$$J_n(x) = q\mu_n n(x)\mathcal{E}(x) + qD_n\frac{dn(x)}{dx}$$

Conduction Current: drift   diffusion   (4-23)

$$J_p(x) = q\mu_p p(x)\mathcal{E}(x) - qD_p\frac{dp(x)}{dx}$$

$$J_{\text{total}} = J_{\text{conduction}} + J_{\text{displacement}} = J_n + J_p + C\frac{dV}{dt}$$

Continuity: $$\frac{\partial p(x, t)}{\partial t} = \frac{\partial \delta p}{\partial t} = -\frac{1}{q}\frac{\partial J_p}{\partial x} - \frac{\delta p}{\tau_p} \qquad \frac{\partial \delta n}{\partial t} = \frac{1}{q}\frac{\partial J_n}{\partial x} - \frac{\delta n}{\tau_n} \quad (4\text{-}31)$$

For steady state diffusion: $$\frac{d^2\delta n}{dx^2} = \frac{\delta n}{D_n\tau_n} \equiv \frac{\delta n}{L_n^2} \qquad \frac{d^2\delta p}{dx^2} = \frac{\delta p}{L_p^2} \quad (4\text{-}34)$$

Diffusion length: $L \equiv \sqrt{D\tau}$ Einstein relation: $\dfrac{D}{\mu} = \dfrac{kT}{q}$ (4-29)

p-n JUNCTIONS

Equilibrium: $$V_0 = \frac{kT}{q}\ln\frac{p_p}{p_n} = \frac{kT}{q}\ln\frac{N_a}{n_i^2/N_d} = \frac{kT}{q}\ln\frac{N_aN_d}{n_i^2} \quad (5\text{-}8)$$

$$\frac{p_p}{p_n} = \frac{n_n}{n_p} = e^{qV_0/kT} \quad (5\text{-}10) \qquad W = \left[\frac{2\epsilon(V_0 - V)}{q}\left(\frac{N_a + N_d}{N_a N_d}\right)\right]^{1/2} \quad (5\text{-}57)$$

One-sided abrupt p^+-n: $x_{n0} = \dfrac{WN_a}{N_a + N_d} \simeq W$ (5-23) $V_0 = \dfrac{qN_dW^2}{2\epsilon}$

$$\Delta p_n = p(x_{n0}) - p_n = p_n(e^{qV/kT} - 1) \quad (5\text{-}29)$$

$$\delta p(x_n) = \Delta p_n e^{-x_n/L_p} = p_n(e^{qV/kT} - 1)e^{-x_n/L_p} \quad (5\text{-}31b)$$

Ideal diode: $$I = qA\left(\frac{D_p}{L_p}p_n + \frac{D_n}{L_n}n_p\right)(e^{qV/kT} - 1) = I_0(e^{qV/kT} - 1) \quad (5\text{-}36)$$

Non-ideal: $\begin{array}{l} I = I_0'(e^{qV/\mathbf{n}kT} - 1) \\ (\mathbf{n} = 1 \text{ to } 2) \end{array}$ (5-74)

With light: $I_{\text{op}} = qAg_{\text{op}}(L_p + L_n + W)$ (8-1)

Solid State Electronic Devices

Multi-level copper metallization of a complementary metal oxide semiconductor (CMOS) chip. This scanning electron micrograph (scale: 1 cm = 3.5 microns) of a CMOS integrated circuit shows six levels of copper metallization that are used to carry electrical signals on the chip. The inter-metal dielectric insulators have been chemically etched away here to reveal the copper interconnects. (Photograph courtesy of IBM.)

FIFTH EDITION

Solid State Electronic Devices

BEN G. STREETMAN AND SANJAY BANERJEE

Microelectronics Research Center
Department of Electrical and Computer Engineering
The University of Texas at Austin

Prentice
Hall

PRENTICE HALL
Upper Saddle River, New Jersey 07458

Library of Congress Cataloging-in-Publication Data
Streetman, Ben G.
    Solid state electronic devices / Ben G. Streetman and Sanjay
Banerjee -- 5th ed.
        p. cm. -- (Prentice Hall series in solid state physical
electronics)
    Includes bibliographical references and index.
    ISBN 0-13-025538-6 (casebound)
    1. Semiconductors. I. Banerjee, Sanjay. II. Title.
III. Series
    TK7871.85.S77   2000
    621.3815'2--dc21
                                        99-16963
                                        CIP

Publisher: **Tom Robbins**
Assistant Vice President of Production and Manufacturing: **David W. Riccardi**
Editor-in-Chief: **Marcia Horton**
Associate Editor: **Alice Dworkin**
Cover Design: **Joseph Sengotta**
Cover: **Fourth Generation PowerPC Microprocessor chip courtesy of Motorola, Inc.**
        Photograph by Kobi Benzvi and Pradipto Mukherjee
Manufacturing Manager: **Trudy Pisciotti**

©2000, 1995, 1990, 1980, 1972 by Prentice Hall, Inc.
A Simon & Schuster Company
Upper Saddle River, New Jersey 07458

Printed in the United States of America

10   9   8   7   6   5   4   3   2

ISBN 0-13-025538-6

Prentice Hall International (UK) Limited, *London*
Prentice Hall of Australia Pty. Limited, *Sydney*
Prentice Hall Canada Inc., *Toronto*
Prentice Hall Hispanoamericana, S.A., *Mexico*
Prentice Hall of India Private Limited, *New Delhi*
Prentice Hall of Japan, Inc., *Tokyo*
Simon & Schuster Asia Pte. Ltd., *Singapore*
Editora Prentice Hall do Brasil, Ltda., *Rio de Janeiro*

CONTENTS

PREFACE   XV

ABOUT THE AUTHORS   XIX

1 CRYSTAL PROPERTIES AND GROWTH OF
SEMICONDUCTORS   1

  1.1 Semiconductor Materials   1
      1.2.1 Periodic Structures   3
  1.2 Crystal Lattices   3
      1.2.2 Cubic Lattices   5
      1.2.3 Planes and Directions   7
      1.2.4 The Diamond Lattice   9
  1.3 Bulk Crystal Growth   12
      1.3.1 Starting Materials   12
      1.3.2 Growth of Single Crystal Ingots   13
      1.3.3 Wafers   14
      1.3.4 Doping   16
  1.4 Epitaxial Growth   17
      1.4.1 Lattice Matching in Epitaxial Growth   18
      1.4.2 Vapor-Phase Epitaxy   21
      1.4.3 Molecular Beam Epitaxy   23

2 ATOMS AND ELECTRONS   28

  2.1 Introduction to Physical Models   28
  2.2 Experimental Observations   30
      2.2.1 The Photoelectric Effect   30
      2.2.2 Atomic Spectra   31
  2.3 The Bohr Model   33
  2.4 Quantum Mechanics   36
      2.4.1 Probability and the Uncertainty Principle   36
      2.4.2 The Schrödinger Wave Equation   38
      2.4.3 Potential Well Problem   40
      2.4.4 Tunneling   42
  2.5 Atomic Structure and the Periodic Table   43
      2.5.1 The Hydrogen Atom   43
      2.5.2 The Periodic Table   46

3   ENERGY BANDS AND CHARGE
    CARRIERS IN SEMICONDUCTORS   55

    3.1   Bonding Forces and Energy Bands in Solids   55
          3.1.1   Bonding Forces in Solids   55
          3.1.2   Energy Bands   58
          3.1.3   Metals, Semiconductors, and Insulators   61
          3.1.4   Direct and Indirect Semiconductors   62
          3.1.5   Variation of Energy Bands with Alloy Composition   64
    3.2   Charge Carriers in Semiconductors   66
          3.2.1   Electrons and Holes   67
          3.2.2   Effective Mass   70
          3.2.3   Intrinsic Material   74
          3.2.4   Extrinsic Material   75
          3.2.5   Electrons and Holes in Quantum Wells   79
    3.3   Carrier Concentrations   80
          3.3.1   The Fermi Level   80
          3.3.2   Electron and Hole Concentrations at Equilibrium   83
          3.3.3   Temperature Dependence of Carrier Concentrations   88
          3.3.4   Compensation and Space Charge Neutrality   90
    3.4   Drift of Carriers in Electric and Magnetic Fields   92
          3.4.1   Conductivity and Mobility   92
          3.4.2   Drift and Resistance   96
          3.4.3   Effects of Temperature and Doping on Mobility   97
          3.4.4   High-Field Effects   99
          3.4.5   The Hall Effect   100
    3.5   Invariance of the Fermi Level at Equilibrium   102

4   EXCESS CARRIERS IN SEMICONDUCTORS   108

    4.1   Optical Absorption   108
    4.2   Luminescence   111
          4.2.1   Photoluminescence   111
          4.2.2   Electroluminescence   114
    4.3   Carrier Lifetime and Photoconductivity   114
          4.3.1   Direct Recombination of Electrons and Holes   115
          4.3.2   Indirect Recombination; Trapping   117
          4.3.3   Steady State Carrier Generation; Quasi-Fermi Levels   120
          4.3.4   Photoconductive Devices   123
    4.4   Diffusion of Carriers   124
          4.4.1   Diffusion Processes   124
          4.4.2   Diffusion and Drift of Carriers; Built-in Fields   127

4.4.3 Diffusion and Recombination; The Continuity Equation 130
4.4.4 Steady State Carrier Injection; Diffusion Length 132
4.4.5 The Haynes-Shockley Experiment 134
4.4.6 Gradients in the Quasi-Fermi Levels 137

5 JUNCTIONS 142

5.1 Fabrication of p-n Junctions 142
5.1.1 Thermal Oxidation 142
5.1.2 Diffusion 144
5.1.3 Rapid Thermal Processing 146
5.1.4 Ion Implantation 147
5.1.5 Chemical Vapor Deposition (CVD) 150
5.1.6 Photolithography 151
5.1.7 Etching 155
5.1.8 Metallization 156

5.2 Equilibrium Conditions 157
5.2.1 The Contact Potential 159
5.2.2 Equilibrium Fermi Levels 163
5.2.3 Space Charge at a Junction 164

5.3 Forward- and Reverse-Biased Junctions;
Steady State Conditions 169
5.3.1 Qualitative Description of Current Flow at a Junction 169
5.3.2 Carrier Injection 174
5.3.3 Reverse Bias 183

5.4 Reverse-Bias Breakdown 185
5.4.1 Zener Breakdown 186
5.4.2 Avalanche Breakdown 188
5.4.3 Rectifiers 190
5.4.4 The Breakdown Diode 193

5.5 Transient and A-C Conditions 194
5.5.1 Time Variation of Stored Charge 195
5.5.2 Reverse Recovery Transient 198
5.5.3 Switching Diodes 201
5.5.4 Capacitance of p-n Junctions 202
5.5.5 The Varactor Diode 210

5.6 Deviations from the Simple Theory 211
5.6.1 Effects of Contact Potential on Carrier Injection 212
5.6.2 Recombination and Generation in the Transition Region 214
5.6.3 Ohmic Losses 217
5.6.4 Graded Junctions 218

5.7 Metal-Semiconductor Junctions 220
5.7.1 Schottky Barriers 220

5.7.2   Rectifying Contacts   222
5.7.3   Ohmic Contacts   224
5.7.4   Typical Schottky Barriers   226
5.8   Heterojunctions   227

6   FIELD-EFFECT TRANSISTORS   241

6.1   Transistor Operation   242
6.1.1   The Load Line   242
6.1.2   Amplification and Switching   244
6.2   The Junction FET   244
6.2.1   Pinch-off and Saturation   245
6.2.2   Gate Control   247
6.2.3   Current-Voltage Characteristics   249
6.3   The Metal-Semiconductor FET   251
6.3.1   The GaAs MESFET   251
6.3.2   The High Electron Mobility Transistor (HEMT)   252
6.3.3   Short Channel Effects   254
6.4   The Metal-Insulator-Semiconductor FET   255
6.4.1   Basic Operation and Fabrication   256
6.4.2   The Ideal MOS Capacitor   260
6.4.3   Effects of Real Surfaces   272
6.4.4   Threshold Voltage   275
6.4.5   MOS Capacitance-Voltage Analysis   277
6.4.6   Time-dependent Capacitance Measurements   280
6.4.7   Current-Voltage Characteristics of MOS Gate Oxides   283
6.5   The MOS Field-Effect Transistor   286
6.5.1   Output Characteristics   286
6.5.2   Transfer Characteristics   288
6.5.3   Mobility Models   290
6.5.4   Short Channel MOSFET I-V Characteristics   293
6.5.5   Control of Threshold Voltage   293
6.5.6   Substrate Bias Effects   300
6.5.7   Subthreshold Characteristics   301
6.5.8   Equivalent Circuit for the MOSFET   304
6.5.9   MOSFET Scaling and Hot Electron Effects   307
6.5.10 Drain-Induced Barrier Lowering   311
6.5.11 Short Channel and Narrow Width Effect   313
6.5.12 Gate-Induced Drain Leakage   315

7   BIPOLAR JUNCTION TRANSISTORS   322

7.1   Fundamentals of BJT Operation   322
7.2   Amplification with BJTs   325

7.3   BJT Fabrication   329
7.4   Minority Carrier Distributions and Terminal Currents   332
    7.4.1   Solution of the Diffusion Equation in the Base Region   333
    7.4.2   Evaluation of the Terminal Currents   334
    7.4.3   Approximations of the Terminal Currents   337
    7.4.4   Current Transfer Ratio   339
7.5   Generalized Biasing   340
    7.5.1   The Coupled-Diode Model   340
    7.5.2   Charge Control Analysis   344
7.6   Switching   346
    7.6.1   Cutoff   347
    7.6.2   Saturation   348
    7.6.3   The Switching Cycle   349
    7.6.4   Specifications for Switching Transistors   350
7.7   Other Important Effects   351
    7.7.1   Drift in the Base Region   352
    7.7.2   Base Narrowing   353
    7.7.3   Avalanche Breakdown   354
    7.7.4   Injection Level; Thermal Effects   356
    7.7.5   Base Resistance and Emitter Crowding   357
    7.7.6   Gummel-Poon Model   359
    7.7.7   Kirk Effect   363
7.8   Frequency Limitations of Transistors   365
    7.8.1   Capacitance and Charging Times   365
    7.8.2   Transit Time Effects   368
    7.8.3   Webster Effect   369
    7.8.4   High-Frequency Transistors   369
7.9   Heterojunction Bipolar Transistors   371

8   OPTOELECTRONIC DEVICES   379
8.1   Photodiodes   379
    8.1.1   Current and Voltage in an Illuminated Junction   379
    8.1.2   Solar Cells   382
    8.1.3   Photodetectors   384
    8.1.4   Noise and Bandwidth of Photodetectors   386
8.2   Light-Emitting Diodes   390
    8.2.1   Light-Emitting Materials   390
    8.2.2   Fiber Optic Communications   392
    8.2.3   Multilayer Heterojunctions for LEDs   395
8.3   Lasers   396

8.4   Semiconductor Lasers   400
   8.4.1   Population Inversion at a Junction   400
   8.4.2   Emission Spectra for p-n Junction Lasers   403
   8.4.3   The Basic Semiconductor Laser   404
   8.4.4   Heterojunction Lasers   405
   8.4.5   Materials for Semiconductor Lasers   408

9   INTEGRATED CIRCUITS   415

9.1   Background   415
   9.1.1   Advantages of Integration   416
   9.1.2   Types of Integrated Circuits   418
   9.1.3   Monolithic and Hybrid Circuits   418
9.2   Evolution of Integrated Circuits   420
9.3   Monolithic Device Elements   423
   9.3.1   CMOS Process Integration   423
   9.3.2   Silicon-on-Insulator (SOI)   437
   9.3.3   Integration of Other Circuit Elements   439
9.4   Charge Transfer Devices   444
   9.4.1   Dynamic Effects in MOS Capacitors   444
   9.4.2   The Basic CCD   446
   9.4.3   Improvements on the Basic Structure   447
   9.4.4   Applications of CCDs   448
9.5   Ultra Large-Scale Integration (ULSI)   449
   9.5.1   Logic Devices   452
   9.5.2   Semiconductor Memories   461
9.6   Testing, Bonding, and Packaging   474
   9.6.1   Testing   474
   9.6.2   Wire Bonding   476
   9.6.3   Flip-Chip Techniques   478
   9.6.4   Packaging   479

10   NEGATIVE CONDUCTANCE MICROWAVE DEVICES   486

10.1 Tunnel Diodes   486
   10.1.1 Degenerate Semiconductors   487
   10.1.2 Tunnel Diode Operation   487
   10.1.3 Circuit Applications   490
10.2 The IMPATT Diode   490
10.3 The Gunn Diode   494
   10.3.1 The Transferred Electron Mechanism   494
   10.3.2 Formation and Drift of Space Charge Domains   496
   10.3.3 Fabrication   499

11 POWER DEVICES 504

11.1 The p-n-p-n Diode 504
  11.1.1 Basic Structure 505
  11.1.2 The Two-Transistor Analogy 506
  11.1.3 Variation of α with Injection 507
  11.1.4 Forward-Blocking State 507
  11.1.5 Conducting State 508
  11.1.6 Triggering Mechanisms 509
11.2 The Semiconductor Controlled Rectifier 511
  11.2.1 Gate Control 511
  11.2.2 Turning off the SCR 512
  11.2.3 Bilateral Devices 513
  11.2.4 Fabrication and Applications 514
11.3 Insulated Gate Bipolar Transistor 515

APPENDICES

  I. Definitions of Commonly Used Symbols 519
  II. Physical Constants and Conversion Factors 523
  III. Properties of Semiconductor Materials 524
  IV. Derivation of the Density of States in the Conduction Band 525
  V. Derivation of Fermi–Dirac Statistics 530
  VI. Dry and Wet Thermal Oxide Thickness as a Function of Time and Temperature 534
  VII. Solid Solubilities of Impurities in Si 536
  VIII. Diffusivities of Dopants in Si and SiO_2 538
  IX. Projected Range and Straggle as a Function of Implant Energy in Si 540

INDEX 543

PREFACE

This book is an introduction to semiconductor devices for undergraduate electrical engineers, other interested students, and practicing engineers and scientists whose understanding of modern electronics needs updating. The book is organized to bring students with a background in sophomore physics to a level of understanding which will allow them to read much of the current literature on new devices and applications.

GOALS

An undergraduate course in electronic devices has two basic purposes: (1) to provide students with a sound understanding of existing devices, so that their studies of electronic circuits and systems will be meaningful; and (2) to develop the basic tools with which they can later learn about newly developed devices and applications. Perhaps the second of these objectives is the more important in the long run; it is clear that engineers and scientists who deal with electronics will continually be called upon to learn about new devices and processes in the future. For this reason, we have tried to incorporate the basics of semiconductor materials and conduction processes in solids, which arise repeatedly in the literature when new devices are explained. Some of these concepts are often omitted in introductory courses, with the view that they are unnecessary for understanding the fundamentals of junctions and transistors. We believe this view neglects the important goal of equipping students for the task of understanding a new device by reading the current literature. Therefore, in this text most of the commonly used semiconductor terms and concepts are introduced and related to a broad range of devices.

READING LISTS

As a further aid in developing techniques for independent study, the reading list at the end of each chapter includes a few articles which students can read comfortably as they study this book. Some of these articles have been selected from periodicals such as *Scientific American* and *Physics Today*, which specialize in introductory presentations. Other articles chosen from books and the professional literature provide a more quantitative treatment of the material. We do not expect that students will read all articles recommended in the reading lists; nevertheless, some exposure to periodicals is useful in laying the foundation for a career of constant updating and self-education.

PROBLEMS

One of the keys to success in understanding this material is to work problems that exercise the concepts. The problems at the end of each chapter are designed to facilitate learning the material. Very few are simple "plug-in" problems. Instead, they are chosen to reinforce or extend the material presented in the chapter.

UNITS   In keeping with the goals described above, examples and problems are stated in terms of units commonly used in the semiconductor literature. The basic system of units is rationalized MKS, although cm is often used as a convenient unit of length. Similarly, electron volts (eV) are often used rather than joules (J) to measure the energy of electrons. Units for various quantities are given in Appendices I and II.

PRESENTATION   In presenting this material at the undergraduate level, one must anticipate a few instances which call for a phrase such as "It can be shown . . ." This is always disappointing; on the other hand, the alternative is to delay study of solid state devices until the graduate level, where statistical mechanics, quantum theory, and other advanced background can be freely invoked. Such a delay would result in a more elegant treatment of certain subjects, but it would prevent undergraduate students from enjoying the study of some very exciting devices.

The discussion includes both silicon and compound semiconductors, to reflect the continuing growth in importance for compounds in optoelectronic and high-speed device applications. Topics such as heterojunctions, lattice matching using ternary and quaternary alloys, variation of band gap with alloy composition, and properties of quantum wells add up to the breadth of the discussion. Not to be outdone by the compounds, silicon-based devices have continued their dramatic record of advancement. The discussion of FET structures and Si integrated circuits reflects these advancements. Our objective is not to cover all the latest devices, which can only be done in the journal and conference literature. Instead, we have chosen devices to discuss which are broadly illustrative of important principles.

The first four chapters of the book provide background on the nature of semiconductors and conduction processes in solids. Included is a brief introduction to quantum concepts (Chapter 2) for those students who do not already have this background from other courses. Chapter 5 describes the p-n junction and some of its applications. Chapters 6 and 7 deal with the principles of transistor operation. Chapter 8 covers optoelectronics and Chapter 9 discusses integrated circuits. Chapters 10 and 11 apply the theory of junctions and conduction processes to microwave and power devices. All of the devices covered are important in today's electronics; furthermore, learning about these devices should be an enjoyable and rewarding experience. We hope this book provides that kind of experience for its readers.

The fifth edition benefits greatly from comments and suggestions provided by students and teachers of the first four editions. The book's readers have generously provided comments which have been invaluable in developing the present version. We remain indebted to those persons mentioned in the Preface of the first four editions, who contributed so much to the development of the book. In particular, Nick Holonyak has been a source of continuing information and inspiration for all five editions. Additional thanks go to our colleagues at UT-Austin who have provided special assistance, particularly Joe Campbell, Ray Chen, Dennis Deppe, Russ Dupuis, Archie Holmes, Dim-Lee Kwong, Jack Lee, Christine Maziar, Dean Neikirk, and Al Tasch. Kay Shores and Qingyou Lu provided useful assistance with the typing. We thank the many companies and organizations cited in the figure captions for generously providing photographs and illustrations of devices and fabrication processes. Kobi Benzvi and Pradipto Mukherjee at Motorola, Shubneesh Batra and Mary Miller at Micron, and Tom Way at IBM deserve special mention. Finally, we recall with gratitude many years of association with the late Greg Stillman, a valued colleague and friend.

Ben G. Streetman
Sanjay Banerjee

ACKNOW-
LEDGMENTS

PRENTICE HALL SERIES
IN SOLID STATE PHYSICAL ELECTRONICS
Nick Holonyak Jr., Editor

Cheo   FIBER OPTICS: DEVICES AND SYSTEMS 2/E
Haus   WAVES AND FIELDS IN OPTOELECTRONICS
Kroemer   QUANTUM MECHANICS FOR ENGINEERING, MATERIALS SCIENCE,
            AND APPLIED PHYSICS
Nussbaum   CONTEMPORARY OPTICS FOR SCIENTISTS AND ENGINEERS
Peyghambarian/Koch/Mysyrowicz   INTRODUCTION TO SEMICONDUCTOR OPTICS
Shur   PHYSICS OF SEMICONDUCTOR DEVICES
Soclof   DESIGN AND APPLICATIONS OF ANALOG INTEGRATED CIRCUITS
Streetman   SOLID STATE ELECTRONIC DEVICES 5/E
Verdeyen   LASER ELECTRONICS 3/E
Wolfe/Holonyak/Stillman   PHYSICAL PROPERTIES OF SEMICONDUCTORS

Ben G. Streetman is Dean of the College of Engineering at The University of Texas at Austin and holds the Dula D. Cockrell Centennial Chair in Engineering. He is a Professor of Electrical and Computer Engineering and was the founding Director of the Microelectronics Research Center (1984–96). His teaching and research interests involve semiconductor materials and devices. After receiving a Ph.D. from The University of Texas at Austin (1966) he was on the faculty (1966–1982) of the University of Illinois at Urbana-Champaign. He returned to The University of Texas at Austin in 1982. His honors include the Education Medal of the Institute of Electrical and Electronics Engineers (IEEE), the Frederick Emmons Terman Medal of the American Society for Engineering Education (ASEE), and the Heinrich Welker Medal from the International Conference on Compound Semiconductors. He is a member of the National Academy of Engineering. He is a Fellow of the IEEE and the Electrochemical Society. He has been honored as a Distinguished Alumnus of The University of Texas at Austin and as a Distinguished Graduate of the UT College of Engineering. He has received the General Dynamics Award for Excellence in Engineering Teaching, and was honored by the Parents' Association as a Teaching Fellow for outstanding teaching of undergraduates. He has served on numerous panels and committees in industry and government, and several corporate boards. He has published more than 270 articles in the technical literature. Thirty-three students of Electrical Engineering, Materials Science, and Physics have received their Ph.D.s under his direction.

Sanjay Kumar Banerjee is Professor of Electrical and Computer Engineering, and Director of the Microelectronics Research Center at The University of Texas at Austin, where he currently holds the Cullen Trust Endowed Professorship in Engineering No. 1 as well as being a Fellow of the Cockrell Family Regent's Chair. His research interests include silicon-based heterostructure devices, device modeling and ultra-large-scale IC technology. He received his B.Tech from the Indian Institute of Technology, Kharagpur, and his M.S. and Ph.D. from the University of Illinois at Urbana-Champaign in 1979, 1981 and 1983, respectively, all in electrical engineering. At Texas Instruments from 1983–1987, he worked on polysilicon transistors and dynamic random access memory cells used in the world's first 4Megabit DRAM, for which he was the co-recipient of the Best Paper Award at the 1986 IEEE International Solid State Circuits Conference. His honors include the NSF Presidential Young Investigator Award (1988), Engineering Foundation Advisory Council Halliburton Award (1991) for teaching excellence, and the Texas Atomic Energy Centennial Fellowship (1990–1997). He has more than 225 archival referred publications, has presented over 200 talks at conferences, and has 12 U.S. patents. He has supervised 18 Ph.D. and 35 MS students. He is a Distinguished National Lecturer for the IEEE Electron Devices Society (1997–), and Fellow of the IEEE (1996).

Chapter 1

Crystal Properties and Growth of Semiconductors

In studying solid state electronic devices we are interested primarily in the electrical behavior of solids. However, we shall see in later chapters that the transport of charge through a metal or a semiconductor depends not only on the properties of the electron but also on the arrangement of atoms in the solid. In the first chapter we shall discuss some of the physical properties of semiconductors compared with other solids, the atomic arrangements of various materials, and some methods of growing semiconductor crystals. Topics such as crystal structure and crystal growth technology are often the subjects of books rather than introductory chapters; thus we shall consider only a few of the more important and fundamental ideas that form the basis for understanding electronic properties of semiconductors and device fabrication.

Semiconductors are a group of materials having electrical conductivities intermediate between metals and insulators. It is significant that the conductivity of these materials can be varied over orders of magnitude by changes in temperature, optical excitation, and impurity content. This variability of electrical properties makes the semiconductor materials natural choices for electronic device investigations.

1.1
SEMICONDUCTOR
MATERIALS

Semiconductor materials are found in column IV and neighboring columns of the periodic table (Table 1–1). The column IV semiconductors, silicon and germanium, are called *elemental* semiconductors because they are composed of single species of atoms. In addition to the elemental materials, compounds of column III and column V atoms, as well as certain combinations from II and VI, and from IV, make up the *compound* semiconductors.

As Table 1–1 indicates, there are numerous semiconductor materials. As we shall see, the wide variety of electronic and optical properties of these semiconductors provides the device engineer with great flexibility in the design of electronic and optoelectronic functions. The elemental semiconductor Ge was

---

Table 1–1. Common semiconductor materials: (a) the portion of the periodic table where semiconductors occur; (b) elemental and compound semiconductors.

(a)	II	III	IV	V	VI
		B	C	N	
		Al	Si	P	S
	Zn	Ga	Ge	As	Se
	Cd	In		Sb	Te

(b)	Elemental	IV compounds	Binary III–V compounds	Binary II–VI compounds
	Si	SiC	AlP	ZnS
	Ge	SiGe	AlAs	ZnSe
			AlSb	ZnTe
			GaN	CdS
			GaP	CdSe
			GaAs	CdTe
			GaSb	
			InP	
			InAs	
			InSb	

widely used in the early days of semiconductor development for transistors and diodes. Silicon is now used for the majority of rectifiers, transistors, and integrated circuits. However, the compounds are widely used in high-speed devices and devices requiring the emission or absorption of light. The two-element (*binary*) III–V compounds such as GaN, GaP, and GaAs are common in light-emitting diodes (LEDs). As discussed in Section 1.2.4, three-element (*ternary*) compounds such as GaAsP and four-element (*quaternary*) compounds such as InGaAsP can be grown to provide added flexibility in choosing materials properties.

Fluorescent materials such as those used in television screens usually are II–VI compound semiconductors such as ZnS. Light detectors are commonly made with InSb, CdSe, or other compounds such as PbTe and HgCdTe. Si and Ge are also widely used as infrared and nuclear radiation detectors. An important microwave device, the Gunn diode, is usually made of GaAs or InP. Semiconductor lasers are made using GaAs, AlGaAs, and other ternary and quaternary compounds.

One of the most important characteristics of a semiconductor, which distinguishes it from metals and insulators, is its *energy band gap*. This property, which we will discuss in detail in Chapter 3, determines among other things the wavelengths of light that can be absorbed or emitted by the semiconductor. For example, the band gap of GaAs is about 1.43 electron volts (eV), which corresponds to light wavelengths in the near infrared. In contrast, GaP has a band gap of about 2.3 eV, corresponding to wavelengths in the green portion of the spectrum.[1] The band gap E_g for various semiconductor materials is listed along with other properties in Appendix III. As a result of the wide variety of semiconductor band gaps, light-emitting diodes and lasers can be constructed with wavelengths over a broad range of the infrared and visible portions of the spectrum.

The electronic and optical properties of semiconductor materials are strongly affected by impurities, which may be added in precisely controlled amounts. Such impurities are used to vary the conductivities of semiconductors over wide ranges and even to alter the nature of the conduction processes from conduction by negative charge carriers to positive charge carriers. For example, an impurity concentration of one part per million can change a sample of Si from a poor conductor to a good conductor of electric current. This process of controlled addition of impurities, called *doping*, will be discussed in detail in subsequent chapters.

To investigate these useful properties of semiconductors, it is necessary to understand the atomic arrangements in the materials. Obviously, if slight alterations in purity of the original material can produce such dramatic changes in electrical properties, then the nature and specific arrangement of atoms in each semiconductor must be of critical importance. Therefore, we begin our study of semiconductors with a brief introduction to crystal structure.

In this section we discuss the arrangements of atoms in various solids. We shall distinguish between single crystals and other forms of materials and then investigate the periodicity of crystal lattices. Certain important crystallographic terms will be defined and illustrated in reference to crystals having a basic cubic structure. These definitions will allow us to refer to certain planes and directions within a lattice. Finally, we shall investigate the diamond lattice; this structure, with some variations, is typical of most of the semiconductor materials used in electronic devices.

**1.2
CRYSTAL LATTICES**

1.2.1 Periodic Structures

A crystalline solid is distinguished by the fact that the atoms making up the crystal are arranged in a periodic fashion. That is, there is some basic arrangement of atoms that is repeated throughout the entire solid. Thus the crystal appears exactly the same at one point as it does at a series of other equivalent points, once the basic periodicity is discovered. However, not all solids are crystals (Fig. 1–1); some have no periodic structure at all (*amorphous* solids), and others are composed of many small regions of single-crystal material (*polycrystalline* solids). The high-resolution micrograph shown in Fig.

---

[1]The conversion between the energy E of a photon of light (eV) and its wavelength $\lambda(\mu m)$ is $\lambda = 1.24/E$. For GaAs, $\lambda = 1.24/1.43 = 0.87$ μm.

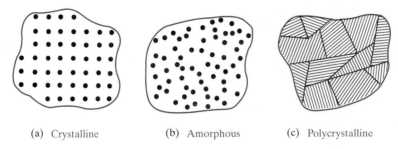

(a) Crystalline   (b) Amorphous   (c) Polycrystalline

Figure 1–1
Three types of solids, classified according to atomic arrangement: (a) crystalline and (b) amorphous materials are illustrated by microscopic views of the atoms, whereas (c) polycrystalline structure is illustrated by a more macroscopic view of adjacent single-crystalline regions, such as (a).

6–33 illustrates the periodic array of atoms in the single-crystal silicon of a transistor channel compared with the amorphous SiO_2 (glass) of the oxide layer.

The periodic arrangement of atoms in a crystal is called the *lattice*. Since there are many different ways of placing atoms in a volume, the distances and orientation between atoms can take many forms. However, in every case the lattice contains a volume, called a *unit cell*, which is representative of the entire lattice and is regularly repeated throughout the crystal. As an example of such a lattice, Fig. 1–2 shows a two-dimensional arrangement of atoms with a unit cell ODEF. This cell has an atom at each corner shared with adjacent cells. Notice that we can define vectors **a** and **b** such that if the unit cell is translated by integral multiples of these vectors, a new unit cell identical to the original is found (e.g., O′D′E′F′). These vectors **a** and **b** (and **c** if the lattice is three dimensional) are called the *basis vectors* for the lattice. Points within the lattice are indistinguishable if the vector between the points is

$$\mathbf{r} = p\mathbf{a} + q\mathbf{b} + s\mathbf{c} \qquad (1-1)$$

where p, q, and s are integers.

The smallest unit cell that can be repeated to form the lattice is called a *primitive cell*. In many lattices, however, the primitive cell is not the most convenient to work with. The importance of the unit cell lies in the fact that we can analyze the crystal as a whole by investigating a representative volume. For example, from the unit cell we can find the distances between nearest atoms and next nearest atoms for calculation of the forces holding the lattice together; we can look at the fraction of the unit cell volume filled by atoms and relate the density of the solid to the atomic arrangement. But even more important for our interest in electronic devices, the properties of the periodic crystal lattice determine the allowed energies of electrons that participate in the conduction process. Thus the lattice determines not only the mechanical properties of the crystal but also its electrical properties.

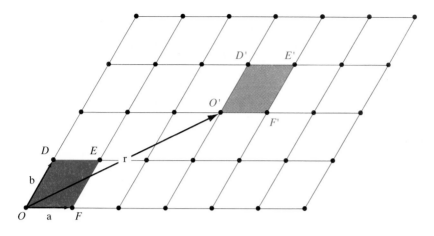

Figure 1–2
A two-dimensional
lattice showing
translation of a
unit cell by
$\mathbf{r} = 3\mathbf{a} + 2\mathbf{b}$.

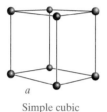 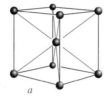 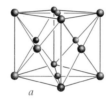

Simple cubic   Body-centered cubic   Face-centered cubic

Figure 1–3
Unit cells for three
types of cubic lat-
tice structures.

1.2.2 Cubic Lattices

The simplest three-dimensional lattice is one in which the unit cell is a cubic volume, such as the three cells shown in Fig. 1–3. The *simple cubic* structure (abbreviated *sc*) has an atom located at each corner of the unit cell. The *body-centered cubic* (*bcc*) lattice has an additional atom at the center of the cube, and the *face-centered cubic* (*fcc*) unit cell has atoms at the eight corners and centered on the six faces.

    As atoms are packed into the lattice in any of these arrangements, the distances between neighboring atoms will be determined by a balance between the forces that attract them together and other forces that hold them apart. We shall discuss the nature of these forces for particular solids in Section 3.1.1. For now, we can calculate the maximum fraction of the lattice volume that can be filled with atoms by approximating the atoms as hard spheres. For example, Fig. 1–4 illustrates the packing of spheres in a face-centered cubic cell of side *a*, such that the nearest neighbors touch. The dimension *a* for a cubic unit cell is called the *lattice constant*. For the fcc lattice the nearest neighbor distance is one-half the diagonal of a face, or $\frac{1}{2}(a\sqrt{2})$. Therefore, for the atom centered on the face to just touch the atoms at each corner of the face, the radius of the sphere must be one-half the nearest neighbor distance, or $\frac{1}{4}(a\sqrt{2})$.

Figure 1–4
Packing of hard
spheres in an fcc
lattice.

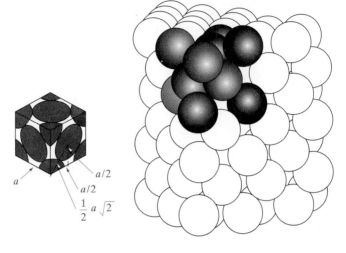

EXAMPLE 1–1   Find the fraction of the fcc unit cell volume filled with hard spheres as in Fig. 1–4.

SOLUTION   Each corner atom in a cubic unit cell is shared with seven neighboring cells; thus each unit cell contains $\frac{1}{8}$ of a sphere at each of the eight corners for a total of one atom. Similarly, the fcc cell contains half an atom at each of the six faces for a total of three. Thus we have

$$\text{Atoms per cell} = 1 \text{ (corners)} + 3 \text{ (faces)} = 4$$

$$\text{Nearest neighbor distance} = \tfrac{1}{2}(a\sqrt{2})$$

$$\text{Radius of each sphere} = \tfrac{1}{4}(a\sqrt{2})$$

$$\text{Volume of each sphere} = \tfrac{4}{3}\pi \, [\tfrac{1}{4}(a\sqrt{2})]^3 = \frac{\pi a^3 \sqrt{2}}{24}$$

Maximum fraction of cell filled

$$= \frac{\text{no. of spheres} \times \text{vol. of each sphere}}{\text{total vol. of each cell}}$$

$$= \frac{4 \times (\pi a^3 \sqrt{2})/24}{a^3}$$

$$= \frac{\pi\sqrt{2}}{6} = 74 \text{ percent filled}$$

Therefore, if the atoms in an fcc lattice are packed as densely as possible, with no distance between the outer edges of nearest neighbors, 74 percent of the volume is filled. This is a relatively high percentage compared with some other lattice structures (Prob. 1.14).

1.2.3 Planes and Directions

In discussing crystals it is very helpful to be able to refer to planes and directions within the lattice. The notation system generally adopted uses a set of three integers to describe the position of a plane or the direction of a vector within the lattice. The three integers describing a particular plane are found in the following way:

1. Find the intercepts of the plane with the crystal axes and express these intercepts as integral multiples of the basis vectors (the plane can be moved in and out from the origin, retaining its orientation, until such an integral intercept is discovered on each axis).

2. Take the reciprocals of the three integers found in step 1 and reduce these to the smallest set of integers h, k, and l, which have the same relationship to each other as the three reciprocals.

3. Label the plane (hkl).

The plane illustrated in Fig. 1–5 has intercepts at **2a**, **4b**, and **1c** along the three crystal axes. Taking the reciprocals of these intercepts, we get $\frac{1}{2}$, $\frac{1}{4}$, and 1. These three fractions have the same relationship to each other as the integers 2, 1, and 4 (obtained by multiplying each fraction by 4). Thus the plane can be referred to as a (214) plane.

EXAMPLE 1–2

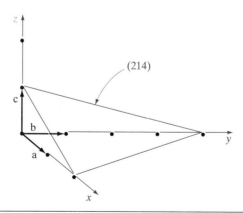

Figure 1–5
A (214) crystal plane.

The three integers h, k, and l are called the *Miller indices*; these three numbers define a set of parallel planes in the lattice. One advantage of taking the reciprocals of the intercepts is avoidance of infinities in the notation. One intercept is infinity for a plane parallel to an axis; however, the reciprocal of such an intercept is taken as zero. If a plane contains one of the axes, it is parallel to that axis and has a zero reciprocal intercept. If a plane passes through the origin, it can be translated to a parallel position for calculation of the Miller indices. If an intercept occurs on the negative branch of an axis, the minus sign is placed above the Miller index for convenience, such as $(h\bar{k}l)$.

From a crystallographic point of view, many planes in a lattice are equivalent; that is, a plane with given Miller indices can be shifted about in the lattice simply by choice of the position and orientation of the unit cell. The indices of such equivalent planes are enclosed in braces { } instead of parentheses. For example, in the cubic lattice of Fig. 1–6 all the cube faces are crystallographically equivalent in that the unit cell can be rotated in various directions and still appear the same. The six equivalent faces are collectively designated as {100}.

A direction in a lattice is expressed as a set of three integers with the same relationship as the components of a vector in that direction. The three vector components are expressed in multiples of the basis vectors, and the three integers are reduced to their smallest values while retaining the relationship among them. For example, the body diagonal in the cubic lattice (Fig. 1–7a) is composed of the components 1**a**, 1**b**, and 1**c**; therefore, this diagonal is the [111] direction. (Brackets are used for direction indices.) As in

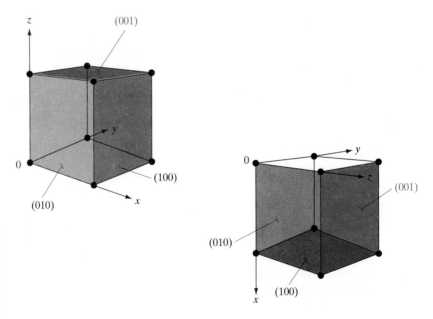

Figure 1–6
Equivalence of the cube faces ({100} planes) by rotation of the unit cell within the cubic lattice.

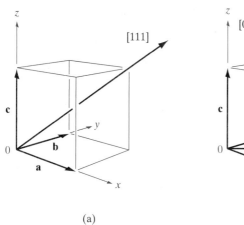

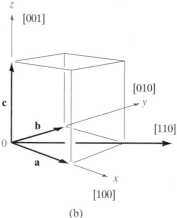

Figure 1–7
Crystal directions in the cubic lattice.

(a)   (b)

the case of planes, many directions in a lattice are equivalent, depending only on the arbitrary choice of orientation for the axes. Such equivalent direction indices are placed in angular brackets ⟨ ⟩. For example, the crystal axes in the cubic lattice [100], [010], and [001] are all equivalent and are called ⟨100⟩ directions (Fig. 1–7b).

Comparing Figs. 1–6 and 1–7, we notice that in cubic lattices a direction [*hkl*] is perpendicular to the plane (*hkl*). This is convenient in analyzing lattices with cubic unit cells, but it should be remembered that it is not necessarily true in noncubic systems.

1.2.4   The Diamond Lattice

The basic lattice structure for many important semiconductors is the *diamond* lattice, which is characteristic of Si and Ge. In many compound semiconductors, atoms are arranged in a basic diamond structure but are different on alternating sites. This is called a *zincblende* lattice and is typical of the III–V compounds. One of the simplest ways of stating the construction of the diamond lattice is the following:

> The diamond lattice can be thought of as an fcc structure with an extra atom placed at **a**/4 + **b**/4 + **c**/4 from each of the fcc atoms.

Figure 1–8a illustrates the construction of a diamond lattice from an fcc unit cell. We notice that when the vectors are drawn with components one-fourth of the cube edge in each direction, only four additional points within the same unit cell are reached. Vectors drawn from any of the other fcc atoms simply determine corresponding points in adjacent unit cells. This method of constructing the diamond lattice implies that the original fcc has associated with it a second interpenetrating fcc displaced by ¼, ¼, ¼. The two interpenetrating fcc *sublattices* can be visualized by looking down on the unit

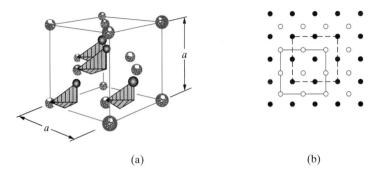

(a)   (b)

Figure 1–8

Diamond lattice structure: (a) a unit cell of the diamond lattice constructed by placing atoms $\frac{1}{4}, \frac{1}{4}, \frac{1}{4}$ from each atom in an fcc; (b) top view (along any ⟨100⟩ direction) of an extended diamond lattice. The colored circles indicate one fcc sublattice and the black circles indicate the interpenetrating fcc.

cell of Fig. 1–8a from the top (or along any ⟨100⟩ direction). In the top view of Fig. 1–8b, atoms belonging to the original fcc are represented by open circles, and the interpenetrating sublattice is shaded. If the atoms are all similar, we call this structure a diamond lattice; if the atoms differ on alternating sites, it is a zincblende structure. For example, if one fcc sublattice is composed of Ga atoms and the interpenetrating sublattice is As, the zincblende structure of GaAs results. Most of the compound semiconductors have this type of lattice, although some of the II–VI compounds are arranged in a slightly different structure called the *wurtzite* lattice. We shall restrict our discussion here to the diamond and zincblende structures, since they are typical of most of the commonly used semiconductors.

EXAMPLE 1–3   Calculate the densities of Si and GaAs from the lattice constants (Appendix III), atomic weights, and Avogadro's number. Compare the results with densities given in Appendix III. The atomic weights of Si, Ga, and As are 28.1, 69.7, and 74.9, respectively.

SOLUTION   For Si: $a = 5.43 \times 10^{-8}$ cm, 8 atoms/cell,

$$\frac{8}{a^3} = \frac{8}{(5.43 \times 10^{-8})^3} = 5 \times 10^{22} \text{ atoms/cm}^3$$

$$\text{density} = \frac{5 \times 10^{22}(\text{atoms/cm}^3) \times 28.1(\text{g/mole})}{6.02 \times 10^{23}(\text{atoms/mole})} = 2.33 \text{ g/cm}^3$$

For GaAs: $a = 5.65 \times 10^{-8}$ cm, 4 each Ga, As atoms/cell

$$\frac{4}{a^3} = \frac{4}{(5.65 \times 10^{-8})^3} = 2.22 \times 10^{22} \text{ atoms/cm}^3$$

$$\text{density} = \frac{2.22 \times 10^{22}(69.7 + 74.9)}{6.02 \times 10^{23}} = 5.33 \text{ g/cm}^3$$

A particularly interesting and useful feature of the III–V compounds is the ability to vary the mixture of elements on each of the two interpenetrating fcc sublattices of the zincblende crystal. For example, in the ternary compound AlGaAs, it is possible to vary the composition of the ternary alloy by choosing the fraction of Al or Ga atoms on the column III sublattice. It is common to represent the composition by assigning subscripts to the various elements. For example, $Al_xGa_{1-x}As$ refers to a ternary alloy in which the column III sublattice in the zincblende structure contains a fraction x of Al atoms and $1-x$ of Ga atoms. The composition $Al_{0.3}Ga_{0.7}As$ has 30 percent Al and 70 percent Ga on the column III sites, with the interpenetrating column V sublattice occupied entirely by As atoms. It is extremely useful to be able to grow ternary alloy crystals such as this with a given composition. For the $Al_xGa_{1-x}As$ example we can grow crystals over the entire composition range from $x = 0$ to $x = 1$, thus varying the electronic and optical properties of the material from that of GaAs ($x = 0$) to that of AlAs ($x = 1$). To vary the properties even further, it is possible to grow four-element (quaternary) compounds such as $In_xGa_{1-x}As_yP_{1-y}$ having a very wide range of properties.

It is important from an electronic point of view to notice that each atom in the diamond and zincblende structures is surrounded by four nearest neighbors (Fig. 1–9). The importance of this relationship of each atom to its neighbors will become evident in Section 3.1.1 when we discuss the bonding forces which hold the lattice together.

The fact that atoms in a crystal are arranged in certain planes is important to many of the mechanical, metallurgical, and chemical properties of the material. For example, crystals often can be cleaved along certain atomic planes, resulting in exceptionally planar surfaces. This is a familiar result in cleaved diamonds for jewelry; the facets of a diamond reveal clearly the triangular, hexagonal, and rectangular symmetries of intersecting planes in various crystallographic directions. Semiconductors with diamond and zincblende lattices have similar cleavage planes. Chemical reactions, such as etching of the crystal, often take place preferentially along certain directions. These properties serve as interesting illustrations of crystal symmetry, but in addition, each plays an important role in fabrication processes for many semiconductor devices.

Figure 1–9
Diamond lattice
unit cell, showing
the four nearest
neighbor struc-
ture. (From *Elec-
trons and Holes in
Semiconductors*
by W. Shockley,
© 1950 by Litton
Educational Pub-
lishing Co., Inc.;
by permission of
Van Nostrand
Reinhold Co.,
Inc.)

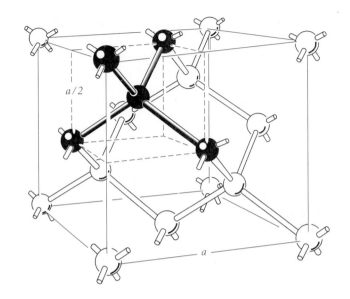

**1.3
BULK CRYSTAL
GROWTH**

The progress of solid state device technology since the invention of the transistor in 1948 has depended not only on the development of device concepts
but also on the improvement of materials. For example, the fact that integrated circuits can be made today is the result of a considerable breakthrough
in the growth of pure, single-crystal Si in the early and mid−1950s. The requirements on the growing of device-grade semiconductor crystals are more
stringent than those for any other materials. Not only must semiconductors
be available in large single crystals, but also the purity must be controlled
within extremely close limits. For example, Si crystals now being used in devices are grown with concentrations of most impurities of less than one part
in ten billion. Such purities require careful handling and treatment of the
material at each step of the manufacturing process.

1.3.1 Starting Materials

The raw feedstock for Si crystal is silicon dioxide (SiO_2). We react SiO_2 with
C in the form of coke in an arc furnace at very high temperatures (~1800
°C) to reduce SiO_2 according to the following reaction:

$$SiO_2 + 2C \rightarrow Si + 2CO \qquad (1–2)$$

This forms metallurgical grade Si (MGS) which has impurities such as
Fe, Al and heavy metals at levels of several hundred to several thousand
parts per million (ppm). Refer back to Example 1–3 to see that 1 ppm of Si
corresponds to an impurity level of $5 \times 10^{16} cm^{-3}$. While MGS is clean enough

for metallurgical applications such as using Si to make stainless steel, it is not pure enough for electronic applications; it is also not single-crystal.

The MGS is refined further to yield semiconductor-grade or electronic-grade Si (EGS), in which the levels of impurities are reduced to parts per billion or ppb (1 ppb=$5 \times 10^{13} cm^{-3}$). This involves reacting the MGS with dry HCl according to the following reaction to form trichlorosilane, $SiHCl_3$, which is a liquid with a boiling point of 32 ° C.

$$Si + 3HCl \rightarrow SiHCl_3 + H_2 \qquad (1-3)$$

Along with $SiHCl_3$, chlorides of impurities such as $FeCl_3$ are formed which fortunately have boiling points that are different from that of $SiHCl_3$. This allows a technique called fractional distillation to be used, in which we heat up the mixture of $SiHCl_3$ and the impurity chlorides, and condense the vapors in different distillation towers held at appropriate temperatures. We can thereby separate pure $SiHCl_3$ from the impurities. $SiHCl_3$ is then converted to highly pure EGS by reaction with H_2,

$$2SiHCl_3 + 2H_2 \rightarrow 2Si + 6HCl \qquad (1-4)$$

1.3.2 Growth of Single Crystal Ingots

Next, we have to convert the high purity but still polycrystalline EGS to single-crystal Si ingots or boules. This is generally done today by a process commonly called the *Czochralski* method. In order to grow single-crystal material, it is necessary to have a seed crystal which can provide a template for growth. We melt the EGS in a quartz-lined graphite crucible by resistively heating it to the melting point of Si (1412°C).

A seed crystal is lowered into the molten material and then is raised slowly, allowing the crystal to grow onto the seed (Fig. 1-10). Generally, the crystal is rotated slowly as it grows to provide a slight stirring of the melt and to average out any temperature variations that would cause inhomogeneous solidification. This technique is widely used in growing Si, Ge, and some of the compound semiconductors.

In pulling compounds such as GaAs from the melt, it is necessary to prevent volatile elements (e.g., As) from vaporizing. In one method a layer of B_2O_3, which is dense and viscous when molten, floats on the surface of the molten GaAs to prevent As evaporation. This growth method is called *liquid-encapsulated Czochralski (LEC)* growth.

In Czochralski crystal growth, the shape of the ingot is determined by a combination of the tendency of the cross section to assume a polygonal shape due to the crystal structure and the influence of surface tension, which encourages a circular cross section. The crystal facets are noticeable in the initial growth near the seed crystal in Fig. 1–10(b). However, the cross section of the large ingot in Fig. 1–11 is almost circular.

Figure 1–10
Pulling of a Si crystal from the melt (Czochralski method): (a) schematic diagram of the crystal growth process; (b) an 8-in. diameter, ⟨100⟩ oriented Si crystal being pulled from the melt. (Photograph courtesy of MEMC Electronics Intl.)

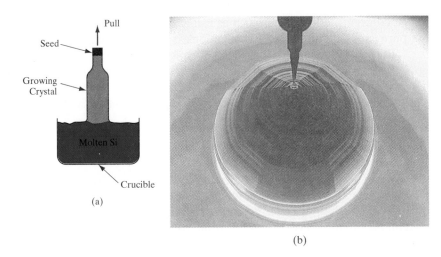

(a)

(b)

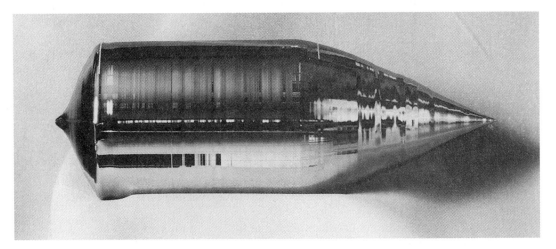

Figure 1–11
Silicon crystal grown by the Czochralski method. This large single-crystal ingot provides 300 mm (12-in.) diameter wafers when sliced using a saw. The ingot is about 1.5 m long (excluding the tapered regions), and weighs about 275 kg. (Photograph courtesy of MEMC Electronics Intl.)

In the fabrication of Si integrated circuits (Chapter 9) it is economical to use very large Si wafers, so that many IC chips can be made simultaneously. As a result, considerable research and development have gone into methods for growing very large Si crystals. For example, Fig. 1–11 illustrates a 12-inch-diameter Si ingot, 1.5 m long, weighing 275 kg.

1.3.3 Wafers

After the single-crystal ingot is grown, it is then mechanically processed to manufacture wafers. The first step involves mechanically grinding the more-

or-less cylindrical ingot into a perfect cylinder with a precisely controlled diameter. This is important because in a modern integrated circuit fabrication facility many processing tools and wafer handling robots require tight tolerances on the size of the wafers. Using X-ray crystallography, crystal planes in the ingot are identified. For reasons discussed in Section 6.4.3, most Si ingots are grown along the ⟨100⟩ direction (Fig. 1–10). For such ingots, a small notch is ground on one side of the cylinder to delineate a {110} face of the crystal. This is useful because for ⟨100⟩ Si wafers, the {110} cleavage planes are orthogonal to each other. This notch then allows the individual integrated circuit chips to be made oriented along {110} planes so that when the chips are sawed apart, there is less chance of spurious cleavage of the crystal, which could cause good chips to be lost.

Next, the Si cylinder is sawed into individual wafers about 775 μm thick, by using a diamond-tipped inner-hole blade saw, or a wire saw (Fig. 1–12a).

Figure 1–12
Steps involved in manufacturing Si wafers: (a) A 300 mm Si cylindrical ingot, with a notch on one side, being loaded into a wire saw to produce Si wafers; (b) a technician holding a cassette of 300 mm wafers. (Photographs courtesy of MEMC Electronics Intl.)

The resulting wafers are mechanically lapped and ground on both sides to achieve a flat surface, and to remove the mechanical damage due to sawing. Such damage would have a detrimental effect on devices. The flatness of the wafer is critical from the point of view of "depth of focus" or how sharp an image can be focussed on the wafer surface during photolithography, as discussed in Chapter 5. The Si wafers are then rounded or "chamfered" along the edges to minimize the likelihood of chipping the wafers during processing. Finally, the wafers undergo chemical-mechanical polishing using a slurry of very fine SiO_2 particles in a basic NaOH solution to give the front surface of the wafer a mirror-like finish. The wafers are now ready for integrated circuit fabrication (Fig. 1–12b). The economic value added in this process is impressive. From sand (SiO_2) costing pennies, we can obtain Si wafers costing a few hundred dollars, on which we can make hundreds of microprocessors, for example, each costing several hundred dollars.

1.3.4 Doping

As previously mentioned, there are some impurities in the molten EGS. We may also add intentional impurities or dopants to the Si melt to change its electronic properties. At the solidifying interface between the melt and the solid, there will be a certain distribution of impurities between the two phases. An important quantity that identifies this property is the *distribution coefficient k_d*, which is the ratio of the concentration of the impurity in the solid C_S to the concentration in the liquid C_L at equilibrium:

$$k_d = \frac{C_S}{C_L} \tag{1–5}$$

The distribution coefficient is a function of the material, the impurity, the temperature of the solid–liquid interface, and the growth rate. For an impurity with a distribution coefficient of one-half, the relative concentration of the impurity in the molten liquid to that in the refreezing solid is two to one. Thus the concentration of impurities in that portion of material that solidifies first is one-half the original concentration C_0. The distribution coefficient is thus important during growth from a melt. This can be illustrated by an example involving Czochralski growth:

EXAMPLE 1–4   A Si crystal is to be grown by the Czochralski method, and it is desired that the ingot contain 10^{16} phosphorus atoms/cm^3.

   (a) What concentration of phosphorus atoms should the melt contain to give this impurity concentration in the crystal during the initial growth? For P in Si, $k_d = 0.35$.

(b) If the initial load of Si in the crucible is 5 kg, how many grams of phosphorus should be added? The atomic weight of phosphorus is 31.

(a) Assume that $C_S = k_d C_L$ throughout the growth. Thus the initial concentration of P in the melt should be

SOLUTION

$$\frac{10^{16}}{0.35} = 2.86 \times 10^{16}\, \text{cm}^{-3}$$

(b) The P concentration is so small that the volume of melt can be calculated from the weight of Si. From Example 1–3 the density of Si is 2.33 g/cm³. In this example we will neglect the difference in density between solid and molten Si.

$$\frac{5000\ \text{g of Si}}{2.33\ \text{g/cm}^3} = 2146\ \text{cm}^3\ \text{of Si}$$

$$2.86 \times 10^{16}\ \text{cm}^{-3} \times 2146\ \text{cm}^3 = 6.14 \times 10^{19}\ \text{P atoms}$$

$$\frac{6.14 \times 10^{19}\ \text{atoms} \times 31\ \text{g/mole}}{6.02 \times 10^{23}\ \text{atoms/mole}} = 3.16 \times 10^{-3}\ \text{g of P}$$

Since the P concentration in the growing crystal is only about one-third of that in the melt, Si is used up more rapidly than P in the growth. Thus the melt becomes richer in P as the growth proceeds, and the crystal is doped more heavily in the latter stages of growth. This assumes that k_d is not varied; a more uniformly doped ingot can be grown by varying the pull rate (and therefore k_d) appropriately. Modern Czochralski growth systems use computer controls to vary the temperature, pull rate, and other parameters to achieve fairly uniformly doped ingots.

One of the most important and versatile methods of crystal growth for device applications is the growth of a thin crystal layer on a wafer of a compatible crystal. The substrate crystal may be a wafer of the same material as the grown layer or a different material with a similar lattice structure. In this process the substrate serves as the seed crystal onto which the new crystalline material grows. The growing crystal layer maintains the crystal structure and orientation of the substrate. The technique of growing an oriented single-crystal layer on a substrate wafer is called *epitaxial growth*, or *epitaxy*. As we shall see in this section, epitaxial growth can be performed at temperatures considerably below the melting point of the substrate crystal. A variety of methods are used to provide the appropriate atoms to the surface of the

**1.4
EPITAXIAL
GROWTH**

growing layer. These methods include *chemical vapor deposition (CVD)*,[2] growth from a melt (*liquid-phase epitaxy, LPE*), and evaporation of the elements in a vacuum (*molecular beam epitaxy, MBE*). With this wide range of epitaxial growth techniques, it is possible to grow a variety of crystals for device applications, having properties specifically designed for the electronic or optoelectronic device being made.

1.4.1 Lattice Matching in Epitaxial Growth

When Si epitaxial layers are grown on Si substrates, there is a natural matching of the crystal lattice, and high-quality single-crystal layers result. On the other hand, it is often desirable to obtain epitaxial layers that differ somewhat from the substrate, which is known as *heteroepitaxy*. This can be accomplished easily if the lattice structure and lattice constant *a* match for the two materials. For example, GaAs and AlAs both have the zincblende structure, with a lattice constant of about 5.65 Å. As a result, epitaxial layers of the ternary alloy AlGaAs can be grown on GaAs substrates with little lattice mismatch. Similarly, GaAs can be grown on Ge substrates (see Appendix III).

Since AlAs and GaAs have similar lattice constants, it is also true that the ternary alloy AlGaAs has essentially the same lattice constant over the entire range of compositions from AlAs to GaAs. As a result, one can choose the composition *x* of the ternary compound $Al_xGa_{1-x}As$ to fit the particular device requirement, and grow this composition on a GaAs wafer. The resulting epitaxial layer will be lattice-matched to the GaAs substrate.

Figure 1–13 illustrates the energy band gap E_g as a function of lattice constant *a* for several III–V ternary compounds as they are varied over their composition ranges. For example, as the ternary compound InGaAs is varied by choice of composition on the column III sublattice from InAs to GaAs, the band gap changes from 0.36 to 1.43 eV while the lattice constant of the crystal varies from 6.06 Å for InAs to 5.65 Å for GaAs. Clearly, we cannot grow this ternary compound over the entire composition range on a particular binary substrate, which has a fixed lattice constant. As Fig. 1–13 illustrates, however, it is possible to grow a specific composition of InGaAs on an InP substrate. The vertical (invariant lattice constant) line from InP to the InGaAs curve shows that a midrange ternary composition (actually, $In_{0.53}Ga_{0.47}As$) can be grown lattice-matched to an InP substrate. Similarly, a ternary InGaP alloy with about 50 percent Ga and 50 percent In on the column III sublattice can be grown lattice-matched to a GaAs substrate. To achieve a broader range of alloy compositions, grown lattice-matched on particular substrates, it is helpful to

[2]The generic term *chemical vapor deposition* includes deposition of layers that may be polycrystalline or amorphous. When a CVD process results in a single-crystal epitaxial layer, a more specific term is *vapor-phase epitaxy (VPE)*.

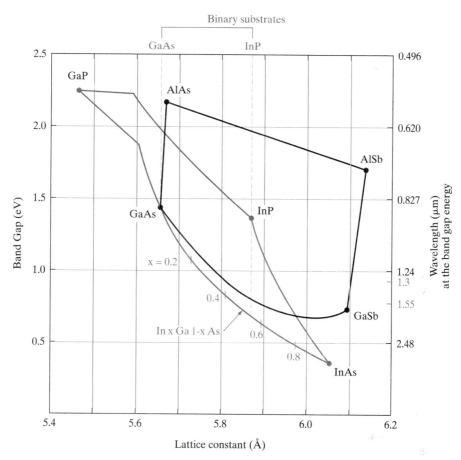

Figure 1–13

Relationship between band gap and lattice constant for alloys in the InGaAsP and AlGaAsSb systems. The dashed vertical lines show the lattice constants for the commercially available binary substrates GaAs and InP. For the marked example of $In_xGa_{1-x}As$, the ternary composition $x = 0.53$ can be grown lattice-matched on InP, since the lattice constants are the same. For quaternary alloys, the compositions on both the III and V sublattices can be varied to grow lattice-matched epitaxial layers along the dashed vertical lines between curves. For example, $In_xGa_{1-x}As_yP_{1-y}$ can be grown on InP substrates, with resulting band gaps ranging from 0.75 eV to 1.35 eV. In using this figure, assume the lattice constant a of a ternary alloy varies linearly with the composition x.

use quaternary alloys such as InGaAsP. The variation of compositions on both the column III and column V sublattices provides additional flexibility in choosing a particular band gap while providing lattice-matching to convenient binary substrates such as GaAs or InP.

In the case of GaAsP, the lattice constant is intermediate between that of GaAs and GaP, depending upon the composition. For example, GaAsP crystals used in red LEDs have 40 percent phosphorus and 60 percent arsenic on the column V sublattice. Since such a crystal cannot be grown directly on either a GaAs or a GaP substrate, it is necessary to gradually change the lattice constant as the crystal is grown. Using a GaAs or Ge wafer as a substrate, the growth is begun at a composition near GaAs. A region ~25 μm thick is grown while gradually introducing phosphorus until the desired As/P ratio is achieved. The desired epitaxial layer (e.g., 100 μm thick) is then grown on this graded layer. By this method epitaxial growth always occurs on a crystal of similar lattice constant. Although some crystal dislocations occur due to lattice strain in the graded region, such crystals are of high quality and can be used in LEDs.

In addition to the widespread use of lattice-matched epitaxial layers, the advanced epitaxial growth techniques described in the following sections allow the growth of very thin (~100Å) layers of lattice-mismatched crystals. If the mismatch is only a few percent and the layer is thin, the epitaxial layer grows with a lattice constant in compliance with that of the seed crystal (Fig. 1–14). The resulting layer is in compression or tension along the surface plane

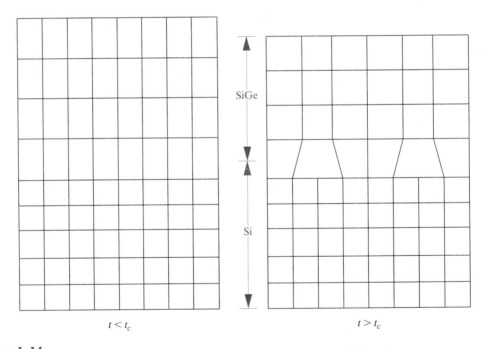

$t < t_c$ $t > t_c$

Figure 1–14
Heteroepitaxy and misfit dislocations. For example, in heteroepitaxy of a SiGe layer on Si, the lattice mismatch between SiGe and Si leads to compressive strain in the SiGe layer. The amount of strain depends on the mole fraction of Ge. (a) For layer thicknesses less than the critical layer thickness, t_c, pseudomorphic growth occurs. (b) However, above t_c, misfit dislocations form at the interface which may reduce the usefulness of the layers in device applications.

as its lattice constant adapts to the seed crystal (Fig. 1–14). Such a layer is called *pseudomorphic* because it is not lattice-matched to the substrate without strain. However, if the epitaxial layer exceeds a critical layer thickness, t_c, which depends on the lattice mismatch, the strain energy leads to formation of defects called *misfit dislocations.* Using thin alternating layers of slightly mismatched crystal layers, it is possible to grow a *strained-layer superlattice* (*SLS*) in which alternate layers are in tension and compression. The overall SLS lattice constant is an average of that of the two bulk materials.

1.4.2 Vapor-Phase Epitaxy

The advantages of low temperature and high purity epitaxial growth can be achieved by crystallization from the vapor phase. Crystalline layers can be grown onto a seed or substrate from a chemical vapor of the semiconductor material or from mixtures of chemical vapors containing the semiconductor. *Vapor-phase epitaxy* (*VPE*) is a particularly important source of semiconductor material for use in devices. Some compounds such as GaAs can be grown with better purity and crystal perfection by vapor epitaxy than by other methods. Furthermore, these techniques offer great flexibility in the actual fabrication of devices. When an epitaxial layer is grown on a substrate, it is relatively simple to obtain a sharp demarcation between the type of impurity doping in the substrate and in the grown layer. The advantages of this freedom to vary the impurity will be discussed in subsequent chapters. We point out here, however, that Si integrated-circuit devices (Chapter 9) are usually built in layers grown by VPE on Si wafers.

Epitaxial layers are generally grown on Si substrates by the controlled deposition of Si atoms onto the surface from a chemical vapor containing Si. In one method, a gas of silicon tetrachloride reacts with hydrogen gas to give Si and anhydrous HCl:

$$SiCl_4 + 2H_2 \rightleftharpoons Si + 4HCl \qquad (1-6)$$

If this reaction occurs at the surface of a heated crystal, the Si atoms released in the reaction can be deposited as an epitaxial layer. The HCl remains gaseous at the reaction temperature and does not disturb the growing crystal. As indicated, this reaction is reversible. This is very important because it implies that by adjusting the process parameters, the reaction in Eq. (1–6) can be driven to the left (providing etching of the Si rather than deposition). This etching can be used for preparing an atomically clean surface on which epitaxy can occur.

This vapor epitaxy technique requires a chamber into which the gases can be introduced and a method for heating the Si wafers. Since the chemical reactions take place in this chamber, it is called a *reaction chamber* or, more simply, a *reactor*. Hydrogen gas is passed through a heated vessel in which $SiCl_4$

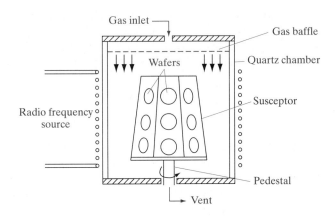

Figure 1–15
A barrel-type reactor for Si VPE. These are atmospheric pressure systems. The Si wafers are held in slots cut on the sides of a SiC-coated graphite susceptor that flares out near the base to promote gas flow patterns conducive to uniform epitaxy.

is evaporated; then the two gases are introduced into the reactor over the substrate crystal, along with other gases containing the desired doping impurities. The Si slice is placed on a graphite susceptor or some other material that can be heated to the reaction temperature with an rf heating coil or tungsten halogen lamps. This method can be adapted to grow epitaxial layers of closely controlled impurity concentration on many Si slices simultaneously (Fig. 1–15).

The reaction temperature for the hydrogen reduction of $SiCl_4$ is approximately 1150–1250°C. Other reactions may be employed at somewhat lower temperatures, including the use of dichlorosilane (SiH_2Cl_2) at 1000–1100 °C, or the pyrolysis of silane (SiH_4) at 1000°C. Pyrolysis involves the breaking up of the silane at the reaction temperature:

$$SiH_4 \rightarrow Si + 2H_2 \qquad (1\text{--}7)$$

There are several advantages of the lower reaction temperature processes, including the fact that they reduce migration of impurities from the substrate to the growing epitaxial layer.

In some applications it is useful to grow thin Si layers on insulating substrates. For example, vapor-phase epitaxial techniques can be used to grow ~1μm Si films on sapphire and other insulators. This application of VPE is discussed in Section 9.3.2.

Vapor-phase epitaxial growth is also important in the III–V compounds, such as GaAs, GaP, and the ternary alloy GaAsP, which is widely used in the fabrication of LEDs. Substrates are held at about 800°C on a rotating wafer holder while phosphine, arsine, and gallium chloride gases are mixed and passed over the samples. The GaCl is obtained by reacting anhydrous HCl with molten Ga within the reactor. Variation of the crystal composition for GaAsP can be controlled by altering the mixture of arsine and phosphine gases.

Another useful method for epitaxial growth of compound semiconductors is called *metal-organic vapor-phase epitaxy (MOVPE)*, or *organometallic vapor-phase epitaxy (OMVPE)*. For example, the organometallic compound trimethylgallium can be reacted with arsine to form GaAs and methane:

$$(CH_3)_3Ga + AsH_3 \rightarrow GaAs + 3CH_4 \qquad (1-8)$$

This reaction takes place at about 700°C, and epitaxial growth of high-quality GaAs layers can be obtained. Other compound semiconductors can also be grown by this method. For example, trimethylaluminum can be added to the gas mixture to grow AlGaAs. This growth method is widely used in the fabrication of a variety of devices, including solar cells and lasers. The convenient variability of the gas mixture allows the growth of multiple thin layers similar to those discussed below for molecular beam epitaxy.

1.4.3 Molecular Beam Epitaxy

One of the most versatile techniques for growing epitaxial layers is called *molecular beam epitaxy (MBE)*. In this method the substrate is held in a high vacuum while molecular or atomic beams of the constituents impinge upon its surface (Fig. 1–16a). For example, in the growth of AlGaAs layers on GaAs substrates, the Al, Ga, and As components, along with dopants, are heated in separate cylindrical cells. Collimated beams of these constituents escape into the vacuum and are directed onto the surface of the substrate. The rates at which these atomic beams strike the surface can be closely controlled, and growth of very high quality crystals results. The sample is held at a relatively low temperature (about 600° C for GaAs) in this growth procedure. Abrupt changes in doping or in crystal composition (e.g., changing from GaAs to AlGaAs) can be obtained by controlling shutters in front of the individual beams. Using slow growth rates (≤ 1 μm/h), it is possible to control the shutters to make composition changes on the scale of the lattice constant. For example, Fig. 1–16b illustrates a portion of a crystal grown with alternating layers of GaAs and AlGaAs only four monolayers thick. Because of the high vacuum and close controls involved, MBE requires a rather sophisticated setup (Fig. 1–17). However, the versatility of this growth method makes it very attractive for many applications.

As MBE has developed in recent years, it has become common to replace some of the solid sources shown in Fig. 1–16 with gaseous chemical sources. This approach, called *chemical beam epitaxy*, or *gas-source MBE*, combines many of the advantages of MBE and VPE.

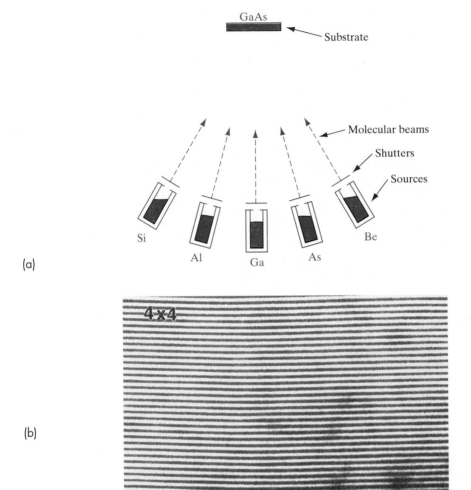

Figure 1–16

Crystal growth by molecular beam epitaxy (MBE): (a) evaporation cells inside a high -vacuum chamber directing beams of Al, Ga, As, and dopants onto a GaAs substrate; (b) scanning electron micrograph of the cross section of an MBE-grown crystal having alternating layers of GaAs (dark lines) and AlGaAs (light lines). Each layer is four monolayers ($4 \times a/2 = 11.3 \text{Å}$) thick. (Photograph courtesy of Bell Laboratories.)

Figure 1–17
Molecular beam epitaxy facility in the Microelectronics Research Center at the University of Texas at Austin.

1.1 Using Appendix III, which of the listed semiconductors in Table 1–1 has the largest band gap? The smallest? What are the corresponding wavelengths if light is emitted at the energy E_g? Is there a noticeable pattern in the band gap energy of III–V compounds related to the column III element?

PROBLEMS

1.2 For a bcc lattice of identical atoms with a lattice constant of 5Å, calculate the maximum packing fraction and the radius of the atoms treated as hard spheres with nearest neighbors touching.

1.3 (a) Label the planes illustrated in Fig. P1–3.

Figure P1–3

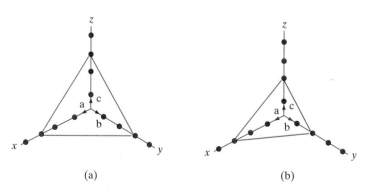

(a)   (b)

(b) Draw equivalent $\langle 111 \rangle, \langle 100 \rangle, \langle 110 \rangle$ directions in a cubic lattice; use a unit cube for illustrating each set of equivalent directions.

1.4 Calculate the volume density of Si atoms (number of atoms/cm³) given that the lattice constant of Si is 5.43 Å. Calculate the areal density of atoms (number/cm²) on the (110) plane. Calculate the distance between two adjacent (111) planes in Si passing through nearest-neighbor atoms.

1.5 The atomic radii of In and Sb atoms are approximately 1.44 Å and 1.36 Å, respectively. Using the hard-sphere approximation, find the lattice constant of InSb (zincblende structure), and the volume of the primitive cell. What is the atomic density on the (110) planes? (Hint: The volume of the primitive cell is $^1/_4$ the fcc unit cell volume.)

1.6 Sodium chloride (NaCl) is a cubic crystal that differs from a sc in that alternating atoms are different; each Na is surrounded by six Cl nearest neighbors and vice versa in the three-dimensional lattice. Draw a two-dimensional NaCl lattice looking down a ⟨100⟩ direction and indicate a unit cell. Remember the unit cell must be repetitive upon displacement by the basis vectors.

1.7 Sketch a view down a ⟨110⟩ direction of a diamond lattice, using Fig. 1–9 as a guide. Include lines connecting nearest neighbors.

1.8 Show by a sketch that the bcc lattice can be represented by two interpenetrating sc lattices. To simplify the sketch, show a ⟨100⟩ view of the lattice.

1.9 (a) Find the number of atoms/cm^2 on the (100) surface of a Si wafer.

   (b) What is the distance (in Å) between nearest In neighbors in InP?

1.10 The ionic radii of Na$^+$ (atomic weight 23) and Cl$^-$ (atomic weight 35.5) are 1.0 and 1.8 Å, respectively. Treating the ions as hard spheres, calculate the density of NaCl. Compare this with the measured density of 2.17 g/cm^3.

1.11 The atoms seen in Fig. 1–8b along a ⟨100⟩ direction of the diamond lattice are not all coplanar. Taking the top plane of colored atoms in Fig. 1–8a to be (0), the parallel plane a/4 down to be $(\frac{1}{4})$, the plane through the center to be $(\frac{1}{2})$, and the second plane of black atoms to be $(\frac{3}{4})$, label the plane of each atom in Fig. 1–8b.

1.12 How many atoms are found inside a unit cell of a simple cubic, body-centered cubic, and face-centered cubic crystal? How far apart in terms of lattice constant a are nearest-neighbor atoms in each case, measured from center to center?

1.13 Draw a cube such as Fig. 1–7 and show four {111} planes with different orientations. Repeat for {110} planes.

1.14 Find the maximum fractions of the unit cell volume that can be filled by hard spheres in the sc, bcc, and diamond lattices.

1.15 Calculate the densities of Ge and InP from the lattice constants (Appendix III), atomic weights, and Avogadro's number. Compare the results with the densities given in Appendix III.

1.16 Beginning with a sketch of an fcc lattice, add atoms at $(\frac{1}{4}, \frac{1}{4}, \frac{1}{4})$ from each fcc atom to obtain the diamond lattice. Show that only the four added atoms in Fig. 1-8a appear in the diamond unit cell.

1.17 Assuming the lattice constant varies linearly with composition x for a ternary alloy (e.g., see the variation for InGaAs in Fig. 1–13), what composition of AlSb$_x$As$_{1-x}$ is lattice matched to InP? What composition of In$_x$Ga$_{1-x}$P is lattice-matched to GaAs? What is the band gap energy in each case?

1.18 A Si crystal is to be pulled from the melt and doped with arsenic ($k_d = 0.3$). If the Si weighs 1 kg, how many grams of arsenic should be introduced to achieve 10^{15} cm^{-3} doping during the initial growth?

Ashcroft, N. W., and N. D. Mermin. *Solid State Physics.* Philadelphia: W.B. Saunders, 1976.

Buckley, D. N. "The Light Fantastic: Materials and Processing Technologies for Photonics." *The Electrochemical Society Interface* 1 (Winter 1992): 41+.

Capasso, F. "Bandgap and Interface Engineering for Advanced Electronic and Photonic Devices." *Thin Solid Films* 216 (28 August 1992): 59–67.

Denbaars, S.P. "Gallium–Nitride-Based Materials for Blue to Ultraviolet Optoelectronic Devices," *Proc. IEEE* 85(11) (November 1997): 1740–1749.

Hammond, M. L. "Epitaxial Silicon Reactor Technology—A Review: I. Reactor Technology." *Solid State Technology* 31 (May 1988): 159–64.

Herman, M. A. *Molecular Beam Epitaxy: Fundamentals and Current Status.* Berlin: Springer-Verlag, 1989.

Houng, Y. M. "Chemical Beam Epitaxy." *Critical Reviews in Solid State and Materials Sciences* 17 (1992): 277–306.

Jungbluth, E. D. "Crystal Growth Methods Shape Communications Lasers." *Laser Focus World* 29 (February 1993): 61–72.

Kasper, E., and J. F. Luy. "Molecular Beam Epitaxy of Silicon Based Electronic Structures." *Microelectronics Journal* 22 (May 1992): 5–16.

Kittel, C. *Introduction to Solid State Physics*, 6th ed. New York: Wiley, 1986.

Kuphal, E. "Liquid Phase Epitaxy." *Applied Physics* A52 (June 1991): 380–409.

Levi, B. G. "What's the Shape of Things to Come in Semiconductors." *Physics Today* 45 (September 1992): 17+.

Li, S. S. *Semiconductor Physical Electronics.* New York: Plenum Press, 1993.

Liaw, H. M., and J. W. Rose. "Silicon Vapor-Phase Epitaxy." In *Epitaxial Silicon Technology*, ed. B. J. Baliga. New York: Academic Press, 1986.

Narayanamurti, V. "Artificially Structured Thin-Film Materials and Interfaces." *Science* 235 (27 February 1987): 1023+.

Schubert, E. F. *Doping in III-V Semiconductors.* Cambridge: Cambridge University Press, 1993.

Speier, P. "MOVPE for Optoelectronics." *Microelectronic Engineering* 18 (May 1992): 1–31.

Stringfellow, G. B. *Organometallic Vapor-Phase Epitaxy.* New York: Academic Press, 1989.

Swaminathan, V., and Macrander, A. T. *Material Aspects of GaAs and InP Based Structures.* Englewood Cliffs, NJ: Prentice Hall, 1991.

Tsang, W. T. "Advances in MOVPE, MBE, and CBE." *Journal of Crystal Growth* 120 (May 1992): 1–24.

Chapter 2
Atoms and Electrons

Since this book is primarily an introduction to solid state devices, it would be preferable not to delay this discussion with subjects such as atomic theory, quantum mechanics, and electron models. However, the behavior of solid state devices is directly related to these subjects. For example, it would be difficult to understand how an electron is transported through a semiconductor device without some knowledge of the electron and its interaction with the crystal lattice. Therefore, in this chapter we shall investigate some of the important properties of electrons, with special emphasis on two points: (1) the electronic structure of atoms, and (2) the interaction of atoms and electrons with excitation, such as the absorption and emission of light. By studying electron energies in an atom, we lay the foundation for understanding the influence of the lattice on electrons participating in current flow through a solid. Our discussions concerning the interaction of light with electrons form the basis for later descriptions of changes in the conductivity of a semiconductor with optical excitation, properties of light-sensitive devices, and lasers.

First, we shall investigate some of the experimental observations which led to the modern concept of the atom, and then we shall give a brief introduction to the theory of quantum mechanics. Several important concepts will emerge from this introduction: the electrons in atoms are restricted to certain energy levels by quantum rules; the electronic structure of atoms is determined from these quantum conditions; and this "quantization" defines certain allowable transitions involving absorption and emission of energy by the electrons.

2.1 INTRODUCTION TO PHYSICAL MODELS

The main effort of science is to describe what happens in nature, in as complete and concise a form as possible. In physics this effort involves observing natural phenomena, relating these observations to previously established theory, and finally establishing a physical model for the observations. The primary purpose of the model is to allow the information obtained in present observations to be used to understand new experiments. Therefore, the most useful models are expressed mathematically, so that quantitative explanations of new experiments can be made succinctly in terms of established principles. For example, we can explain the behavior of a spring-supported weight moving up and down periodically after an initial displacement, because the differential equations describing such a simple harmonic motion have been established and are understood by students of elementary physics. But the physical model upon which these equations of motion are based arises from serious study of natural phenomena such as gravitational force, the response of bodies to accelerating forces, the relationship of kinetic and potential energy, and the properties of springs. The mass and spring problem is relatively easy to solve because each of these properties of nature is well understood.

When a new physical phenomenon is observed, it is necessary to find out how it fits into the established models and "laws" of physics. In the vast majority of cases this involves a direct extension of the mathematics of well-established models to the particular conditions of the new problem. In fact, it is not uncommon for a scientist or engineer to predict that a new phenomenon should occur before it is actually observed, simply by a careful study and extension of existing models and laws. The beauty of science is that natural phenomena are not isolated events but are related to other events by a few analytically describable laws. However, it does happen occasionally that a set of observations cannot be described in terms of existing theories. In such cases it is necessary to develop models which are based as far as possible on existing laws, but which contain new aspects arising from the new phenomena. Postulating new physical principles is a serious business, and it is done only when there is no possibility of explaining the observations with established theory. When new assumptions and models are made, their justification lies in the following question: "Does the model describe precisely the observations, and can reliable predictions be made based on the model?" The model is good or poor depending on the answer to this question.

In the 1920s it became necessary to develop a new theory to describe phenomena on the atomic scale. A long series of careful observations had been made that clearly indicated that many events involving electrons and atoms did not obey the classical laws of mechanics. It was necessary, therefore, to develop a new kind of mechanics to describe the behavior of particles on this small scale. This new approach, called *quantum mechanics,* describes atomic phenomena very well and also properly predicts the way in which electrons behave in solids—our primary interest here. Through the years, quantum mechanics has been so successful that now it stands beside the classical laws as a valid description of nature.

A special problem arises when students first encounter the theory of quantum mechanics. The problem is that quantum concepts are largely mathematical in nature and do not involve the "common sense" quality associated with classical mechanics. At first, many students find quantum concepts difficult, not so much because of the mathematics involved, but because they feel the concepts are somehow divorced from "reality." This is a reasonable reaction, since ideas which we consider to be real or intuitively satisfying are usually based on our own observation. Thus the classical laws of motion are easy to understand because we observe bodies in motion every day. On the other hand, we observe the effects of atoms and electrons only indirectly, and naturally we have very little feeling for what is happening on the atomic scale. It is necessary, therefore, to depend on the facility of the theory to predict experimental results rather than to attempt to force classical analogues onto the nonclassical phenomena of atoms and electrons.

Our approach in this chapter will be to investigate the important experimental observations that led to the quantum theory, and then to indicate how the theory accounts for these observations. Discussions of quantum theory must necessarily be largely qualitative in such a brief presentation, and those topics that are most important to solid state theory will be emphasized here. Several good references for further individual study are given at the end of this chapter.

**2.2
EXPERIMENTAL
OBSERVATIONS**

The experiments that led to the development of quantum theory were concerned with the nature of light and the relation of optical energy to the energies of electrons within atoms. These experiments supplied only indirect evidence of the nature of phenomena on the atomic scale; however, the cumulative results of a number of careful experiments showed clearly that a new theory was needed.

2.2.1 The Photoelectric Effect

An important observation by Planck indicated that radiation from a heated sample is emitted in discrete units of energy, called *quanta;* the energy units were described by $h\nu$, where ν is the frequency of the radiation, and h is a quantity now called Planck's constant ($h = 6.63 \times 10^{-34}$ J-s). Soon after Planck developed this hypothesis, Einstein interpreted an important experiment that clearly demonstrated the discrete nature (*quantization*) of light. This experiment involved absorption of optical energy by the electrons in a metal and the relationship between the amount of energy absorbed and the frequency of the light (Fig. 2–1). Let us suppose that monochromatic light is incident on the surface of a metal plate in a vacuum. The electrons in the metal absorb energy from the light, and some of the electrons receive enough energy to be ejected from the metal surface into the vacuum. This phenomenon is called the *photoelectric effect*. If the energy of the escaping electrons is measured, a plot can be made of the maximum energy as a function of the frequency ν of the incident light (Fig. 2–1b).

One simple way of finding the maximum energy of the ejected electrons is to place another plate above the one shown in Fig. 2–1a and then create an electric field between the two plates. The potential necessary to retard all electron flow between the plates gives the energy E_m. For a particular frequency of light incident on the sample, a maximum energy E_m is observed for the emitted electrons. The resulting plot of E_m vs. ν is linear, with a slope equal to Planck's constant. The equation of the line shown in Fig. 2–1b is

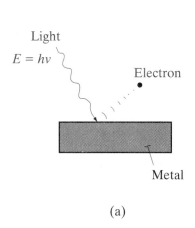

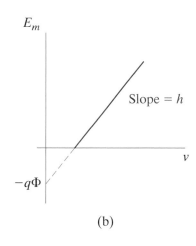

Figure 2–1
The photoelectric
effect: (a) elec-
trons are ejected
from the surface
of a metal when
exposed to light
of frequency v in
a vacuum; (b) plot
of the maximum
kinetic energy of
ejected electrons
vs. frequency of
the incoming
light.

(a)   (b)

$$E_m = hv - q\Phi \qquad\qquad (2\text{–}1)$$

where q is the magnitude of the electronic charge. The quantity Φ (volts) is a characteristic of the particular metal used. When Φ is multiplied by the electronic charge, an energy (joules) is obtained which represents the minimum energy required for an electron to escape from the metal into a vacuum. The energy $q\Phi$ is called the *work function* of the metal. These results indicate that the electrons receive an energy hv from the light and lose an amount of energy $q\Phi$ in escaping from the surface of the metal.

This experiment demonstrates clearly that Planck's hypothesis was correct —light energy is contained in discrete units rather than in a continuous distribution of energies. Other experiments also indicate that, in addition to the wave nature of light, the quantized units of light energy can be considered as localized packets of energy, called *photons*. Some experiments emphasize the wave nature of light, while other experiments reveal the discrete nature of photons. This duality is fundamental to quantum processes and does not imply an ambiguity in the theory.

2.2.2 Atomic Spectra

One of the most valuable experiments of modern physics is the analysis of absorption and emission of light by atoms. For example, an electric discharge can be created in a gas, so that the atoms begin to emit light with wavelengths characteristic of the gas. We see this effect in a neon sign, which is typically a glass tube filled with neon or a gas mixture, with electrodes for creating a discharge. If the intensity of the emitted light is measured as a function of wavelength, one finds a series of sharp lines rather than a continuous distribution of wavelengths. By the early 1900s the characteristic spectra for several atoms were well known. A portion of the measured emission spectrum for hydrogen is shown in Fig. 2–2, in which the vertical lines represent the positions of observed emission peaks on the wavelength scale. Wavelength (λ) is usually measured in angstroms (1 Å $= 10^{-10}$ m) and is related (in meters) to frequency by $\lambda = c/v$, where c is the speed of light (3×10^8 m/s). Photo energy hv is then related to wavelength by

$$E = hv = \frac{hc}{\lambda} \tag{2–2}$$

The lines in Fig. 2–2 appear in several groups labeled the *Lyman, Balmer,* and *Paschen* series after their early investigators. Once the hydrogen spectrum was established, scientists noticed several interesting relationships among the lines. The various series in the spectrum were observed to follow certain empirical forms:

$$\text{Lyman:} \quad v = cR\left(\frac{1}{1^2} - \frac{1}{n^2}\right), \quad n = 2, 3, 4, \ldots \tag{2–3a}$$

$$\text{Balmer:} \quad v = cR\left(\frac{1}{2^2} - \frac{1}{n^2}\right), \quad n = 3, 4, 5, \ldots \tag{2–3b}$$

$$\text{Paschen:} \quad v = cR\left(\frac{1}{3^2} - \frac{1}{n^2}\right), \quad n = 4, 5, 6, \ldots \tag{2–3c}$$

where R is a constant called the Rydberg constant ($R = 109{,}678$ cm^{-1}). If the photon energies hv are plotted for successive values of the integer **n,** we notice that each energy can be obtained by taking sums and differences of other photon energies in the spectrum (Fig. 2–3). For example, E_{42} in the Balmer series is the difference between E_{41} and E_{21} in the Lyman series. This relationship among the various series is called the *Ritz combination*

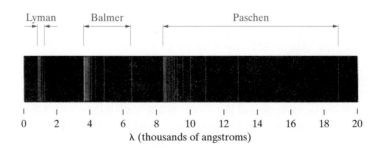

Figure 2–2
Some important lines in the emission spectrum of hydrogen.

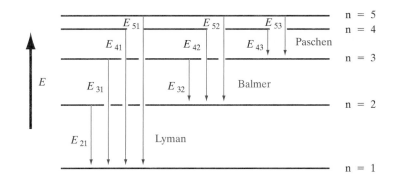

Figure 2–3
Relationships
among photon en-
ergies in the hy-
drogen spectrum.

principle. Naturally, these empirical observations stirred a great deal of interest in constructing a comprehensive theory for the origin of the photons given off by atoms.

The results of emission spectra experiments led Niels Bohr to construct a model for the hydrogen atom, based on the mathematics of planetary systems. If the electron in the hydrogen atom has a series of planetary-type orbits available to it, it can be excited to an outer orbit and then can fall to any one of the inner orbits, giving off energy corresponding to one of the lines of Fig. 2–3. To develop the model, Bohr made several postulates:

**2.3
THE BOHR MODEL**

1. Electrons exist in certain stable, circular orbits about the nucleus. This assumption implies that the orbiting electron does not give off radiation as classical electromagnetic theory would normally require of a charge experiencing angular acceleration; otherwise, the electron would not be stable in the orbit but would spiral into the nucleus as it lost energy by radiation.

2. The electron may shift to an orbit of higher or lower energy, thereby gaining or losing energy equal to the difference in the energy levels (by absorption or emission of a photon of energy $h\nu$).

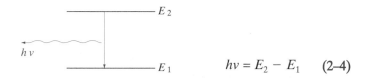

$$h\nu = E_2 - E_1 \qquad (2\text{--}4)$$

3. The angular momentum p_θ of the electron in an orbit is always an integral multiple of Planck's constant divided by 2π ($h/2\pi$ is often abbreviated $\hbar$ for convenience). This assumption,

$$p_\theta = n\hbar, \quad n = 1, 2, 3, 4, \ldots \tag{2-5}$$

is necessary to obtain the observed results of Fig. 2–3.

If we visualize the electron in a stable orbit of radius r about the proton of the hydrogen atom, we can equate the electrostatic force between the charges to the centripetal force:

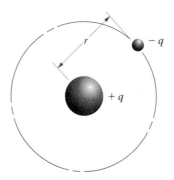

$$-\frac{q^2}{Kr^2} = -\frac{mv^2}{r} \tag{2-6}$$

where $K = 4\pi\epsilon_0$ in MKS units, m is the mass of the electron, and v is its velocity. From assumption 3 we have

$$p_\theta = mvr = n\hbar \tag{2-7}$$

Since n takes on integral values, r should be denoted by r_n to indicate the nth orbit. Then Eq. (2–7) can be written

$$m^2v^2 = \frac{n^2\hbar^2}{r_n^2} \tag{2-8}$$

Substituting Eq. (2–8) in Eq. (2–6) we find that

$$\frac{q^2}{Kr_n^2} = \frac{1}{mr_n} \cdot \frac{n^2\hbar^2}{r_n^2} \tag{2-9}$$

$$r_n = \frac{Kn^2\hbar^2}{mq^2} \tag{2-10}$$

for the radius of the nth orbit of the electron. Now we must find the expression for the total energy of the electron in this orbit, so that we can calculate the energies involved in transitions between orbits.

From Eqs. (2–7) and (2–10) we have

$$v = \frac{n\hbar}{mr_n} \qquad (2\text{-}11)$$

$$v = \frac{n\hbar q^2}{Kn^2\hbar^2} = \frac{q^2}{Kn\hbar} \qquad (2\text{-}12)$$

Therefore, the kinetic energy of the electron is

$$\text{K. E.} = \frac{1}{2}mv^2 = \frac{mq^4}{2K^2n^2\hbar^2} \qquad (2\text{-}13)$$

The potential energy is the product of the electrostatic force and the distance between the charges:

$$\text{P. E.} = -\frac{q^2}{Kr_n} = -\frac{mq^4}{K^2n^2\hbar^2} \qquad (2\text{-}14)$$

Thus the total energy of the electron in the **n**th orbit is

$$E_n = \text{K. E.} + \text{P. E.} = -\frac{mq^4}{2K^2n^2\hbar^2} \qquad (2\text{-}15)$$

The critical test of the model is whether energy differences between orbits correspond to the observed photon energies of the hydrogen spectrum. The transitions between orbits corresponding to the Lyman, Balmer, and Paschen series are illustrated in Fig. 2–4. The energy difference between orbits $\mathbf{n}_1$ and $\mathbf{n}_2$ is given by

$$E_{n2} - E_{n1} = \frac{mq^4}{2K^2\hbar^2}\left(\frac{1}{n_1^2} - \frac{1}{n_2^2}\right) \qquad (2\text{-}16)$$

The frequency of light given off by a transition between these orbits is

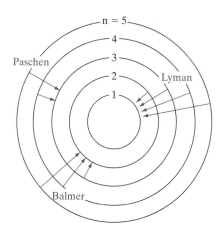

Figure 2–4
Electron orbits and transitions in the Bohr model of the hydrogen atom. Orbit spacing is not drawn to scale.

$$v_{21} = \left[\frac{mq^4}{2K^2\hbar^2 h}\right]\left(\frac{1}{\mathbf{n}_1^2} - \frac{1}{\mathbf{n}_2^2}\right) \qquad (2\text{-}17)$$

The factor in brackets is essentially the Rydberg constant R times the speed of light c. A comparison of Eq. (2–17) with the experimental results summed up by Eq. (2–3) indicates that the Bohr theory provides a good model for electronic transitions within the hydrogen atom, as far as the early experimental evidence is concerned.

Whereas the Bohr model accurately describes the gross features of the hydrogen spectrum, it does not include many fine points. For example, experimental evidence indicates some splitting of levels in addition to the levels predicted by the theory. Also, difficulties arise in extending the model to atoms more complicated than hydrogen. Attempts were made to modify the Bohr model for more general cases, but it soon became obvious that a more comprehensive theory was needed. However, the partial success of the Bohr model was an important step toward the eventual development of the quantum theory. The concept that electrons are quantized in certain allowed energy levels, and the relationship of photon energy and transitions between levels had been established firmly by the Bohr theory.

**2.4
QUANTUM
MECHANICS**

The principles of quantum mechanics were developed from two different points of view at about the same time (the late 1920s). One approach, developed by Heisenberg, utilizes the mathematics of matrices and is called *matrix mechanics*. Independently, Schrödinger developed an approach utilizing a wave equation, now called *wave mechanics*. These two mathematical formulations appear to be quite different. However, closer examination reveals that beyond the formalism, the basic principles of the two approaches are the same. It is possible to show, for example, that the results of matrix mechanics reduce to those of wave mechanics after mathematical manipulation. We shall concentrate here on the wave mechanics approach, since solutions to a few simple problems can be obtained with it, involving less mathematical discussion.

2.4.1 Probability and the Uncertainty Principle

It is impossible to describe with absolute precision events involving individual particles on the atomic scale. Instead, we must speak of the average values (*expectation values*) of position, momentum, and energy of a particle such as an electron. It is important to note, however, that the uncertainties revealed in quantum calculations are not based on some shortcoming of the theory. In fact, a major strength of the theory is that it describes the probabilistic nature of events involving atoms and electrons. The fact is that such quantities as the position and momentum of an electron *do not exist* apart from a particular uncertainty. The magnitude of this inherent uncertainty is described by the *Heisenberg uncertainty principle:*[1]

> In any measurement of the position and momentum of a particle, the uncertainties in the two measured quantities will be related by

$$(\Delta x)\,(\Delta p_x) \geq \hbar \qquad\qquad (2\text{--}18)$$

Similarly, the uncertainties in an energy measurement will be related to the uncertainty in the time at which the measurement was made by

$$(\Delta E)\,(\Delta t) \geq \hbar \qquad\qquad (2\text{--}19)$$

These limitations indicate that simultaneous measurement of position and momentum or of energy and time are inherently inaccurate to some degree. Of course, Planck's constant h is a rather small number (6.63×10^{-34} J-s), and we are not concerned with this inaccuracy in the measurement of x and p_x for a truck, for example. On the other hand, measurements of the position of an electron and its speed are seriously limited by the uncertainty principle.

One implication of the uncertainty principle is that we cannot properly speak of *the* position of an electron, for example, but must look for the "probability" of finding an electron at a certain position. Thus one of the important results of quantum mechanics is that a *probability density function* can be obtained for a particle in a certain environment, and this function can be used to find the expectation value of important quantities such as position, momentum, and energy. We are familiar with the methods for calculating discrete (single-valued) probabilities from common experience. For example, it is clear that the probability of drawing a particular card out of a random deck is $^1/_{52}$, and the probability that a tossed coin will come up heads is $^1/_2$. The techniques for making predictions when the probability varies are less familiar, however. In such cases it is common to define a probability of finding a particle within a certain volume. Given a probability density function $P(x)$ for a one-dimensional problem, the probability of finding the particle in a range from x to $x + dx$ is $P(x)dx$. Since the particle will be *somewhere*, this definition implies that

$$\int_{-\infty}^{\infty} P(x)dx = 1 \qquad\qquad (2\text{--}20)$$

if the function $P(x)$ is properly chosen. Equation (2–20) is implied by stating that the function $P(x)$ is *normalized* (i.e., the integral equals unity).

---

[1]This is often called the *principle of indeterminacy.*

To find the average value of a function of x, we need only multiply the value of that function in each increment dx by the probability of finding the particle in that dx and sum over all x. Thus the average value of $f(x)$ is

$$\langle f(x)\rangle = \int_{-\infty}^{\infty} f(x)P(x)dx \qquad (2\text{--}21a)$$

If the probability density function is not normalized, this equation should be written

$$\langle f(x)\rangle = \frac{\int_{-\infty}^{\infty} f(x)P(x)dx}{\int_{-\infty}^{\infty} P(x)dx} \qquad (2\text{--}21b)$$

2.4.2 The Schrödinger Wave Equation

There are several ways to develop the wave equation by applying quantum concepts to various classical equations of mechanics. One of the simplest approaches is to consider a few basic postulates, develop the wave equation from them, and rely on the accuracy of the results to serve as a justification of the postulates. In more advanced texts these assumptions are dealt with in more convincing detail.

Basic Postulates

1. Each particle in a physical system is described by a wave function $\Psi(x, y, z, t)$. This function and its space derivative ($\partial\Psi/\partial x + \partial\Psi/\partial y + \partial\Psi/\partial z$) are continuous, finite, and single valued.

2. In dealing with classical quantities such as energy E and momentum p, we must relate these quantities with abstract quantum mechanical operators defined in the following way:

Classical variable	Quantum operator
x	x
$f(x)$	$f(x)$
p(x)	$\dfrac{\hbar}{j}\dfrac{\partial}{\partial x}$
E	$-\dfrac{\hbar}{j}\dfrac{\partial}{\partial t}$

and similarly for the other two directions.

3. The probability of finding a particle with wave function Ψ in the volume $dx\, dy\, dz$ is $\Psi^*\Psi\, dx\, dy\, dz$.[2] The product $\Psi^*\Psi$ is normalized according to Eq. (2–20) so that

$$\int_{-\infty}^{\infty} \Psi^*\Psi\, dx\, dy\, dz = 1$$

and the average value $\langle Q \rangle$ of any variable Q is calculated from the wave function by using the operator form Q_{op} defined in postulate 2:

$$\langle Q \rangle = \int_{-\infty}^{\infty} \Psi^* Q_{op}\, \Psi\, dx\, dy\, dz$$

Once we find the wave function Ψ for a particle, we can calculate its average position, energy, and momentum, within the limits of the uncertainty principle. Thus, a major part of the effort in quantum calculations involves solving for Ψ within the conditions imposed by a particular physical system. We notice from assumption 3 that the probability density function is $\Psi^*\Psi$, or $|\Psi|^2$.

The classical equation for the energy of a particle can be written:

$$\text{Kinetic energy} + \text{potential energy} = \text{total energy} \tag{2–22}$$

$$\frac{1}{2m}\,\mathsf{p}^2 \quad + \qquad \mathsf{V} \qquad = \qquad E$$

In quantum mechanics we use the operator form for these variables (postulate 2); the operators are allowed to operate on the wave function Ψ. For a one-dimensional problem Eq. (2–22) becomes[3]

$$-\frac{\hbar^2}{2m}\frac{\partial^2\Psi(x, t)}{\partial x^2} + \mathsf{V}(x)\Psi(x, t) = -\frac{\hbar}{j}\frac{\partial\Psi(x, t)}{\partial t} \tag{2–23}$$

which is the Schrödinger wave equation. In three dimensions the equation is

$$\boxed{-\frac{\hbar^2}{2m}\nabla^2\Psi + \mathsf{V}\Psi = -\frac{\hbar}{j}\frac{\partial\Psi}{\partial t}} \tag{2–24}$$

where $\nabla^2\Psi$ is

$$\frac{\partial^2\Psi}{\partial x^2} + \frac{\partial^2\Psi}{\partial y^2} + \frac{\partial^2\Psi}{\partial z^2}$$

---

[2] Ψ^* is the *complex conjugate* of Ψ, obtained by reversing the sign on each j. Thus $(e^{jx})^* = e^{-jx}$.
[3] The operational interpretation of $(\partial/\partial x)^2$ is the second derivative form $\partial^2/\partial x^2$; the square of j is -1.

The wave function Ψ in Eqs. (2–23) and (2–24) includes both space and time dependencies. It is common to calculate these dependencies separately and combine them later. Furthermore, many problems are time independent, and only the space variables are necessary. Thus we try to solve the wave equation by breaking it into two equations by the technique of separation of variables. Let $\Psi(x, t)$ be represented by the product $\psi(x)\phi(t)$. Using this product in Eq. (2–23) we obtain

$$-\frac{\hbar^2}{2m}\frac{\partial^2\psi(x)}{\partial x^2}\phi(t) + V(x)\psi(x)\phi(t) = -\frac{\hbar}{j}\psi(x)\frac{\partial\phi(t)}{\partial t} \qquad (2\text{–}25)$$

Now the variables can be separated to obtain the time-dependent equation in one dimension,

$$\boxed{\frac{d\phi(t)}{dt} + \frac{jE}{\hbar}\phi(t) = 0} \qquad (2\text{–}26)$$

and the time-independent equation,

$$\boxed{\frac{d^2\psi(x)}{dx^2} + \frac{2m}{\hbar^2}[E - V(x)]\psi(x) = 0} \qquad (2\text{–}27)$$

We can show that the separation constant E corresponds to the energy of the particle when particular solutions are obtained, such that a wave function ψ_n corresponds to a particle energy E_n.

These equations are the basis of wave mechanics. From them we can determine the wave functions for particles in various simple systems. For calculations involving electrons, the potential term $V(x)$ usually results from an electrostatic or magnetic field.

2.4.3 Potential Well Problem

It is quite difficult to find solutions to the Schrödinger equation for most realistic potential fields. One can solve the problem with some effort for the hydrogen atom, for example, but solutions for more complicated atoms are hard to obtain. There are several important problems, however, which illustrate the theory without complicated manipulation. The simplest problem is the potential energy well with infinite boundaries. Let us assume a particle is trapped in a potential well with $V(x)$ zero except at the boundaries $x = 0$ and L, where it is infinitely large (Fig. 2–5a)

$$V(x) = 0, \quad 0 < x < L$$
$$V(x) = \infty, \quad x = 0, L \qquad (2\text{–}28)$$

Inside the well we set $V(x) = 0$ in Eq. (2–27)

$$\frac{d^2\psi(x)}{dx^2} + \frac{2m}{\hbar^2}E\psi(x) = 0, \quad 0 < x < L \qquad (2\text{–}29)$$

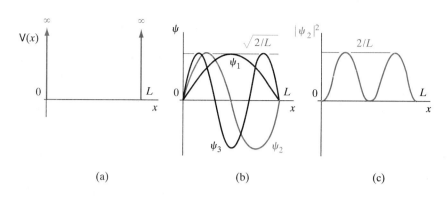

Figure 2–5
The problem of a
particle in a po-
tential well: (a)
potential energy
diagram; (b)
wave functions in
the first three
quantum states;
(c) probability
density distribu-
tion for the
second state.

This is the wave equation for a free particle; it applies to the potential well problem in the region with no potential V(x).

Possible solutions to Eq. (2–29) are sin kx and cos kx, where k is $\sqrt{2mE}/\hbar$ In choosing a solution, however, we must examine the boundary conditions. The only allowable value of ψ at the walls is zero. Otherwise, there would be a nonzero $|\psi|^2$ outside the potential well, which is impossible because a particle cannot penetrate an infinite barrier. Therefore, we must choose only the sine solution and define k such that sin kx goes to zero at $x = L$.

$$\psi = A \sin kx, \quad k = \frac{\sqrt{2mE}}{\hbar} \qquad (2-30)$$

The constant A is the amplitude of the wave function and will be evaluated from the normalization condition (postulate 3). If ψ is to be zero at $x = L$, then k must be some integral multiple of π/L.

$$k = \frac{n\pi}{L}, \quad n = 1, 2, 3, \dots \qquad (2-31)$$

From Eqs. (2–30) and (2–31) we can solve for the total energy E_n for each value of the integer **n.**

$$\frac{\sqrt{2mE_n}}{\hbar} = \frac{n\pi}{L} \qquad (2-32)$$

$$E_n = \frac{n^2\pi^2\hbar^2}{2mL^2} \qquad (2-33)$$

Thus for each allowable value of **n** the particle energy is described by Eq. (2–33). We notice that the *energy is quantized.* Only certain values of energy are allowed. The integer **n** is called a *quantum number;* the particular wave function ψ_n and corresponding energy state E_n describe the *quantum state* of the particle.

The quantized energy levels described by Eq. (2–33) appear in a variety of small-geometry structures encountered in semiconductor devices. We shall return to this potential well problem (often called the "particle in a box" problem) in later discussions.

The constant A is found from postulate 3.

$$\int_{-\infty}^{\infty} \psi^* \psi \, dx = \int_{0}^{L} A^2 \left(\sin \frac{\mathbf{n}\pi}{L} x \right)^2 dx = A^2 \frac{L}{2} \qquad (2\text{–}34)$$

Setting Eq. (2–34) equal to unity we obtain.

$$A = \sqrt{\frac{2}{L}}, \quad \psi_n = \sqrt{\frac{2}{L}} \sin \frac{\mathbf{n}\pi}{L} x \qquad (2\text{–}35)$$

The first three wave functions ψ_1, ψ_2, ψ_3, are sketched in Fig. 2–5b. The probability density function $\psi^* \psi$, or $|\psi|^2$, is sketched for ψ_2 in Fig. 2–5c.

2.4.4 Tunneling

The wave functions are relatively easy to obtain for the potential well with infinite walls, since the boundary conditions force ψ to zero at the walls. A slight modification of this problem illustrates a principle that is very important in some solid state devices—the quantum mechanical *tunneling* of an electron through a barrier of finite height and thickness. Let us consider the potential barrier of Fig. 2–6. If the barrier is not infinite, the boundary conditions do not force ψ to zero at the barrier. Instead, we must use the condition that ψ and its slope $d\psi/dx$ are continuous at each boundary of the barrier (postulate 1). Thus ψ must have a nonzero value within the barrier and also on the other side. Since ψ has a value to the right of the barrier, $\psi^* \psi$ exists

Figure 2–6
Quantum mechanical tunneling: (a) potential barrier of height V_0 and thickness W; (b) probability density for an electron with energy $E < V_0$, indicating a nonzero value of the wave function beyond the barrier.

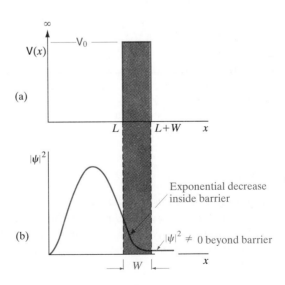

there also, implying that there is some probability of finding the particle beyond the barrier. We notice that the particle does not go over the barrier; its total energy is assumed to be less than the barrier height V_0. The mechanism by which the particle "penetrates" the barrier is called tunneling. However, no classical analogue, including classical descriptions of tunneling through barriers, in appropriate for this effect. Quantum mechanical tunneling is intimately bound to the uncertainty principle. If the barrier is sufficiently thin, we cannot say with certainty that the particle exists only on one side. However, the wave function amplitude for the particle is reduced by the barrier as Fig. 2–6 indicates, so that by making the thickness W greater, we can reduce ψ on the right-hand side to the point that negligible tunneling occurs. Tunneling is important only over very small dimensions, but it can be of great importance in the conduction of electrons in solids, as we shall see in Chapters 5, 6 and 11.

Recently, a novel electronic device called the resonant tunneling diode was developed. This device operates by tunneling electrons through "particle in a potential well" energy levels of the type described in Section 2.4.3.

The Schrödinger equation describes accurately the interactions of particles with potential fields, such as electrons within atoms. Indeed, the modern understanding of atomic theory (the modern atomic *models*) comes from the wave equation and from Heisenberg's matrix mechanics. It should be pointed out, however, that the problem of solving the Schrödinger equation directly for complicated atoms is extremely difficult. In fact, only the hydrogen atom is generally solved directly; atoms of atomic number greater than one are usually handled by techniques involving approximations. Many atoms such as the alkali metals (Li, Na, etc.), which have a neutral core with a single electron in an outer orbit, can be treated by a rather simple extension of the hydrogen atom results. The hydrogen atom solution is also important in identifying the basic selection rules for describing allowed electron energy levels. These quantum mechanical results must coincide with the experimental spectra, and we expect the energy levels to include those predicted by the Bohr model. Without actually working through the mathematics for the hydrogen atom, in this section we shall investigate the energy level schemes dictated by the wave equation.

2.5
ATOMIC STRUCTURE AND THE PERIODIC TABLE

2.5.1 The Hydrogen Atom

Finding the wave functions for the hydrogen atom requires a solution of the Schrödinger equation in three dimensions for a coulombic potential field. Since the problem is spherically symmetric, the spherical coordinate system is used in the calculation (Fig. 2–7). The term $V(x, y, z)$ in Eq. (2–24) must be replaced by $V(r, \theta, \phi)$, representing the Coulomb potential which the electron

Figure 2–7
The spherical co-
ordinate system.

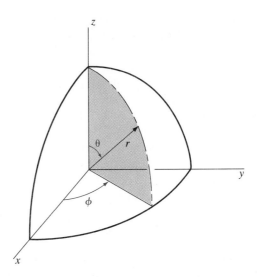

experiences in the vicinity of the proton. The Coulomb potential varies only
with r in spherical coordinates

$$V(r, \theta, \phi) = V(r) = -\left(4\pi\epsilon_0\right)^{-1}\frac{q^2}{r} \tag{2–36}$$

as in Eq. (2–14).

When the separation of variables is made, the time-independent equation can be written as

$$\psi(r, \theta, \phi) = R(r)\Theta(\theta)\Phi(\phi) \tag{2–37}$$

Thus the wave functions are found in three parts. Separate solutions
must be obtained for the r-dependent equation, the θ-dependent equation,
and the ϕ-dependent equation. After these three equations are solved, the
total wave function ψ is obtained from the product.

As in the simple potential well problem, each of the three hydrogen
atom equations gives a solution which is quantized. Thus we would expect a
quantum number to be associated with *each* of the three parts of the wave
equation. As an illustration, the ϕ-dependent equation obtained after separation of variables is

$$\frac{d^2\Phi}{d\phi^2} + \mathbf{m}^2\Phi = 0 \tag{2–38}$$

where $\mathbf{m}$ is a quantum number. The solution to this equation is

$$\Phi_m(\phi) = Ae^{j\mathbf{m}\phi} \tag{2–39}$$

where A can be evaluated by the normalization condition, as before:

$$\int_0^{2\pi} \Phi_m^*(\phi)\Phi_m(\phi)d\phi = 1 \qquad (2\text{–}40)$$

$$A^2\int_0^{2\pi} e^{-jm\phi}e^{jm\phi}\, d\phi = A^2\int_0^{2\pi} d\phi = 2\pi A^2 \qquad (2\text{–}41)$$

Thus the value of A is

$$A = \frac{1}{\sqrt{2\pi}} \qquad (2\text{–}42)$$

and the ϕ-dependent wave function is

$$\Phi_m(\phi) = \frac{1}{\sqrt{2\pi}}\, e^{jm\phi} \qquad (2\text{–}43)$$

Since values of ϕ repeat every 2π radians, Φ should repeat also. This occurs if **m** is an integer, including negative integers and zero. Thus the wave functions for the ϕ-dependent equation are quantized with the following selection rule for the quantum numbers:

$$\mathbf{m} = \ldots, -3, -2, -1, 0, +1, +2, +3, \ldots \qquad (2\text{–}44)$$

By similar treatments, the functions $R(r)$ and $\Theta(\theta)$ can be obtained, each being quantized by its own selection rule. For the r-dependent equation, the quantum number **n** can be any positive integer (not zero), and for the θ-dependent equation the quantum number l can be zero or a positive integer. However, there are interrelationships among the equations which restrict the various quantum numbers used with a single wave function ψ_{nlm}:

$$\psi_{nlm}(r, \theta, \phi) = R_n(r)\Theta_l(\theta)\Phi_m(\phi) \qquad (2\text{–}45)$$

These restrictions are summarized as follows:

$$\mathbf{n} = 1, 2, 3, \ldots \qquad (2\text{–}46\text{a})$$

$$l = 0, 1, 2, \ldots, (\mathbf{n} - 1) \qquad (2\text{–}46\text{b})$$

$$\mathbf{m} = -l, \ldots, -2, -1, 0, +1, +2, \ldots, +l \qquad (2\text{–}46\text{c})$$

In addition to the three quantum numbers arising from the three parts of the wave equation, there is an important quantization condition on the "spin" of the electron. Investigations of electron spin employ the theory of relativity as well as quantum mechanics; therefore, we shall simply state that the intrinsic angular momentum **s** of an electron with ψ_{nlm} specified is

$$\mathbf{s} = \pm\frac{\hbar}{2} \qquad (2\text{–}47)$$

That is, in units of $\hbar$, the electron has a spin of $\frac{1}{2}$, and the angular momentum produced by this spin is positive or negative depending on whether the electron is "spin up" or "spin down." The important point for our discussion is that each allowed energy state of the electron in the hydrogen atom is uniquely described by four quantum numbers: $\mathbf{n}$, l, $\mathbf{m}$ and $\mathbf{s}$.[4]

Using these four quantum numbers, we can identify the various states which the electron can occupy in a hydrogen atom. The number $\mathbf{n}$, called the *principal* quantum number, specifies the "orbit" of the electron in Bohr terminology. Of course, the concept of orbit is replaced by probability density functions in quantum mechanical calculations. It is common to refer to states with a given principal quantum number as belonging to a *shell* rather than an orbit.

There is considerable fine structure in the energy levels about the Bohr orbits, due to the dictates of the other three quantum conditions. For example, an electron with $\mathbf{n} = 1$ (the first Bohr orbit) can have only $l = 0$ and $\mathbf{m} = 0$ according to Eq. (2–46), but there are two spin states allowed from Eq. (2–47). For $\mathbf{n} = 2$, l can be 0 or 1, and $\mathbf{m}$ can be $-1, 0$, or $+1$. The various allowed combinations of quantum numbers appear in the first four columns of Table 2–1. From these combinations it is apparent that the electron in a hydrogen atom can occupy any one of a large number of excited states in addition to the lowest (*ground*) state ψ_{100}. Energy differences between the various states properly account for the observed lines in the hydrogen spectrum.

2.5.2 The Periodic Table

The quantum numbers discussed in Section 2.5.1 arise from the solutions to the hydrogen atom problem. Thus the energies obtainable from the wave functions are unique to the hydrogen atom and cannot be extended to more complicated atoms without appropriate alterations. However, the quantum number selection rules are valid for more complicated structures, and we can use these rules to gain an understanding of the arrangement of atoms in the periodic table of chemical elements. Without these selection rules, it is difficult to understand why only two electrons fit into the first Bohr orbit of an atom, whereas eight electrons are allowed in the second orbit. After even the brief discussion of quantum numbers given above, we should be able to answer these questions with more insight.

Before discussing the periodic table, we must be aware of an important principle of quantum theory, the *Pauli exclusion principle*. This rule states that no two electrons in an interacting system[5] can have the same set of quantum numbers $\mathbf{n}$, l, $\mathbf{m}$, $\mathbf{s}$. In other words, only two electrons can have the same three quantum numbers $\mathbf{n}$, l, $\mathbf{m}$, and those two must have opposite spin. The importance of this principle cannot be

---

[4]In many texts the numbers we have called $\mathbf{m}$ and $\mathbf{s}$ are referred to as $\mathbf{m}_l$ and $\mathbf{m}_s$, respectively.

[5]An interacting system is one in which electron wave functions overlap—in this case an atom with two or more electrons.

overemphasized; it is basic to the electronic structure of all atoms in the periodic table. One implication of this principle is that by listing the various combinations of quantum numbers, we can determine into which shell each electron of a complicated atom fits, and how many electrons are allowed per shell. The quantum states summarized in Table 2–1 can be used to indicate the electronic configurations for atoms in the lowest energy state.

In the first electronic shell ($\mathbf{n} = 1$), l can be only zero since the maximum value of l is always $\mathbf{n} - 1$. Similarly, $\mathbf{m}$ can be only zero since $\mathbf{m}$ runs from the negative value of l to the positive value of l. Two electrons with opposite spin can fit in this ψ_{100} state; therefore, the first shell can have at most two electrons. For the helium atom (atomic number $Z = 2$) in the ground state, both electrons will be in the first Bohr orbit ($\mathbf{n} = 1$), both will have $l = 0$ and $\mathbf{m} = 0$, and they will have opposite spin. Of course, one or both of the He atom electrons can be excited to one of the higher energy states of Table 2–1 and subsequently relax to the ground state, giving off a photon characteristic of the He spectrum.

Table 2–1 Quantum numbers to $\mathbf{n} = 3$ and allowable states for the electron in a hydorgen atom: The first four columns show the various combinations of quantum numbers allowed by the selection rules of Eq. (2–46); the last two columns indicate the number of allowed states (combinations of $\mathbf{n}$, l, $\mathbf{m}$, and $\mathbf{s}$) for each l (subshell) and $\mathbf{n}$ (shell, or Bohr orbit).

n	l	m	s/$\hbar$	Allowable states in subshell	Allowable states in complete shell
1	0	0	$\pm\frac{1}{2}$	2	2
2	0	0	$\pm\frac{1}{2}$	2	
	1	−1	$\pm\frac{1}{2}$		8
		0	$\pm\frac{1}{2}$	6	
		1	$\pm\frac{1}{2}$		
3	0	0	$\pm\frac{1}{2}$	2	
	1	−1	$\pm\frac{1}{2}$		
		0	$\pm\frac{1}{2}$	6	
		1	$\pm\frac{1}{2}$		18
	2	−2	$\pm\frac{1}{2}$		
		−1	$\pm\frac{1}{2}$		
		0	$\pm\frac{1}{2}$	10	
		1	$\pm\frac{1}{2}$		
		2	$\pm\frac{1}{2}$		

As Table 2–1 indicates, there can be two electrons in the $l = 0$ subshell, six electrons when $l = 1$, and ten electrons for $l = 2$. The electronic configurations of various atoms in the periodic table can be deduced from this list of allowed states. The ground state electron structures for a number of atoms are listed in Table 2–2. There is a simple shorthand notation for electronic structures which is commonly used instead of such a table. The only new convention to remember in this notation is the naming of the l values:

$$l = 0, 1, 2, 3, 4, \ldots$$

$$s, p, d, f, g, \ldots$$

This convention was created by early spectroscopists who referred to the first four spectral groups as *sharp*, *principal*, *diffuse*, and *fundamental*. Alphabetical order is used beyond *f*. With this convention for l, we can write an electron state as follows:

$$\underset{(\mathbf{n}\,=\,3)}{\overset{\displaystyle 3p^6}{\diagup}}\,\overset{\displaystyle \text{6 electrons in the } 3p \text{ subshell}}{\underset{(l\,=\,1)}{\big\lfloor}}$$

For example, the total electronic configuration for Si ($Z = 14$) in the ground state is

$$1s^2 2s^2 2p^6 3s^2 3p^2$$

We notice that Si has a closed Ne configuration (see Table 2–2) plus four electrons in an outer $\mathbf{n} = 3$ orbit ($3s^2 3p^2$). These are the four valence electrons of Si; two valence electrons are in an s state and two are in a p state. The Si electronic configuration can be written [Ne] $3s^2 3p^2$ for convenience, since the Ne configuration $1s^2 2s^2 2p^6$ forms a closed shell (typical of the inert elements).

Figure 2–8a shows the orbital model of a Si atom, which has a nucleus consisting of 14 protons (with a charge of $+14$) and neutrons, 10 core electrons in shells $n = 1$ and 2, and 4 valence electrons in the $3s$ and $3p$ subshells. Figure 2–8b shows the energy levels of the various electrons in the coulombic potential well of the nucleus. Since unlike charges attract each other, there is an attractive potential between the negatively charged electrons and the positively charged nucleus. As indicated in Eq. (2–36), a Coulomb potential varies as $1/r$ as a function of distance from the charge, in this case the Si nucleus. The potential energy gradually goes to zero when we approach infinity. We end up getting "particle-in-a-box" states for these electrons in this potential well, as discussed in Section 2.4.3 and Eq. (2–33). Of course, in this case the shape of the potential well is not rectangular, as shown in Fig. 2–5a, but coulombic, as shown in Fig. 2–8b. Therefore, the energy levels have a form closer to those of the H atom as shown in Eq. (2–15), rather than in Eq. (2–33).

Table 2–2 Electronic configurations for atoms in the ground state.

Atomic number (Z)	Element	$n=1$ / $l=0$ / $1s$	2 / 0 / $2s$	2 / 1 / $2p$	3 / 0 / $3s$	3 / 1 / $3p$	3 / 2 / $3d$	4 / 0 / $4s$	4 / 1 / $4p$	Shorthand notation
1	H	1								$1s^1$
2	He	2								$1s^2$
3	Li		1							$1s^2\ 2s^1$
4	Be		2							$1s^2\ 2s^2$
5	B	helium core,	2	1						$1s^2\ 2s^2\ 2p^1$
6	C	2 electrons	2	2						$1s^2\ 2s^2\ 2p^2$
7	N		2	3						$1s^2\ 2s^2\ 2p^3$
8	O		2	4						$1s^2\ 2s^2\ 2p^4$
9	F		2	5						$1s^2\ 2s^2\ 2p^5$
10	Ne		2	6						$1s^2\ 2s^2\ 2p^6$
11	Na				1					[Ne] $3s^1$
12	Mg				2					$3s^2$
13	Al				2	1				$3s^2\ 3p^1$
14	Si	neon core,			2	2				$3s^2\ 3p^2$
15	P	10 electrons			2	3				$3s^2\ 3p^3$
16	S				2	4				$3s^2\ 3p^4$
17	Cl				2	5				$3s^2\ 3p^5$
18	Ar				2	6				$3s^2\ 3p^6$
19	K							1		[Ar] $4s^1$
20	Ca							2		$4s^2$
21	Sc						1	2		$3d^1\ 4s^2$
22	Ti						2	2		$3d^2\ 4s^2$
23	V						3	2		$3d^3\ 4s^2$
24	Cr						5	1		$3d^5\ 4s^1$
25	Mn						5	2		$3d^5\ 4s^2$
26	Fe						6	2		$3d^6\ 4s^2$
27	Co	argon core,					7	2		$3d^7\ 4s^2$
28	Ni	18 electrons					8	2		$3d^8\ 4s^2$
29	Cu						10	1		$3d^{10}\ 4s^1$
30	Zn						10	2		$3d^{10}\ 4s^2$
31	Ga						10	2	1	$3d^{10}\ 4s^2\ 4p^1$
32	Ge						10	2	2	$3d^{10}\ 4s^2\ 4p^2$
33	As						10	2	3	$3d^{10}\ 4s^2\ 4p^3$
34	Se						10	2	4	$3d^{10}\ 4s^2\ 4p^4$
35	Br						10	2	5	$3d^{10}\ 4s^2\ 4p^5$
36	Kr						10	2	6	$3d^{10}\ 4s^2\ 4p^6$

Number of electrons

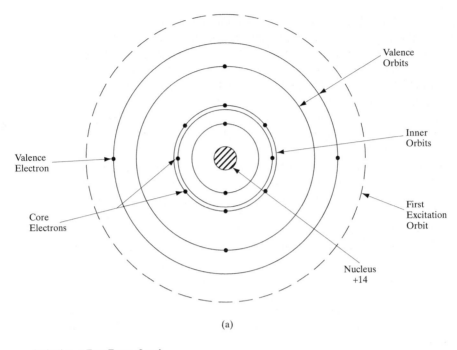

(a)

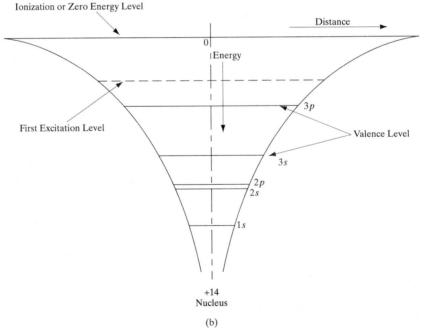

(b)

Figure 2–8
Electronic structure and energy levels in a Si atom: (a) The orbital model of a Si atom showing the 10 core electrons (n = 1 and 2), and the 4 valence electrons (n = 3); (b) energy levels in the coulombic potential of the nucleus are also shown schematically.

If we solve the Schrödinger equation for the Si atom as we did in Section 2.5.1 for the H atom, we can get the radial and angular dependence of the wavefunctions or "orbitals" of the electrons. Let us focus on the valence shell, $\mathbf{n} = 3$, where we have two $3s$ and two $3p$ electrons. It turns out that the $3s$ orbital is spherically symmetric with no angular dependence, and is positive everywhere. It can hold 2 electrons with opposite spin according to the Pauli principle. There are 3 p-orbitals which are mutually perpendicular. These are shaped like dumb-bells with a positive lobe and a negative lobe (Fig. 2–9). The $3p$ subshell can hold up to 6 electrons, but in the case of Si has only 2. Interestingly, in a Si crystal when we bring individual atoms very close together, the s- and p-orbitals overlap so much that they lose their distinct character, and lead to four mixed sp^3 orbitals. The negative part of the p orbital cancels the s-type wavefunction, while the positive part enhances it, thereby leading to a "directed" bond in space. As shown in Fig. 2–9, these *linear combinations of atomic orbitals (LCAO) or* "hybridized" sp^3 orbitals point symmetrically in space along the 4 tetragonal directions (See Fig. 1–9). In Chapter 3 we shall see that these "directed" chemical bonds are responsible for the tetragonal diamond or zincblende lattice structure in most semiconductors. They are also very important in the understanding of energy bands, and in the conduction of charges in these semiconductors.

The column IV semiconductor Ge ($Z = 32$) has an electronic structure similar to Si, except that the four valence electrons are outside a closed $\mathbf{n} = 3$

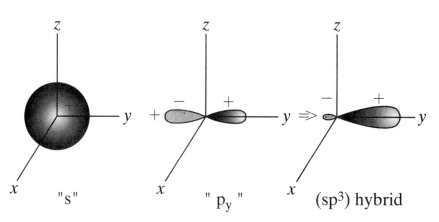

"s"   " p_y " (sp^3) hybrid

Figure 2–9
Orbitals in a Si atom: The spherically symmetric "s" type wave functions or orbitals are positive everywhere, while the three mutually perpendicular "p" type orbitals (p_x, p_y, p_z) are dumb-bell shaped and have a positive lobe and a negative lobe. The four sp^3 "hybridized" orbitals, only one of which is shown here, point symmetrically in space and lead to the diamond lattice in Si.

shell. Thus the Ge configuration is $[Ar]\, 3d^{10}4s^{2}4p^{2}$. There are several cases in Table 2–2 that do not follow the most straight-forward choice of quantum numbers. For example, we notice that in K ($Z = 19$) and Ca ($Z = 20$) the $4s$ state is filled before the $3d$ state; in Cr ($Z = 24$) and Cu ($Z = 29$) there is a transfer of an electron back to the $3d$ state. These exceptions, required by minimum energy considerations, are discussed more fully in most atomic physics texts.

PROBLEMS

2.1 (a) Sketch a simple vacuum tube device and the associated circuitry for measuring E_m in the photoelectric effect experiment. The electrodes can be placed in a sealed glass envelope.

(b) Sketch the photocurrent I vs. retarding voltage V that you would expect to measure for a given electrode material and configuration. Make the sketch for several intensities of light at a given wavelength.

(c) The work function of platinum is 4.09 eV. What retarding potential will be required to reduce the photocurrent to zero in a photoelectric experiment with Pt electrodes if the wavelength of incident light is 2440 Å? Remember that an energy of $q\Phi$ is lost by each electron in escaping the surface.

2.2 Point A is at an electrostatic potential of +1V relative to point B in a vacuum. An electron initially at rest at B moves to A. What energy (expressed in J and eV) does the electron have at A? What is its velocity (m/s)?

2.3 (a) Show that the various lines in the hydrogen spectrum can be expressed in angstroms as

$$\lambda(\text{Å}) = \frac{911 n_1^2 n^2}{n^2 - n_1^2}$$

where $n_1 = 1$ for the Lyman series, 2 for the Balmer series, and 3 for the Paschen series. The integer n is larger than n_1.

(b) Calculate λ for the Lyman series to $n = 5$, the Balmer series to $n = 7$, and the Paschen series to $n = 10$. Plot the results as in Fig. 2–2. What are the wavelength limits for each of the three series?

2.4 Show that the calculated Bohr expression for frequency of emitted light in the hydrogen spectrum, Eq. (2–17), corresponds to the experimental expressions, Eq. (2–3).

2.5 (a) The position of an electron is determined to within 1 Å. What is the minimum uncertainty in its momentum?

(b) An electron's energy is measured with an uncertainty of 1 eV. What is the minimum uncertainty in the time over which the measurement was made?

2.6 The de Broglie wavelength of a particle $\lambda = h/mv$ describes the wave-particle duality for small particles such as electrons. What is the de Broglie wavelength (in Å) of an electron at 100 eV? What is the wavelength for electrons at 12 keV, which is typical of electron microscopes? Comparing this to visible light, comment on the advantages of electron microscopes.

2.7 A sample of radioactive material undergoes decay such that the number of atoms $N(t)$ remaining in the unstable state at time t is related to the number N_0 at $t = 0$ by the relation $N(t) = N_0 \exp(-t/\tau)$. Show that τ is the average lifetime $\langle t \rangle$ of an atom in the unstable state before it spontaneously decays. Equation (2–21b) can be used with t substituted for x.

2.8 Given a plane wave $\psi = A exp(jk_x x)$ what is the expectation value for p_x^2 and p_z where p is momentum?

2.9 A free electron traveling in the x-direction can be described by a plane wave, with a wave function of the form $\psi_k(x) = A e^{j\mathbf{k}_x x}$, where $\mathbf{k}_x$ is a wave vector, or propagation constant. Use postulate 3 and the momentum operator to relate the electron momentum $\langle \mathrm{p}_x \rangle$ to $\mathbf{k}_x$.

2.10 An electron is described by a plane-wave wavefunction $\psi(x, t) = A e^{j(10x - 7t)}$. Calculate the expectation value of the x-component of momentum, the y-component of momentum and the energy of the electron. (Give values in MKS units.)

2.11 Calculate the first three energy levels for an electron in a quantum well of width 10Å with infinite walls.

2.12 What do Li, Na, and K have in common? What do F, Cl, and Br have in common? What are the electron configurations for ionized Na and Cl?

READING LIST

Ashcroft, N. W., and N. D. Mermin. *Solid State Physics.* Philadelphia: W.B. Saunders, 1976.

Baggot, J. "Beating the Uncertainty Principle." *New Scientist* 133 (15 February 1992): 36–40.

Bate, R. T. "The Quantum-Effect Device: Tomorrow's Transistor?" *Scientific American* 258 (March 1988): 96–100.

Brehm, J. J., and W. J. Mullin. *Introduction to the Structure of Matter.* New York: Wiley, 1989.

Bube, R. H. *Electrons in Solids,* 3rd ed. Boston: Harcourt Brace Jovanovich, 1992.

Capasso, F., and S. Datta. "Quantum Electron Devices." *Physics Today* 43 (February 1990): 74–82.

Cassidy, D. C. "Heisenberg: Uncertainty and the Quantum Revolution." *Scientific American* 266 (May 1992): 106–12.

Chang, L. L., and L. Esaki. "Semiconductor Quantum Heterostructures." *Physics Today* 45 (October 1992): 36–43.

Cohen-Tannouoji, C., B. Diu, and F. Laloe. *Quantum Mechanics.* New York: Wiley, 1977.

Corcoran, E. "Diminishing Dimensions." *Scientific American* 263 (November 1990): 122–6+.

Datta, S. *Modular Series on Solid State Devices: Vol. 8. Quantum Phenomena.* Reading. MA: Addison-Wesley, 1989.

Feynman, R. P. *The Feynman Lectures on Physics, Vol. 3. Quantum Mechanics.* Reading, MA: Addison-Wesley, 1965.

Hummel, R. E. *Electronic Properties of Materials,* 2nd ed. Berlin: Springer-Verlag, 1993.

Kroemer, H. *Quantum Mechanics.* Englewood Cliffs, NJ: Prentice Hall, 1994.

Park, D. *Introduction to the Quantum Theory.* New York: McGraw-Hill, 1992.

Sakurai, J. J. *Modern Quantum Mechanics.* Reading, MA: Addison-Wesley, 1994.

Singh, J. *Semiconductor Devices.* New York: McGraw-Hill, 1994.

Sundaram, M., S. A. Chalmers, and P. F. Hopkins. "New Quantum Structures." *Science* 254 (29 November 1991): 1326–35.

Weisbuch, C., and B. Vinter. *Quantum Semiconductor Structures.* Boston: Academic Press, 1991.

Chapter 3

Energy Bands and Charge Carriers in Semiconductors

In this chapter we begin to discuss the specific mechanisms by which current flows in a solid. In examining these mechanisms we shall learn why some materials are good conductors of electric current, whereas others are poor conductors. We shall see how the conductivity of a semiconductor can be varied by changing the temperature or the number of impurities. These fundamental concepts of charge transport form the basis for later discussions of solid state device behavior.

In Chapter 2 we found that electrons are restricted to sets of discrete energy levels within atoms. Large gaps exist in the energy scale in which no energy states are available. In a similar fashion, electrons in solids are restricted to certain energies and are not allowed at other energies. The basic difference between the case of an electron in a solid and that of an electron in an isolated atom is that in the solid the electron has a *range,* or *band,* of available energies. The discrete energy levels of the isolated atom spread into bands of energies in the solid because in the solid the wave functions of electrons in neighboring atoms overlap, and an electron is not necessarily localized at a particular atom. Thus, for example, an electron in the outer orbit of one atom feels the influence of neighboring atoms, and its overall wave function is altered. Naturally, this influence affects the potential energy term and the boundary conditions in the Schrödinger equation, and we would expect to obtain different energies in the solution. Usually, the influence of neighboring atoms on the energy levels of a particular atom can be treated as a small perturbation, giving rise to shifting and splitting of energy states into energy bands.

3.1.1 Bonding Forces in Solids

The interaction of electrons in neighboring atoms of a solid serves the very important function of holding the crystal together. For example, alkali halides such as NaCl are typified by *ionic bonding.* In the NaCl lattice, each Na atom

55

Figure 3–1
Different types of
chemical bonding
in solids (a) an
example of ionic
bonding in NaCl;
(b) covalent
bonding in the Si
crystal, veiwed
along a <100>
direction (see
also Figs. 1–8
and 1–9).

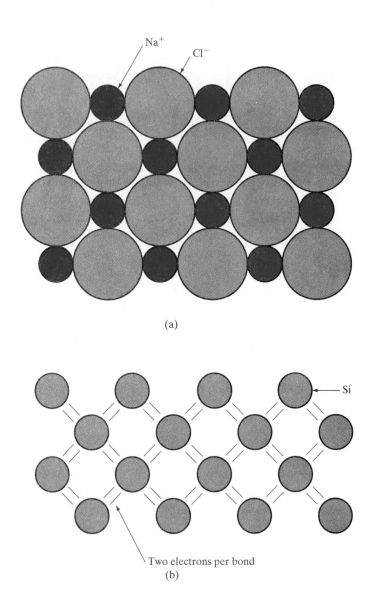

(a)

(b)

Two electrons per bond

is surrounded by six nearest neighbor Cl atoms, and vice versa. Four of the
nearest neighbors are evident in the two-dimensional representation shown
in Fig. 3–1a. The electronic structure of Na ($Z = 11$) is [Ne] $3s^1$, and Cl ($Z = 17$)
has the structure [Ne]$3s^2 3p^5$. In the lattice each Na atom gives up its outer $3s$
electron to a Cl atom, so that the crystal is made up of ions with the electronic
structures of the inert atoms Ne and Ar (Ar has the electronic structure
[Ne]$3s^2 3p^6$). However, the ions have net electric charges after the electron exchange. The Na$^+$ ion has a net positive charge, having lost an electron, and
the Cl$^-$ ion has a net negative charge, having gained an electron.

Each Na^+ ion exerts an electrostatic attractive force upon its six Cl^- neighbors, and vice versa. These coulombic forces pull the lattice together until a balance is reached with repulsive forces. A reasonably accurate calculation of the atomic spacing can be made by considering the ions as hard spheres being attracted together (Example 1–1).

An important observation in the NaCl structure is that all electrons are tightly bound to atoms. Once the electron exchanges have been made between the Na and Cl atoms to form the Na^+ and Cl^- ions, the outer orbits of all atoms are completely filled. Since the ions have the closed-shell configurations of the inert atoms Ne and Ar, there are no loosely bound electrons to participate in current flow; as a result, NaCl is a good insulator.

In a metal atom the outer electronic shell is only partially filled, usually by no more than three electrons. We have already noted that the alkali metals (e.g., Na) have only one electron in the outer orbit. This electron is loosely bound and is given up easily in ion formation. This accounts for the great chemical activity in the alkali metals, as well as for their high electrical conductivity. In the metal the outer electron of each alkali atom is contributed to the crystal as a whole, so that the solid is made up of ions with closed shells immersed in a sea of free electrons. The forces holding the lattice together arise from an interaction between the positive ion cores and the surrounding free electrons. This is one type of *metallic bonding.* Obviously, there are complicated differences in the bonding forces for various metals, as evidenced by the wide range of melting temperatures (234 K for Hg, 3643 K for W). However, the metals have the sea of electrons in common, and these electrons are free to move about the crystal under the influence of an electric field.

A third type of bonding is exhibited by the diamond lattice semiconductors. We recall that each atom in the Ge, Si, or C diamond lattice is surrounded by four nearest neighbors, each with four electrons in the outer orbit. In these crystals each atom shares its valence electrons with its four neighbors (Fig. 3–1b). Bonding between nearest neighbor atoms is illustrated in the diamond lattice diagram of Fig. 1–9. The bonding forces arise from a quantum mechanical interaction between the shared electrons. This is known as *covalent bonding;* each electron pair constitutes a covalent bond. In the sharing process it is no longer relevant to ask which electron belongs to a particular atom—both belong to the bond. The two electrons are indistinguishable, except that they must have opposite spin to satisfy the Pauli exclusion principle. Covalent bonding is also found in certain molecules, such as H_2.

As in the case of the ionic crystals, no free electrons are available to the lattice in the covalent diamond structure of Fig. 3–1b. By this reasoning Ge and Si should also be insulators. However, we have pictured an idealized lattice at 0 K in this figure. As we shall see in subsequent sections, an electron can be thermally or optically excited out of a covalent bond and thereby become free to participate in conduction. This is an important feature of semiconductors.

Compound semiconductors such as GaAs have mixed bonding, in which both ionic and covalent bonding forces participate. Some ionic bonding is to be expected in a crystal such as GaAs because of the difference in placement of the Ga and As atoms in the periodic table. The ionic character of the bonding becomes more important as the atoms of the compound become further separated in the periodic table, as in the II–VI compounds.

3.1.2 Energy Bands

As isolated atoms are brought together to form a solid, various interactions occur between neighboring atoms, including those described in the preceding section. The forces of attraction and repulsion between atoms will find a balance at the proper interatomic spacing for the crystal. In the process, important changes occur in the electron energy level configurations, and these changes result in the varied electrical properties of solids.

In Fig. 2–8, we showed the orbital model of a Si atom, along with the energy levels of the various electrons in the coulombic potential well of the nucleus. Let us focus on the outermost shell or valence shell, $\mathbf{n} = 3$, where two $3s$ and two $3p$ electrons interact to form the four "hybridized" sp^3 electrons when the atoms are brought close together. In Fig. 3–2, we schematically show the coulombic potential wells of two atoms close to each other, along with the wave functions of two electrons centered on the two nuclei. By solving the Schrödinger equation for such an interacting system, we find that the composite two-electron wave functions are *linear combinations* of the individual *atomic orbitals* (LCAO). The odd or anti-symmetric combination is called the anti-bonding orbital, while the even or symmetric combination is the bonding orbital. It can be seen that the bonding orbital has a higher value of the wave function (and therefore the electron probability density) than the anti-bonding state in the region between the two nuclei. This corresponds to the covalent bond between the atoms.

To determine the energy levels of the bonding and the anti-bonding states, it is important to recognize that in the region between the two nuclei the coulombic potential energy $V(r)$ is lowered (solid line in Fig. 3–2) compared to isolated atoms (dashed lines). It is easy to see why the potential energy would be lowered in this region, because an electron here would be attracted by two nuclei, rather than just one. For the bonding state the electron probability density is higher in this region of lowered potential energy than for the anti-bonding state. As a result, the original isolated atomic energy level would be split into two, a lower bonding energy level and a higher anti-bonding level. It is the lowering of the energy of the bonding state that gives rise to cohesion of the crystal. For even smaller inter-atomic spacings, the energy of the crystal goes up because of repulsion between the nuclei, and other electronic interactions. Since the probability density is given by the

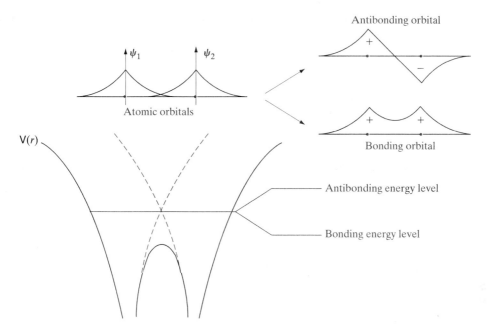

Figure 3–2
Linear combinations of atomic orbitals (LCAO): The LCAO when 2 atoms are brought together leads to 2 distinct "normal" modes—a higher energy anti-bonding orbital, and a lower energy bonding orbital. Note that the electron probability density is high in the region between the ion cores (covalent "bond"), leading to lowering of the bonding energy level and the cohesion of the crystal. If instead of 2 atoms, one brings together N atoms, there will be N distinct LCAO, and N closely-spaced energy levels in a band.

square of the wave function, if the entire wave function is multiplied by –1, it does not lead to a different LCAO. The important point to note in this discussion is that the number of distinct LCAO, and the number of distinct energy levels depends on the number of atoms that are brought together. The lowest energy level corresponds to the totally symmetric LCAO, the highest corresponds to the totally anti-symmetric case and the other combinations lead to energy levels in between.

Qualitatively, we can see that as atoms are brought together, the application of the Pauli exclusion principle becomes important. When two atoms are completely isolated from each other so that there is no interaction of electron wave functions between them, they can have identical electronic structures. As the spacing between the two atoms becomes smaller, however, electron wave functions begin to overlap. The exclusion principle dictates that no two electrons in a given interacting system may have the same quantum state; thus there must be at most one electron per level after there is a splitting of the discrete energy levels of the isolated atoms into new levels belonging to the pair rather than to individual atoms.

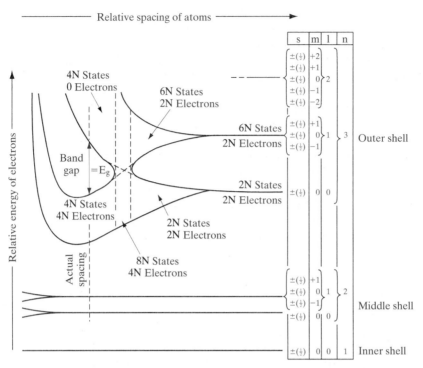

Figure 3–3
Energy levels in Si as a function of inter-atomic spacing. The core levels (**n** = 1,2) in Si are completely filled with electrons. At the actual atomic spacing of the crystal, the $2N$ electrons in the $3s$ sub-shell and the $2N$ electrons in the $3p$ sub-shell undergo sp^3 hybridization, and all end up in the lower $4N$ states (valence band), while the higher lying $4N$ states (conduction band) are empty, separated by a bandgap.

In a solid, many atoms are brought together, so that the split energy levels form essentially continuous *bands* of energies. As an example, Fig. 3–3 illustrates the imaginary formation of a silicon crystal from isolated silicon atoms. Each isolated silicon atom has an electronic structure $1s^2 2s^2 2p^6 3s^2 3p^2$ in the ground state. Each atom has available two $1s$ states, two $2s$ states, six $2p$ states, two $3s$ states, six $3p$ states, and higher states (see Tables 2–1 and 2–2). If we consider N atoms, there will be $2N$, $2N$, $6N$, $2N$, and $6N$ states of type $1s$, $2s$, $2p$, $3s$, and $3p$, respectively. As the interatomic spacing decreases, these energy levels split into bands, beginning with the outer (**n** = 3) shell. As the "$3s$" and "$3p$" bands grow, they merge into a single band composed of a mixture of energy levels. This band of "$3s$–$3p$" levels contains $8N$ available states. As the distance between atoms approaches the equilibrium interatomic spacing of silicon, this band splits into two bands separated by an *energy gap* E_g. The upper band (called the *conduction band*) contains $4N$ states, as does the lower (*valence*) band. Thus, apart from the low-lying and tightly bound "core" levels, the silicon crystal has two bands of available energy levels separated by an energy gap E_g wide, which contains no allowed energy levels for electrons to occupy. This gap is sometimes called a "forbidden band," since in a perfect crystal it contains no electron energy states.

We should pause at this point and count electrons. The lower "1s" band is filled with the $2N$ electrons which originally resided in the collective $1s$ states of the isolated atoms. Similarly, the $2s$ band and the $2p$ bands will have $2N$ and $6N$ electrons in them, respectively. However, there were $4N$ electrons in the original isolated **n** = 3 shells ($2N$ in $3s$ states and $2N$ in $3p$ states). These $4N$ electrons must occupy states in the valence band or the conduction band in the crystal. At 0 K the electrons will occupy the lowest energy states available to them. In the case of the Si crystal, there are exactly $4N$ states in the valence band available to the $4N$ electrons. Thus at 0 K, *every* state in the valence band will be filled, while the conduction band will be completely empty of electrons. As we shall see, this arrangement of completely filled and empty energy bands has an important effect on the electrical conductivity of the solid.

3.1.3 Metals, Semiconductors, and Insulators

Every solid has its own characteristic energy band structure. This variation in band structure is responsible for the wide range of electrical characteristics observed in various materials. The silicon band structure of Fig. 3–3, for example, can give a good picture of why silicon in the diamond lattice is a good insulator. To reach such a conclusion, we must consider the properties of completely filled and completely empty energy bands in the current conduction process.

Before discussing the mechanisms of current flow in solids further, we can observe here that for electrons to experience acceleration in an applied electric field, they must be able to move into new energy states. This implies there must be empty states (allowed energy states which are not already occupied by electrons) available to the electrons. For example, if relatively few electrons reside in an otherwise empty band, ample unoccupied states are available into which the electrons can move. On the other hand, the silicon band structure is such that the valence band is completely filled with electrons at 0 K and the conduction band is empty. There can be no charge transport within the valence band, since no empty states are available into which electrons can move. There are no electrons in the conduction band, so no charge transport can take place there either. Thus silicon has a high resistivity typical of insulators.

Semiconductor materials at 0 K have basically the same structure as insulators—a filled valence band separated from an empty conduction band by a band gap containing no allowed energy states (Fig. 3–4). The difference lies in the size of the band gap E_g, which is much smaller in semiconductors than in insulators. For example, the semiconductor Si has a band gap of about 1.1 eV compared with 5 eV for diamond. The relatively small band gaps of semiconductors (Appendix III) allow for excitation of electrons from the

Figure 3–4
Typical band
structures at 0 K.

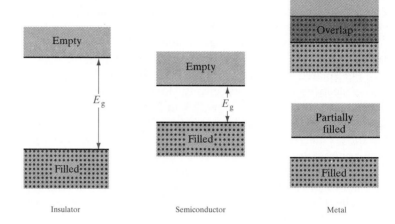

lower (valence) band to the upper (conduction) band by reasonable amounts of thermal or optical energy. For example, at room temperature a semiconductor with a 1-eV band gap will have a significant number of electrons excited thermally across the energy gap into the conduction band, whereas an insulator with E_g = 10 eV will have a negligible number of such excitations. Thus an important difference between semiconductors and insulators is that the number of electrons available for conduction can be increased greatly in semiconductors by thermal or optical energy.

In metals the bands either overlap or are only partially filled. Thus electrons and empty energy states are intermixed within the bands so that electrons can move freely under the influence of an electric field. As expected from the metallic band structures of Fig. 3–4, metals have a high electrical conductivity.

3.1.4 Direct and Indirect Semiconductors

The "thought experiment" of Section 3.1.2, in which isolated atoms were brought together to form a solid, is useful in pointing out the existence of energy bands and some of their properties. Other techniques are generally used, however, when quantitative calculations are made of band structures. In a typical calculation, a single electron is assumed to travel through a perfectly periodic lattice. The wave function of the electron is assumed to be in the form of a plane wave[1] moving, for example, in the x- direction with propagation constant **k,** also called a *wave vector*. The space-dependent wave function for the electron is

[1]Discussions of plane waves are available in most sophomore physics texts or in introductory electromagnetics texts.

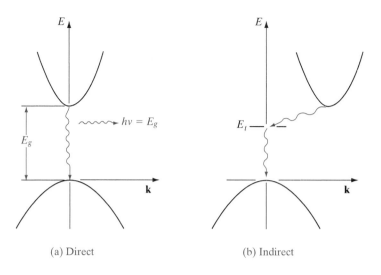

Figure 3–5
Direct and indirect electron transitions in semiconductors: (a) direct transition with accompanying photon emission; (b) indirect transition via a defect level.

(a) Direct

(b) Indirect

$$\psi_{\mathbf{k}}(x) = U(\mathbf{k}_x, x)e^{j\mathbf{k}_x x} \qquad (3\text{–}1)$$

where the function $U(\mathbf{k}_x, x)$ modulates the wave function according to the periodicity of the lattice.

In such a calculation, allowed values of energy can be plotted vs. the propagation constant $\mathbf{k}$. Since the periodicity of most lattices is different in various directions, the $(E, \mathbf{k})$ diagram must be plotted for the various crystal directions, and the full relationship between E and $\mathbf{k}$ is a complex surface which should be visualized in three dimensions.

The band structure of GaAs has a minimum in the conduction band and a maximum in the valence band for the same $\mathbf{k}$ value ($\mathbf{k} = 0$). On the other hand, Si has its valence band maximum at a different value of $\mathbf{k}$ than its conduction band minimum. Thus an electron making a smallest-energy transition from the conduction band to the valence band in GaAs can do so without a change in $\mathbf{k}$ value; on the other hand, a transition from the minimum point in the Si conduction band to the maximum point of the valence band requires some change in $\mathbf{k}$. Thus there are two classes of semiconductor energy bands; *direct* and *indirect* (Fig. 3–5). We can show that an indirect transition, involving a change in $\mathbf{k}$, requires a change of momentum for the electron.

Assuming that U is constant in Eq. (3–1) for an essentially free electron, show that the x-component of the electron momentum in the crystal is given by $\langle p_x \rangle = \hbar \mathbf{k}_x$.

EXAMPLE 3–1

SOLUTION   From Eq. (3–1)

$$\psi_{\mathbf{k}}(x) = Ue^{j\mathbf{k}_x x}$$

Using Eq. (2–21b) and the momentum operator,

$$\langle p_x \rangle = \frac{\displaystyle\int_{-\infty}^{\infty} U^2 e^{-j\mathbf{k}_x x} \frac{\hbar}{j} \frac{\partial}{\partial x} \left(e^{j\mathbf{k}_x x}\right) dx}{\displaystyle\int_{-\infty}^{\infty} U^2 \, dx}$$

$$= \frac{\hbar \mathbf{k}_x \displaystyle\int_{-\infty}^{\infty} U^2 \, dx}{\displaystyle\int_{-\infty}^{\infty} U^2 \, dx} = \hbar \mathbf{k}_x$$

This result implies that $(E, \mathbf{k})$ diagrams such as shown in Fig. 3–5 can be considered plots of electron energy vs. momentum, with a scaling factor $\hbar$.

The direct and indirect semiconductors are identified in Appendix III. In a direct semiconductor such as GaAs, an electron in the conduction band can fall to an empty state in the valence band, giving off the energy difference E_g as a photon of light. On the other hand, an electron in the conduction band minimum of an indirect semiconductor such as Si cannot fall directly to the valence band maximum but must undergo a momentum change as well as changing its energy. For example, it may go through some defect state (E_t) within the band gap. We shall discuss such defect states in Sections 4.2.1 and 4.3.2. In an indirect transition which involves a change in **k**, the energy is generally given up as heat to the lattice rather than as an emitted photon. This difference between direct and indirect band structures is very important for deciding which semiconductors can be used in devices requiring light output. For example, semiconductor light emitters and lasers (Chapter 8) generally must be made of materials capable of direct band-to-band transitions or of indirect materials with vertical transitions between defect states.

Band diagrams such as those shown in Fig. 3–5 are cumbersome to draw in analyzing devices, and do not provide a view of the variation of electron energy with distance in the sample. Therefore, in most discussions we shall use simple band pictures such as those shown in Fig. 3–4, remembering that electron transitions across the band gap may be direct or indirect.

3.1.5   Variation of Energy Bands with Alloy Composition

As III–V ternary and quaternary alloys are varied over their composition ranges (see Sections 1.2.4 and 1.4.1), their band structures change. For example, Fig. 3–6 illustrates the band structure of GaAs and AlAs, and the way in which the bands change with composition x in the ternary compound $Al_xGa_{1-x}As$.

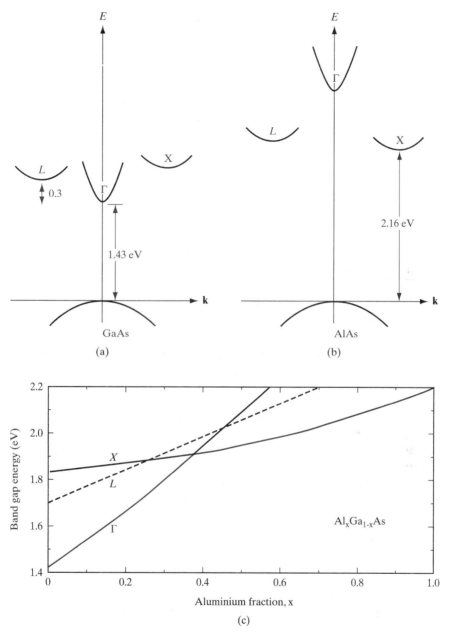

Figure 3–6
Variation of direct and indirect conduction bands in AlGaAs as a function of composition: (a) the $(E,\mathbf{k})$ diagram for GaAs, showing three minima in the conduction band; (b) AlAs band diagram; (c) positions of the three conduction band minima in $Al_xGa_{1-x}As$ as x varies over the range of compositions from GaAs $(x = 0)$ to AlAs $(x = 1)$. The smallest band gap, E_g (shown in color), follows the direct Γ band to $x = 0.38$, and then follows the indirect X band.

The binary compound GaAs is a direct material, with a band gap of 1.43 eV at room temperature. For reference, we call the direct ($\mathbf{k} = 0$) conduction band minimum Γ. There are also two higher-lying indirect minima in the GaAs conduction band, but these are sufficiently far above Γ that few electrons reside there (we discuss an important exception in Chapter 10 in which high-field excitation of electrons into the indirect minima leads to the Gunn effect). We call the lowest-lying GaAs indirect minimum L and the other X. In AlAs the direct Γ minimum is much higher than the indirect X minimum, and this material is therefore indirect with a band gap of 2.16 eV at room temperature.

In the ternary alloy $Al_xGa_{1-x}As$ all of these conduction band minima move up relative to the valence band as the composition x varies from 0 (GaAs) to 1 (AlAs). However, the indirect minimum X moves up less than the others, and for compositions above about 38 percent Al this indirect minimum becomes the lowest-lying conduction band. Therefore, the ternary alloy AlGaAs is a direct semiconductor for Al compositions on the column III sublattice up to about 38 percent, and is an indirect semiconductor for higher Al mole fractions. The band gap energy E_g is shown in color on Fig. 3–6(c).

The variation of energy bands for the ternary alloy $GaAs_{1-x}P_x$ is generally similar to that of AlGaAs shown in Fig. 3–6. GaAsP is a direct semiconductor from GaAs to about $GaAs_{.55}P_{.45}$ and is indirect from this composition to GaP (see Fig. 8–11). This material is often used in visible LEDs.

Since light emission is most efficient for direct materials, in which electrons can drop from the conduction band to the valence band without changing $\mathbf{k}$ (and therefore momentum), LEDs in GaAsP are generally made in material grown with a composition less than $x = 0.45$. For example, most red LEDs in this material are made at about $x = 0.4$, where the Γ minimum is still the lowest-lying conduction band edge, and where the photon resulting from a direct transition from this band to the valence band is in the red portion of the spectrum (about 1.9 eV). The use of impurities to enhance radiative recombination in indirect material will be discussed in Section 8.2.

**3.2
CHARGE CARRIERS
IN SEMI-
CONDUCTORS**

The mechanism of current conduction is relatively easy to visualize in the case of a metal; the metal atoms are imbedded in a "sea" of relatively free electrons, and these electrons can move as a group under the influence of an electric field. This free electron view is oversimplified, but many important conduction properties of metals can be derived from just such a model. However, we cannot account for all of the electrical properties of semiconductors in this way. Since the semiconductor has a filled valence band and an empty conduction band at 0 K, we must consider the increase in conduction band electrons by thermal excitations across the band gap as the temperature is raised. In addition, after electrons are excited to the conduction band, the empty states left in the valence band can contribute to the conduction process.

The introduction of impurities has an important effect on the energy band structure and on the availability of charge carriers. Thus there is considerable flexibility in controlling the electrical properties of semiconductors.

3.2.1   Electrons and Holes

As the temperature of a semiconductor is raised from 0 K, some electrons in the valence band receive enough thermal energy to be excited across the band gap to the conduction band. The result is a material with some electrons in an otherwise empty conduction band and some unoccupied states in an otherwise filled valence band (Fig. 3–7).[2] For convenience, an empty state in the valence band is referred to as a *hole*. If the conduction band electron and the hole are created by the excitation of a valence band electron to the conduction band, they are called an *electron–hole pair* (abbreviated EHP).

After excitation to the conduction band, an electron is surrounded by a large number of unoccupied energy states. For example, the equilibrium number of electron–hole pairs in pure Si at room temperature is only about 10^{10} EHP/cm^3, compared to the Si atom density of 5×10^{22} atoms/cm^3. Thus the few electrons in the conduction band are free to move about via the many available empty states.

The corresponding problem of charge transport in the valence band is somewhat more complicated. However, it is possible to show that the effects of current in a valence band containing holes can be accounted for by simply keeping track of the holes themselves.

In a filled band, all available energy states are occupied. For every electron moving with a given velocity, there is an equal and opposite electron motion elsewhere in the band. If we apply an electric field, the net current is zero because for every electron j moving with velocity v_j there is a corresponding electron j' with velocity $-v_j$. Figure 3–8 illustrates this effect in terms of the electron energy vs. wave vector plot for the valence band. Since **k** is proportional

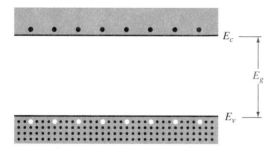

Figure 3–7
Electron-hole pairs in a semiconductor.

[2] In Fig. 3–7 and in subsequent discussions, we refer to the bottom of the conduction band as E_c and the top of the valence band as E_v.

to electron momentum, it is clear the two electrons have oppositely directed velocities. With N electrons/cm^3 in the band we express the current density using a sum over all of the electron velocities, and including the charge $-q$ on each electron. In a unit volume,

$$J = (-q)\sum_{i}^{N} v_i = 0 \quad \text{(\textit{filled band})} \tag{3-2a}$$

Now if we create a hole by removing the jth electron, the net current density in the valence band involves the sum over all velocities, minus the contribution of the electron we have removed.

$$J = (-q)\sum_{i}^{N} v_i - (-q)v_j \quad \text{(\textit{jth electron missing})} \tag{3-2b}$$

But the first term is zero, from Eq. (3–2a). Thus the net current is $+qv_j$. In other words, the current contribution of the hole is equivalent to that of a positively charged particle with velocity v_j, that of the missing electron. Of course, the charge transport is actually due to the motion of the new uncompensated electron (j'). Its current contribution $(-q)(-v_j)$ is equivalent to that of a positively charged particle with velocity $+v_j$. For simplicity, it is customary to treat empty states in the valence band as charge carriers with positive charge and positive mass.

A simple analogy may help in understanding the behavior of holes. If we have two bottles, one completely filled with water and one completely empty, we can ask ourselves "Will there be any net transport of water when we tilt the bottles?" The answer is "no". In the case of the empty bottle, the answer is obvious. In the case of the completely full bottle also, there cannot be any net motion of water because there is no empty space for water to move into. Similarly, an empty conduction band completely devoid of electrons or a valence band completely full of electrons cannot give rise to a net motion of electrons, and thus to current conduction.

Next, we imagine transferring some water droplets from the full bottle into the empty bottle, leaving behind some air bubbles, and ask ourselves the same question. Now when we tilt the bottles there will be net transport of water: the water droplets will roll downhill in one bottle and the air bubbles will move uphill in the other. Similarly, a few electrons in an otherwise empty conduction band move opposite to an electric field, while holes in an otherwise filled valence band move in the direction of the field. The bubble analogy is imperfect, but it may provide a physical feel for why the charge and mass of a hole have opposite signs from those of an electron.

In all the following discussions we shall concentrate on the electrons in the conduction band and on the holes in the valence band. We can account for the current flow in a semiconductor by the motion of these two types of charge carriers. We draw valence and conduction bands on an electron energy scale E, as in Fig. 3–8. However, we should remember that in the valence

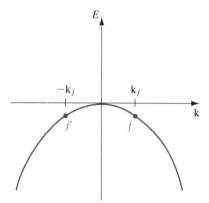

Figure 3–8
A valence band with all states filled, including states j and j', marked for discussion. The jth electron with wave vector k_j is matched by an electron at j' with the opposite wave vector $-k_j$. There is no net current in the band unless an electron is removed. For example, if the jth electron is removed, the motion of the electron at j' is no longer compensated.

band, hole energy increases oppositely to electron energy, because the two carriers have opposite charge. Thus hole energy increases downward in Fig. 3–8 and holes, seeking the lowest energy state available, are generally found at the *top* of the valence band. In contrast, conduction band electrons are found at the bottom of the conduction band.

It would be instructive to compare the $(E, \mathbf{k})$ band diagrams with the "simplified" band diagrams that are used for routine device analysis (Fig. 3–9). As discussed in Examples 3–1 and 3–2, an $(E, \mathbf{k})$ diagram is a plot of the total electron energy (potential plus kinetic) as a function of the crystal-direction–dependent electron wave vector (which is proportional to the momentum and therefore the velocity) at some point in space. Hence, the bottom of the conduction band corresponds to zero electron velocity or kinetic energy, and simply gives us the potential energy at that point in space. For holes, the top of the valence band corresponds to zero kinetic energy. For simplified band diagrams, we plot the edges of the conduction and valence bands (i.e., the potential energy) as a function of position in the device. Energies higher in the band correspond to additional kinetic energy of the electron. Also, the fact that the band edge corresponds to the electron potential energy tells us that the variation of the band edge in space is related to the electric field at different points in the semiconductor. We will show this relationship explicitly in Section 4.4.2.

In Fig. 3–9, an electron at location A sees an electric field given by the slope of the band edge (potential energy), and gains kinetic energy (at the expense of potential energy) by moving to point B. Correspondingly, in the $(E, \mathbf{k})$ diagram, the electron starts at $k = 0$, but moves to a non-zero wave vector $\mathbf{k}_\mathrm{B}$.

Figure 3–9
Superimposition of the $(E,\mathbf{k})$ bandstructure on the E-versus-position simplified band diagram for a semiconductor in an electric field. Electron energies increase going up, while hole energies increase going down. Similarly, electron and hole wavevectors point in opposite directions and these charge carriers move opposite to each other, as shown.

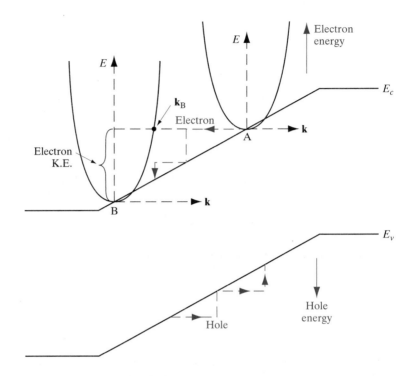

The electron then loses kinetic energy to heat by scattering mechanisms (discussed in Section 3.4.3) and returns to the bottom of the band at B. The slopes of the (E, x) band edges at different points in space reflect the local electric fields at those points. In practice, the electron may lose its kinetic energy in stages by a series of scattering events, as shown by the dashed lines.

3.2.2 Effective Mass

The electrons in a crystal are not completely free, but instead interact with the periodic potential of the lattice. As a result, their "wave-particle" motion cannot be expected to be the same as for electrons in free space. Thus, in applying the usual equations of electrodynamics to charge carriers in a solid, we must use altered values of particle mass. In doing so, we account for most of the influences of the lattice, so that the electrons and holes can be treated as "almost free" carriers in most computations. The calculation of effective mass must take into account the shape of the energy bands in three-dimensional $\mathbf{k}$- space, taking appropriate averages over the various energy bands.

EXAMPLE 3–2   Find the $(E, \mathbf{k})$ relationship for a free electron and relate it to the electron mass.

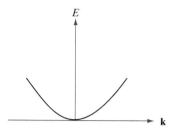

From Example 3–1, the electron momentum is $p = mv = \hbar\mathbf{k}$. Then

$$E = \frac{1}{2}mv^2 = \frac{1}{2}\frac{p^2}{m} = \frac{\hbar^2}{2m}\mathbf{k}^2$$

Thus the electron energy is parabolic with wave vector $\mathbf{k}$. The electron mass is inversely related to the curvature (second derivative) of the $(E, \mathbf{k})$ relationship, since

$$\frac{d^2E}{d\mathbf{k}^2} = \frac{\hbar^2}{m}$$

Although electrons in solids are not free, most energy bands are close to parabolic at their minima (for conduction bands) or maxima (for valence bands). We can also approximate effective mass near those band extrema from the curvature of the band.

The effective mass of an electron in a band with a given $(E, \mathbf{k})$ relationship is found in Example 3–2 to be

$$m* = \frac{\hbar^2}{d^2E/d\mathbf{k}^2} \tag{3-3}$$

Thus the curvature of the band determines the electron effective mass. For example, in Fig. 3–6a it is clear that the electron effective mass in GaAs is much smaller in the direct Γ conduction band (strong curvature) than in the L or X minima (weaker curvature, smaller value in the denominator of the $m*$ expression).

A particularly interesting feature of Figs. 3–5 and 3–6 is that the curvature of $d^2E/d\mathbf{k}^2$ is positive at the conduction band minima, but is negative at the valence band maxima. Thus, the electrons near the top of the valence band have *negative effective mass*, according to Eq. (3–3). Valence band electrons with negative charge and negative mass move in an electric field in the same direction as holes with positive charge and positive mass. As discussed in Section 3.2.1, we can fully account for charge transport in the valence band by considering hole motion.

For a band centered at $\mathbf{k} = 0$ (such as the Γ band in GaAs), the $(E, \mathbf{k})$ relationship near the minimum is usually parabolic:

$$E = \frac{\hbar^2}{2m*} \mathbf{k}^2 + E_c \qquad (3\text{-}4)$$

Comparing this relation to Eq. (3–3) indicates that the effective mass $m*$ is constant in a parabolic band. On the other hand, many conduction bands have complex $(E, \mathbf{k})$ relationships that depend on the direction of electron transport with respect to the principal crystal directions. In this case, the effective mass is a tensor quantity. However, we can use appropriate averages over such bands in most calculations.

Figure 3–10a shows the bandstructures for Si and GaAs viewed along two major directions. While the shape is parabolic near the band edges (as indicated in Figure 3–5, and Example 3–2), there are significant non-parabolicities at higher energies. The energies are plotted along the high symmetry [111] and [100] directions in the crystal. The $\mathbf{k} = 0$ point is denoted as Γ. When we go along

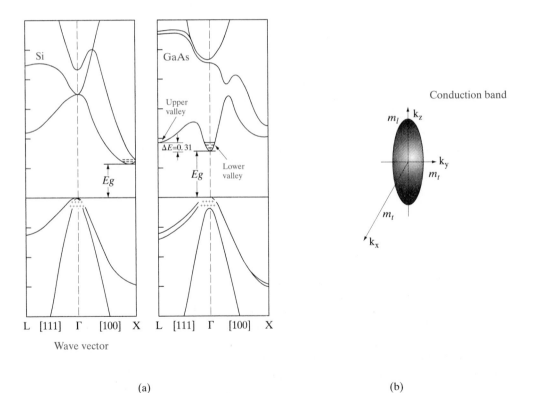

(a)   (b)

Figure 3–10
Realistic bandstructures in semiconductors: (a) Conduction and valence bands in Si and GaAs along [111] and [100]; (b) ellipsoidal constant energy surface for Si, near the 6 conduction band minima along the X directions. (From Chelikowsky and Cohen, Phys. Rev. B14, 556, 1976).

the [100] direction, we reach a valley near X, while we reach the L valley along the [111] direction. (Since the energies are plotted along different directions, the curves do not look symmetric.) The valence band maximum in most semiconductors is at the Γ point. It has three branches: the *heavy hole* band with the smallest curvature, a *light hole band* with a larger curvature, and a *split-off band* at a different energy. We notice that for GaAs the conduction band minimum and the valence band maximum are both at $\mathbf{k} = 0$; therefore it is direct bandgap. Silicon, on the other hand, has 6 equivalent conduction minima at X along the 6 equivalent $\langle 100 \rangle$ directions; therefore, it is indirect.

Figure 3–10b shows the constant energy surface for electrons in one of the six conduction bands for Si. The way to relate these surfaces to the bandstructures shown in Fig. 3–10a is to consider a certain value of energy, and determine all the $\mathbf{k}$ vectors in 3 dimensions for which we get this energy. We find that for Si we have 6 cigar-shaped ellipsoidal equi-energy surfaces near the conduction band minima along the six equivalent X-directions, with a longitudinal effective mass, m_l, along the major axis, and two transverse effective masses, m_t, along the minor axes. For GaAs, the conduction band is more or less spherical for low energies. On the other hand, we have warped spherical surfaces in the valence band. The importance of these surfaces will be clear in Section 3. 4.1 when we consider different types of effective masses in semiconductors.

In any calculation involving the mass of the charge carriers, we must use effective mass values for the particular material involved. In all subsequent discussions, the electron effective mass is denoted by m_n^* and the hole effective mass by m_p^*. The n subscript indicates the electron as a negative charge carrier, and the p subscript indicates the hole as a positive charge carrier.

There is nothing mysterious about the concept of an "effective" mass, m_n^*, and about the fact that it is different in different semiconductors. Indeed, the "true" mass of an electron, m, is the same in Si, Ge or GaAs—it is the same as for a free electron in vacuum. To understand why the effective mass is different from the true mass, consider Newton's second law which states that the time rate of change of momentum is the force.

$$dp/dt = d(mv)/dt = \text{Force} \qquad (3\text{--}5a)$$

An electron in a crystal experiences a total force $F_{int} + F_{ext}$, where F_{int} is the collection of internal periodic crystal forces, and F_{ext} is the externally applied force. It is inefficient to solve this complicated problem involving the periodic crystal potential (which is obviously different in different semiconductors) every time we try to solve a semiconductor device problem. It is better to solve the complicated problem of carrier motion in the periodic crystal potential just once, and encapsulate that information in what is called the bandstructure, $(E, \mathbf{k})$, whose curvature gives us the effective mass, m_n^*. The electron then responds to external forces with this new m_n^*. Newton's law is then written as:

$$d(m_n^* v)/dt = F_{ext} \qquad (3\text{--}5b)$$

This is clearly an enormous simplification compared to the more detailed problem. Obviously, the periodic crystal forces depend on the details of a specific semiconductor; therefore, the effective mass is different in different materials.

Once we determine the band curvature effective mass components from the orientation-dependent $(E, \mathbf{k})$, we have to combine them appropriately for different types of calculations. We shall see in Section 3.3.2 that when we are interested in determining the numbers of carriers in the bands, we have to use a "density-of-states" effective mass by taking the geometric mean of the band curvature effective masses, and the number of equivalent band extrema. On the other hand we will find in Section 3.4.1 that in problems involving the motion of carriers, one must take the harmonic mean of the band curvature effective masses to get the "conductivity" effective mass.

3.2.3 Intrinsic Material

A perfect semiconductor crystal with no impurities or lattice defects is called an *intrinsic* semiconductor. In such material there are no charge carriers at 0 K, since the valence band is filled with electrons and the conduction band is empty. At higher temperatures electron–hole pairs are generated as valence band electrons are excited thermally across the band gap to the conduction band. These EHPs are the only charge carriers in intrinsic material.

The generation of EHPs can be visualized in a qualitative way by considering the breaking of covalent bonds in the crystal lattice (Fig. 3–11). If one of the Si valence electrons is broken away from its position in the bonding structure such that it becomes free to move about in the lattice, a conduction electron is created and a broken bond (hole) is left behind. The energy required to break the bond is the band gap energy E_g. This model helps in visualizing the physical mechanism of EHP creation, but the energy band model is more productive for purposes of quantitative calculation. One important difficulty in the "broken bond" model is that the free electron and the hole seem deceptively localized in the lattice. Actually, the positions of the free electron and the hole are spread out over several lattice spacings and should be considered quantum mechanically by probability distributions (see Section 2.4).

Figure 3–11
Electron–hole
pairs in the
covalent bonding
model of the
Si crystal.

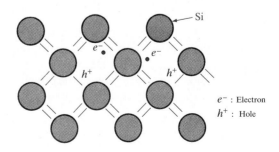

e^- : Electron
h^+ : Hole

Since the electrons and holes are created in pairs, the conduction band electron concentration n (electrons per cm^3) is equal to the concentration of holes in the valence band p (holes per cm^3). Each of these intrinsic carrier concentrations is commonly referred to as n_i. Thus *for intrinsic material*

$$n = p = n_i \qquad (3\text{--}6)$$

At a given temperature there is a certain concentration of electron–hole pairs n_i. Obviously, if a steady state carrier concentration is maintained, there must be *recombination* of EHPs at the same rate at which they are generated. Recombination occurs when an electron in the conduction band makes a transition (direct or indirect) to an empty state (hole) in the valence band, thus annihilating the pair. If we denote the generation rate of EHPs as g_i (EHP/cm^3-s) and the recombination rate as r_i, equilibrium requires that

$$r_i = g_i \qquad (3\text{--}7a)$$

Each of these rates is temperature dependent. For example, $g_i(T)$ increases when the temperature is raised, and a new carrier concentration n_i is established such that the higher recombination rate $r_i(T)$ just balances generation. At any temperature, we can predict that the rate of recombination of electrons and holes r_i is proportional to the equilibrium concentration of electrons n_0 and the concentration of holes p_0:

$$r_i = \alpha_r n_0 p_0 = \alpha_r n_i^2 = g_i \qquad (3\text{--}7b)$$

The factor α_r is a constant of proportionality which depends on the particular mechanism by which recombination takes place. We shall discuss the calculation of n_i as a function of temperature in Section 3.3.3; recombination processes will be discussed in Chapter 4.

3.2.4 Extrinsic Material

In addition to the intrinsic carriers generated thermally, it is possible to create carriers in semiconductors by purposely introducing impurities into the crystal. This process, called *doping*, is the most common technique for varying the conductivity of semiconductors. By doping, a crystal can be altered so that it has a predominance of either electrons or holes. Thus there are two types of doped semiconductors, n-type (mostly electrons) and p-type (mostly holes). When a crystal is doped such that the equilibrium carrier concentrations n_0 and p_0 are different from the intrinsic carrier concentration n_i, the material is said to be *extrinsic*.

When impurities or lattice defects are introduced into an otherwise perfect crystal, additional levels are created in the energy band structure, usually within the band gap. For example, an impurity from column V of the periodic table (P, As, and Sb) introduces an energy level very near the conduction band in Ge or Si. This level is filled with electrons at 0 K, and very little thermal energy is required to excite these electrons to the conduction band (Fig. 3–12a). Thus at about 50–100 K virtually all of the electrons in the

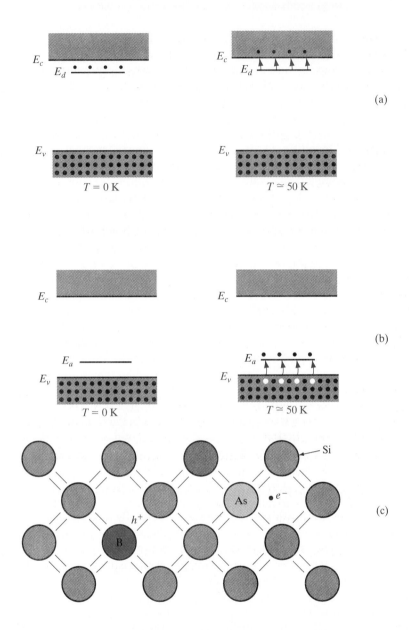

impurity level are "donated" to the conduction band. Such an impurity level
is called a *donor* level, and the column V impurities in Ge or Si are called
donor impurities. From Fig. 3–12a we note that the material doped with donor
impurities can have a considerable concentration of electrons in the conduction band, even when the temperature is too low for the intrinsic EHP
concentration to be appreciable. Thus semiconductors doped with a significant number of donor atoms will have $n_0 \gg (n_i, p_0)$ at room temperature.
This is n-type material.

Atoms from column III (B, Al, Ga, and In) introduce impurity levels in Ge or Si near the valence band. These levels are empty of electrons at 0 K Fig. 3–12b). At low temperatures, enough thermal energy is available to excite electrons from the valence band into the impurity level, leaving behind holes in the valence band. Since this type of impurity level "accepts" electrons from the valence band, it is called an *acceptor* level, and the column III impurities are acceptor impurities in Ge and Si. As Fig. 3–12b indicates, doping with acceptor impurities can create a semiconductor with a hole concentration p_0 much greater than the conduction band electron concentration n_0 (this type is p-type material).

In the covalent bonding model, donor and acceptor atoms can be visualized as shown in Fig. 3–12c. An As atom (column V) in the Si lattice has the four necessary valence electrons to complete the covalent bonds with the neighboring Si atoms, plus one extra electron. This fifth electron does not fit into the bonding structure of the lattice and is therefore loosely bound to the As atom. A small amount of thermal energy enables this extra electron to overcome its coulombic binding to the impurity atom and be donated to the lattice as a whole. Thus it is free to participate in current conduction. This process is a qualitative model of the excitation of electrons out of a donor level and into the conduction band (Fig. 3–12a). Similarly, the column III impurity B has only three valence electrons to contribute to the covalent bonding (Fig. 3–12c), thereby leaving one bond incomplete. With a small amount of thermal energy, this incomplete bond can be transferred to other atoms as the bonding electrons exchange positions. Again, the idea of an electron "hopping" from an adjacent bond into the incomplete bond at the B site provides some physical insight into the behavior of an acceptor, but the model of Fig. 3–12b is preferable for most discussions.

We can calculate rather simply the approximate energy required to excite the fifth electron of a donor atom into the conduction band (the donor *binding energy*). Let us assume for rough calculations that the As atom of Fig. 3–12c has its four covalent bonding electrons rather tightly bound and the fifth "extra" electron loosely bound to the atom. We can approximate this situation by using the Bohr model results, considering the loosely bound electron as ranging about the tightly bound "core" electrons in a hydrogen-like orbit. From Eq. (2–15) the magnitude of the ground-state energy ($\mathbf{n} = 1$) of such an electron is

$$E = \frac{mq^4}{2K^2\hbar^2} \tag{3–8}$$

The value of K must be modified from the free-space value $4\pi\epsilon_0$ used in the hydrogen atom problem to

$$K = 4\pi\epsilon_0\,\epsilon_r \tag{3–9}$$

where ϵ_r is the relative dielectric constant of the semiconductor material. In addition, we must use the conductivity effective mass m_n^* typical of the semiconductor, discussed in more detail in Section 3.4.1.

EXAMPLE 3–3   Calculate the approximate donor binding energy for GaAs ($\epsilon_r = 13.2$, m_n^* $= 0.067m_0$).

SOLUTION   From Eq. (3–8) and Appendix II we have

$$E = \frac{m_n^* q^4}{8(\epsilon_0 \epsilon_r)^2 h^2} = \frac{0.067(9.11 \times 10^{-31})(1.6 \times 10^{-19})^4}{8(8.85 \times 10^{-12} \times 13.2)^2(6.63 \times 10^{-34})^2}$$

$$= 8.34 \times 10^{-22}\,\text{J} = 0.0052\ \text{eV}$$

Thus the energy required to excite the donor electron from the **n** = 1 state to the free state (**n** = ∞) is ≃ 5.2 meV. This corresponds to the energy difference $E_c - E_d$ in Fig. 3–10a and is in very close agreement with actual measured values.

Generally, the column V donor levels lie approximately 0.01 eV below the conduction band in Ge, and the column III acceptor levels lie about 0.01 eV above the valence band. In Si the usual donor and acceptor levels lie about 0.03–0.06 eV from a band edge.

In III–V compounds, column VI impurities occupying column V sites serve as donors. For example, S, Se, and Te are donors in GaAs, since they substitute for As and provide an extra electron compared with the As atom. Similarly, impurities from column II (Be, Zn, Cd) substitute for column III atoms to form acceptors in the III—V compounds. A more ambiguous case arises when a III–V material is doped with Si or Ge, from column IV. These impurities are called *amphoteric*, meaning that Si or Ge can serve as donors or acceptors depending on whether they reside on the column III or column V sublattice of the crystal. In GaAs it is common for Si impurities to occupy Ga sites. Since the Si has an extra electron compared with the Ga it replaces, it serves as a donor. However, an excess of As vacancies arising during growth or processing of the GaAs can cause Si impurities to occupy As sites, where they serve as acceptors.

The importance of doping will become obvious when we discuss electronic devices made from junctions between p-type and n-type semiconductor material. The extent to which doping controls the electronic properties of semiconductors can be illustrated here by considering changes in the sample resistance which occur with doping. In Si, for example, the intrinsic carrier concentration n_i is about 10^{10} cm^{-3} at room temperature. If we dope Si with 10^{15} As atoms/cm^3, the conduction electron concentration changes by five orders of magnitude. The resistivity of Si changes from about 2×10^5 Ω-cm to 5 Ω-cm with this doping.

When a semiconductor is doped n-type or p-type, one type of carrier dominates. In the example given above, the conduction band electrons outnumber the holes in the valence band by many orders of magnitude. We refer to the small number of holes in n-type material as *minority carriers* and the relatively large number of conduction band electrons as *majority carriers.* Similarly, electrons are the minority carriers in p-type material, and holes are the majority carriers.

3.2.5 Electrons and Holes in Quantum Wells

We have discussed single-valued (*discrete*) energy levels in the band gap arising from doping, and a *continuum* of allowed states in the valence and conduction bands. A third possibility is the formation of discrete levels for electrons and holes as a result of quantum-mechanical confinement.

One of the most useful applications of MBE or OMVPE growth of multi-layer compound semiconductors, as described in Section 1.4, is the fact that a continuous single crystal can be grown in which adjacent layers have different band gaps. For example, Fig. 3–13 shows the spatial variation in conduction and valence bands for a multilayer structure in which a very thin layer of GaAs is sandwiched between two layers of AlGaAs, which has a wider band gap that the GaAs. We will discuss the details of such *heterojunctions* (junctions between dissimilar materials) in Section 5.8. It is interesting to point out here, however, that a consequence of confining electrons and holes in a very thin layer is that these particles behave according to the *particle in a potential well* problem, with quantum states calculated in Section

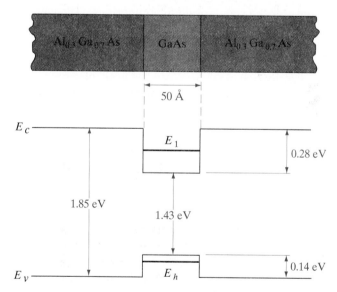

Figure 3–13
Energy band discontinuities for a thin layer of GaAs sandwiched between layers of wider band gap AlGaAs. In this case, the GaAs region is so thin that quantum states are formed in the valence and conduction bands. Electrons in the GaAs conduction band reside on "particle in a potential well" states such a E_1 shown here, rather than in the usual conduction band states. Holes in the quantum well occupy similar discrete states, such a E_h.

2.4.3. Therefore, instead of having the continuum of states normally available in the conduction band, the conduction band electrons in the narrow-gap material are confined to discrete quantum states as described by Eq. (2–33), modified for effective mass and finite barrier height. Similarly, the states in the valence band available for holes are restricted to discrete levels in the quantum well. This is one of the clearest demonstrations of the quantum mechanical results discussed in Chapter 2. From a practical device point of view, the formation of discrete quantum states in the GaAs layer of Fig. 3–13 changes the energy at which photons can be emitted. An electron on one of the discrete conduction band states (E_1 in Fig. 3–13) can make a transition to an empty discrete valence band state in the GaAs quantum well (such as E_h), giving off a photon of energy $E_g + E_1 + E_h$, greater than the GaAs band gap. Semiconductor lasers have been made in which such a quantum well is used to raise the energy of the transition from the infrared, typical of GaAs, to the red portion of the spectrum. We will see other examples of quantum wells in semiconductor devices in later chapters.

3.3
CARRIER
CONCENTRATIONS

In calculating semiconductor electrical properties and analyzing device behavior, it is often necessary to know the number of charge carriers per cm^3 in the material. The majority carrier concentration is usually obvious in heavily doped material, since one majority carrier is obtained for each impurity atom (for the standard doping impurities). The concentration of minority carriers is not obvious, however, nor is the temperature dependence of the carrier concentrations.

To obtain equations for the carrier concentrations we must investigate the distribution of carriers over the available energy states. This type of distribution is not difficult to calculate, but the derivation requires some background in statistical methods. Since we are primarily concerned here with the application of these results to semiconductor materials and devices, we shall accept the distribution function as given.

3.3.1 The Fermi Level

Electrons in solids obey *Fermi–Dirac* statistics.[3] In the development of this type of statistics, one must consider the indistinguishability of the electrons,

[3]Examples of other types of statistics are *Maxwell–Boltzmann* for classical particles (e.g., gas) and *Bose–Einstein* for photons. For two discrete energy levels, E_2 and E_1 (with $E_2 > E_1$), classical gas atoms follow a Boltzmann distribution; the number n_2 of atoms in state E_2 is related to the number n_1 in E_1 at thermal equilibrium by

$$\frac{n_2}{n_1} = \frac{N_2 e^{-E_2/kT}}{N_1 e^{-E_1/kT}} = \frac{N_2}{N_1} e^{-(E_2-E_1)/kT}$$

assuming the two levels have N_2 and N_1 number of states, respectively. The exponential term $\exp(-\Delta E/kT)$ is commonly called the *Boltzmann factor*. It appears also in the denominator of the Fermi–Dirac distribution function. We shall return to the Boltzmann distribution in Chapter 8 in discussions of the properties of lasers.

their wave nature, and the Pauli exclusion principle. The rather simple result of these statistical arguments is that the distribution of electrons over a range of allowed energy levels at thermal equilibrium is

$$f(E) = \frac{1}{1 + e^{(E-E_F)/kT}} \qquad (3\text{--}10)$$

where k is Boltzmann's constant ($k = 8.62 \times 10^{-5}$ eV/K $= 1.38 \times 10^{-23}$ J/K). The function $f(E)$, the *Fermi–Dirac distribution function,* gives the probability that an available energy state at E will be occupied by an electron at absolute temperature T. The quantity E_F is called the *Fermi level,* and it represents an important quantity in the analysis of semiconductor behavior. We notice that, for an energy E equal to the Fermi level energy E_F, the occupation probability is

$$f(E_F) = [1 + e^{(E_F - E_F)/kT}]^{-1} = \frac{1}{1 + 1} = \frac{1}{2} \qquad (3\text{--}11)$$

Thus an energy state at the Fermi level has a probability of $1/2$ of being occupied by an electron.

A closer examination of $f(E)$ indicates that at 0 K the distribution takes the simple rectangular form shown in Fig. 3–14. With $T = 0$ in the denominator of the exponent, $f(E)$ is $1/(1 + 0) = 1$ when the exponent is negative ($E < E_F$), and is $1/(1 + \infty) = 0$ when the exponent is positive ($E > E_F$). This rectangular distribution implies that at 0 K every available energy state up to E_F is filled with electrons, and all states above E_F are empty.

At temperatures higher than 0 K, some probability exists for states above the Fermi level to be filled. For example, at $T = T_1$ in Fig. 3–14 there is some probability $f(E)$ that states above E_F are filled, and there is a corresponding probability $[1 - f(E)]$ that states below E_F are empty. The Fermi function is symmetrical about E_F for all temperatures; that is, the probability $f(E_F + \Delta E)$ that a state ΔE above E_F is filled is the same as the probability $[1 - f(E_F - \Delta E)]$ that a state ΔE below E_F is empty. The symmetry of the

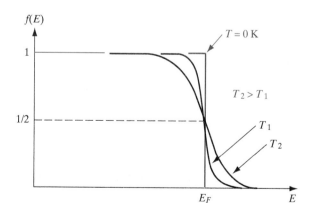

Figure 3–14
The Fermi-Dirac distribution function.

distribution of empty and filled states about E_F makes the Fermi level a natural reference point in calculations of electron and hole concentrations in semiconductors.

In applying the Fermi–Dirac distribution to semiconductors, we must recall that $f(E)$ is the probability of occupancy of an *available* state at E. Thus if there is no available state at E (e.g., in the band gap of a semiconductor), there is no possibility of finding an electron there. We can best visualize the relation between $f(E)$ and the band structure by turning the $f(E)$ vs. E diagram on its side so that the E scale corresponds to the energies of the band diagram (Fig. 3–15). For intrinsic material we know that the concentration of holes in the valence band is equal to the concentration of electrons in the conduction band. Therefore, the Fermi level E_F must lie at the middle of the band gap in intrinsic material.[4] Since $f(E)$ is symmetrical about E_F, the electron probability "tail" of $f(E)$ extending into the conduction band of Fig. 3–15a is symmetrical with the hole probability tail $[1 - f(E)]$ in the valence band. The distribution function has values within the band gap between

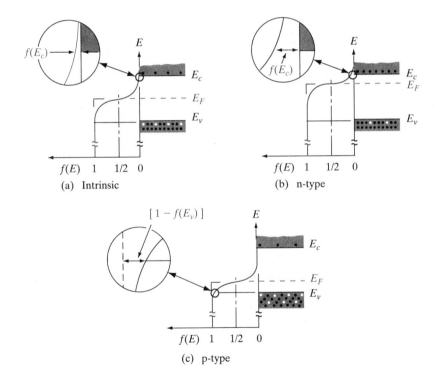

Figure 3–15
The Fermi distribution function applied to semiconductors: (a) intrinsic material; (b) n-type material; (c) p-type material.

[4]Actually the intrinsic E_F is displaced slightly from the middle of the gap, since the densities of available states in the valence and conduction bands are not equal (Section 3.3.2).

E_v and E_c, but there are no energy states available, and no electron occupancy results from $f(E)$ in this range.

The tails in $f(E)$ are exaggerated in Fig. 3–15 for illustrative purposes. Actually, the probability values at E_v and E_c are quite small for intrinsic material at reasonable temperatures. For example, in Si at 300 K, $n_i = p_i \approx 10^{10}$ cm^{-3}, whereas the densities of available states at E_v and E_c are on the order of 10^{19} cm^{-3}. Thus the probability of occupancy $f(E)$ for an individual state in the conduction band and the hole probability $[1 - f(E)]$ for a state in the valence band are quite small. Because of the relatively large density of states in each band, small changes in $f(E)$ can result in significant changes in carrier concentration.

In n-type material there is a high concentration of electrons in the conduction band compared with the hole concentration in the valence band (recall Fig. 3–12a). Thus in n-type material the distribution function $f(E)$ must lie above its intrinsic position on the energy scale (Fig. 3–15b). Since $f(E)$ retains its shape for a particular temperature, the larger concentration of electrons at E_c in n-type material implies a correspondingly smaller hole concentration at E_v. We notice that the value of $f(E)$ for each energy level in the conduction band (and therefore the total electron concentration n_0) increases as E_F moves closer to E_c. Thus the energy difference $(E_c - E_F)$ gives a measure of n; we shall express this relation mathematically in the following section.

For p-type material the Fermi level lies near the valence band (Fig. 3–15c) such that the $[1- f(E)]$ tail below E_v is larger than the $f(E)$ tail above E_c. The value of $(E_F - E_v)$ indicates how strongly p-type the material is.

It is usually inconvenient to draw $f(E)$ vs. E on every energy band diagram to indicate the electron and hole distributions. Therefore, it is common practice merely to indicate the position of E_F in band diagrams. This is sufficient information, since for a particular temperature the position of E_F implies the distributions in Fig. 3–15.

3.3.2 Electron and Hole Concentrations at Equilibrium

The Fermi distribution function can be used to calculate the concentrations of electrons and holes in a semiconductor, if the densities of available states in the valence and conduction bands are known. For example, the concentration of electrons in the conduction band is

$$n_0 = \int_{E_c}^{\infty} f(E)N(E)dE \qquad (3\text{–}12)$$

where $N(E)dE$ is the density of states (cm^{-3}) in the energy range dE. The subscript 0 used with the electron and hole concentration symbols (n_0, p_0) indicates equilibrium conditions. The number of electrons per unit volume in the energy range dE is the product of the density of states and the probability of occupancy $f(E)$. Thus the total electron concentration is the integral over the entire conduction band, as in Eq. (3–12).[5] The function $N(E)$ can be calculated by using quantum mechanics and the Pauli exclusion principle (Appendix IV).

It is shown in Appendix IV that $N(E)$ is proportional to $E^{1/2}$, so the density of states in the conduction band increases with electron energy. On the other hand, the Fermi function becomes extremely small for large energies. The result is that the product $f(E)N(E)$ decreases rapidly above E_c, and very few electrons occupy energy states far above the conduction band edge. Similarly, the probability of finding an empty state (hole) in the valence band $[1 - f(E)]$ decreases rapidly below E_v, and most holes occupy states near the top of the valence band. This effect is demonstrated in Fig. 3–16, which shows the density of available states, the Fermi function, and the resulting number of electrons and holes occupying available energy states in the conduction and valence bands at thermal equilibrium (i.e., with no excitations except thermal energy). For holes, increasing energy points down in Fig. 3–16, since the E scale refers to electron energy.

The result of the integration of Eq. (3–12) is the same as that obtained if we represent all of the distributed electron states in the conduction band by an *effective density of states* N_c located at the conduction band edge E_c. Therefore, the conduction band electron concentration is simply the effective density of states at E_c times the probability of occupancy at E_c[6]

$$n_0 = N_c f(E_c) \tag{3–13}$$

In this expression we assume the Fermi level E_F lies at least several kT below the conduction band. Then the exponential term is large compared with unity, and the Fermi function $f(E_c)$ can be simplified as

$$f(E_c) = \frac{1}{1 + e^{(E_c - E_F)/kT}} \simeq e^{-(E_c - E_F)/kT} \tag{3–14}$$

Since kT at room temperature is only 0.026 eV, this is generally a good approximation. For this condition the concentration of electrons in the conduction band is

$$\boxed{n_0 = N_c e^{-(E_c - E_F)/kT}} \tag{3–15}$$

The effective density of states N_c is shown in Appendix IV to be

$$N_c = 2\left(\frac{2\pi m_n^* kT}{h^2}\right)^{3/2} \tag{3–16}$$

---

[5]The upper limit is actually improper in Eq. (3–12), since the conduction band does not extend to infinite energy. This is unimportant in the calculation of n_0, however, since $f(E)$ becomes negligibly small for large values of E. Most electrons occupy states near the bottom of the conduction band at equilibrium.

[6]The simple expression for n_0 obtained in Eq. (3–13) is the direct result of integrating Eq. (3–12), as in Appendix IV. Equations (3–15) and (3–19) properly include the effects of the conduction and valence bands through the density-of-states terms.

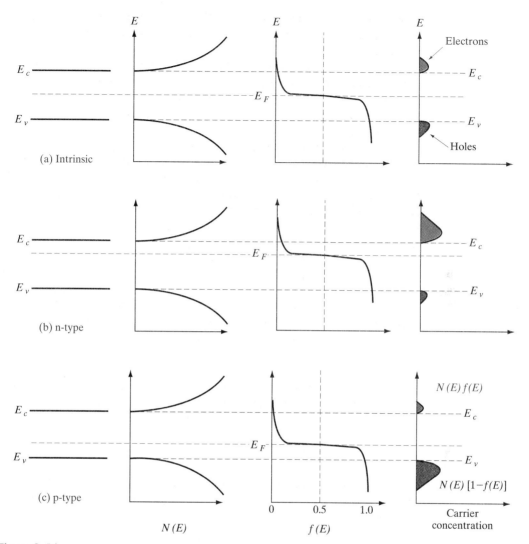

Figure 3–16
Schematic band diagram, density of states, Fermi–Dirac distribution, and the carrier concentrations for (a) intrinsic, (b) n-type, and (c) p-type semiconductors at thermal equilibrium.

Since the quantities in Eq. (3–16a) are known, values of N_c can be tabulated as a function of temperature. As Eq. (3–15) indicates, the electron concentration increases as E_F moves closer to the conduction band. This is the result we would predict from Fig. 3–15b.

In Eq. (3–16a), m_n^* is the density-of-states effective mass for electrons. To illustrate how it is obtained from the band curvature effective masses mentioned in Section 3.2.2, let us consider the 6 equivalent conduction band minima along the X-directions for Si. Looking at the cigar-shaped equi-energy surfaces in Fig. 3–10b, we find that we have more than one band curvature to deal with in calculating effective masses. There is a longitudinal effective mass m_l along the major axis of the ellipsoid, and the transverse effective mass m_t along the two minor axes. Since we have $(m_n^*)^{3/2}$ appearing in the density-of-states expression Eq. (3–16a), by using dimensional equivalence and adding contributions from all 6 valleys, we get

$$(m_n^*)^{3/2} = 6(m_l m_t^2)^{1/2} \tag{3–16b}$$

It can be seen that this is the geometric mean of the effective masses.

EXAMPLE 3–4   Calculate the density-of-states effective mass of electrons in Si.

SOLUTION   For Si, $m_l = 0.98\ m_0$; $m_t = 0.19\ m_0$ from Appendix III.
There are six equivalent X valleys in the conduction band.

$$m_n^* = 6^{2/3}[0.98(0.19)^2]^{1/3} m_0 = 1.1\ m_0$$

Note: For GaAs, the conduction band equi-energy surfaces are spherical. So there is only one band curvature effective mass, and it is equal to the density-of-states effective mass ($= 0.067\ m_0$).

By similar arguments, the concentration of holes in the valence band is

$$p_0 = N_v[1 - f(E_v)] \tag{3–17}$$

where N_v is the effective density of states in the valence band. The probability of finding an empty state at E_v is

$$1 - f(E_v) = 1 - \frac{1}{1 + e^{(E_v - E_F)/kT}} \simeq e^{-(E_F - E_v)/kT} \tag{3–18}$$

for E_F larger than E_v by several kT. From these equations, the concentration of holes in the valence band is

$$\boxed{p_0 = N_v e^{-(E_F - E_v)/kT}} \tag{3–19}$$

The effective density of states in the valence band reduced to the band edge is

$$N_v = 2\left(\frac{2\pi m_p^* kT}{h^2}\right)^{3/2} \tag{3–20}$$

As expected from Fig. 3–15c, Eq. (3–19) predicts that the hole concentration increases as E_F moves closer to the valence band.

The electron and hole concentrations predicted by Eqs. (3–15) and (3–19) are valid whether the material is intrinsic or doped, provided thermal equilibrium is maintained. Thus *for intrinsic material, E_F lies at some intrinsic level E_i near the middle of the band gap* (Fig. 3–15a), and the intrinsic electron and hole concentrations are

$$n_i = N_c e^{-(E_c - E_i)/kT}, \quad p_i = N_v e^{-(E_i - E_v)/kT} \qquad (3\text{–}21)$$

The product of n_0 and p_0 at equilibrium is a constant for a particular material and temperature, even if the doping is varied:

$$n_0 p_0 = \left(N_c e^{-(E_c - E_F)/kT}\right)\left(N_v e^{-(E_F - E_v)/kT}\right) = N_c N_v e^{-(E_c - E_v)/kT} \quad (3\text{–}22a)$$

$$= N_c N_v e^{-E_g/kT}$$

$$n_i p_i = \left(N_c e^{-(E_c - E_i)/kT}\right)\left(N_v e^{-(E_i - E_v)/kT}\right) = N_c N_v e^{-E_g/kT} \qquad (3\text{–}22b)$$

The intrinsic electron and hole concentrations are equal (since the carriers are created in pairs), $n_i = p_i$; thus the intrinsic concentration is

$$n_i = \sqrt{N_c N_v}\, e^{-E_g/2kT} \qquad (3\text{–}23)$$

The constant product of electron and hole concentrations in Eq. (3–22) can be written conveniently as

$$\boxed{n_0 p_0 = n_i^2} \qquad (3\text{–}24)$$

This is an important relation, and we shall use it extensively in later calculations. The intrinsic concentration for Si at room temperature is approximately $n_i = 1.5 \times 10^{10}$ cm^{-3}.

Comparing Eqs. (3–21) and (3–23), we note that the intrinsic level E_i is the middle of the band gap ($E_c - E_i = E_g/2$), if the effective densities of states N_c and N_v are equal. There is usually some difference in effective mass for electrons and holes, however, and N_c and N_v are slightly different as Eqs. (3–16) and (3–20) indicate. The intrinsic level E_i is displaced from the middle of the band gap, more for GaAs than for Ge or Si.

Another convenient way of writing Eqs. (3–15) and (3–19) is

$$\boxed{\begin{aligned} n_0 &= n_i e^{(E_F - E_i)/kT} \\ p_0 &= n_i e^{(E_i - E_F)/kT} \end{aligned}} \qquad \begin{aligned} (3\text{–}25a) \\ (3\text{–}25b) \end{aligned}$$

obtained by the application of Eq. (3–21). This form of the equations indicates directly that the electron concentrations is n_i when E_F is at the intrinsic level E_i, and that n_0 increases exponentially as the Fermi level moves away from E_i toward the conduction band. Similarly, the hole concentration p_0 varies from n_i to larger values as E_F moves from E_i toward the valence band. Since these equations reveal the qualitative features of carrier concentration so directly, they are particularly convenient to remember.

EXAMPLE 3–5   A Si sample is doped with 10^{17} As atoms/cm^3. What is the equilibrium hole concentration p_0 at 300 K? Where is E_F relative to E_i?

SOLUTION   Since $N_d \gg n_i$, we can approximate $n_0 = N_d$ and

$$p_0 = \frac{n_i^2}{n_0} = \frac{2.25 \times 10^{20}}{10^{17}} = 2.25 \times 10^3 \text{ cm}^{-3}$$

From Eq. (3–25a), we have

$$E_F - E_i = kT \ln \frac{n_0}{n_i} = 0.0259 \ln \frac{10^{17}}{1.5 \times 10^{10}} = 0.407 \text{ eV}$$

The resulting band diagram is:

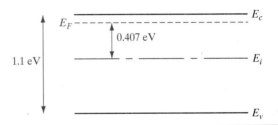

3.3.3 Temperature Dependence of Carrier Concentrations

The variation of carrier concentration with temperature is indicated by Eq. (3–25). Initially, the variation of n_0 and p_0 with T seems relatively straightforward in these relations. The problem is complicated, however, by the fact that n_i has a strong temperature dependence [Eq. (3–23)] and that E_F can also vary with temperature. Let us begin by examining the intrinsic carrier concentration. By combining Eqs. (3–23), (3–16a), and (3–20) we obtain

$$n_i(T) = 2\left(\frac{2\pi kT}{h^2}\right)^{3/2} (m_n^* m_p^*)^{3/4} e^{-E_g/2kT} \qquad (3\text{–}26)$$

The exponential temperature dependence dominates $n_i(T)$, and a plot of $\ln n_i$ vs. $10^3/T$ appears linear (Fig. 3–17).[7] In this figure we neglect variations due to the $T^{3/2}$ dependence of the density-of-states function and the fact

---

[7] When plotting quantities such as carrier concentration, which involve a Boltzmann factor, it is common to use an inverse temperature scale. This allows terms which are exponential in $1/T$ to appear linear in the semilogarithmic plot. When reading such graphs, remember that temperature increases from right to left.

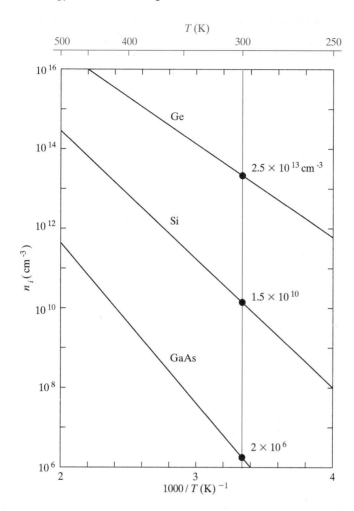

T (K)

n_i (cm^{-3})

Ge

2.5×10^{13} cm^{-3}

Si

1.5×10^{10}

GaAs

2×10^6

$1000 / T$ (K)$^{-1}$

Figure 3–17
Intrinsic carrier concentration for Ge, Si, and GaAs as a function of inverse temperature. The room temperature values are marked for reference.

that E_g varies somewhat with temperature.[8] The value of n_i at any temperature is a definite number for a given semiconductor, and is known for most materials. Thus we can take n_i as given in calculating n_0 or p_0 from Eq. (3–25).[9]

With n_i and T given, the unknowns in Eq. (3–25) are the carrier concentrations and the Fermi level position relative to E_i. One of these two

---

[8]For Si the band gap E_g varies from about 1.11 eV at 300 K to about 1.16 eV at 0 K.

[9]Care must be taken to use consistent units in these calculations. For example, if an energy such as E_g is expressed in electron volts (eV), it should be multiplied by q (1.6 × 10^{-19} C) to convert to joules if k is in J/K; alternatively, E_g can be kept in eV and the value of k in eV/K can be used. At 300 K we can use kT = 0.0259 eV and E_g in eV.

quantities must be given if the other is to be found. If the carrier concentration is held at a certain value, as in heavily doped extrinsic material, E_F can be obtained from Eq. (3–25). The temperature dependence of electron concentration in a doped semiconductor can be visualized as shown in Fig. 3–18. In this example, Si is doped n-type with a donor concentration N_d of 10^{15} cm^{-3}. At very low temperatures (large $1/T$), negligible intrinsic EHPs exist, and the donor electrons are bound to the donor atoms. As the temperature is raised, these electrons are donated to the conduction band, and at about 100 K ($1000/T = 10$) all the donor atoms are ionized. This temperature range is called the *ionization* region. Once the donors are ionized, the conduction band electron concentration is $n_0 \simeq N_d = 10^{15}$ cm^{-3}, since one electron is obtained for each donor atom. When every available extrinsic electron has been transferred to the conduction band, n_0 is virtually constant with temperature until the concentration of intrinsic carriers n_i becomes comparable to the extrinsic concentration N_d. Finally, at higher temperatures n_i is much greater than N_d, and the intrinsic carriers dominate. In most devices it is desirable to control the carrier concentration by doping rather than by thermal EHP generation. Thus one usually dopes the material such that the extrinsic range extends beyond the highest temperature at which the device is to be used.

3.3.4 Compensation and Space Charge Neutrality

When the concept of doping was introduced, we assumed the material contained either N_d donors or N_a acceptors, so that the extrinsic majority carrier concentrations were $n_0 \simeq N_d$ or $p_0 \simeq N_a$, respectively, for the n-type or p-type

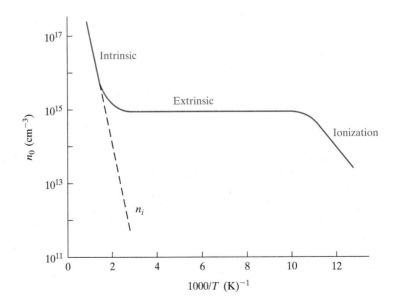

Figure 3–18
Carrier concentration vs. inverse temperature for Si doped with 10^{15} donors/cm^3.

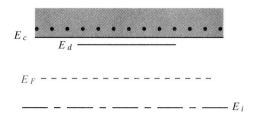

E_c

E_d

E_F

E_i

E_a

E_v

Figure 3–19
Compensation in an n-type semiconductor $(N_d > N_a)$.

material. If often happens, however, that a semiconductor contains both donors and acceptors. For example, Fig. 3–19 illustrates a semiconductor for which both donors and acceptors are present, but $N_d > N_a$. The predominance of donors makes the material n-type, and the Fermi level is therefore in the upper part of the band gap. Since E_F is well above the acceptor level E_a, this level is essentially filled with electrons. However, with E_F above E_i, we cannot expect a hole concentration in the valence band commensurate with the acceptor concentration. In fact, the filling of the E_a states occurs at the expense of the donated conduction band electrons. The mechanism can be visualized as follows: Assume an acceptor state is filled with a valence band electron as described in Fig. 3–12b, with a hole resulting in the valence band. This hole is then filled by recombination with one of the conduction band electrons. Extending this logic to all the acceptor atoms, we expect the resultant concentration of electrons in the conduction band to be $N_d - N_a$ instead of the total N_d. This process is called *compensation*. By this process it is possible to begin with an *n*-type semiconductor and add acceptors until $N_a = N_d$ and no donated electrons remain in the conduction band. In such compensated material, $n_0 = n_i = p_0$ and intrinsic conduction is obtained. With further acceptor doping the semiconductor becomes p-type with a hole concentration of essentially $N_a - N_d$.

The exact relationship among the electron, hole, donor, and acceptor concentrations can be obtained by considering the requirements for *space charge neutrality*. If the material is to remain electrostatically neutral, the sum of the positive charges (holes and ionized donor atoms) must balance the sum of the negative charges (electrons and ionized acceptor atoms):

$$p_0 + N_d^+ = n_0 + N_a^- \qquad (3\text{–}27)$$

Thus in Fig. 3–19 the net electron concentration in the conduction band is

$$n_0 = p_0 + (N_d^+ - N_a^-) \qquad (3\text{–}28)$$

If the material is doped n-type ($n_0 \gg p_0$) and all the impurities are ionized, we can approximate Eq. (3–28) by $n_0 \approx N_d - N_a$.

Since the intrinsic semiconductor itself is electrostatically neutral and the doping atoms we add are also neutral, the requirement of Eq. (3–27) must be maintained at equilibrium. The electron and hole concentrations and the Fermi level adjust such at Eqs. (3–27) and (3–25) are satisfied.

3.4 DRIFT OF CARRIERS IN ELECTRIC AND MAGNETIC FIELDS

Knowledge of carrier concentrations in a solid is necessary for calculating current flow in the presence of electric or magnetic fields. In addition to the values of *n* and *p*, we must be able to take into account the collisions of the charge carriers with the lattice and with the impurities. These processes will affect the ease with which electrons and holes can flow through the crystal, that is, their *mobility* within the solid. As should be expected, these collision and scattering processes depend on temperature, which affects the thermal motion of the lattice atoms and the velocity of the carriers.

3.4.1 Conductivity and Mobility

The charge carriers in a solid are in constant motion, even at thermal equilibrium. At room temperature, for example, the thermal motion of an individual electron may be visualized as random scattering from lattice vibrations, impurities, other electrons, and defects (Fig. 3–20). Since the scattering is random, there is no net motion of the group of *n* electrons/cm^3 over any period of time. This is not true of an individual electron, of course. The probability of the electron in Fig. 3–20 returning to its starting point after some time *t* is negligibly small. However, if a large number of electrons is considered (e.g., 10^{16} cm^{-3} in an n-type semiconductor), there will be no preferred direction of motion for the group of electrons and no net current flow.

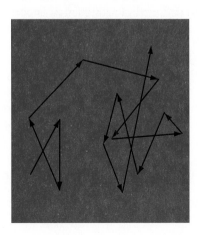

Figure 3–20
Thermal motion of an electron in a solid.

If an electric field $\mathscr{E}_x$ is applied in the x-direction, each electron experiences a net force $-q\mathscr{E}_x$ from the field. This force may be insufficient to alter appreciably the random path of an individual electron; the effect when averaged over all the electrons, however, is a net motion of the group in the $-x$-direction. If p_x is the x-component of the total momentum of the group, the force of the field on the n electrons/cm^3 is

$$-nq\mathscr{E}_x = \frac{d\mathsf{p}_x}{dt}\bigg|_{field} \tag{3-29}$$

Initially, Eq. (3–29) seems to indicate a continuous acceleration of the electrons in the $-x$-direction. This is not the case, however, because the net acceleration of Eq. (3–29) is just balanced in steady state by the decelerations of the collision processes. Thus while the steady field $\mathscr{E}_x$ does produce a net momentum p_{-x}, the net rate of change of momentum when collisions are included must be zero in the case of steady state current flow.

To find the total rate of momentum change from collisions, we must investigate the collision probabilities more closely. If the collisions are truly random, there will be a constant probability of collision at any time for each electron. Let us consider a group of N_0 electrons at time $t = 0$ and define $N(t)$ as the number of electrons that *have not* undergone a collision by time t. The rate of decrease in $N(t)$ at any time t is proportional to the number left unscattered at t,

$$-\frac{dN(t)}{dt} = \frac{1}{\bar{t}}N(t) \tag{3-30}$$

where $\bar{t}^{-1}$ is a constant of proportionality.

The solution to Eq. (3–30) is an exponential fuction

$$N(t) = N_0 e^{-t/\bar{t}} \tag{3-31}$$

and $\bar{t}$ represents the mean time between scattering events,[10] called the *mean free time*. The probability that any electron has a collision in the time interval dt is $dt/\bar{t}$. Thus the differential change in p_x due to collisions in time dt is

$$d\mathsf{p}_x = -\mathsf{p}_x\frac{dt}{\bar{t}} \tag{3-32}$$

The rate of change of p_x due to the decelerating effect of collisions is

$$\frac{d\mathsf{p}_x}{dt}\bigg|_{collisions} = -\frac{\mathsf{p}_x}{\bar{t}} \tag{3-33}$$

[10]Equations (3–30) and (3–31) are typical of events dominated by random processes, and the forms of these equations occur often in many branches of physics and engineering. For example, in the radioactive decay of unstable nuclear isotopes, N_0 nuclides decay exponentially with a mean lifetime $\bar{t}$. Other examples will be found in this text, including the absorption of light in a semiconductor and the recombination of excess EHPs.

The sum of acceleration and deceleration effects must be zero for steady state. Taking the sum of Eqs. (3–29) and (3–33), we have

$$-\frac{p_x}{t} - nq\mathscr{E}_x = 0 \tag{3–34}$$

The average momentum per electron is

$$\langle p_x \rangle = \frac{p_x}{n} = -q\bar{t}\mathscr{E}_x \tag{3–35}$$

where the angular brackets indicate an average over the entire group of electrons. As expected for steady state, Eq. (3–35) indicates that the electrons have *on the average* a constant net velocity in the negative *x*-direction:

$$\langle v_x \rangle = \frac{\langle p_x \rangle}{m_n^*} = -\frac{q\bar{t}}{m_n^*}\mathscr{E}_x \tag{3–36}$$

Actually, the individual electrons move in many directions by thermal motion during a given time period, but Eq. (3–36) tells us the *net drift* of an average electron in response to the electric field. The drift speed described by Eq. (3–36) is usually much smaller than the random speed due to the thermal motion v_{th}.

The current density resulting from this net drift is just the number of electrons crossing a unit area per unit time $(n\langle v_x \rangle)$ multiplied by the charge on the electron $(-q)$:

$$\boxed{\begin{array}{c} J_x = -qn\langle v_x \rangle \\[4pt] \dfrac{\text{ampere}}{\text{cm}^2} = \dfrac{\text{coulomb}}{\text{electron}} \cdot \dfrac{\text{electrons}}{\text{cm}^3} \cdot \dfrac{\text{cm}}{\text{s}} \end{array}} \tag{3–37}$$

Using Eq. (3–36) for the average velocity, we obtain

$$J_x = \frac{nq^2\bar{t}}{m_n^*}\mathscr{E}_x \tag{3–38}$$

Thus the current density is proportional to the electric field, as we expect from Ohm's law:

$$J_x = \sigma\mathscr{E}_x, \quad \text{where } \sigma \equiv \frac{nq^2\bar{t}}{m_n^*} \tag{3–39}$$

The conductivity $\sigma(\Omega\text{-cm})^{-1}$ can be written

$$\sigma = qn\mu_n, \quad \text{where } \mu_n \equiv \frac{q\bar{t}}{m_n^*} \tag{3–40a}$$

The quantity μ_n, called the *electron mobility*, describes the ease with which electrons drift in the material. Mobility is a very important quantity in characterizing semiconductor materials and in device development.

Here m_n^* is the conductivity effective mass for electrons, different from the density-of-states effective mass mentioned in Eq. (3–16b). While we use the density-of-states effective mass to count the number of carriers in bands, we must use the conductivity effective mass for charge transport problems. To illustrate how it is obtained from the band curvature effective masses mentioned in Section 3.2.2, once again let us consider the 6 equivalent conduction band minima along the X-directions for Si, with the band curvature longitudinal effective mass, m_l, along the major axis of the ellipsoid, and the transverse effective mass, m_t, along the two minor axes (Fig. 3–10b). Since we have $1/m_n^*$ in the mobility expression Eq. (3–40a), by using dimensional equivalence, we can write the conductivity effective mass as the harmonic mean of the band curvature effective masses.

$$\frac{1}{m_n^*} = \frac{1}{3}\left(\frac{1}{m_l} + \frac{2}{m_t}\right) \qquad (3\text{–}40\text{b})$$

Calculate the conductivity effective mass of electrons in Si. **EXAMPLE 3–6**

SOLUTION

For Si, $m_l = 0.98\, m_0$; $m_t = 0.19\, m_0$ (Appendix III)
There are 6 equivalent X valleys in the conduction band.

$$1/m_n^* = 1/3(1/m_x + 1/m_y + 1/m_z) = 1/3(1/m_l + 2/m_t)$$

$$1/m_n^* = \frac{1}{3}\left(\frac{1}{0.98\, m_0} + \frac{2}{0.19\, m_0}\right)$$

$$m_n^* = 0.26\, m_0$$

Note: For GaAs, the conduction band equi-energy surfaces are spherical. So there is only one band curvature effective mass. (The density of states effective mass and the conductivity effective mass are both 0.067 m_0.)

The mobility defined in Eq. (3–40a) can be expressed as the average particle drift velocity per unit electric field. Comparing Eqs. (3–36) and (3–40a), we have

$$\mu_n = -\frac{\langle v_x\rangle}{\mathscr{E}_x} \qquad (3\text{–}41)$$

The units of mobility are (cm/s)/(V/cm) = cm²/V-s, as Eq. (3–41) suggests. The minus sign in the definition results in a positive value of mobility, since electrons drift opposite to the field.
The current density can be written in terms of mobility as

$$J_x = qn\mu_n\mathscr{E}_x \qquad (3\text{–}42)$$

This derivation has been based on the assumption that the current is carried primarily by electrons. For hole conduction we change n to p, $-q$ to $+q$, and μ_n to μ_p, where $\mu_p = +\langle v_x \rangle / \mathscr{E}_x$ is the mobility for holes. If both electrons and holes participate, we must modify Eq. (3–42) to

$$\boxed{J_x = q(n\mu_n + p\mu_p)\mathscr{E}_x = \sigma\mathscr{E}_x} \qquad (3\text{–}43)$$

Values of μ_n and μ_p are given for many of the common semiconductor materials in Appendix III. According to Eq. (3–40), the parameters determining mobility are m^* and mean free time $\bar{t}$. Effective mass is a property of the material's band structure, as described by Eq. (3–3). Thus we expect m_n^* to be small in the strongly curved Γ minimum of the GaAs conduction band (Fig. 3–6), with the result that μ_n is very high. In a more gradually curved band, a larger m^* in the denominator of Eq. (3–40) leads to a smaller value of mobility. It is reasonable to expect that lighter particles are more mobile than heavier particles (which is satisfying, since the common-sense value of effective mass is not always apparent). The other parameter determining mobility is the mean time between scattering events, $\bar{t}$. In Section 3.4.3 we shall see that this is determined primarily by temperature and impurity concentration in the semiconductor.

3.4.2 Drift and Resistance

Let us look more closely at the drift of electrons and holes. If the semiconductor bar of Fig. 3–21 contains both types of carrier, Eq. (3–43) gives the conductivity of the material. The resistance of the bar is then

$$R = \frac{\rho L}{wt} = \frac{L}{wt}\frac{1}{\sigma} \qquad (3\text{–}44)$$

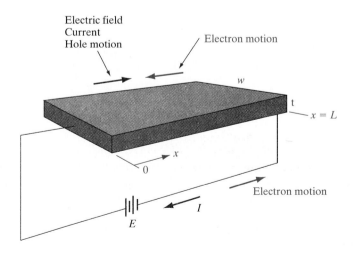

Electric field
Current
Hole motion

Electron motion

w

t

$x = L$

x

0

Electron motion

I

E

Figure 3–21
Drift of electrons and holes in a semiconductor bar.

where ρ is the resistivity (Ω-cm). The physical mechanism of carrier drift requires that the holes in the bar move as a group in the direction of the electric field and that the electrons move as a group in the opposite direction. Both the electron and the hole components of current are in the direction of the $\mathcal{E}$ field, since conventional current is positive in the direction of hole flow and opposite to the direction of electron flow. The drift current described by Eq. (3–43) is constant throughout the bar. A valid question arises, therefore, concerning the nature of the electron and hole flow at the contacts and in the external circuit. We should specify that the contacts to the bar of Fig. 3–21 are *ohmic,* meaning that they are perfect sources and sinks of both carrier types and have no special tendency to inject or collect either electrons or holes.

If we consider that current is carried around the external circuit by electrons, there is no problem in visualizing electrons flowing into the bar at one end and out at the other (always opposite to I). Thus for every electron leaving the left end ($x = 0$) of the bar in Fig. 3–21, there is a corresponding electron entering at $x = L$, so that the electron concentration in the bar remains constant at n. But what happens to the holes at the contacts? As a hole reaches the ohmic contact at $x = L$, it recombines with an electron, which must be supplied through the external circuit. As this hole disappears, a corresponding hole must appear at $x = 0$ to maintain space charge neutrality. It is reasonable to consider the source of this hole as the generation of an EHP at $x = 0$, with the hole flowing into the bar and the electron flowing into the external circuit.

Find the resistivity of intrinsic Si at 300 K.   **EXAMPLE 3–7**

From Appendix III, $\mu_n = 1350$ and $\mu_p = 480$ cm²/V-s for intrinsic Si. Thus,   **SOLUTION**
since $n_0 = p_0 = n_i$,

$$\sigma_i = q(\mu_n + \mu_p)n_i = 1.6 \times 10^{-19}(1830)(1.5 \times 10^{10})$$
$$= 4.39 \times 10^{-6} \, (\Omega - \text{cm})^{-1}$$
$$\rho_i = \sigma_i^{-1} = 2.28 \times 10^5 \, \Omega - \text{cm}$$

3.4.3 Effects of Temperature and Doping on Mobility

The two basic types of scattering mechanisms that influence electron and hole mobility are *lattice scattering* and *impurity scattering.* In lattice scattering a carrier moving through the crystal is scattered by a vibration of the lattice, resulting from the temperature.[11] The frequency of such scattering events

---

[11]Collective vibrations of atoms in the crystal are called *phonons.* Thus lattice scattering is also known as *phonon scattering.*

Figure 3–22
Approximate tem-
perature depen-
dence of mobility
with both lattice
and impurity
scattering.

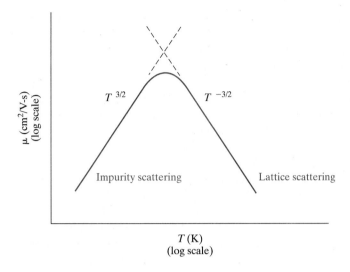

increases as the temperature increases, since the thermal agitation of the lattice becomes greater. Therefore, we should expect the mobility to decrease as the sample is heated (Fig. 3–22). On the other hand, scattering from crystal defects such as ionized impurities becomes the dominant mechanism at low temperatures. Since the atoms of the cooler lattice are less agitated, lattice scattering is less important; however, the thermal motion of the carriers is also slower. Since a slowly moving carrier is likely to be scattered more strongly by an interaction with a charged ion than is a carrier with greater momentum, impurity scattering events cause a decrease in mobility with decreasing temperature. As Fig. 3–22 indicates, the approximate temperature dependencies are $T^{-3/2}$ for lattice scattering and $T^{3/2}$ for impurity scattering. Since the scattering probability of Eq. (3–32) is inversely proportional to the mean free time and therefore to mobility, the mobilities due to two or more scattering mechanisms add inversely:

$$\frac{1}{\mu} = \frac{1}{\mu_1} + \frac{1}{\mu_2} + \dots \qquad (3\text{–}45)$$

As a result, the mechanism causing the lowest mobility value dominates, as shown in Fig. 3–22.

As the concentration of impurities increases, the effects of impurity scattering are felt at higher temperatures. For example, the electron mobility μ_n of intrinsic silicon at 300 K is 1350 cm^2/(V-s). With a donor doping concentration of 10^{17} cm^{-3}, however, μ_n is 700 cm^2/(V-s). Thus the presence of the 10^{17} ionized donors/cm^3 introduces a significant amount of impurity scattering. This effect is illustrated in Fig. 3–23, which shows the variation of mobility with doping concentration at room temperature.

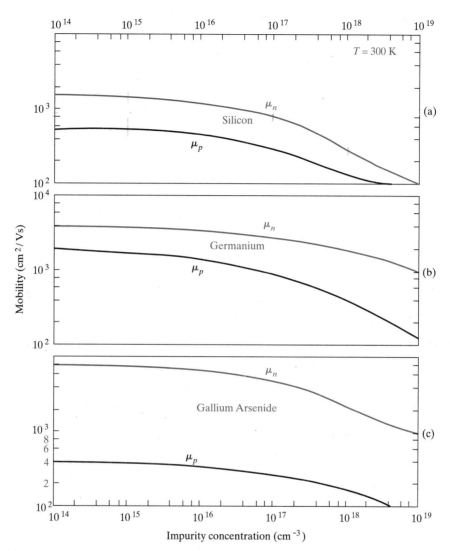

Figure 3–23
Variation of mobility with total doping impurity concentration ($N_a + N_d$) for Ge, Si, and GaAs at 300 K.

3.4.4 High-Field Effects

One assumption implied in the derivation of Eq. (3–39) was that Ohm's law is valid in the carrier drift processes. That is, it was assumed that the drift current is proportional to the electric field and that the proportionality constant (σ) is not a function of field $\mathscr{E}$. This assumption is valid over a wide range of $\mathscr{E}$. However, large electric fields ($> 10^3$V/cm) can cause the drift velocity and

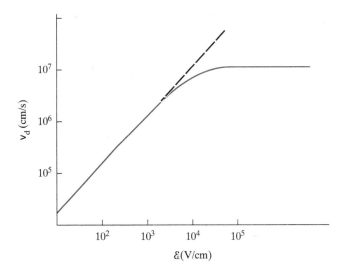

Figure 3–24
Saturation of electron drift velocity at high electric fields for Si.

therefore the current $J = -qn\mathbf{v}_d$ to exhibit a sublinear dependence on the electric field. This dependence of σ upon $\mathscr{E}$ is an example of a *hot carrier* effect, which implies that the carrier drift velocity $\mathbf{v}_d$ is comparable to the thermal velocity $\mathbf{v}_{th}$.

In many cases an upper limit is reached for the carrier drift velocity in a high field (Fig. 3–24). This limit occurs near the mean thermal velocity ($\simeq 10^7$ cm/s) and represents the point at which added energy imparted by the field is transferred to the lattice rather than increasing the carrier velocity. The result of this *scattering limited velocity* is a fairly constant current at high field. This behavior is typical of Si, Ge, and some other semiconductors. However, there are other important effects in some materials; for example, in Chapter 10 we shall discuss a *decrease* in electron velocity at high fields for GaAs and certain other materials, which results in negative conductivity and current instabilities in the sample. Another important high-field effect is avalanche multiplication, which we shall discuss in Section 5.4.2.

3.4.5 The Hall Effect

If a magnetic field is applied perpendicular to the direction in which holes drift in a p-type bar, the path of the holes tends to be deflected (Fig. 3–25). Using vector notation, the total force on a single hole due to the electric and magnetic fields is

$$\mathbf{F} = q(\mathscr{E} + \mathbf{v} \times \mathscr{B}) \tag{3–46}$$

In the *y*-direction the force is

$$F_y = q(\mathscr{E}_y - \mathbf{v}_x \mathscr{B}_z) \tag{3–47}$$

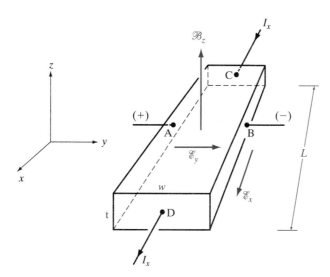

Figure 3–25
The Hall effect.

The important result of Eq. (3–47) is that unless an electric field $\mathscr{E}_y$ is established along the width of the bar, each hole will experience a net force (and therefore an acceleration) in the $-y$-direction due to the $qv_x\mathscr{B}_z$ product. Therefore, to maintain a steady state flow of holes down the length of the bar, the electric field $\mathscr{E}_y$ must just balance the product $v_x\mathscr{B}_z$:

$$\mathscr{E}_y = v_x\mathscr{B}_z \tag{3–48}$$

so that the net force F_y is zero. Physically, this electric field is set up when the magnetic field shifts the hole distribution slightly in the $-y$-direction. Once the electric field $\mathscr{E}_y$ becomes as large as $v_x\mathscr{B}_z$, no net lateral force is experienced by the holes as they drift along the bar. The establishment of the electric field $\mathscr{E}_y$ is known as the *Hall effect,* and the resulting voltage $V_{AB} = \mathscr{E}_y w$ is called the *Hall voltage.* If we use the expression derived in Eq. (3–37) for the drift velocity (using $+q$ and p_0 for holes), the field $\mathscr{E}_y$ becomes

$$\mathscr{E}_y = \frac{J_x}{qp_0}\mathscr{B}_z = R_H J_x \mathscr{B}_z, \quad R_H \equiv \frac{1}{qp_0} \tag{3–49}$$

Thus the Hall field is proportional to the product of the current density and the magnetic flux density. The proportionality constant $R_H = (qp_0)^{-1}$ is called the *Hall coefficient.* A measurement of the Hall voltage for a known current and magnetic field yields a value for the hole concentration p_0

$$p_0 = \frac{1}{qR_H} = \frac{J_x\mathscr{B}_z}{q\mathscr{E}_y} = \frac{(I_x/wt)\mathscr{B}_z}{q(V_{AB}/w)} = \frac{I_x\mathscr{B}_z}{qtV_{AB}} \tag{3–50}$$

Since all of the quantities in the right-hand side of Eq. (3–50) can be measured, the Hall effect can be used to give quite accurate values for carrier concentration.

If a measurement of resistance R is made, the sample resistivity ρ can be calculated:

$$\rho(\Omega-cm) = \frac{Rwt}{L} = \frac{V_{CD}/I_x}{L/wt} \tag{3-51}$$

Since the conductivity $\sigma = 1/\rho$ is given by $q\mu_p p_0$, the mobility is simply the ratio of the Hall coefficient and the resistivity:

$$\mu_p = \frac{\sigma}{qp_0} = \frac{1/\rho}{q(1/qR_H)} = \frac{R_H}{\rho} \tag{3-52}$$

Measurements of the Hall coefficient and the resistivity over a range of temperatures yield plots of majority carrier concentration and mobility vs. temperature. Such measurements are extremely useful in the analysis of semiconductor materials. Although the discussion here has been related to p-type material, similar results are obtained for n-type material. A negative value of q is used for electrons, and the Hall voltage V_{AB} and Hall coefficient R_H are negative. In fact, measurement of the sign of the Hall voltage is a common technique for determining if an unknown sample is p-type or n-type.

EXAMPLE 3-8

A sample of Si is doped with 10^{17} phosphorus atoms/cm³. What would you expect to measure for its resistivity? What Hall voltage would you expect in a sample 100 μm thick if $I_x = 1mA$ and $\mathcal{B}_z = 1\ kG = 10^{-5}\ Wb/cm^2$?

SOLUTION

From Fig. 3-23, the mobility is 700 cm²/(V-s). Thus the conductivity is

$$\sigma = q\mu_n n_0 = (1.6 \times 10^{-19})\,(700)\,(10^{17}) = 11.2(\Omega-cm)^{-1}$$

since p_0 is negligible. The resistivity is

$$\rho = \sigma^{-1} = 0.0893\ \Omega-cm$$

The Hall coefficient is

$$R_H = -(qn_0)^{-1} = -62.5\ cm^3/C$$

from Eq. (3-49), or we could use Eq. (3-52). The Hall voltage is

$$V_{AB} = \frac{I_x \mathcal{B}_z}{t} R_H = \frac{(10^{-3}\ A)(10^{-5}\ Wb/cm^2)}{10^{-2}\ cm}\,(-62.5\ cm^3/C) = -62.5\ \mu V$$

3.5
INVARIANCE OF
THE FERMI LEVEL
AT EQUILIBRIUM

In this chapter we have discussed homogeneous semiconductors, without variations in doping and without junctions between dissimilar materials. In the following chapters we will be considering cases in which nonuniform doping occurs in a given semiconductor, or junctions occur between different semiconductors or a semiconductor and a metal. These cases are crucial to the various types of electronic and optoelectronic devices made in semiconductors. In anticipation of those discussions, an important concept should be established here regarding the demands of equilibrium. That concept can be summarized by noting that *no discontinuity or gradient can arise in the equilibrium Fermi level E_F.*

To demonstrate this assertion, let us consider two materials in intimate contact such that electrons can move between the two (Fig. 3–26). These may be, for example, dissimilar semiconductors, n- and p-type regions, a metal and a semiconductor, or simply two adjacent regions of a nonuniformly doped semiconductor. Each material is described by a Fermi–Dirac distribution function and some distribution of available energy states that electrons can occupy.

There is no current, and therefore no net charge transport, at thermal equilibrium. There is also no net transfer of energy. Therefore, for each energy E in Fig. 3–26 any transfer of electrons from material 1 to material 2 must be exactly balanced by the opposite transfer of electrons from 2 to 1. We will let the density of states at energy E in material 1 be called $N_1(E)$ and in material 2 we will call it $N_2(E)$. At energy E the rate of transfer of electrons from 1 to 2 is proportional to the number of filled states at E in material 1 times the number of empty states at E in material 2:

$$\text{rate from 1 to 2} \propto N_1(E)f_1(E) \cdot N_2(E)[1 - f_2(E)] \qquad (3\text{–}53)$$

where $f(E)$ is the probability of a state being filled at E in each material, i.e., the Fermi–Dirac distribution function given by Eq. (3–10). Similarly,

$$\text{rate from 2 to 1} \propto N_2(E)f_2(E) \cdot N_1(E)[1 - f_1(E)] \qquad (3\text{–}54)$$

At equilibrium these must be equal:

$$N_1(E)f_1(E) \cdot N_2(E)[1 - f_2(E)] = N_2(E)f_2(E) \cdot N_1(E)[1 - f_1(E)] \qquad (3\text{–}55)$$

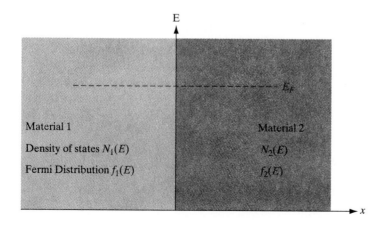

Material 1

Density of states $N_1(E)$

Fermi Distribution $f_1(E)$

Material 2

$N_2(E)$

$f_2(E)$

Figure 3–26
Two materials in intimate contact at equilibrium. Since the net motion of electrons is zero, the equilibrium Fermi level must be constant throughout.

Rearranging terms, we have, at energy E,

$$N_1 f_1 N_2 - N_1 f_1 N_2 f_2 = N_2 f_2 N_1 - N_2 f_2 N_1 f_1 \qquad (3-56)$$

which results in

$$f_1(E) = f_2(E), \quad \text{that is,} \quad [1 + e^{(E-E_{F1})/kT}]^{-1} = [1 + e^{(E-E_{F2})/kT}]^{-1} \qquad (3-57)$$

Therefore, we conclude that $E_{F1} = E_{F2}$. That is, there is no discontinuity in the equilibrium Fermi level. More generally, we can state that the Fermi level at equilibrium must be constant throughout materials in intimate contact. One way of stating this is that no gradient exists in the Fermi level at equilibrium:

$$\boxed{\frac{dE_F}{dx} = 0} \qquad (3-58)$$

We will make considerable use of this result in the chapters to follow.

PROBLEMS

3.1 It was mentioned in Section 3.2 that the covalent bonding model gives a false impression of the localization of carriers. As an illustration, calculate the radius of the electron orbit around the donor in Fig. 3–12c, assuming a ground state hydrogen-like orbit in Si. Compare with the Si lattice constant. Use $m_n^* = 0.26 m_0$ for Si.

3.2 Calculate values for the Fermi function $f(E)$ at 300 K and plot vs. energy in eV as in Fig. 3–14. Choose $E_F = 1$ eV and make the calculated points closer together near the Fermi level to obtain a smooth curve. Notice that $f(E)$ varies quite rapidly within a few kT of E_F.

3.3 A semiconductor such as Si has a bandstructure about the minimum along [100] described approximately by $E = E_0 - A\cos(\alpha k_x) - B\{\cos(\beta k_y) + \cos(\beta k_z)\}$. What is the density-of-states effective mass associated with the X minimum? [Hint: $\cos(2x) = 1 - 2x^2$ for small x.]

3.4 At room temperature, an unknown, intrinsic, cubic semiconductor has the following bandstructure: there are 6 X minima along the <100> directions. If $m_n^*(\Gamma) = 0.065 m_0$, $m_n^*(X) = 0.30 m_0$ (for each of the X minima and $m_p^* = 0.47 m_0$, at what temperature is the number of electrons in the Γ minima and the X minima equal if the Γ to X energy separation is 0.35 eV, and the bandgap is 1.7 eV (m_0 = free electron mass)?

3.5 Consider n-type GaAs and assume that the total number of conduction electrons, n, is independent of temperature. The density-of-states effective mass, m_L^*, in the L valley is 15 times larger than in the Γ valley. Also the energy separation, E_s, between the Γ and L minima is 0.35 eV, and the mobility in the Γ minimum is 50 times that in L. Calculate and sketch how the conductivity varies from low $T(\ll E_s/k)$ to high $T(\gg E_s/k)$. What is the ratio of the conductivities at 1000°C and 300°C?

3.6 Calculate the band gap of Si from Eq. (3–23) and the plot of n_i vs. $1000/T$ (Fig. 3–17). *Hint:* the slope cannot be measured directly from a semilogarithmic plot; read the values from two points on the plot and take the natural logarithm as needed for the solution.

3.7 Show that Eq. (3–25) results from Eqs. (3–15) and (3–19). If $n_0 = 10^{16}$ cm^{-3}, where is the Fermi level relative to E_i in Si at 300 K?

3.8 Derive an expression relating the intrinsic level E_i to the center of the band gap $E_g/2$. Calculate the displacement of E_i from $E_g/2$ for Si at 300 K, assuming the effective mass values for electrons and holes are $1.1m_0$ and $0.56m_0$, respectively.

3.9 (a) Explain why holes are found at the *top* of the valence band, whereas electrons are found at the *bottom* of the conduction band.

    (b) Explain why Si doped with 10^{14} cm^{-3} Sb is n-type at 400 K but similarly doped Ge is not.

3.10 A Si sample is doped with 6×10^{15} cm^{-3} donors and 2×10^{15} cm^{-3} acceptors. Find the position of the Fermi level with respect to E_i at 300 K. What is the value and sign of the Hall coefficient?

3.11 (a) Show that the minimum conductivity of a semiconductor sample occurs when $n_0 = n_i \sqrt{\mu_p/\mu_n}$. Hint: begin with Eq. (3–43) and apply Eq. (3–24).

    (b) What is the expression for the minimum conductivity σ_{min}?

    (c) Calculate σ_{min} for Si at 300 K and compare with the intrinsic conductivity.

3.12 (a) A Si bar 0.1 cm long and 100 μm^2 in cross-sectional area is doped with 10^{17} cm^{-3} phosphorus. Find the current at 300 K with 10V applied. Repeat for a Si bar 1 μm long.

    (b) How long does it take an average electron to drift 1 μm in pure Si at an electric field of 100 V/cm? Repeat for 10^5 V/cm.

3.13 A perfect III–V semiconductor (relative dielectric constant = 13) is doped with column VI and column II impurities. Given that $\mu_n = 1000$ cm^2/V-s, $\mu_p = 500$ cm^2/V-s, what energy levels are introduced in the bandgap? (The mean free time = 0.1 ps for electrons and 0.4 ps for holes.)

3.14 In soldering wires to a sample such as that shown in Fig. 3–25, it is difficult to align the Hall probes A and B precisely. If B is displaced slightly down the length of the bar from A, an erroneous Hall voltage results. Show that the true Hall voltage V_H can be obtained from two measurements of V_{AB}, with the magnetic field first in the $+z$-direction and then in the $-z$-direction.

3.15 We put 11 electrons in an infinite 1-D potential well of size 100 Å. What is the Fermi level at 0 K? What is the probability of exciting a carrier to the first excited state at $T = 300$ K? Use the free electron mass in this problem.

3.16 Use Eq. (3–45) to calculate and plot the mobility vs. temperature $\mu(T)$ from 10 K to 500 K for Si doped with $N_d = 10^{14}, 10^{16}$, and 10^{18} donors cm^{-3}. Consider the mobility to be determined by impurity and phonon (lattice) scattering. Impurity scattering limited mobility can be described by

$$\mu_I = 3.29 \times 10^{15} \frac{\epsilon_r^2 T^{3/2}}{N_d^+(m_n^*/m_0)^{1/2}\left[\ln(1+z) - \dfrac{z}{1+z}\right]}$$

where

$$z = 1.3 \times 10^{13}\epsilon_r T^2 (m_n^*/m_0)(N_d^+)^{-1}$$

Assume that the ionized impurity concentration N_d^+ is equal to N_d at all temperatures.

The conductivity effective mass m_n^* for Si is 0.26 m_0. Acoustic phonon (lattice) scattering limited mobility can be described by

$$\mu_{AC} = 1.18 \times 10^{-5}c_1(m_n^*/m_0)^{-5/2}T^{-3/2}(E_{AC})^{-2}$$

where the stiffness (c_1) is given by

$$c_1 = 1.9 \times 10^{12} \text{ dyne cm}^{-2} \text{ for Si}$$

and the conduction band acoustic deformation potential (E_{AC}) is

$$E_{AC} = 9.5 \text{ eV for Si}$$

3.17 Rework Prob. 3.16 considering carrier freeze-out onto donors at low T. That is, consider

$$N_d^+ = \frac{N_d}{1 + \exp(E_d/kT)}$$

as the ionized impurity concentration. Consider the donor ionization energy (E_d) to be 45 meV for Si.

3.18 Hall measurements are made on a p-type semiconductor bar 500 μm wide and 20 μm thick. The Hall contacts A and B are displaced 2 μm with respect to each other in the direction of current flow of 3 mA. The voltage between A and B with a magnetic field of 10 kG ($1\text{kG} = 10^{-5}$ Wb/cm^2) pointing out of the plane of the sample is 3.2 mV. When the magnetic field direction is reversed the voltage changes to -2.8 mV. What is the hole concentration and mobility?

3.19 For a hypothetical semiconductor, we have $\mu_n = \mu_p = 1000$ cm^2/V-s and $N_c = N_v = 10^{19}$ cm^{-3}. If the conductivity of the intrinsic semiconductor at 300 K is 4×10^{-6} $(\Omega\text{-cm})^{-1}$, what is the conductivity at 600 K?

3.20 An unknown semiconductor has $E_g = 1.1$ eV and $N_c = N_v$. It is doped with 10^{15} cm^{-3} donors where the donor level is 0.2 eV below E_C. Given that E_F is 0.25 eV below E_c, calculate n_i and the concentration of electrons and holes in the semiconductor at 300 K.

3.21 Referring to Fig. 3.25, consider a semiconductor bar with $w = 0.1$ mm, $t = 10$ μm and $L = 5$ mm. For $\mathcal{B} = 10$ kG in the direction shown (1 kG $= 10^{-5}$ Wb/cm^2) and a current of 1mA, we have $V_{AB} = -2$ mV, $V_{CD} = 100$ mV. Find the type, concentration and mobility of the majority carrier.

Ashcroft, N. W., and N. D. Mermin. *Solid State Physics.* Philadelphia: W.B. Saunders, 1976.

Blakemore, J. S. *Semiconductor Statistics.* New York: Dover Publications, 1987.

Bube, R. H. *Electrons in Solids,* 3rd ed. Boston: Harcourt Brace Jovanovich, 1992.

Burns, G. *Solid State Physics.* San Diego: Academic Press, 1985.

Capasso, F. "Bandgap and Interface Engineering for Advanced Electronic and Photonic Devices." *Thin Solid Films* 216 (28 August 1992): 59–67.

Capasso, F., and S. Datta. "Quantum Electron Devices." *Physics Today* 43 (February 1990): 74–82.

Drummund, T. J., P. L. Gourley, and T. E. Zipperian. "Quantum-Tailored Solid-State Devices." *IEEE Spectrum* 25 (June 1988): 33–37.

Hummel, R. E. *Electronic Properties of Materials,* 2nd ed. Berlin: Springer-Verlag, 1993.

Kittel, C. *Introduction to Solid State Physics,* 7th ed. New York: Wiley, 1996.

Li, S. S. *Semiconductor Physical Electronics.* New York: Plenum Press, 1993.

Muller, R. S., and T. I. Kamins. *Device Electronics for Integrated Circuits.* New York: Wiley, 1986.

Neamen, D. A. *Semiconductor Physics and Devices: Basic Principles.* Homewood, IL: Irwin, 1992.

Pierret, R. F. *Advanced Semiconductor Fundamentals.* Reading, MA: Addison-Wesley, 1987.

Schubert, E. F. *Doping in III–V Semiconductors.* Cambridge: Cambridge University Press, 1993.

Singh, J. *Physics of Semiconductors and Their Heterostructures.* New York: McGraw-Hill, 1993.

Singh, J. *Semiconductor Devices.* New York: McGraw-Hill, 1994.

Sundaram, M., S. A. Chalmers, and P. F. Hopkins. "New Quantum Structures." *Science* 254 (29 November 1991): 1326–35.

Swaminathan, V., and Macrander, A. T. *Material Aspects of GaAs and InP Based Structures.* Englewood Cliffs, NJ: Prentice Hall, 1991.

Wang, S. *Fundamentals of Semiconductor Theory and Device Physics.* Englewood Cliffs, NJ: Prentice Hall, 1989.

Wolfe, C. M., G. E. Stillman, and N. Holonyak, Jr. *Physical Properties of Semiconductors.* Englewood Cliffs, NJ: Prentice Hall, 1989.

READING LIST

Chapter 4
Excess Carriers in Semiconductors

Most semiconductor devices operate by the creation of charge carriers in excess of the thermal equilibrium values. These excess carriers can be created by optical excitation or electron bombardment, or as we shall see in Chapter 5, they can be injected across a forward-biased p-n junction. However the excess carriers arise, they can dominate the conduction processes in the semiconductor material. In this chapter we shall investigate the creation of excess carriers by optical absorption and the resulting properties of photoluminescence and photoconductivity. We shall study more closely the mechanism of electron–hole pair recombination and the effects of carrier trapping. Finally, we shall discuss the diffusion of excess carriers due to a carrier gradient, which serves as a basic mechanism of current conduction along with the mechanism of drift in an electric field.

4.1 OPTICAL ABSORPTION[1]

An important technique for measuring the band gap energy of a semiconductor is the absorption of incident photons by the material. In this experiment photons of selected wavelengths are directed at the sample, and relative transmission of the various photons is observed. Since photons with energies greater than the band gap energy are absorbed while photons with energies less than the band gap are transmitted, this experiment gives an accurate measure of the band gap energy.

It is apparent that a photon with energy $hv \geq E_g$ can be absorbed in a semiconductor (Fig. 4–1). Since the valence band contains many electrons and the conduction band has many empty states into which the electrons may be excited, the probability of photon absorption is high. As Fig. 4–1 indicates, an electron excited to the conduction band by optical absorption may initially have more energy than is common for conduction band electrons (almost all electrons are near E_c unless the sample is very heavily doped). Thus the excited electron loses energy to the lattice in scattering events until its velocity reaches the thermal equilibrium velocity of other conduction band elec-

[1]In this context the word "optical" does not necessarily imply that the photons absorbed are in the visible part of the spectrum. Many semiconductors absorb photons in the infrared region, but this is included in the term "optical absorption."

108

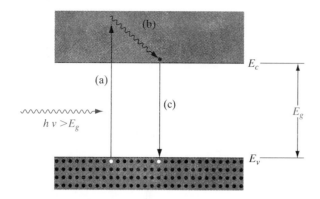

Figure 4–1
Optical absorption of a photon with $h\nu > E_g$: (a) an EHP is created during photon absorption; (b) the excited electron gives up energy to the lattice by scattering events; (c) the electron recombines with a hole in the valence band.

trons. The electron and hole created by this absorption process are *excess carriers;* since they are out of balance with their environment, they must eventually recombine. While the excess carriers exist in their respective bands, however, they are free to contribute to the conductivity of the material.

A photon with energy less than E_g is unable to excite an electron from the valence band to the conduction band. Thus in a pure semiconductor, there is negligible absorption of photons with $h\nu < E_g$. This explains why some materials are transparent in certain wavelength ranges. We are able to "see through" certain insulators, such as a good NaCl crystal, because a large energy gap containing no electron states exists in the material. If the band gap is about 2 eV wide, only long wavelengths (infrared) and the red part of the visible spectrum are transmitted; on the other hand, a band gap of about 3 eV allows infrared and the entire visible spectrum to be transmitted.

If a beam of photons with $h\nu > E_g$ falls on a semiconductor, there will be some predictable amount of absorption, determined by the properties of the material. We would expect the ratio of transmitted to incident light intensity to depend on the photon wavelength and the thickness of the sample. To calculate this dependence, let us assume that a photon beam of intensity $\mathbf{I}_0$ (photons/cm^2-s) is directed at a sample of thickness l (Fig. 4–2). The beam contains only photons of wavelength λ, selected by a monochromator. As the beam passes through the sample, its intensity at a distance x from the surface can be calculated by considering the probability of absorption within any increment dx. Since a photon which has survived to x without absorption has no memory of how far it has traveled, its probability of absorption in any dx is constant. Thus the degradation of the intensity $-d\mathbf{I}(x)/dx$ is proportional to the intensity remaining at x:

$$-\frac{d\mathbf{I}(x)}{dx} = \alpha\mathbf{I}(x) \qquad (4\text{–}1)$$

The solution to this equation is

$$\mathbf{I}(x) = \mathbf{I}_0 e^{-\alpha x} \qquad (4\text{–}2)$$

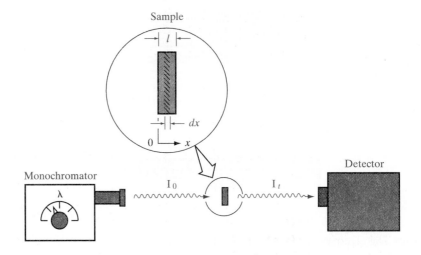

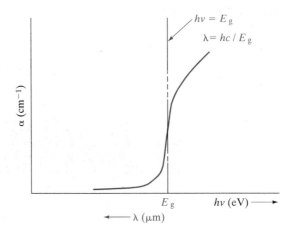

and the intensity of light transmitted through the sample thickness l is

$$\mathbf{I}_t = \mathbf{I}_0 e^{-\alpha l} \tag{4–3}$$

The coefficient α is called the *absorption coefficient* and has units of cm^{-1}. This coefficient will of course vary with the photon wavelength and with the material. In a typical plot of α vs. wavelength (Fig. 4–3), there is negligible absorption at long wavelengths ($h\nu$ small) and considerable absorption of photons with energies larger than E_g. According to Eq. (2–2), the relation between photon energy and wavelength is $E = hc/\lambda$. If E is given in electron volts and λ in micrometers, this becomes $E = 1.24/\lambda$.

Figure 4–4 indicates the band gap energies of some of the common semiconductors, relative to the visible, infrared, and ultraviolet portions of the spectrum. We observe that GaAs, Si, Ge, and InSb lie outside the vis-

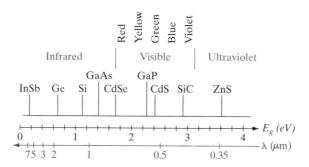

Figure 4–4
Band gaps of
some common
semiconductors
relative to the op-
tical spectrum.

ible region, in the infrared. Other semiconductors, such as GaP and CdS, have band gaps wide enough to pass photons in the visible range. It is important to note here that a semiconductor absorbs photons with energies equal to the band gap, *or larger*. Thus Si absorbs not only band gap light (~1 μm) but also shorter wavelengths, including those in the visible part of the spectrum.

When electron–hole pairs are generated in a semiconductor, or when carriers are excited into higher impurity levels from which they fall to their equilibrium states, light can be given off by the material. Many of the semiconductors are well suited for light emission, particularly the compound semiconductors with direct band gaps. The general property of light emission is called *luminescence.*[2] This overall category can be subdivided according to the excitation mechanism: If carriers are excited by photon absorption, the radiation resulting from the recombination of the excited carriers is called *photoluminescence;* if the excited carriers are created by high-energy electron bombardment of the material, the mechanism is called *cathodoluminescence;* if the excitation occurs by the introduction of current into the sample, the resulting luminescence is called *electroluminescence.* Other types of excitation are possible, but these three are the most important for device applications.

**4.2
LUMINESCENCE**

4.2.1 Photoluminescence

The simplest example of light emission from a semiconductor occurs for direct excitation and recombination of an EHP, as depicted in Fig. 3–5a. If the recombination occurs directly rather than via a defect level, band gap light is given off in the process. For steady state excitation, the recombination of EHPs occurs at the same rate at the generation, and one photon is emitted

---

[2]The emission processes considered here should not be confused with radiation due to incandescence which occurs in heated materials. The various luminescent mechanisms can be considered "cold" processes as compared to the "hot" process of incandescence, which increases with temperature. In fact, most luminescent processes become more efficient as the temperature is lowered.

for each photon absorbed. Direct recombination is a fast process; the mean lifetime of the EHP is usually on the order of 10^{-8} s or less. Thus the emission of photons stops within approximately 10^{-8} s after the excitation is turned off. Such fast luminescent processes are often referred to as *fluorescence*. In some materials, however, emission continues for periods up to seconds or minutes after the excitation is removed. These slow processes are called *phosphorescence*, and the materials are called *phosphors*. An example of a slow process is shown in Fig. 4–5. This material contains a defect level (perhaps due to an impurity) in the band gap which has a strong tendency to temporarily capture (*trap*) electrons from the conduction band. The events depicted in the figure are as follows; (a) An incoming photon with $hv_1 > E_g$ is absorbed, creating an EHP; (b) the excited electron gives up energy to the lattice by scattering until it nears the bottom of the conduction band; (c) the electron is *trapped* by the impurity level E_t and remains trapped until it can be thermally reexcited to the conduction band (d); (e) finally direct recombination occurs as the electron falls to an empty state in the valence band, giving off a photon (hv_2) of approximately the band gap energy. The delay time between excitation and recombination can be relatively long if the probability of thermal reexcitation from the trap (d) is small. Even longer delay times result if the electron is retrapped several times before recombination. If the trapping probability is greater than the probability of recombination, an electron may make several trips between the trap and the conduction band before recombination finally occurs. In such material the emission of phosphorescent light persists for a relatively long time after the excitation is removed.

The color of light emitted by a phosphor such as ZnS depends primarily on the impurities present, since many radiative transitions involve impurity levels within the band gap. This selection of colors is particularly useful in the fabrication of a color television screen.

One of the most common examples of photoluminescence is the fluorescent lamp. Typically such a lamp is composed of a glass tube filled with gas

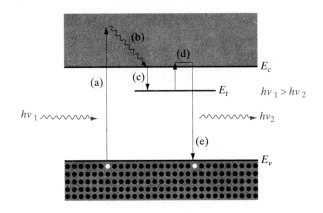

Figure 4–5

Excitation and recombination mechanisms in photoluminescence with a trapping level for electrons.

(e.g., a mixture of argon and mercury), with a fluorescent coating on the inside of the tube. When an electric discharge is induced between electrodes in the tube, the excited atoms of the gas emit photons, largely in the visible and ultra-violet regions of the spectrum. This light is absorbed by the luminescent coating, and the visible photons are emitted. The efficiency of such a lamp is considerably better than that of an incandescent bulb, and the wavelength mixture in light given off can be adjusted by proper selection of the fluorescent material.

A 0.46-μm-thick sample of GaAs is illuminated with monochromatic light of $hv = 2$ eV. The absorption coefficient α is 5×10^4 cm^{-1}. The power incident on the sample is 10 mW.   **EXAMPLE 4–1**

(a) Find the total energy absorbed by the sample per second (J/s).

(b) Find the rate of excess thermal energy given up by the electrons to the lattice before recombination (J/s).

(c) Find the number of photons per second given off from recombination events, assuming perfect quantum efficiency.

(a) From Eq. (4–3),   **SOLUTION**

$$\mathbf{I}_t = \mathbf{I}_0 e^{-\alpha l} = 10^{-2} \exp(-5 \times 10^4 \times 0.46 \times 10^{-4})$$

$$= 10^{-2} e^{-2.3} = 10^{-3} \text{ W}$$

Thus the absorbed power is

$$10 - 1 = 9 \text{ mW} = 9 \times 10^{-3} \text{ J/s}$$

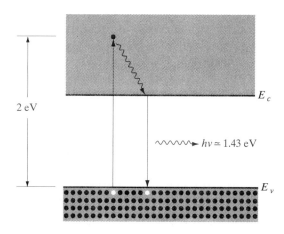

Figure 4–6
Excitation and band-to-band recombination leading to photoluminescence.

(b)   The fraction of each photon energy unit which is converted to heat is

$$\frac{2 - 1.43}{2} = 0.285$$

Thus the amount of energy converted to heat per second is

$$0.285 \times 9 \times 10^{-3} = 2.57 \times 10^{-3} \, \text{J/s}$$

(c)   Assuming one emitted photon for each photon absorbed (perfect quantum efficiency), we have

$$\frac{9 \times 10^{-3} \, \text{J/s}}{1.6 \times 10^{-19} \, \text{J/eV} \times 2 \, \text{eV/photon}} = 2.81 \times 10^{16} \, \text{photons/s}$$

Alternative solution: Recombination radiation accounts for $9 - 2.57 = 6.43$ mW at 1.43 eV/photon.

$$\frac{6.43 \times 10^{-3}}{1.6 \times 10^{-19} \times 1.43} = 2.81 \times 10^{16} \, \text{photons/s}$$

4.2.2   Electroluminescence

There are many ways by which electrical energy can be used to generate photon emission in a solid. In LEDs an electric current causes the injection of minority carriers into regions of the crystal where they can recombine with majority carriers, resulting in the emission of recombination radiation. This important effect (*injection electroluminescence*) will be discussed in Chapter 8 in terms of p-n junction theory.

The first electroluminescent effect to be observed was the emission of photons by certain phosphors in an alternating electric field (the Destriau effect). In this device, a phosphor powder such as ZnS is held in a binder material (often a plastic) of a high dielectric constant. When an a-c electric field is applied, light is given off by the phosphor. Such cells can be useful as lighting panels, although their efficiency has thus far been too low for most applications and their reliability is poor.

4.3
CARRIER LIFETIME
AND PHOTO-
CONDUCTIVITY

When excess electrons and holes are created in a semiconductor, there is a corresponding increase in the conductivity of the sample as indicated by Eq. (3–43). If the excess carriers arise from optical luminescence, the resulting increase in conductivity is called *photoconductivity*. This is an important effect, with useful applications in the analysis of semiconductor materials and in the operation of several types of devices. In this section we shall examine the mechanisms by which excess electrons and holes recombine and apply the recombination kinetics to the analysis of photoconductive devices. However, the importance of recombination is not limited to cases in which the excess carriers are created optically. In fact, virtually every semiconductor device depends in some way on the recombination of excess electrons and holes. Therefore, the concepts developed in this section will be used extensively in the analyses of diodes, transistors, lasers, and other devices in later chapters.

4.3.1 Direct Recombination of Electrons and Holes

It was pointed out in Section 3.1.4 that electrons in the conduction band of a semiconductor may make transitions to the valence band (i.e., recombine with holes in the valence band) either directly or indirectly. In direct recombination, an excess population of electrons and holes decays by electrons falling from the conduction band to empty states (holes) in the valence band. Energy lost by an electron in making the transition is given up as a photon. Direct recombination occurs *spontaneously;* that is, the probability that an electron and a hole will recombine is constant in time. As in the case of carrier scattering, this constant probability leads us to expect an exponential solution for the decay of the excess carriers. In this case the rate of decay of electrons at any time t is proportional to the number of electrons remaining at t and the number of holes, with some constant of proportionality for recombination, α_r. The *net* rate of change in the conduction band electron concentration is the thermal generation rate $\alpha_r n_i^2$ from Eq. (3–7) minus the recombination rate

$$\frac{dn(t)}{dt} = \alpha_r n_i^2 - \alpha_r n(t)p(t) \qquad (4\text{–}4)$$

Let us assume the excess electron–hole population is created at $t = 0$, for example by a short flash of light, and the initial excess electron and hole concentrations Δn and Δp are equal.[3] Then as the electrons and holes recombine in pairs, the instantaneous concentrations of excess carriers $\delta n(t)$ and $\delta p(t)$ are also equal. Thus we can write the total concentrations of Eq. (4–4) in terms of the equilibrium values n_0 and p_0 and the excess carrier concentrations $\delta n(t) = \delta p(t)$. Using Eq. (3–24) we have

$$\frac{d\delta n(t)}{dt} = \alpha_r n_i^2 - \alpha_r[n_0 + \delta n(t)][p_0 + \delta p(t)]$$

$$= -\alpha_r[(n_0 + p_0)\delta n(t) + \delta n^2(t)] \qquad (4\text{–}5)$$

This nonlinear equation would be difficult to solve in its present form. Fortunately, it can be simplified for the case of low-level injection. If the excess

---

[3]We will use $\delta n(t)$ and $\delta p(t)$ to mean instantaneous excess carrier concentrations, and Δn, Δp for their values at $t = 0$. Later we will use similar symbolism for spatial distributions, such as $\delta n(x)$ and $\Delta n(x = 0)$.

carrier concentrations are small, we can neglect the δn^2 term. Furthermore, if the material is extrinsic, we can usually neglect the term representing the equilibrium minority carriers. For example, if the material is p-type ($p_0 \gg n_0$), Eq. (4–5) becomes

$$\frac{d\delta n(t)}{dt} = -\alpha_r p_0 \delta n(t) \tag{4–6}$$

The solution to this equation is an exponential decay from the original excess carrier concentration Δn:

$$\delta n(t) = \Delta n e^{-\alpha_r p_0 t} = \Delta n e^{-t/\tau_n} \tag{4–7}$$

Excess electrons in a p-type semiconductor recombine with a decay constant $\tau_n = (\alpha_r p_0)^{-1}$, called the *recombination lifetime*. Since the calculation is made in terms of the minority carriers, τ_n is often called the *minority carrier lifetime*. The decay of excess holes in n-type material occurs with $\tau_p = (\alpha_r n_0)^{-1}$. In the case of direct recombination, the excess majority carriers decay at exactly the same rate as the minority carriers.

There is a large percentage change in the minority carrier electron concentration in Example 4–2 and a small percentage change in the majority hole concentration. Basically, the approximations of extrinsic material and low-level injection allow us to represent $n(t)$ in Eq. (4–4) by the excess concentration $\delta n(t)$ and $p(t)$ by the equilibrium value p_0. Figure 4–7 indicates that this is a good approximation for the example. A more general expression for the carrier lifetime is

$$\tau_n = \frac{1}{\alpha_r(n_0 + p_0)} \tag{4–8}$$

This expression is valid for n- or p-type material if the injection level is low.

EXAMPLE 4–2   A numerical example may be helpful in visualizing the approximations made in the analysis of direct recombination. Let us assume a sample of GaAs is doped with 10^{15} acceptors/cm^3. The intrinsic carrier concentration of GaAs is approximately 10^6 cm^{-3}; thus the minority electron concentration is $n_0 = n_i^2/p_0 = 10^{-3}$ cm^{-3}. Certainly the approximation of $p_0 \gg n_0$ is valid in this case. Now if 10^{14} EHP/cm^3 are created at $t = 0$, we can calculate the decay of these carriers in time. The approximation of $\delta n \ll p_0$ is reasonable, as Fig. 4–7 indicates. This figure shows the decay in time of the excess populations for a carrier recombination lifetime of $\tau_n = \tau_p = 10^{-8}$ s.

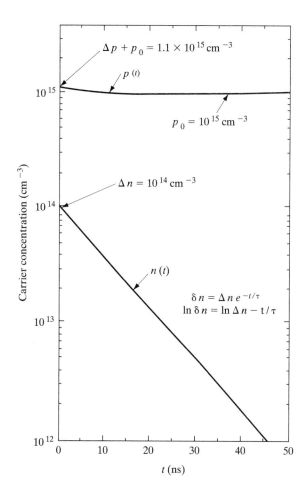

Figure 4–7
Decay of excess electrons and holes by recombination, for $\Delta n = \Delta p = 0.1 p_0$, with n_0 negligible, and $\tau = 10$ ns (Example 4–2). The exponential decay of $\delta n(t)$ is linear on this semilogarithmic graph.

4.3.2 Indirect Recombination; Trapping

In column IV semiconductors and in certain compounds, the probability of direct electron–hole recombination is very small (Appendix III). There is some band gap light given off by materials such as Si and Ge during recombination, but this radiation is very weak and may be detected only by sensitive equipment. The vast majority of the recombination events in indirect materials occur via *recombination levels* within the band gap, and the resulting energy loss by recombining electrons is usually given up to the lattice as heat rather than by the emission of photons. Any impurity or lattice defect can serve as a recombination center if it is capable of receiving a carrier of one type and subsequently capturing the opposite type of carrier, thereby annihilating the pair. For example, Fig. 4–8 illustrates a recombination level E_r which is below E_F at equilibrium and therefore is substantially filled with

Figure 4–8
Capture processes at a recombination level: (a) hole capture at a filled recombination center; (b) electron capture at an empty center.

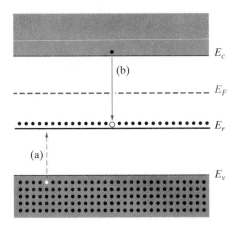

Figure 4–8 Capture processes at a recombination level: (a) hole capture at a filled recombination center; (b) electron capture at an empty center.

electrons. When excess electrons and holes are created in this material, each EHP recombines at E_r in two steps: (a) hole capture and (b) electron capture.

Since the recombination centers in Fig. 4–8 are filled at equilibrium, the first event in the recombination process is hole capture. It is important to note that this event is equivalent to an electron at E_r falling to the valence band, leaving behind an empty state in the recombination level. Thus in hole capture, energy is *given up* as heat to the lattice. Similarly, energy is given up when a conduction band electron subsequently falls to the empty state in E_r. When both of these events have occurred, the recombination center is back to its original state (filled with an electron), but an EHP is missing. Thus one EHP recombination has taken place, and the center is ready to participate in another recombination event by capturing a hole.

The carrier lifetime resulting from indirect recombination is somewhat more complicated than is the case for direct recombination, since it is necessary to account for unequal times required for capturing each type of carrier. In particular, recombination is often delayed by the tendency for a captured carrier to be thermally reexcited to its original band before capture of the opposite type of carrier can occur (Section 4.2.1). For example, if electron capture (b) does not follow immediately after hole capture (a) in Fig. 4–8, the hole may be thermally reexcited to the valence band. Energy is required for this process, which is equivalent to a valence band electron being raised to the empty state in the recombination level. This process delays the recombination, since the hole must be captured again before recombination can be completed.

When a carrier is trapped temporarily at a center and then is reexcited without recombination taking place, the process is often called *temporary trapping*. Although the nomenclature varies somewhat, it is common to refer to an impurity or defect center as a *trapping center* (or simply *trap*) if, after capture of one type of carrier, the most probable next event is reexcitation. If the most probable next event is capture of the opposite type of carrier, the center is predominately a recombination center. The recombination can be slow or fast, depending on the average time the first carrier is held before the second carrier is captured. In general, trapping levels located deep in the band gap are slower in releasing trapped carriers than are the levels located near one of the bands. This results from the fact that more energy is required, for example, to reexcite a trapped electron from a center near the middle of the gap to the conduction band than is required to reexcite an electron from a level closer to the conduction band.

As an example of impurity levels in semiconductors, Fig. 4–9[4] shows the energy level positions of various impurities in Si. In this diagram a superscript indicates whether the impurity is positive (donor) or negative (acceptor) when ionized. Some impurities introduce multiple levels in the band gap; for example, Zn introduces a level (Zn^-) located 0.31 eV above the valence band and a second level ($Zn^=$) near the middle of the gap. Each Zn impurity atom is capable of accepting two electrons from the semiconductor, one in the lower level and then one in the upper level.

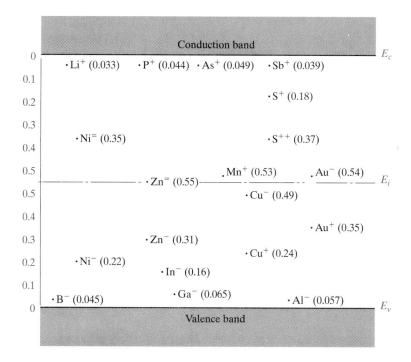

Figure 4–9
Energy levels of impurities in Si. The energies are measured from the nearest band edge (E_v or E_c); donor levels are designated by a plus sign and acceptors by a minus sign.

[4]References: S. M. Sze and J. C. Irvin, "Resistivity, Mobility, and Impurity Levels in GaAs, Ge and Si at 300 K," *Solid State Electronics*, vol. 11, pp. 599–602 (June 1968); E. Schibli and A. G. Milnes, "Deep Impurities in Silicon," *Materials Science and Engineering*, vol. 2, pp. 173–180 (1967).

Figure 4–10
Experimental
arrangement for
photoconductive
decay measure-
ments, and a typi-
cal oscilloscope
trace.

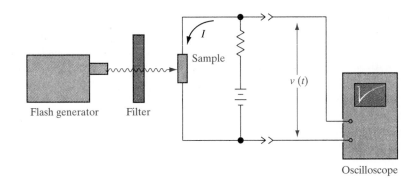

The effects of recombination and trapping can be measured by a *pho-toconductive decay* experiment. As Fig. 4–7 shows, a population of excess electrons and holes disappears with a decay constant characteristic of the particular recombination process. The conductivity of the sample during the decay is

$$\sigma(t) = q[n(t)\mu_n + p(t)\mu_p] \tag{4–9}$$

Therefore, the time dependence of the carrier concentrations can be monitored by recording the sample resistance as a function of time. A typical experimental arrangement is shown schematically in Fig. 4–10. A source of short pulses of light is required, along with an oscilloscope for displaying the sample voltage as the resistance varies. Microsecond light pulses can be obtained by periodically discharging a capacitor through a flash tube containing a gas such as xenon. For shorter pulses, special techniques such as the use of a pulsed laser must be used.

4.3.3 Steady State Carrier Generation; Quasi-Fermi Levels

In the previous discussion we emphasized the transient decay of an excess EHP population. However, the various recombination mechanisms are also important in a sample at thermal equilibrium or with a steady state EHP generation-recombination balance.[5] For example, a semiconductor at equilibrium experiences thermal generation of EHPs at a rate $g(T) = g_i$ described by Eq. (3–7). This generation is balanced by the recombination rate so that the equilibrium concentrations of carriers n_0 and p_0 are maintained:

$$g(T) = \alpha_r n_i^2 = \alpha_r n_0 p_0 \tag{4–10}$$

This equilibrium rate balance can include generation from defect centers as well as band-to-band generation.

---

[5]The term *equilibrium* refers to a condition of no external excitation except for temperature, and no net motion of charge (e.g., a sample at a constant temperature, in the dark, with no fields applied). *Steady state* refers to a nonequilibrium condition in which all processes are constant and are balanced by opposing processes (e.g., a sample with a constant current or a constant optical generation of EHPs just balanced by recombination).

If a steady light is shone on the sample, an optical generation rate g_{op} will be added to the thermal generation, and the carrier concentrations n and p will increase to new steady state values. We can write the balance between generation and recombination in terms of the equilibrium carrier concentrations and the departures from equilibrium δn and δp:

$$g(T) + g_{op} = \alpha_r np = \alpha_r(n_0 + \delta n)(p_0 + \delta p) \tag{4-11}$$

For steady state recombination and no trapping, $\delta n = \delta p$; thus Eq. (4–11) becomes

$$g(T) + g_{op} = \alpha_r n_0 p_0 + \alpha_r[(n_0 + p_0)\delta n + \delta n^2] \tag{4-12}$$

The term $\alpha_r n_0 p_0$ is just equal to the thermal generation rate $g(T)$. Thus, neglecting the δn^2 term for low-level excitation, we can rewrite Eq. (4–12) as

$$g_{op} = \alpha_r(n_0 + p_0)\delta n = \frac{\delta n}{\tau_n} \tag{4-13}$$

The excess carrier concentration can be written as

$$\delta n = \delta p = g_{op}\tau_n \tag{4-14}$$

More general expressions are given in Eq. (4–16), which allow for the case $\tau_p \neq \tau_n$, when trapping is present.

As a numerical example, let us assume that 10^{13} EHP/cm^3 are created optically every microsecond in a Si sample with $n_0 = 10^{14}$ cm^{-3} and $\tau_n = \tau_p = 2$ μsec. The steady state excess electron (or hole) concentration is then 2×10^{13} cm^{-3} from Eq. (4–14). While the percentage change in the majority electron concentration is small, the minority carrier concentration changes from **EXAMPLE 4–3**

$$p_0 = n_i^2/n_0 = (2.25 \times 10^{20})/10^{14} = 2.25 \times 10^6 \text{ cm}^{-3} \quad (equilibrium)$$

to

$$p = 2 \times 10^{13} \text{ cm}^{-3} \quad (steady\ state)$$

Note that the equilibrium equation $n_0 p_0 = n_i^2$ cannot be used with the subscripts removed; that is, $np \neq n_i^2$ when excess carriers are present.

It is often desirable to refer to the steady state electron and hole concentrations in terms of Fermi levels, which can be included in band diagrams for various devices. The Fermi level E_F used in Eq. (3–25) is meaningful only when no excess carriers are present. However, we can write expressions for the steady state concentrations in the same *form* as the equilibrium expressions by defining separate *quasi-Fermi levels* F_n and F_p for electrons and holes. The resulting carrier concentration equations

$$\boxed{\begin{array}{l} n = n_i e^{(F_n - E_i)/kT} \\ p = n_i e^{(E_i - F_p)/kT} \end{array}}$$

(4–15)

can be considered as defining relations for the quasi-Fermi levels.[6]

EXAMPLE 4–4

In Example 4–3, the steady state electron concentration is

$$n = n_0 + \delta n = 1.2 \times 10^{14} = (1.5 \times 10^{10})e^{(F_n - E_i)/0.0259}$$

where $kT \simeq 0.0259$ eV at room temperature. Thus the electron quasi-Fermi level position $F_n - E_i$ is found from

$$F_n - E_i = 0.0259 \ln(8 \times 10^3) = 0.233 \text{ eV}$$

and F_n lies 0.233 eV above the intrinsic level. By a similar calculation, the hole quasi-Fermi level lies 0.186 eV below E_i (Fig. 4–11). In this example, the equilibrium Fermi level is $0.0259 \ln(6.67 \times 10^3) = 0.228$ eV above the intrinsic level.

Figure 4–11
Quasi-Fermi levels F_n and F_p for a Si sample with $n_0 = 10^{14}$ cm^{-3}, $\tau_p = 2$ μs, and $g_{op} = 10^{19}$ EHP/cm^3-s (Example 4–4).

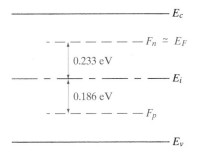

The quasi-Fermi levels of Fig. 4–11 illustrate dramatically the deviation from equilibrium caused by the optical excitation; the steady state F_n is only slightly above the equilibrium E_F, whereas F_p is greatly displaced below E_F. From the figure it is obvious that the excitation causes a large percentage change in minority carrier hole concentration and a relatively small change in the electron concentration.

In summary, the quasi-Fermi levels F_n and F_p are the steady state analogues of the equilibrium Fermi level E_F. When excess carriers are present, the deviations of F_n and F_p from E_F indicate how far the electron and hole populations are from the equilibrium values n_0 and p_0. A given concentration of excess EHPs causes a large shift in the minority carrier quasi-Fermi level

[6]In some texts the quasi-Fermi level is called *IMREF*, which is Fermi spelled backward.

compared with that for the majority carriers. The separation of the quasi-Fermi levels $F_n - F_p$ is a direct measure of the deviation from equilibrium (at equilibrium $F_n = F_p = E_F$). The concept of quasi-Fermi levels is very useful in visualizing minority and majority carrier concentrations in devices where these quantities vary with position.

4.3.4 Photoconductive Devices

There are a number of applications for devices which change their resistance when exposed to light. For example, such light detectors can be used in the home to control automatic night lights which turn on at dusk and turn off at dawn. They can also be used to measure illumination levels, as in exposure meters for cameras. Many systems include a light beam aimed at the photoconductor, which signals the presence of an object between the source and detector. Such systems are useful in moving-object counters, burglar alarms, and many other applications. Detectors are used in optical signaling systems in which information is transmitted by a light beam and is received at a photoconductive cell.

Considerations in choosing a photoconductor for a given application include the sensitive wavelength range, time response, and optical sensitivity of the material. In general, semiconductors are most sensitive to photons with energies equal to the band gap or slightly more energetic than band gap. Less energetic photons are not absorbed, and photons with $hv \gg E_g$ are absorbed at the surface and contribute little to the bulk conductivity. Therefore, the table of band gaps (Appendix III) indicates the photon energies to which most semiconductor photodetectors respond. For example, CdS ($E_g = 2.42$ eV) is commonly used as a photoconductor in the visible range, and narrow-gap materials such as Ge (0.67 eV) and InSb (0.18 eV) are useful in the infrared portion of the spectrum. Some photoconductors respond to excitations of carriers from impurity levels within the band gap and therefore are sensitive to photons of less than band gap energy.

The optical sensitivity of a photoconductor can be evaluated by examining the steady state excess carrier concentrations generated by an optical generation rate g_{op}. If the mean time each carrier spends in its respective band before capture is τ_n and τ_p, we have

$$\delta n = \tau_n g_{op} \quad \text{and} \quad \delta p = \tau_p g_{op} \qquad (4\text{--}16)$$

and the photoconductivity change is

$$\Delta\sigma = qg_{op}(\tau_n\mu_n + \tau_p\mu_p) \qquad (4\text{--}17)$$

For simple recombination, τ_n and τ_p will be equal. If trapping is present, however, one of the carriers may spend little time in its band before being trapped. From Eq. (4–17) it is obvious that for maximum photoconductive response, we want high mobilities and long lifetimes. Some semiconductors are especially good candidates for photoconductive devices because of their high mobility; for example, InSb has an electron mobility of about 10^5 cm^2/V-s and therefore is used as a sensitive infrared detector in many applications.

The time response of a photoconductive cell is limited by the recombination times, the degree of carrier trapping, and the time required for carriers to drift through the device in an electric field. Often these properties can be adjusted by proper choice of material and device geometry, but in some cases improvements in response time are made at the expense of sensitivity. For example, the drift time can be reduced by making the device short, but this substantially reduces the responsive area of the device. In addition, it is often desirable that the device have a large dark resistance, and for this reason, shortening the length may not be practical. There is usually a compromise between sensitivity, response time, dark resistance, and other requirements in choosing a device for a particular application.

**4.4
DIFFUSION OF
CARRIERS**

When excess carriers are created nonuniformly in a semiconductor, the electron and hole concentrations vary with position in the sample. Any such spatial variation (*gradient*) in *n* and *p* calls for a net motion of the carriers from regions of high carrier concentration to regions of low carrier concentration. This type of motion is called *diffusion* and represents an important charge transport process in semiconductors. The two basic processes of current conduction are diffusion due to a carrier gradient and drift in an electric field.

4.4.1 Diffusion Processes

When a bottle of perfume is opened in one corner of a closed room, the scent is soon detected throughout the room. If there is no convection or other net motion of air, the scent spreads by diffusion. The diffusion is the natural result of the *random motion* of the individual molecules. Consider, for example, a volume of arbitrary shape with scented air molecules inside and unscented molecules outside the volume. All the molecules undergo random thermal motion and collisions with other molecules. Thus each molecule moves in an arbitrary direction until it collides with another air molecule, after which it moves in a new direction. If the motion is truly random, a molecule at the edge of the volume has equal probabilities of moving into or out of the volume on its next step (assuming the curvature of the surface is negligible on the molecular scale). Therefore, after a mean free time $\bar{t}$, half the molecules at the edge will have moved into the volume and half will have moved out of the volume. The net effect is that the volume containing scented molecules has increased. This process will continue until the molecules are uniformly distributed in the room. Only then will a given volume gain as many molecules as it loses in a given time. In other words, net diffusion will continue as long as gradients exist in the distribution of scented molecules.

Carriers in a semiconductor diffuse in a carrier gradient by random thermal motion and scattering from the lattice and impurities. For example, a pulse of excess electrons injected at $x = 0$ at time $t = 0$ will spread out in time as shown in Fig. 4–12. Initially, the excess electrons are concentrated at $x = 0$; as time passes, however, electrons diffuse to regions of low electron concentration until finally $n(x)$ is constant.

We can calculate the rate at which the electrons diffuse in a one-dimensional problem by considering an arbitrary distribution $n(x)$ such as Fig. 4–13a. Since the mean free path $\bar{l}$ between collisions is a small incremental distance, we can divide x into segments $\bar{l}$ wide, with $n(x)$ evaluated at the center of each segment (Fig. 4–13b).

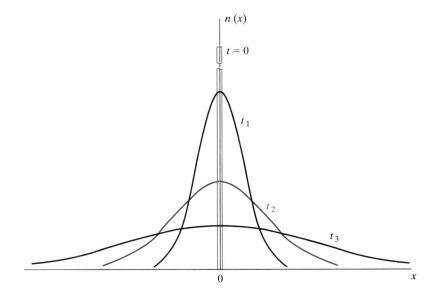

Figure 4–12
Spreading of a pulse of electrons by diffusion.

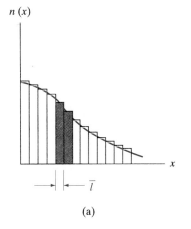

(a)

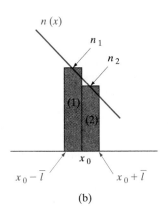

(b)

Figure 4–13
An arbitrary electron concentration gradient in one dimension: (a) division of $n(x)$ into segments of length equal to a mean free path for the electrons; (b) expanded view of two of the segments centered at x_0.

The electrons in segment (1) to the left of x_0 in Fig. 4–13b have equal chances of moving left or right, and in a mean free time $\bar{t}$ one-half of them will move into segment (2). The same is true of electrons within one mean free path of x_0 to the right; one-half of these electrons will move through x_0 from right to left in a mean free time. Therefore, the *net* number of electrons passing x_0 from left to right in one mean free time is $\frac{1}{2}(n_1 \bar{l} A) - \frac{1}{2}(n_2 \bar{l} A)$, where the area perpendicular to x is A. The rate of electron flow in the $+x$-direction per unit area (the electron flux density ϕ_n) is given by

$$\phi_n(x_0) = \frac{\bar{l}}{2\bar{t}}(n_1 - n_2) \tag{4–18}$$

Since the mean free path $\bar{l}$ is a small differential length, the difference in electron concentration $(n_1 - n_2)$ can be written as

$$n_1 - n_2 = \frac{n(x) - n(x + \Delta x)}{\Delta x} \bar{l} \tag{4–19}$$

where x is taken at the center of segment (1) and $\Delta x = \bar{l}$. In the limit of small Δx (i.e., small mean free path $\bar{l}$ between scattering collisions), Eq. (4–18) can be written in terms of the carrier gradient $dn(x)/dx$:

$$\phi_n(x) = \frac{\bar{l}^2}{2\bar{t}} \lim_{\Delta x \to 0} \frac{n(x) - n(x + \Delta x)}{\Delta x} = \frac{-\bar{l}^2}{2\bar{t}} \frac{dn(x)}{dx} \tag{4–20}$$

The quantity $\bar{l}^2/2\bar{t}$ is called the *electron diffusion coefficient*[7] D_n, with units cm²/s. The minus sign in Eq. (4–20) arises from the definition of the derivative; it simply indicates that the net motion of electrons due to diffusion is in the direction of *decreasing* electron concentration. This is the result we expect, since net diffusion occurs from regions of high particle concentration to regions of low particle concentration. By identical arguments, we can show that holes in a hole concentration gradient move with a diffusion coefficient D_p. Thus

$$\phi_n(x) = -D_n \frac{dn(x)}{dx} \tag{4–21a}$$

$$\phi_p(x) = -D_p \frac{dp(x)}{dx} \tag{4–21b}$$

The diffusion current crossing a unit area (the current density) is the particle flux density multiplied by the charge of the carrier:

$$J_n(\text{diff.}) = -(-q)D_n \frac{dn(x)}{dx} = +qD_n \frac{dn(x)}{dx} \tag{4–22a}$$

---

[7]If motion in three dimensions were included, the diffusion would be smaller in the *x*-direction. Actually, the diffusion coefficient should be calculated from the true energy distributions and scattering mechanisms. Diffusion coefficients are usually determined experimentally for a particular material, as described in Section 4.4.5.

$$J_p(\text{diff.}) = -(+q)D_p\frac{dp(x)}{dx} = -qD_p\frac{dp(x)}{dx} \qquad (4\text{--}22b)$$

It is important to note that electrons and holes move together in a carrier gradient [Eqs. (4–21)], but the resulting currents are in opposite directions [Eqs. (4–22)] because of the opposite charge of electrons and holes.

4.4.2 Diffusion and Drift of Carriers; Built-in Fields

If an electric field is present in addition to the carrier gradient, the current densities will each have a drift component and a diffusion component

$$J_n(x) = q\mu_n n(x)\mathscr{E}(x) + qD_n\frac{dn(x)}{dx} \qquad (4\text{--}23a)$$

$$\underbrace{\phantom{J_n(x) = q\mu_n n(x)\mathscr{E}(x)}}_{\text{drift}} \quad \underbrace{\phantom{qD_n\frac{dn(x)}{dx}}}_{\text{diffusion}}$$

$$J_p(x) = q\mu_p p(x)\mathscr{E}(x) - qD_p\frac{dp(x)}{dx} \qquad (4\text{--}23b)$$

and the total current density is the sum of the contributions due to electrons and holes:

$$J(x) = J_n(x) + J_p(x) \qquad (4\text{--}24)$$

We can best visualize the relation between the particle flow and the current of Eqs. (4–23) by considering a diagram such as shown in Fig. 4–14. In this figure an electric field is assumed to be in the x-direction, along with carrier distributions $n(x)$ and $p(x)$ which decrease with increasing x. Thus the derivatives in Eqs. (4–21) are negative, and diffusion takes place in the +x-direction. The resulting electron and hole diffusion currents [J_n (diff.) and J_p (diff.)] are in opposite directions, according to Eqs. (4–22). Holes drift in the direction of the electric field [ϕ_p (drift)], whereas electrons drift in the opposite direction because of their negative charge. The resulting drift current is in the +x-direction in each case. Note that the drift and diffusion components of the current are additive for holes when the field is in the direction of decreasing hole concentration, whereas the two components are subtractive for electrons under similar conditions. The total current may be due primarily to the flow of electrons or

Figure 4–14
Drift and diffusion directions for electrons and holes in a carrier gradient and an electric field. Particle flow directions are indicated by dashed arrows, and the resulting currents are indicated by solid arrows.

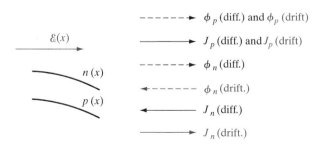

holes, depending on the relative concentrations and the relative magnitudes and directions of electric field and carrier gradients.

An important result of Eqs. (4–23) is that minority carriers can contribute significantly to the current through diffusion. Since the drift terms are proportional to carrier concentration, minority carriers seldom provide much drift current. On the other hand, diffusion current is proportional to the *gradient* of concentration. For example, in n-type material the minority hole concentration *p* may be many orders of magnitude smaller than the electron concentration *n*, but the gradient *dp/dx* may be significant. As a result, minority carrier currents through diffusion can sometimes be as large as majority carrier currents.

In discussing the motion of carriers in an electric field, we should indicate the influence of the field on the energies of electrons in the band diagrams. Assuming an electric field $\mathscr{E}(x)$ in the *x*-direction, we can draw the energy bands as in Fig. 4–15, to include the change in potential energy of electrons in the field. Since electrons drift in a direction opposite to the field, we expect the potential energy for electrons to increase in the direction of the field, as in Fig. 4–15. The electrostatic potential $\mathscr{V}(x)$ varies in the opposite direction, since it is defined in terms of positive charges and is therefore related to the electron potential energy $E(x)$ displayed in the figure by $\mathscr{V}(x) = E(x)/(-q)$.

From the definition of electric field,

$$\mathscr{E}(x) = -\frac{d\mathscr{V}(x)}{dx} \tag{4-25}$$

we can relate $\mathscr{E}(x)$ to the electron potential energy in the band diagram by choosing some reference in the band for the electrostatic potential. We are interested only in the spatial variation $\mathscr{V}(x)$ for Eq. (4–25). Choosing E_i as a convenient reference, we can relate the electric field to this reference by

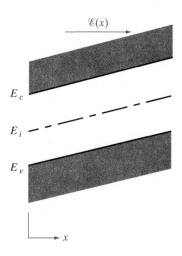

Figure 4–15
Energy band diagram of a semiconductor in an electric field $\mathscr{E}(x)$.

$$\mathscr{E}(x) = -\frac{dV(x)}{dx} = -\frac{d}{dx}\left[\frac{E_i}{(-q)}\right] = \frac{1}{q}\frac{dE_i}{dx} \qquad (4\text{--}26)$$

Therefore, the variation of band energies with $\mathscr{E}(x)$ as drawn in Fig. 4–15 is correct. The direction of the slope in the bands relative to $\mathscr{E}$ is simple to remember: Since the diagram indicates electron energies, we know the slope in the bands must be such that electrons drift "downhill" in the field. Therefore, $\mathscr{E}$ points "uphill" in the band diagram.

At equilibrium, no net current flows in a semiconductor. Thus any fluctuation which would begin a diffusion current also sets up an electric field which redistributes carriers by drift. An examination of the requirements for equilibrium indicates that the diffusion coefficient and mobility must be related. Setting Eq. (4–23b) equal to zero for equilibrium, we have

$$\mathscr{E}(x) = \frac{D_p}{\mu_p}\frac{1}{p(x)}\frac{dp(x)}{dx} \qquad (4\text{--}27)$$

Using Eq. (3–25b) for $p(x)$,

$$\mathscr{E}(x) = \frac{D_p}{\mu_p}\frac{1}{kT}\left(\frac{dE_i}{dx} - \frac{dE_F}{dx}\right) \qquad (4\text{--}28)$$

The equilibrium Fermi level does not vary with x, and the derivative of E_i is given by Eq. (4–26). Thus Eq. (4–28) reduces to

$$\frac{D}{\mu} = \frac{kT}{q} \qquad (4\text{--}29)$$

This result is obtained for either carrier type. This important equation is called the *Einstein relation*. It allows us to calculate either D or μ from a measurement of the other. Table 4–1 lists typical values of D and μ for several semiconductors at room temperature. It is clear from these values that $D/\mu \approx 0.026$ V.

An important result of the balance of drift and diffusion at equilibrium is that *built-in* fields accompany gradients in E_i [see Eq. (4–26)]. Such gradients in the bands at equilibrium (E_F constant) can arise when the band gap varies due to changes in alloy composition. More commonly, built-in fields result from doping gradients. For example, a donor distribution $N_d(x)$

Table 4–1 Diffusion coefficient and mobility of electrons and holes for intrinsic semiconductors at 300 K. *Note:* Use Fig. 3–23 for doped semiconductors.

	D_n (cm^2/s)	D_p (cm^2/s)	μ_n (cm^2/V-s)	μ_p (cm^2/V-s)
Ge	100	50	3900	1900
Si	35	12.5	1350	480
GaAs	220	10	8500	400

causes a gradient in $n_0(x)$, which must be balanced by a built-in electric field $\mathscr{E}(x)$.

EXAMPLE 4–5

An intrinsic Si sample is doped with donors from one side such that $N_d = N_0 \exp(-ax)$. (a) Find an expression for $\mathscr{E}(x)$ at equilibrium over the range for which $N_d \gg n_i$. (b) Evaluate $\mathscr{E}(x)$ when $a = 1(\mu m)^{-1}$. (c) Sketch a band diagram such as in Fig. 4–15 and indicate the direction of $\mathscr{E}$.

SOLUTION

(a) From Eq. (4–23a),

$$\mathscr{E}(x) = -\frac{D_n}{\mu_n}\frac{dn/dx}{n} = -\frac{kT}{q}\frac{N_0(-a)e^{-ax}}{N_0 e^{-ax}} = +\frac{kT}{q}a$$

We notice for this exponential impurity distribution, $\mathscr{E}(x)$ depends on a but not on N_0 or x.

(b) $\mathscr{E}(x) = 0.0259(10^4) = 259 \text{ V/cm}$

(c)

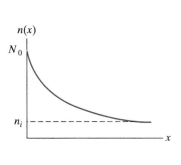

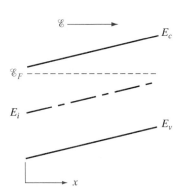

4.4.3 Diffusion and Recombination; The Continuity Equation

In the discussion of diffusion of excess carriers, we have thus far neglected the important effects of recombination. These effects must be included in a description of conduction processes, however, since recombination can cause a variation in the carrier distribution. For example, consider a differential length Δx of a semiconductor sample with area A in the yz-plane (Fig. 4–16). The hole current density leaving the volume, $J_p(x + \Delta x)$, can be larger or smaller than the current density entering, $J_p(x)$, depending on the generation and recombination of carriers taking place within the volume. The net increase in hole concentration per unit time, $\partial p/\partial t$, is the difference between the hole flux per unit volume entering and leaving, minus the recombination rate. We can convert hole current density to hole particle flux density by dividing J_p by q. The current densities are already expressed per unit area; thus dividing $J_p(x)/q$

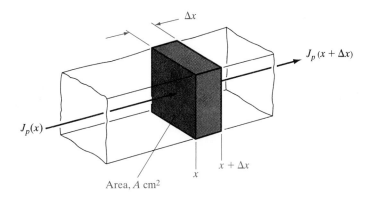

Figure 4–16
Current entering
and leaving a
volume ΔxA.

by Δx gives the number of carriers per unit volume entering ΔxA per unit time, and $(1/q)J_p(x + \Delta x)/\Delta x$ is the number leaving per unit volume and time:

$$\left.\frac{\partial p}{\partial t}\right|_{x \to x + \Delta x} = \frac{1}{q}\frac{J_p(x) - J_p(x + \Delta x)}{\Delta x} - \frac{\delta p}{\tau_p} \qquad (4\text{–}30)$$

$$\underset{\text{hole buildup}}{\text{Rate of}} = \underset{\text{tion in } \Delta xA \text{ per unit time}}{\text{increase of hole concentra}-} \underset{\text{rate}}{\text{recombination}}$$

As Δx approaches zero, we can write the current change in derivative form:

$$\frac{\partial p(x, t)}{\partial t} = \frac{\partial \delta p}{\partial t} = -\frac{1}{q}\frac{\partial J_p}{\partial x} - \frac{\delta p}{\tau_p} \qquad (4\text{–}31a)$$

The expression (4–31a) is called the *continuity equation* for holes. For electrons we can write

$$\frac{\partial \delta n}{\partial t} = \frac{1}{q}\frac{\partial J_n}{\partial x} - \frac{\delta n}{\tau_n} \qquad (4\text{–}31b)$$

since the electronic charge is negative.

When the current is carried strictly by diffusion (negligible drift), we can replace the currents in Eqs. (4–31) by the expressions for diffusion current; for example, for electron diffusion we have

$$J_n(\text{diff.}) = qD_n\frac{\partial \delta n}{\partial x} \qquad (4\text{–}32)$$

Substituting this into Eq. (4–31b) we obtain the *diffusion equation* for electrons,

$$\boxed{\frac{\partial \delta n}{\partial t} = D_n\frac{\partial^2 \delta n}{\partial x^2} - \frac{\delta n}{\tau_n}} \qquad (4\text{–}33a)$$

and similarly for holes,

$$\frac{\partial \delta p}{\partial t} = D_p \frac{\partial^2 \delta p}{\partial x^2} - \frac{\delta p}{\tau_n}$$ (4–33b)

These equations are useful in solving transient problems of diffusion with recombination. For example, a pulse of electrons in a semiconductor (Fig. 4–12) spreads out by diffusion and disappears by recombination. To solve for the electron distribution in time, $n(x, t)$, we would begin with the diffusion equation, Eq. (4–33a).

4.4.4 Steady State Carrier Injection; Diffusion Length

In many problems a steady state distribution of excess carriers is maintained, such that the time derivatives in Eqs. (4–33) are zero. In the steady state case the diffusion equations become

$$\frac{d^2 \delta n}{dx^2} = \frac{\delta n}{D_n \tau_n} \equiv \frac{\delta n}{L_n^2}$$ (4–34a)

$$\frac{d^2 \delta p}{dx^2} = \frac{\delta p}{D_p \tau_p} \equiv \frac{\delta p}{L_p^2}$$ (4–34b)

(steady state)

where $L_n \equiv \sqrt{D_n \tau_n}$ is called the electron *diffusion length* and L_p is the diffusion length for holes. We no longer need partial derivatives, since the time variation is zero for steady state.

The physical significance of the diffusion length can be understood best by an example. Let us assume that excess holes are somehow injected into a semi-infinite semiconductor bar at $x = 0$, and the steady state hole injection maintains a constant excess hole concentration at the injection point $\delta p(x = 0) = \Delta p$. The injected holes diffuse along the bar, recombining with a characteristic lifetime τ_p. In steady state we expect the distribution of excess holes to decay to zero for large values of x, because of the recombination (Fig. 4–17). For this problem we use the steady state diffusion equation for holes, Eq. (4–34b). The solution to this equation has the form

$$\delta p(x) = C_1 e^{x/L_p} + C_2 e^{-x/L_p}$$ (4–35)

We can evaluate C_1 and C_2 from the boundary conditions. Since recombination must reduce $\delta p(x)$ to zero for large values of x, $\delta p = 0$ at $x = \infty$ and therefore $C_1 = 0$. Similarly, the condition $\delta p = \Delta p$ at $x = 0$ gives $C_2 = \Delta p$, and the solution is

$$\delta p(x) = \Delta p e^{-x/L_p}$$ (4–36)

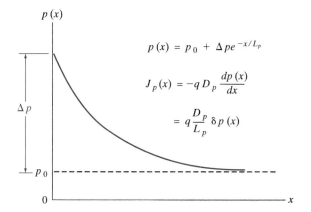

$$p(x) = p_0 + \Delta p e^{-x/L_p}$$

$$J_p(x) = -q D_p \frac{dp(x)}{dx}$$

$$= q \frac{D_p}{L_p} \delta p(x)$$

Figure 4–17
Injection of holes
at x = 0, giving a
steady state hole
distribution p(x)
and a resulting
diffusion current
density $J_p(x)$.

The injected excess hole concentration dies out exponentially in x due to recombination, and the diffusion length L_p represents the distance at which the excess hole distribution is reduced to $1/e$ of its value at the point of injection. We can show that L_p *is the average distance a hole diffuses before recombining*. To calculate an average diffusion length, we must obtain an expression for the probability that an injected hole recombines in a particular interval dx. The probability that a hole injected at $x = 0$ *survives* to x without recombination is $\delta p(x)/\Delta p = \exp(-x/L_p)$, the ratio of the steady state concentrations at x and 0. On the other hand, the probability that a hole at x will *recombine* in the subsequent interval dx is

$$\frac{\delta p(x) - \delta p(x + dx)}{\delta p(x)} = \frac{-(d\delta p(x)/dx)dx}{\delta p(x)} = \frac{1}{L_p} dx \qquad (4\text{–}37)$$

Thus the total probability that a hole injected at $x = 0$ will recombine in a given dx is the product of the two probabilities:

$$\left(e^{-x/L_p}\right)\left(\frac{1}{L_p} dx\right) = \frac{1}{L_p} e^{-x/L_p} dx \qquad (4\text{–}38)$$

Then, using the usual averaging techniques described by Eq. (2–21), the average distance a hole diffuses before recombining is

$$\langle x \rangle = \int_0^\infty x \frac{e^{-x/L_p}}{L_p} dx = L_p \qquad (4\text{–}39)$$

The steady state distribution of excess holes causes diffusion, and therefore a hole current, in the direction of decreasing concentration. From Eqs. (4–22b) and (4–36) we have

$$J_p(x) = -q D_p \frac{dp}{dx} = -q D_p \frac{d\delta p}{dx} = q \frac{D_p}{L_p} \Delta p e^{-x/L_p} = q \frac{D_p}{L_p} \delta p(x) \quad (4\text{–}40)$$

Since $p(x) = p_0 + \delta p(x)$, the space derivative involves only the excess concentration. We notice that since $\delta p(x)$ is proportional to its derivative for an exponential distribution, the diffusion current at any x is just proportional to the excess concentration δp at that position.

Although this example seems rather restricted, its usefulness will become apparent in Chapter 5 in the discussion of p-n junctions. The injection of minority carriers across a junction often leads to exponential distributions as in Eq. (4–36), with the resulting diffusion current of Eq. (4–40).

4.4.5 The Haynes-Shockley Experiment

One of the classic semiconductor experiments is the demonstration of drift and diffusion of minority carriers, first performed by J.R Haynes and W. Shockley in 1951 at the Bell Telephone Laboratories. The experiment allows independent measurement of the minority carrier mobility μ and diffusion coefficient D. The basic principles of the Haynes–Shockley experiment are as follows: A pulse of holes is created in an n-type bar (for example) that contains an electric field (Fig. 4–18); as the pulse drifts in the field and spreads out by diffusion, the excess hole concentration is monitored at some point down the bar; the time required for the holes to drift a given distance in the field gives a measure of the mobility; and the spreading of the pulse during a given time is used to calculate the diffusion coefficient.

In Fig. 4–18 a pulse of excess carriers is created by a light flash at some point $x = 0$ in an n-type semiconductor ($n_0 \gg p_0$). We assume that the excess carriers have a negligible effect on the electron concentration but change the hole concentration significantly. The excess holes drift in the direction of the electric field and eventually reach the point $x = L$, where they are

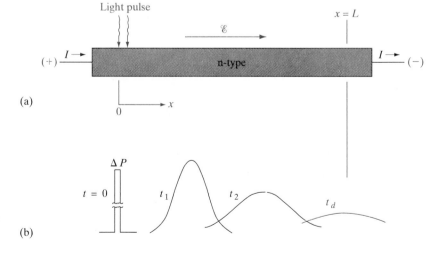

Figure 4–18
Drift and diffusion of a hole pulse in an n-type bar: (a) sample geometry; (b) position and shape of the pulse for several times during its drift down the bar.

monitored. By measuring the drift time t_d, we can calculate the drift velocity v_d and, therefore, the hole mobility:

$$v_d = \frac{L}{t_d} \qquad (4\text{--}41)$$

$$\mu_p = \frac{v_d}{\mathscr{E}} \qquad (4\text{--}42)$$

Thus the hole mobility can be calculated directly from a measurement of the drift time for the pulse as it moves down the bar. In contrast with the Hall effect (Section 3.4.5), which can be used with resistivity to obtain the *majority* carrier mobility, the Haynes–Shockley experiment is used to measure the *minority* carrier mobility.

As the pulse drifts in the $\mathscr{E}$ field it also spreads out by diffusion. By measuring the spread in the pulse, we can calculate D_p. To predict the distribution of holes in the pulse as a function of time, let us first reexamine the case of diffusion of a pulse *without drift, neglecting recombination* (Fig. 4–12). The equation which the hole distribution must satisfy is the time-dependent diffusion equation, Eq. (4–33b). For the case of negligible recombination (τ_p long compared with the times involved in the diffusion), we can write the diffusion equation as

$$\frac{\partial \delta p(x, t)}{\partial t} = D_p \frac{\partial^2 \delta p(x, t)}{\partial x^2} \qquad (4\text{--}43)$$

The function which satisfies this equation is called a *gaussian distribution,*

$$\delta p(x, t) = \left[\frac{\Delta P}{2\sqrt{\pi D_p t}} \right] e^{-x^2/4D_p t} \qquad (4\text{--}44)$$

where ΔP is the number of holes per unit area created over a negligibly small distance at $t = 0$. The factor in brackets indicates that the peak value of the pulse (at $x = 0$) decreases with time, and the exponential factor predicts the spread of the pulse in the positive and negative x-directions (Fig. 4–19). If we

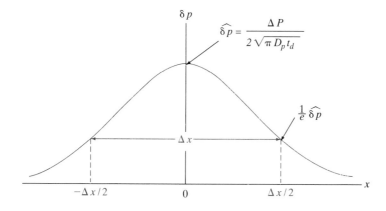

Figure 4–19
Calculation of D_p from the shape of the δp distribution after time t_d. No drift or recombination is included.

designate the peak value of the pulse as $\delta\hat{p}$ at any time (say t_d), we can use Eq. (4–44) to calculate D_p from the value of δp at some point x. The most convenient choice is the point $\Delta x/2$, at which δp is down by $1/e$ of its peak value $\delta\hat{p}$. At this point we can write

$$e^{-1}\delta\hat{p} = \delta\hat{p}e^{-(\Delta x/2)^2/4D_pt_d} \tag{4–45}$$

$$D_p = \frac{(\Delta x)^2}{16t_d} \tag{4–46}$$

Since Δx cannot be measured directly, we use an experimental setup such as Fig. 4–20, which allows us to display the pulse on an oscilloscope as the carriers pass under a detector. As we shall see in Chapter 5, a forward-biased p-n junction serves as an excellent injector of minority carriers, and a reverse-biased junction serves as a detector. The measured quantity in Fig. 4–20 is the pulse width Δt displayed on the oscilloscope in time. It is related to Δx by the drift velocity, as the pulse drifts past the detector point (2).

$$\Delta x = \Delta t v_d = \Delta t \frac{L}{t_d} \tag{4–47}$$

EXAMPLE 4–6

An n-type Ge sample is used in the Haynes–Shockley experiment shown in Fig. 4–20. The length of the sample is 1 cm, and the probes (1) and (2) are separated by 0.95 cm. The battery voltage E_0 is 2 V. A pulse arrives at point (2)

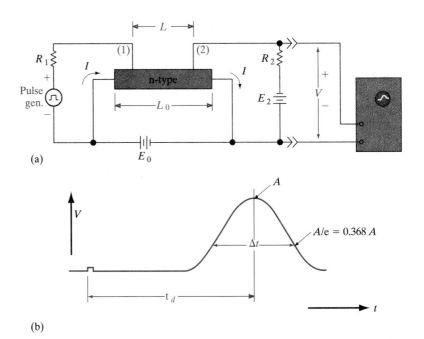

Figure 4–20
The Haynes–Shockley experiment: (a) circuit schematic; (b) typical trace on the oscilloscope screen.

(a)

(b)

0.25 ms after injection at (1); the width of the pulse Δt is 117 μs. Calculate the hole mobility and diffusion coefficient, and check the results against the Einstein relation.

SOLUTION

$$\mu_p = \frac{v_d}{\mathscr{E}} = \frac{0.95/(0.25 \times 10^{-3})}{2/1} = 1900 \text{ cm}^2/(\text{V-s})$$

$$D_p = \frac{(\Delta x)^2}{16t_d} = \frac{(\Delta t L)^2}{16t_d^3}$$

$$= \frac{(117 \times 0.95)^2 \times 10^{-12}}{16(0.25)^3 \times 10^{-9}} = 49.4 \text{ cm}^2/\text{s}$$

$$\frac{D_p}{\mu_p} = \frac{49.4}{1900} = 0.026 = \frac{kT}{q}$$

4.4.6 Gradients in the Quasi-Fermi Levels

In Section 3.5 we saw that equilibrium implies no gradient in the Fermi level E_F. In contrast, any combination of drift and diffusion implies a gradient in the steady state quasi-Fermi level.

We can use the results of Eqs. (4–23), (4–26), and (4–29) to demonstrate the power of the concept of quasi-Fermi levels in semiconductors [see Eq. (4–15)]. If we take the general case of nonequilibrium electron concentration with drift and diffusion, we must write the total electron current as

$$J_n(x) = q\mu_n n(x)\mathscr{E}(x) + qD_n \frac{dn(x)}{dx} \tag{4-48}$$

where the gradient in electron concentration is

$$\frac{dn(x)}{dx} = \frac{d}{dx}[n_i e^{(F_n - E_i)/kT}] = \frac{n(x)}{kT}\left(\frac{dF_n}{dx} - \frac{dE_i}{dx}\right) \tag{4-49}$$

Using the Einstein relation, the total electron current becomes

$$J_n(x) = q\mu_n n(x)\mathscr{E}(x) + \mu_n n(x)\left[\frac{dF_n}{dx} - \frac{dE_i}{dx}\right] \tag{4-50}$$

But Eq. (4–26) indicates that the subtractive term in the brackets is just $q\mathscr{E}(x)$, giving a direct cancellation of $q\mu_n n(x)\mathscr{E}(x)$ and leaving

$$J_n(x) = \mu_n n(x)\frac{dF_n}{dx} \tag{4-51}$$

Thus, the processes of electron drift and diffusion are summed up by the spatial variation of the quasi-Fermi level. The same derivation can be made for holes, and we can write the current due to drift and diffusion in the form of a *modified Ohm's law*

$$J_n(x) = q\mu_n n(x) \frac{d(F_n/q)}{dx} = \sigma_n(x) \frac{d(F_n/q)}{dx} \qquad (4\text{–}52a)$$

$$J_p(x) = q\mu_p p(x) \frac{d(F_p/q)}{dx} = \sigma_p(x) \frac{d(F_p/q)}{dx} \qquad (4\text{–}52b)$$

Therefore, any drift, diffusion, or combination of the two in a semiconductor results in currents proportional to the gradients of the two quasi-Fermi levels. Conversely, a lack of current implies constant quasi-Fermi levels.

PROBLEMS

4.1 With E_F located 0.4 eV above the valence band in a Si sample, what charge state would you expect for most Ga atoms in the sample? What would be the predominant charge state of Zn? Au? *Note:* By charge state we mean neutral, singly positive, doubly negative, etc.

4.2 A Si sample is doped with 10^{16} cm^{-3} Sb. How many Zn atoms/cm^3 must be added to exactly compensate this material ($n_0 = p_0 = n_i$)?

4.3 Construct a semilogarithmic plot such as Fig. 4–7 for GaAs doped with 2×10^{15} donors/cm^3 and having 4×10^{14} EHP/cm^3 created uniformly at $t = 0$. Assume that $\tau_n = \tau_p = 50$ ns.

4.4 Calculate the recombination coefficient α_r, for the low-level excitation described in Prob. 4.3. Assume that this value of α_r applies when the GaAs sample is uniformly exposed to a steady state optical generation rate $g_{op} = 10^{20}$ EHP/cm^3-s. Find the steady state excess carrier concentration $\Delta n = \Delta p$.

4.5 A sample is doped with donors such that $n_0 = Gx$ for $n_0 \gg n_i$, where G is a constant. Find the built-in electric field $\mathscr{E}(x)$.

4.6 A Si sample with 10^{15}/cm^3 donors is uniformly optically excited at room temperature such that 10^{19}/cm^3 electron–hole pairs are generated per second. Find the separation of the quasi-Fermi levels and the change of conductivity upon shining the light. Electron and hole lifetimes are both 10 μs. $D_p = 12$ cm^2/s.

4.7 An n-type Si sample with $N_d = 10^{15}$ cm^{-3} is steadily illuminated such that $g_{op} = 10^{21}$ EHP/cm^3-s. If $\tau_n = \tau_p = 1$μs for this excitation, calculate the separation in the quasi-Fermi levels, $(F_n - F_p)$. Draw a band diagram such as Fig. 4–11.

4.8 For a 2 cm long doped Si bar ($N_d = 10^{16}$ cm^{-3}) with a cross-sectional area = 0.05 cm^2, what is the current if we apply 10V across it? If we generate 10^{20} electron–hole pairs per second per cm^3 uniformly in the bar and the lifetime $\tau_n = \tau_p = 10^{-4}$s, what is the new current? Assume the low level α_r doesn't change for high level injection. If the voltage is then increased to 100,000 V, what is the new current? Assume $\mu_p = 500$ cm^2/V-s, but you must choose the appropriate values for electrons.

4.9 Design and sketch a photoconductor using a 5-μm-thick film of CdS, assuming that $\tau_n = \tau_p = 10^{-6}$ s and $N_d = 10^{14}$ cm^{-3}. The dark resistance (with $g_{op} = 0$) should be 10 MΩ, and the device must fit in a square 0.5 cm on a side; therefore, some sort of folded or zigzag pattern is in order. With an excitation of $g_{op} = 10^{21}$ EHP/cm^3-s, what is the resistance change?

4.10 In a very long p-type Si bar with cross-sectional area = 0.5 cm^2 and $N_a = 10^{17}$ cm^{-3}, we inject holes such that the steady state *excess* hole concentration is 5×10^{16} cm^{-3} at $x = 0$. What is the steady state separation between F_p and E_c at $x = 1000$Å? What is the hole current there? How much is the excess stored hole charge? Assume $\mu_p = 500$ cm^2/V-s and $\tau_p = 10^{-10}$s.

4.11 Assume that a photoconductor in the shape of a bar of length L and area A has a constant voltage V applied, and it is illuminated such that g_{op} EHP/cm^3-s are generated uniformly throughout. If $\mu_n \gg \mu_p$, we can assume the optically induced change in current ΔI is dominated by the mobility μ_n and lifetime τ_n for electrons. Show that $\Delta I = qALg_{op}\tau_n/\tau_t$ for this photoconductor, where τ_t is the transit time of electrons drifting down the length of the bar.

4.12 For the steady state minority hole distribution shown in Fig. 4–17, find the expression for the hole quasi-Fermi level position $E_i - F_p(x)$ while $p(x) \gg p_0$ (i.e., while F_p is below E_F). On a band diagram, draw the variation of $F_p(x)$. Be careful—when the minority carriers are few (e.g., when δp is n_i), F_p still has a long way to go to reach E_F.

4.13 Boron is diffused into an intrinsic Si sample, giving the acceptor distribution shown in Figure P4-13. Sketch the equilibrium band diagram and show the direction of the resulting electric field, for $N_a(x) \gg n_i$. Repeat for phosphorus, with $N_d(x) \gg n_i$.

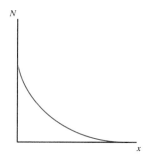

Figure P4–13

4.14 The current required to feed the hole injection at $x = 0$ in Fig. 4–17 is obtained by evaluating Eq. (4–40) at $x = 0$. The result is $I_p(x = 0) = qAD_p\Delta p/L_p$. Show that this current can be calculated by integrating the charge stored in the steady state hole distribution $\delta p(x)$ and then dividing by the average hole lifetime τ_p. Explain why this approach gives $I_p(x = 0)$.

4.15 We wish to use the Haynes–Shockley experiment to calculate the hole lifetime τ_p in an n-type sample. Assume the peak voltage of the pulse displayed on the oscilloscope screen is proportional to the hole concentration under the collector terminal at time t_d, and that the displayed pulse can be approximated as a gaussian, as in Eq. (4–44), which decays due to recombination by e^{-t/τ_p}. The electric field is varied and the following data taken: For $t_d = 200$ μs, the peak is 20 mV; for $t_d = 50$ μs, the peak is 80 mV. What is τ_p?

4.16 Consider a sample of GaAs ($n_i = 10^6$ cm^{-3} at 300 K) doped with 10^{15} donors per cm^3 illuminated with the 5145 Å line of an argon ion laser. For GaAs at 5145 Å, $\alpha = 10^4$ cm^{-1}. Calculate and plot the steady state excess electron profile $\delta n(x)$ in the region within 5 μm of the surface for photon fluxes of 10^{15}, 10^{17}, and 10^{19} photons cm^{-2} s^{-1} using low-level injection assumptions and directly solving Eq. (4–12). For this problem, assume that $\tau_n = \tau_p = 10^{-6}$ s. Neglect diffusion.

4.17 For the sample of Prob. 4–16, calculate and plot the steady state excess electron profile $\delta n(x)$ in the region within 5 μm of the surface for a photon flux of 10^{19} photons cm^{-2} s^{-1} using low-level injection assumptions and directly solving Eq. (4–12) for values of α_r of 10^{-9}, 10^{-7}, and 10^{-5} cm^3 s^{-1}.

4.18 Using the results of Prob. 4–16 obtained for a photon flux of 10^{15} photons cm^{-2} s^{-1}, calculate and plot the transient excess carrier profile, 1, 2, and 5 ns after the laser flux is interrupted, by integrating Eq. (4–5) within each depth interval, using 10^{-6} cm^3 s^{-1} for α_r. In this case, ignore carrier diffusion.

4.19 Assume an n-type semiconductor bar is illuminated over a narrow region of its length, such that $\Delta n = \Delta p$ in the illuminated zone, and excess carriers diffuse away and recombine in both directions along the bar. Assuming $\delta n = \delta p$, sketch the excess carrier distribution and, on a band diagram, sketch the quasi-Fermi levels F_n and F_p over several diffusion lengths from the illuminated zone. See the cautionary note in Prob. 4–12.

READING LIST   Ashcroft, N. W., and N. D. Mermin. *Solid State Physics.* Philadelphia: W.B. Saunders, 1976.

Bhattacharya, P. *Semiconductor Optoelectronic Devices.* Englewood Cliffs, NJ: Prentice Hall, 1994.

Blakemore, J. S. *Semiconductor Statistics.* New York: Dover Publications, 1987.

Collins, R. T., and M. A. Tischler. "Silicon Emits Light, but How?" *Laser Focus World* 28 (February 1992): 18+.

Ghandhi, S. K. *VLSI Fabrication Principles,* 2nd ed. New York: Wiley, 1994.

Gupta, S., S. L. Williamson, and J. F. Whitaker. "Epitaxial Methods Produce Robust Ultrafast Detectors." *Laser Focus World* 28 (June 1992): 97–101+.

Hummel, R. E. *Electronic Properties of Materials,* 2nd ed. Berlin: Springer-Verlag, 1993.

Li, S. S. *Semiconductor Physical Electronics.* New York: Plenum Press, 1993.

Madden M. R., and P. C. Williamson. "Photodetector Hybrids: What Are They and Who Needs Them?" *Laser Focus World* 28 (July 1992): 107–109+.

Muller, R. S., and T. I. Kamins. *Device Electronics for Integrated Circuits.* New York: Wiley, 1986.

Neamen, D. A. *Semiconductor Physics and Devices: Basic Principles.* Homewood, IL: Irwin, 1992.

Neudeck, G. W. *Modular Series on Solid State Devices: Vol II. The PN Junction Diode.* Reading, MA: Addison-Wesley, 1983.

Pankove, J. I. *Optical Processes in Semiconductors.* Englewood Cliffs, NJ: Prentice Hall, 1971.

Pierret, R. F. *Advanced Semiconductor Fundamentals.* Reading, MA: Addison-Wesley, 1987.

Singh, J. *Physics of Semiconductors and Their Heterostructures.* New York: McGraw-Hill, 1993.

Singh, J. *Semiconductor Devices.* New York: McGraw-Hill, 1994.

Swaminathan, V., and Macrander, A. T. *Material Aspects of GaAs and InP Based Structures.* Englewood Cliffs, NJ: Prentice Hall, 1991.

Sze, S. M. *Physics of Semiconductor Devices.* New York: Wiley, 1981.

Thomas, G. A. "An Electron–Hole Liquid." *Scientific American* 234 (June 1976): 28–37.

Wang, S. *Fundamentals of Semiconductor Theory and Device Physics.* Englewood Cliffs, NJ: Prentice Hall, 1989.

Weisbuch, C., and B. Vinter. *Quantum Semiconductor Structures.* Boston: Academic Press, 1991.

Wolfe, C. M., G. E. Stillman, and N. Holonyak, Jr. *Physical Properties of Semiconductors.* Englewood Cliffs, NJ: Prentice Hall, 1989.

Chapter 5

Junctions

Most semiconductor devices contain at least one junction between p-type and n-type material. These p-n junctions are fundamental to the performance of functions such as rectification, amplification, switching, and other operations in electronic circuits. In this chapter we shall discuss the equilibrium state of the junction and the flow of electrons and holes across a junction under steady state and transient conditions. This is followed by a discussion of metal–semiconductor junctions and heterojunctions between semiconductors having different band gaps. With the background provided in this chapter on junction properties, we can then discuss specific devices in later chapters.

5.1 FABRICATION OF p-n JUNCTIONS

Although this book deals primarily with how devices work rather than how they are made, it is instructive to have an overview of the fabrication process in order to appreciate device physics. We have already discussed in Chapter 1 how single-crystal substrates and epitaxial layers needed for high quality devices are grown, and how the doping can be varied as a function of depth. However, we have not discussed how doping can be varied laterally across the surface, which is key to making integrated circuits on a wafer. Hence, it is necessary to be able to form patterned masks on the wafer corresponding to the circuitry, and introduce the dopants selectively through windows in the mask. We will first briefly describe the major process steps that form the underpinnings of modern integrated circuit manufacturing. Relatively few unit process steps can be used in different permutations and combinations to make everything from simple diodes to the most complex microprocessors.

5.1.1 Thermal Oxidation

Many fabrication steps involve heating up the wafer in order to enhance a chemical process. An important example of this is thermal oxidation of Si to form SiO_2. This involves placing a batch of wafers in a clean silica (quartz) tube which can be heated to very high temperatures (~800–1000°C) using heating coils in a furnace with ceramic brick insulating liners. An oxygen-containing gas such as dry O_2 or H_2O is flowed into the tube at atmospheric pressure, and flowed out at the other end. Traditionally, horizontal furnaces

Figure 5–1a
Silicon wafers being loaded into a furnace. For 8-inch and larger wafers, this type of horizontal loading is often replaced by a vertical furnace.

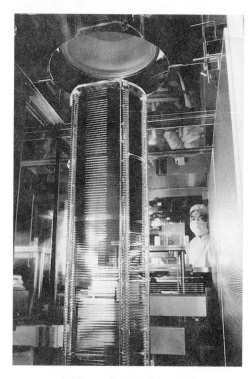

Figure 5–1b
Vertical furnace for large Si wafers. The silica wafer holder is loaded with eight-inch Si wafers and moved into the furnace above for oxidation, diffusion, or deposition operations. (Photograph courtesy of Tokyo Electron Ltd.)

were used (Fig. 5–1a). More recently, it has become common to employ vertical furnaces (Fig. 5-1b). A batch of Si wafers is placed in the silica wafer holders, each facing down to minimize particulate contamination. The wafers are then moved into the furnace. The gases flow in from the top and flow out at the bottom, providing more uniform flow than in conventional horizontal furnaces. The overall reactions that occur during oxidation are:

$$Si + O_2 \rightarrow SiO_2 \text{ (dry oxidation)}$$

$$Si + 2H_2O \rightarrow SiO_2 + 2H_2 \text{ (wet oxidation)}$$

In both cases, Si is consumed from the surface of the substrate. For every micron of SiO_2 grown, 0.44 μm of Si is consumed, leading to a 2.2$\times$ volume expansion of the consumed layer upon oxidation. The oxidation proceeds by having the oxidant (O_2 or H_2O) molecules diffuse through the already grown oxide to the Si–SiO_2 interface, where the above reactions take place. One of the very important reasons why Si integrated circuits exist (and by extension why modern computers exist) is that a stable thermal oxide can be grown on Si with excellent interface electrical properties. Other semiconductor materials do not have such a useful native oxide. We can argue that modern electronics and computer technology owe their existence to this simple oxidation process.

Plots of oxide thickness as a function of time, at different temperatures, are shown for dry and wet oxidation of (100) Si in Appendix VI.

5.1.2 Diffusion

Another thermal process that was used extensively in IC fabrication in the past is thermal in-diffusion of dopants in furnaces such as those shown in Fig. 5–1a. The wafers are first oxidized and windows are opened in the oxide using the photolithography and etching steps described in Sections 5.1.6 and 5.17, respectively. Dopants such as B, P or As are introduced into these patterned wafers in a high temperature (~800–1100°C) diffusion furnace, generally using a gas or vapor source. The dopants are gradually transported from the high concentration region near the surface into the substrate through diffusion, similar to that described for carriers in Section 4.4. The maximum number of impurities that can be dissolved (the solid solubility) in Si is shown for various impurities as a function of temperature in Appendix VII. The diffusivity of dopants in solids, D, has a strong Arrhenius dependence on temperature, T. It is given by $D = D_0 \exp -(E_A/kT)$, where D_0 is a constant depending on the material and the dopant, and E_A is the activation energy. The average distance the dopants diffuse is related to the diffusion length as in Section 4.4.4. In this case, the diffusion length is $\sqrt{Dt}$, where t is the processing time. The product Dt is sometimes called the *thermal budget*. The Arrhenius dependence of diffusivity on temperature explains why high temperatures are required for diffusion; otherwise, the diffusivities are far too low. Since D varies exponentially with T, it is critical to have very precise control over the furnace temperatures, within several degrees, in order to have control over the diffusion profiles (Fig. 5–2). The dopants are effectively blocked or masked by the oxide because their diffusivity in oxide is very low. The diffusivities of various dopants in Si and SiO_2 are shown as a function of temperature in Appendix VIII. Difficulty with profile control and the

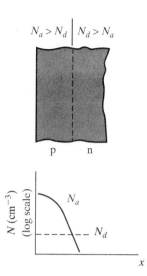

$N_a > N_d$ | $N_d > N_a$

p   n

N (cm^{-3}) (log scale)

N_a

N_d

x

Figure 5–2
Impurity concen-
tration profile for
fabricating a p-n
junction by
diffusion.

very high temperature requirement has led to diffusion being supplanted by ion implantation as a doping technique, as discussed in Section 5.1.4.

The trend of using larger Si wafers has changed many processing steps. For example, eight-inch and larger wafers are best handled in a vertical furnace (Fig. 5–1b) rather than the traditional horizontal furnace (Fig. 5–1a). Also, large wafers are often handled individually for a variety of deposition, etching, and implantation processes. Such single-wafer processing has led to development of robotic systems for fast and accurate wafer handling.

The distribution of impurities in the sample at any time during the diffusion can be calculated from a solution of the diffusion equation with appropriate boundary conditions. If the source of dopant atoms at the surface of the sample is limited (e.g., a given number of atoms deposited on the Si surface before diffusion), a gaussian distribution as described by Eq. (4–44) (for $x > 0$) is obtained. On the other hand, if the dopant atoms are supplied continuously, such that the concentration at the surface is maintained at a constant value, the distribution follows what is called a *complementary error function*. In Fig. 5–2, there is some point in the sample at which the introduced acceptor concentration just equals the background donor concentration in the originally n-type sample. This point is the location of the p-n junction. To the left of this point in the sample of Fig. 5–2, acceptor atoms predominate and the material is p-type, whereas to the right of the junction, the background donor atoms predominate and the material is n-type. The depth of the junction beneath the surface of the sample can be controlled by the time and temperature of the diffusion (Prob. 5.2).

In the horizontal diffusion furnace shown in Fig. 5–1a, Si wafers are placed in the tube during diffusion, and the impurity atoms are introduced into the gas which flows through the silica tube. Common impurity source materials for diffusions in Si are B_2O_3, BBr_3, and BCl_3 for boron; phosphorus sources include PH_3, P_2O_5, and $POCl_3$. Solid sources are placed in the silica tube upstream from the sample or in a separate heating zone of the furnace; gaseous sources can be metered directly into the gas flow system; and with liquid sources inert carrier gas is bubbled through the liquid before being introduced into the furnace tube. The Si wafers are held in a silica "boat" (Fig. 5–1a) which can be pushed into position in the furnace and removed by a silica rod.

It is important to remember the degree of cleanliness required in these processing steps. Since typical doping concentrations represent one part per million or less, cleanliness and purity of materials is critically important. Thus the impurity source and carrier gas must be extremely pure; the silica tube, sample holder and pushrod must be cleaned and etched in hydrofluoric acid (HF) before use (once in use, the tube cleanliness can be maintained if no unwanted impurities are introduced); finally, the Si wafers themselves must undergo an elaborate cleaning procedure before diffusion, including a final etch containing HF to remove any unwanted SiO_2 from the surface.

5.1.3 Rapid Thermal Processing

Increasingly, many thermal steps formerly performed in furnaces are being done using what is called *rapid thermal processing* (RTP). This includes rapid thermal oxidation, annealing of ion implantation, and chemical vapor deposition, which are discussed in the following paragraphs. A simple RTP system is shown in Fig. 5–3. Instead of having a large batch of wafers in a conventional furnace where the temperature cannot be changed rapidly, a single wafer is held (face down to minimize particulates) on low-thermal-mass quartz pins, surrounded by a bank of high-intensity (tens of kW) tung-

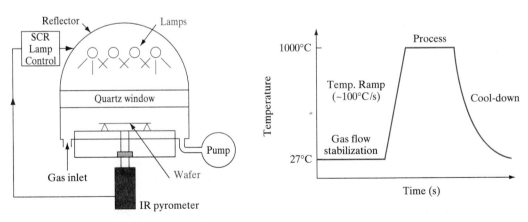

Figure 5–3
Schematic diagram of a rapid thermal processor, and typical time-temperature profile.

sten-halogen infrared lamps, with gold-plated reflectors around them. By turning on the lamps, the high intensity infrared radiation shines through the quartz chamber and is absorbed by the wafer, causing its temperature to rise *very* rapidly (~50–100°C/s). The processing temperature can be reached quickly, after the gas flows have been stabilized in the chamber. At the end of the process, the lamps are turned off, allowing the wafer temperature to drop rapidly, once again because of the much lower thermal mass of an RTP system compared to a furnace. In RTP, therefore, temperature is essentially used as a "switch" to start or quench the reaction. Two critical aspects of RTP are ensuring temperature uniformity across large wafers, and accurate temperature measurement, for example with thermocouples or pyrometers.

A key parameter in all thermal processing steps is the thermal budget, Dt. Generally speaking, we try to minimize this quantity because an excessive Dt product leads to loss of control over compact doping profiles, which is detrimental to ultra-small devices. In furnace processing, thermal budgets are minimized by operating at as low a temperature as feasible so that D is small. On the other hand, RTP operates at higher temperatures (~1000°C) but does so for only a few seconds (compared to minutes or hours in a furnace).

5.1.4 Ion Implantation

A useful alternative to high-temperature diffusion is the direct implantation of energetic ions into the semiconductor. In this process a beam of impurity ions is accelerated to kinetic energies ranging from several keV to several MeV and is directed onto the surface of the semiconductor. As the impurity atoms enter the crystal, they give up their energy to the lattice in collisions and finally come to rest at some average penetration depth, called the *projected range*. Depending on the impurity and its implantation energy, the range in a given semiconductor may vary from a few hundred angstroms to about 1 μm. For most implantations the ions come to rest distributed almost evenly about the projected range R_p, as shown in Fig. 5–4a. An implanted dose of ϕ ions/cm^2 is distributed approximately by a gaussian formula

$$N(x) = \frac{\phi}{\sqrt{2\pi}\Delta R_p} \exp\left[-\frac{1}{2}\left(\frac{x - R_p}{\Delta R_p}\right)^2 \right] \qquad (5\text{--}1\text{a})$$

where ΔR_p, called the *straggle*, measures the half-width of the distribution at $e^{-1/2}$ of the peak Fig. (5–4a). Both R_p and ΔR_p increase with increasing implantation energy. These parameters are shown as a function of energy for various implant species into Si in Appendix IX. By performing several implantations at different energies, it is possible to synthesize a desired impurity distribution, such as the uniformly doped region in Fig. 5–4b.

Figure 5–4
Distributions of implanted impurities: (a) gaussian distribution of boron atoms about a projected range R_p (in this example, a dose of 10^{14} B atoms/cm^2 implanted at 140 keV); (b) a relatively flat distribution obtained by summing four gaussions implanted at selected energies and doses.

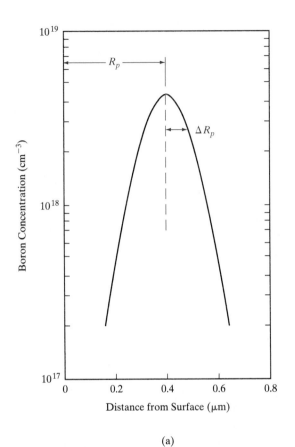

(a)

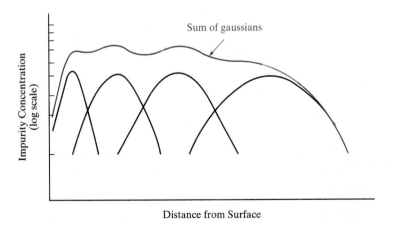

(b)

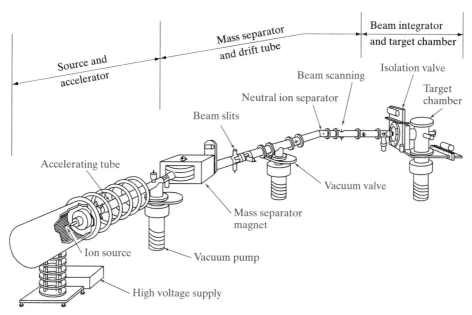

Figure 5–5
Schematic diagram of an ion implantation system.

An ion implanter is shown schematically in Fig. 5–5. A gas containing the desired impurity is ionized within the *source* and is then extracted into the *acceleration tube*. After acceleration to the desired kinetic energy, the ions are passed through a *mass separator* to ensure that only the desired ion species enters the *drift tube*.[1] The ion beam is then focused and scanned electrostatically over the surface of the wafer in the *target chamber*. Repetitive scanning in a raster pattern provides exceptionally uniform doping of the wafer surface. The target chamber commonly includes automatic wafer-handling facilities to speed up the process of implanting many wafers per hour.

An obvious advantage of implantation is that it can be done at relatively low temperatures; this means that doping layers can be implanted without disturbing previously diffused regions. The ions can be blocked by metal or photoresist layers; therefore, the photolithographic techniques described in Section 5.1.6 can be used to define ion implanted doping patterns. Very shallow (tenths of a micron) and well-defined doping layers can be achieved by this method. As we shall see in later chapters, many devices require thin doping regions and may be improved by ion implantation techniques. Furthermore, it is possible to implant impurities which do not diffuse conveniently into semiconductors.

One of the major advantages of implantation is the precise control of doping concentration it provides. Since the ion beam current can be measured accurately during implantation, a precise quantity of impurity can be

---

[1] In many ion implanters the mass separation occurs before the ions are accelerated to high energy.

introduced. This control over doping level, along with the uniformity of the implant over the wafer surface, make ion implantation particularly attractive for the fabrication of Si integrated circuits (Chapter 9).

One problem with this doping method is the lattice damage which results from collisions between the ions and the lattice atoms. However, most of this damage can be removed in Si by heating the crystal after the implantation. This process is called *annealing*. Although Si can be heated to temperatures in excess of 1000°C without difficulty, GaAs and some other compounds tend to dissociate at high temperatures. For example, As evaporation from the surface of GaAs during annealing damages the sample. Therefore, it is common to encapsulate the GaAs with a thin layer of silicon nitride during the anneal. Another approach to annealing either Si or compounds is to heat the sample only briefly (e.g., 10 s) using RTP, rather than a conventional furnace. Annealing leads to some unintended diffusion of the implanted species. It is desirable to minimize this diffusion by optimizing the annealing time and temperature. The profile after annealing is given by

$$N(x) = \frac{\phi}{\sqrt{2\pi}(\Delta R_p^2 + 2Dt)^{1/2}} \exp\left[-\frac{1}{2}\left(\frac{(x - R_p)^2}{\Delta R_p^2 + 2Dt}\right)\right] \quad (5\text{--}1b)$$

5.1.5 Chemical Vapor Deposition (CVD)

At various stages of device fabrication, thin films of dielectrics, semiconductors and metals have to be formed on the wafer and then patterned and etched. We have already discussed one important example of this involving thermal oxidation of Si. SiO_2 films can also be formed by *low pressure* (~100 mTorr)[2] chemical vapor deposition (LPCVD) (Fig. 5–6) or plasma-enhanced

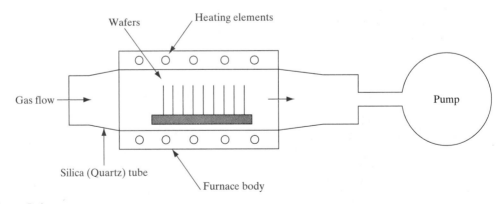

Figure 5–6
Low pressure chemical vapor deposition (LPCVD) reactor.

[2]Torr or Torricelli = 1 mm Hg or 133 Pa.

CVD (PECVD). The key differences are that thermal oxidation consumes Si from the substrate, and very high temperatures are required, whereas CVD of SiO_2 does not consume Si from the substrate and can be done at much lower temperatures. The CVD process reacts a Si-containing gas such as SiH_4 with an oxygen-containing precursor, causing a chemical reaction, leading to the deposition of SiO_2 on the substrate. Being able to deposit SiO_2 is very important in certain applications. As a complicated device structure is built up, the Si substrate may not be available for reaction, or there may be metallization on the wafer that cannot withstand very high temperatures. In such cases, CVD is a necessary alternative.

Although we have used deposition of SiO_2 as an important example, LPCVD is also widely used to deposit other dielectrics such as silicon nitride (Si_3N_4), and polycrystalline or amorphous Si. It should also be clear that the VPE of Si or MOCVD of compound semiconductors discussed in Chapter 1 is really a special, more challenging example of CVD where not only must a film be deposited, but single-crystal growth must also be maintained.

5.1.6 Photolithography

Patterns corresponding to complex circuitry are formed on a wafer using *photolithography*. This involves first generating a *reticle* which is a transparent silica (quartz) plate containing the pattern (Fig. 5–7a). Opaque regions on the mask are made up of an ultraviolet light-absorbing layer, such as iron oxide. The reticle typically contains the patterns corresponding to a single *chip* or *die*, rather than the entire wafer (in which case it would be called a

Figure 5–7a
A photolithographic reticle used for one step in the processing of a 16 Mb dynamic random access memory (DRAM). In a "stepper" projection exposure system, ultraviolet light shines through the glass plate and the image is projected onto the wafer to expose photoresist for one die in the array of circuits, then steps to the next. (Photograph courtesy of IBM Corp.)

Figure 5–7b
Schematic dia-
gram of an opti-
cal stepper.

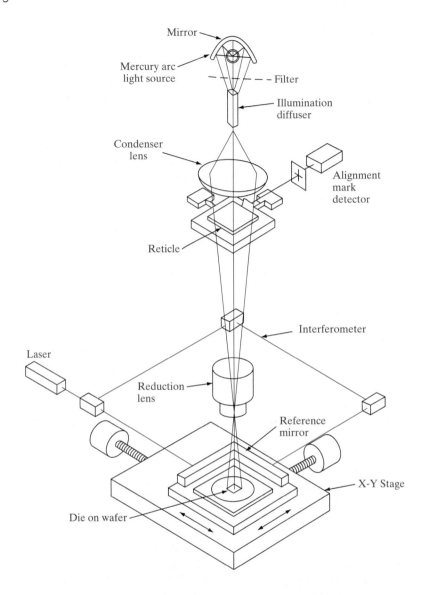

mask). It is usually created by a computer controlled electron beam driven by the circuit layout data, using pattern generation software. A thin layer of electron beam sensitive material called electron beam resist is placed on the iron-oxide–covered quartz plate, and the resist is exposed by the electron beam. A resist is a thin organic polymer layer that undergoes chemical changes if it is exposed to energetic particles such as electrons or photons. The resist is exposed selectively, corresponding to the patterns that are required.

After exposure, the resist is *developed* in a chemical solution. There are two types of resist. The developer is either used to remove the exposed (*positive* resist) or unexposed (*negative* resist) material. The iron oxide layer is then selectively etched off in a plasma to generate the appropriate patterns. The reticle can be used repeatedly to pattern Si wafers. To make a typical integrated circuit, a dozen or more reticles are required, corresponding to different process steps.

The Si wafers are first covered with an ultraviolet light-sensitive organic material or photoemulsion called *photoresist* by dispensing the liquid resist onto the wafer and spinning it rapidly (~3000 rpm) to form a uniform coating (~0.5 μm). As mentioned above, there are two types of resist —negative, which forms the opposite polarity image on the wafer compared to that on the reticle, and positive (same polarity). Currently, positive resist has supplanted negative because it can achieve far better resolution, down to ~0.25 μm using ultraviolet light from mercury lamps (wavelength ~0.365 μm). The light shines on the resist-covered wafer through the reticle, causing the exposed regions to become acidified. Subsequently, the exposed wafers are developed in a basic solution of NaOH, which causes the exposed resist to etch away. Thereby, the pattern on the reticle is transferred to the die on the wafer. After the remaining resist is cured by baking at ~125°C in order to harden it, the appropriate process step can be performed, such as implanting dopants through windows in the resist pattern or plasma etching of the underlying layers.

The exposure of the wafers is achieved die-by-die in a step-and-repeat system called a *stepper* (Fig. 5–7b). As the name implies, the ultraviolet light shines selectively through the reticle onto a single die location. After the photoexposure is done, the wafer mechanically translates on a precisely controlled *x-y* translation stage to the next die location and is exposed again. It is very important to be able to precisely align the patterns on the reticle with respect to pre-existing patterns on the wafer, which is why these tools are also sometimes known as *mask aligners*. An advantage of such a "stepper" projection system is that re-focusing and realignment can be done at each die to accommodate slight variations in surface flatness across the wafer. This is especially important in printing ultra-small linewidths over a very large wafer. The success of modern IC manufacture has depended on numerous advances in deep ultraviolet light sources, precision optical projection systems, techniques for registration between masking layers, and stepper design.

What makes photolithography (along with etching) so critical is that it obviously determines how small and closely packed the individual devices (e.g. transistors) can be made. We shall see that smaller devices operate better in terms of higher speed and lower power dissipation. What makes modern lithography so challenging is the fact that pattern dimensions are comparable to the wavelength of light that is used. Under these circumstances we cannot treat light propagation using simple geometrical ray optics;

rather, the wave nature of light is manifested in terms of diffraction, which makes it harder to control the patterns. The *diffraction-limited minimum geometry* is given by

$$l_{min} = 0.8\ \lambda/NA \qquad (5\text{--}2a)$$

where λ is the wavelength of the light and NA (~0.5) is the numerical aperture or "size" of the lens used in the aligner. This expression implies that for finer patterns, we should work with larger (and, therefore, more expensive) lenses and shorter wavelengths. As a result, smaller geometries require shorter wavelengths. This has led the push to replace UV mercury lamp sources (0.365 µm) with argon fluoride (ArF) excimer lasers (0.193 µm), or extreme ultraviolet (EUV) sources (0.154 µm). Novel exposure techniques employing phase shifting and Fourier optics allow resolutions near the dimension of the wavelength being used. For a common ultraviolet wavelength of 0.365 µm, for example, one may achieve linewidths of about 0.25 µm. An ArF laser can be used for 0.15 µm linewidths.

The other key parameter in lithography is the so-called *depth-of-focus* (DOF), which is given by

$$DOF = \frac{\lambda}{2\ (NA)^2} \qquad (5\text{--}2b)$$

The DOF tells us the range of distances around the focal plane where the image quality is sharp. Unfortunately, this expression implies that exposure with very short wavelengths leads to poor DOF. This is a big challenge because the topography or the "hills and valleys" on a chip during processing can be larger than the DOF allowed by the optics.

We must therefore add steps in the fabrication process to planarize the surface using *chemical mechanical polishing* (CMP). As the name implies, the planarizing process is partly chemical in nature (using a basic solution), and partly mechanical grinding of the layers using an abrasive slurry. As described in Section 1.3.3, CMP can be achieved using a slurry of fine SiO_2 particles in an NaOH solution.

The expression for diffraction-limited geometry (Eq. 5–2a) explains why there is so much interest in X-ray and electron beam lithography. The de Broglie theorem states that the wavelength of a particle varies inversely with its momentum:

$$\lambda = \frac{h}{p} \qquad (5\text{--}2c)$$

Thus more massive particles or energetic photons should be considered to achieve shorter wavelengths. Viable candidates for this application are electrons, ions, or X-rays. For example, electron beams are easily generated, focused, and deflected. The basic technology for this process has been developed over many years in scanning electron microscopy. Since a 10-keV electron has a wavelength of about 0.1 Å, the linewidth limits become the size of the focused beam and its interaction with the photoresist layer. It is possible to achieve linewidths of 0.1 μm by direct electron beam writing on the wafer photoresist. Furthermore, the computer-controlled electron beam exposure requires no masks. This capability allows extremely dense packing of circuit elements on the chip, but direct writing of complex patterns is slow.

Because of the time required for electron beam wafer exposure, it is usually advantageous to use electron beam writing to make the reticle (Fig. 5–7a), but to expose the wafer photoresist using photons. In addition to the advances in deep ultraviolet sources mentioned previously, X-rays offer the promise of even smaller dimensions. For example, if a heavy metal is used in the mask, X-rays ($\lambda \sim 1$ Å) can be used to expose the wafer with 0.1 μm resolution. A particularly high flux of X-rays can be obtained from the synchrotron radiation emitted by electrons accelerated in a storage ring or synchrotron.

5.1.7 Etching

After the photoresist pattern is formed, it can be used as a mask to etch the material underneath. In the early days of Si technology, etching was done using wet chemicals. For example, dilute HF can be used to etch SiO_2 layers grown on a Si substrate with excellent *selectivity*. The term selectivity here refers to the fact that HF attacks SiO_2, but does not affect the Si substrate underneath or the photoresist mask. Although many wet etches are selective, they are unfortunately *isotropic*, which means that they etch as fast laterally as they etch vertically. This is unacceptable for ultra-small features. Hence, wet etching has been largely supplanted by dry, plasma-based etching which can be made both selective and *anisotropic* (etches vertically but not laterally along the surface). In modern IC processing the main use of wet chemical processing is in cleaning the wafers.

Plasmas are ubiquitous in IC processing. The most popular type of plasma-based etching is known as *reactive ion etching* (RIE) (Fig. 5–8). In a typical process, appropriate etch gases such as chlorofluorocarbons (CFCs) flow

Figure 5–8
Reactive ion etcher. Single or multiple wafers are placed on the rf powered cathode to maximize the ion bombardment. Shown in the figure is a simple *diode* etcher in which we have just two electrodes. We can also use a third electrode to supply rf power separately to the etch gases in a *triode* etcher. The most commonly used rf frequency is 13.56 MHz, which is a frequency dedicated to industrial use so that there is minimal interference with radio communications.

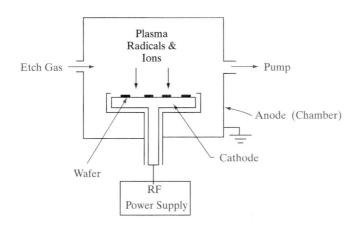

into the chamber at reduced pressure (~1–100 mTorr), and a plasma is struck by applying an rf voltage across a cathode and an anode. The rf voltage accelerates the light electrons in the system to much higher kinetic energies (~10 eV) than the heavier ions. The high energy electrons collide with neutral atoms and molecules to create ions and molecular fragments called radicals. The wafers are held on the rf powered cathode, while the grounded chamber walls act as the anode. From a study of plasma physics, we can show that although the bulk of the plasma is a highly conducting, equi-potential region, less conducting *sheath* regions form next to the two electrodes. It can also be shown that the sheath voltage next to the cathode can be increased by making the (powered) cathode smaller in area than the (grounded) anode. A high d-c voltage (~100–1000 V) develops across the sheath next to the rf powered cathode, such that positive ions gain kinetic energy by being accelerated in this region, and bombard the wafer normal to the surface. This bombardment at normal incidence contributes a physical component to the etch that makes it anisotropic. Physical etching, however, is rather unselective. Simultaneously, the highly reactive radicals in the system give rise to a chemical etch component that is very selective, but not anisotropic. The result is that RIE achieves a good compromise between anisotropy and selectivity, and has become the mainstay of modern IC etch technology.

5.1.8 Metallization

After the semiconductor devices are made by the processing methods described previously, they have to be connected to each other, and ultimately to the IC package, by metallization. Metal films are generally deposited by a physical vapor deposition technique such as evaporation (e.g., Au on GaAs) or sputtering (e.g., Al on Si). Sputtering of Al is achieved by immersing an Al target (typically alloyed with ~1% Si and ~4% Cu to improve the electrical and metallurgical properties of the Al, as described in Section 9.3.1) in an Ar plasma. Argon ions bombard the Al and physically dislodge Al atoms by momentum transfer (Fig. 5–9). Many of the Al atoms ejected from the target deposit on the Si wafers held in close proximity to the target. The Al is then patterned using the metallization reticle and subsequently etched by RIE. Finally, it is sintered at ~450°C for ~30 minutes to form a good electrical, ohmic contact to the Si.

After the interconnection metallization is complete, a protective overcoat of silicon nitride is deposited using plasma-enhanced CVD. Then the individual integrated circuits can be separated by sawing or by scribing and breaking the wafer. The final steps of the process are mounting individual devices in appropriate packages and connecting leads to the Al contact regions. Very precise lead bonders are available for bonding Au or Al wire (about one thousandth of an inch in diameter) to the device and then to the package leads. This phase of device fabrication is called back-end processing, and is discussed in more detail in Chapter 9.

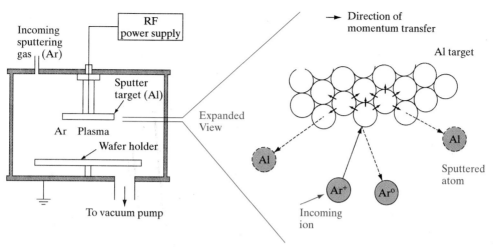

Figure 5–9
Aluminum sputtering by Ar⁺ ions. The Ar⁺ ions with energies of ~1–3 keV physically dislodge Al atoms which end up depositing on the Si wafers held in close proximity. The chamber pressures are kept low such that the mean free path of the ejected Al atoms is long compared to the target-to-wafer separation.

The main steps in making p-n junctions using some of these unit processes are illustrated in Fig. 5–10. Similarly, we will discuss how the key semiconductor devices are made using these same unit processes in subsequent chapters.

In this chapter we wish to develop both a useful mathematical description of the p-n junction and a strong qualitative understanding of its properties. There must be some compromise in these two goals, since a complete mathematical treatment would obscure the essentially simple physical features of junction operation, while a completely qualitative description would not be useful in making calculations. The approach, therefore, will be to describe the junction mathematically while neglecting small effects which add little to the basic solution. In Section 5.6 we shall include several deviations from the simple theory.

    The mathematics of p-n junctions is greatly simplified for the case of the *step junction*, which has uniform p doping on one side of a sharp junction and uniform n doping on the other side. This model represents epitaxial junctions quite well; diffused or implanted junctions, however, are actually *graded* ($N_d - N_a$ varies over a significant distance on either side of the junction). After the basic ideas of junction theory are explored for the step junction, we can make the appropriate corrections to extend the theory to the graded junction. In these discussions we shall assume one-dimensional current flow in samples of uniform cross-sectional area.

**5.2
EQUILIBRIUM
CONDITIONS**

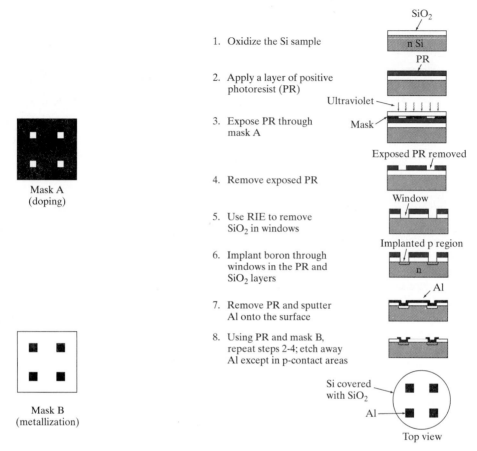

1. Oxidize the Si sample

2. Apply a layer of positive photoresist (PR)

3. Expose PR through mask A

4. Remove exposed PR

Mask A (doping)

5. Use RIE to remove SiO₂ in windows

6. Implant boron through windows in the PR and SiO₂ layers

7. Remove PR and sputter Al onto the surface

8. Using PR and mask B, repeat steps 2-4; etch away Al except in p-contact areas

Mask B (metallization)

Figure 5–10
Simplified description of steps in the fabrication of p-n junctions. For simplicity, only four diodes per wafer are shown, and the relative thicknesses of the oxide, PR, and the Al layers are exaggerated.

In this section we investigate the properties of the step junction at equilibrium (i.e., with no external excitation and no net currents flowing in the device). We shall find that the difference in doping on each side of the junction causes a potential difference between the two types of material. This is a reasonable result, since we would expect some charge transfer because of diffusion between the p material (many holes) and the n material (many electrons). In addition, we shall find that there are four components of current which flow across the junction due to the drift and diffusion of electrons and holes. These four components combine to give zero net current for the equilibrium case. However, the application of bias to the junction increases some of these current components with respect to others, giving net current flow. If we understand the nature of these four current components, a sound view of p-n junction operation, with or without bias, will follow.

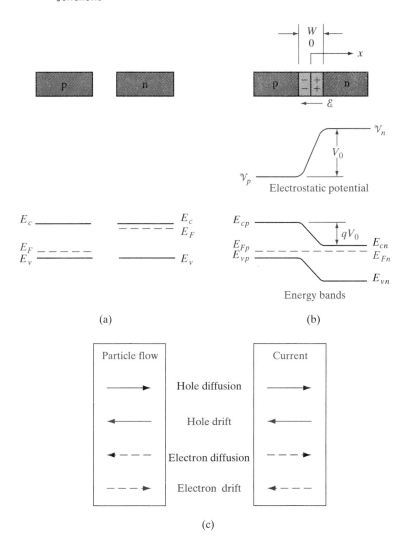

(a)

(b)

(c)

Figure 5–11
Properties of an equilibrium p-n junction: (a) isolated, neutral regions of p-type and n-type material and energy bands for the isolated regions; (b) junction, showing space charge in the transition region W, the resulting electric field ℰ and contact potential V_0, and the separation of the energy bands; (c) directions of the four components of particle flow within the transition region, and the resulting current directions.

5.2.1 The Contact Potential

Let us consider separate regions of p- and n-type semiconductor material, brought together to form a junction (Fig. 5–11). This is not a practical way of forming a device, but this "thought experiment" does allow us to discover the requirements of equilibrium at a junction. Before they are joined, the n material has a large concentration of electrons and few holes, whereas the converse is true for the p material. Upon joining the two regions (Fig. 5–11), we expect diffusion of carriers to take place because of the large carrier concentration gradients at the junction. Thus holes diffuse from the p side into the n side, and electrons diffuse from n to p. The resulting diffusion current cannot build up indefinitely, however, because an opposing electric field is created at the junction (Fig. 5–11b). If the two regions were boxes of red air molecules and green molecules (perhaps due to appropriate types of pollution), eventually there would be a homogeneous mixture of the two after the boxes were joined. This cannot occur in the case of the charged particles in a p-n junction because of the development of space charge and the electric field $\mathcal{E}$. If we consider that electrons diffusing from n to p leave behind uncompensated[3] donor ions (N_d^+) in the n material, and holes leaving the p region leave behind uncompensated acceptors (N_a^-), it is easy to visualize the development of a region of positive space charge near the n side of the junction and negative charge near the p side. The resulting electric field is directed from the positive charge toward the negative charge. Thus $\mathcal{E}$ is in the direction opposite to that of diffusion current for each type of carrier (recall electron current is opposite to the direction of electron flow). Therefore, the field creates a drift component of current from n to p, opposing the diffusion current (Fig. 5–11c).

Since we know that no *net* current can flow across the junction at equilibrium, the current due to the drift of carriers in the $\mathcal{E}$ field must exactly cancel the diffusion current. Furthermore, since there can be no net buildup of electrons or holes on either side as a function of time, the drift and diffusion currents must cancel for *each* type of carrier.

$$J_p(\text{drift}) + J_p(\text{diff.}) = 0 \qquad (5\text{–}3\text{a})$$

$$J_n(\text{drift}) + J_n(\text{diff.}) = 0 \qquad (5\text{–}3\text{b})$$

Therefore, the electric field $\mathcal{E}$ builds up to the point where the net current is zero at equilibrium. The electric field appears in some region W about the junction, and there is an equilibrium potential difference V_0 across W. In the electrostatic potential diagram of Fig. 5–11b, there is a gradient in potential in the direction opposite to $\mathcal{E}$, in accordance with the fundamental relation[4] $\mathcal{E}(x) = -d\mathcal{V}(x)/dx$. We assume the electric field is zero in the neutral regions outside W. Thus there is a constant potential $\mathcal{V}_n$ in the neutral n material, a constant $\mathcal{V}_p$ in the neutral p material, and a potential difference $V_0 = \mathcal{V}_n - \mathcal{V}_p$ between the two. The region W is called the *transition region*,[5] and the potential difference V_0 is called the *contact potential*. The contact potential appearing across W is a *built-in* potential barrier, in that it is necessary to the maintenance

[3]We recall that neutrality is maintained in the bulk materials of Fig. 5–11a by the presence of one electron for each ionized donor ($n = N_d^+$) in the n material and one hole for each ionized acceptor ($p = N_a^-$) in the p material (neglecting minority carriers). Thus, if electrons leave n, some of the positive donor ions near the junction are left uncompensated, as in Fig. 5–11b. The donors and acceptors are fixed in the lattice, in contrast to the mobile electrons and holes.

[4]When we write $\mathcal{E}(x)$, we refer to the value of $\mathcal{E}$ as computed in the x-direction. This value will of course be negative, since it is directed opposite to the true direction of $\mathcal{E}$ as shown in Fig. 5–11b.

[5]Other names for this region are the *space charge region*, since space charge exists within W while neutrality is maintained outside this region, and the *depletion region*, since W is almost depleted of carriers compared with the rest of the crystal. The contact potential V_0 is also called the *diffusion potential*, since it represents a potential barrier which diffusing carriers must surmount in going from one side of the junction to the other.

of equilibrium at the junction; it does not imply any external potential. Indeed, the contact potential cannot be measured by placing a voltmeter across the devices, because new contact potentials are formed at each probe, just canceling V_0. By definition V_0 is an equilibrium quantity, and no net current can result from it.

The contact potential separates the bands as in Fig. 5–11b; the valence and conduction energy bands are higher on the p side of the junction than on the n side[6] by the amount qV_0. The separation of the bands at equilibrium is just that required to make the Fermi level constant throughout the device. We discussed the lack of spatial variation of the Fermi level at equilibrium in Section 3.5. Thus if we know the band diagram, including E_F, for each separate material (Fig. 5–11a), we can find the band separation for the junction at equilibrium simply by drawing a diagram such as Fig. 5–11b with the Fermi levels aligned.

To obtain a quantitative relationship between V_0 and the doping concentrations on each side of the junction, we must use the requirements for equilibrium in the drift and diffusion current equations. For example, the drift and diffusion components of the hole current just cancel at equilibrium:

$$ J_p(x) = q\left[\mu_p p(x)\mathscr{E}(x) - D_p\frac{dp(x)}{dx} \right] = 0 \qquad (5\text{–}4a) $$

This equation can be rearranged to obtain

$$ \frac{\mu_p}{D_p}\mathscr{E}(x) = \frac{1}{p(x)}\frac{dp(x)}{dx} \qquad (5\text{–}4b) $$

where the x-direction is arbitrarily taken from p to n. The electric field can be written in terms of the gradient in the potential, $\mathscr{E}(x) = -d\mathcal{V}(x)/dx$, so that Eq. (5–4b) becomes

$$ -\frac{q}{kT}\frac{d\mathcal{V}(x)}{dx} = \frac{1}{p(x)}\frac{dp(x)}{dx} \qquad (5\text{–}5) $$

with the use of the Einstein relation for μ_p/D_p. This equation can be solved by integration over the appropriate limits. In this case we are interested in the potential on either side of the junction, $\mathcal{V}_p$ and $\mathcal{V}_n$, and the hole concentration just at the edge of the transition region on either side, p_p and p_n. For a step junction it is reasonable to take the electron and hole concentration in the neutral regions outside the transition region as their equilibrium values. Since we have assumed a one-dimensional geometry, p and $\mathcal{V}$ can be taken reasonably as functions of x only. Integration of Eq. (5–5) gives

---

[6]The electron energy diagram of Fig. 5–11b is related to the electrostatic potential diagram by $-q$, the negative charge on the electron. Since $\mathcal{V}_n$ is a higher potential than $\mathcal{V}_p$ by the amount V_0, the electron energies on the n side are *lower* than those on the p side by qV_0.

$$-\frac{q}{kT}\int_{\mathcal{V}_p}^{\mathcal{V}_n} d\mathcal{V} = \int_{p_p}^{p_n} \frac{1}{p} dp$$

$$-\frac{q}{kT}(\mathcal{V}_n - \mathcal{V}_p) = \ln p_n - \ln p_p = \ln\frac{p_n}{p_p} \qquad (5\text{--}6)$$

The potential difference $\mathcal{V}_n - \mathcal{V}_p$ is the contact potential V_0 (Fig. 5–11b). Thus we can write V_0 in terms of the equilibrium hole concentrations on either side of the junction:

$$V_0 = \frac{kT}{q} \ln \frac{p_p}{p_n} \qquad (5\text{--}7)$$

If we consider the step junction to be made up of material with N_a acceptors/cm^3 on the p side and a concentration of N_d donors on the n side, we can write Eq. (5–7) as

$$V_0 = \frac{kT}{q} \ln \frac{N_a}{n_i^2/N_d} = \frac{kT}{q} \ln \frac{N_a N_d}{n_i^2} \qquad (5\text{--}8)$$

by considering the majority carrier concentration to be the doping concentration on each side.

Another useful form of Eq. (5–7) is

$$\frac{p_p}{p_n} = e^{qV_0/kT} \qquad (5\text{--}9)$$

By using the equilibrium condition $p_p n_p = n_i^2 = p_n n_n$, we can extend Eq. (5–9) to include the electron concentrations on either side of the junction:

$$\boxed{\frac{p_p}{p_n} = \frac{n_n}{n_p} = e^{qV_0/kT}} \qquad (5\text{--}10)$$

This relation will be very valuable in calculation of the I–V characteristics of the junction.

EXAMPLE 5–1

An abrupt Si p-n junction has $N_a = 10^{17}$ cm^{-3} on the p side and $N_d = 10^{16}$ cm^{-3} on the n side. At 300 K, (a) calculate the Fermi levels, draw an equilibrium band diagram and find V_0 from the diagram; (b) compare the result from (a) with V_0 calculated from Eq. (5–8).

SOLUTION

(a) Find E_F on each side

$$E_{ip} - E_F = kT \ln\frac{p_p}{n_i} = 0.0259 \ln \frac{10^{17}}{(1.5 \times 10^{10})} = \textbf{0.407 eV}$$

$$E_F - E_{in} = kT \ln\frac{n_n}{n_i} = 0.0259 \ln \frac{10^{16}}{(1.5 \times 10^{10})} = \textbf{0.347 eV}$$

$$qV_0 = 0.407 + 0.347 = \textbf{0.754 eV}$$

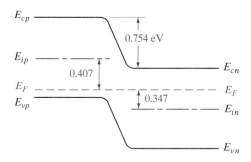

(b) Find V_0 from Eq. (5–8)

$$qV_0 = kT \ln \frac{N_a N_d}{n_i^2} = 0.0259 \ln \frac{10^{33}}{2.25 \times 10^{20}} = \textbf{0.754 eV}$$

5.2.2 Equilibrium Fermi Levels

We have observed that the Fermi level must be constant throughout the device at equilibrium. This observation can be easily related to the results of the previous section. Since we have assumed that p_n and p_p are given by their equilibrium values outside the transition region, we can write Eq. (5–9) in terms of the basic definitions of these quantities using Eq. (3–19):

$$\frac{p_p}{p_n} = e^{qV_0/kT} = \frac{N_v e^{-(E_{Fp} - E_{vp})/kT}}{N_v e^{-(E_{Fn} - E_{vn})/kT}} \qquad (5\text{–}11a)$$

$$e^{qV_0/kT} = e^{(E_{Fn} - E_{Fp})/kT} e^{(E_{vp} - E_{vn})/kT} \qquad (5\text{–}11b)$$

$$qV_0 = E_{vp} - E_{vn} \qquad (5\text{–}12)$$

The Fermi level and valence band energies are written with subscripts to indicate the p side and the n side of the junction.

From Fig. 5–11b the energy bands on either side of the junction are separated by the contact potential V_0 times the electronic charge q; thus the energy difference $E_{vp} - E_{vn}$ is just qV_0. Equation (5–12) results from the fact that the Fermi levels on either side of the junction are equal at equilibrium ($E_{Fn} - E_{Fp} = 0$). When bias is applied to the junction, the potential barrier is raised or lowered from the value of the contact potential, and the Fermi levels on either side of the junction are shifted with respect to each other by an energy in electron volts numerically equal to the applied voltage in volts.

5.2.3 Space Charge at a Junction

Within the transition region, electrons and holes are in transit from one side of the junction to the other. Some electrons diffuse from n to p, and some are swept by the electric field from p to n (and conversely for holes); there are, however, very few carriers within the transition region at any given time, since the electric field serves to sweep out carriers which have wandered into W. To a good approximation, we can consider the space charge within the transition region as due only to the uncompensated donor and acceptor ions. The charge density within W is plotted in Fig. 5–12b. Neglecting carriers within the space charge region, the charge density on the n side is just q

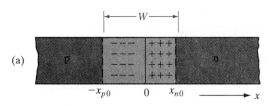

Figure 5–12
Space charge and electric field distribution within the transition region of a p-n junction with $N_d > N_a$: (a) the transition region, with $x = 0$ defined at the metallurgical junction; (b) charge density within the transition region, neglecting the free carriers; (c) the electric field distribution, where the reference direction for $\mathscr{E}$ is arbitrarily taken as the +x-direction.

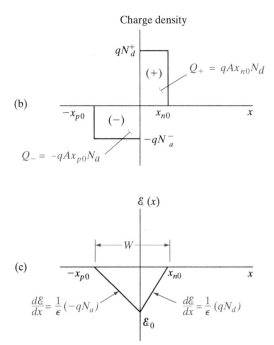

times the concentration of donor ions N_d, and the negative charge density on the p side is $-q$ times the concentration of acceptors N_a. The assumption of carrier depletion within W and neutrality outside W is known as the *depletion approximation*.

Since the dipole about the junction must have an equal number of charges on either side,[7] $(Q_+ = |Q_-|)$, the transition region may extend into the p and n regions unequally, depending on the relative doping of the two sides. For example, if the p side is more lightly doped than the n side $(N_a < N_d)$, the space charge region must extend farther into the p material than into the n, to "uncover" an equivalent amount of charge. For a sample of cross-sectional area A, the total uncompensated charge on either side of the junction is

$$qAx_{p0}N_a = qAx_{n0}N_d \qquad (5\text{--}13)$$

where x_{p0} is the penetration of the space charge region into the p material, and x_{n0} is the penetration into n. The total width of the transition region (W) is the sum of x_{p0} and x_{n0}.

To calculate the electric field distribution within the transition region, we begin with *Poisson's equation*, which relates the gradient of the electric field to the local space charge at any point x:

$$\frac{d\mathscr{E}(x)}{dx} = \frac{q}{\epsilon}(p - n + N_d^+ - N_a^-) \qquad (5\text{--}14)$$

This equation is greatly simplified within the transition region if we neglect the contribution of the carriers $(p - n)$ to the space charge. With this approximation we have two regions of constant space charge:

$$\frac{d\mathscr{E}}{dx} = \frac{q}{\epsilon}N_d, \quad 0 < x < x_{n0} \qquad (5\text{--}15a)$$

$$\frac{d\mathscr{E}}{dx} = -\frac{q}{\epsilon}N_a, \quad -x_{p0} < x < 0 \qquad (5\text{--}15b)$$

assuming complete ionization of the impurities $(N_d^+ = N_d$, and $(N_a^- = N_a)$. We can see from these two equations that a plot of $\mathscr{E}(x)$ vs. x within the transition region has two slopes, positive ($\mathscr{E}$ increasing with x) on the n side and negative ($\mathscr{E}$ becoming more negative as x increases) on the p side. There is some maximum value of the field $\mathscr{E}_0$ at $x = 0$ (the metallurgical junction between the p and n materials), and $\mathscr{E}(x)$ is everywhere negative within the transition region (Fig. 5–12c). These conclusions come from Gauss's law, but we could predict the qualitative features of Fig. 5–12 without equations. We

---

[7]A simple way of remembering this equal charge requirement is to note that electric flux lines must begin and end on charges of opposite sign. Therefore, if Q_+ and Q_- were not of equal magnitude, the electric field would not be contained within W but would extend farther into the p or n regions until the enclosed charges became equal.

expect the electric field $\mathcal{E}(x)$ to be negative throughout W, since we know that the $\mathcal{E}$ field actually points in the $-x$-direction, from n to p (i.e., from the positive charges of the transition region dipole toward the negative charges). The electric field is assumed to go to zero at the edges of the transition region, since we are neglecting any small $\mathcal{E}$ field in the neutral n or p regions. Finally, there must be a maximum $\mathcal{E}_0$ at the junction, since this point is between the charges Q_+ and Q_- on either side of the transition region. All the electric flux lines pass through the $x = 0$ plane, so this is the obvious point of maximum electric field.

The value of $\mathcal{E}_0$ can be found by integrating either part of Eq. (5–15) with appropriate limits (see Fig. 5–12c in choosing the limits of integration).

$$\int_{\mathcal{E}_0}^{0} d\mathcal{E} = \frac{q}{\epsilon} N_d \int_{0}^{x_{n0}} dx, \qquad 0 < x < x_{n0} \qquad (5\text{–}16a)$$

$$\int_{0}^{\mathcal{E}_0} d\mathcal{E} = -\frac{q}{\epsilon} N_a \int_{-x_{p0}}^{0} dx, \qquad -x_{p0} < x < 0 \qquad (5\text{–}16b)$$

Therefore, the maximum value of the electric field is

$$\mathcal{E}_0 = -\frac{q}{\epsilon} N_d x_{n0} = -\frac{q}{\epsilon} N_a x_{p0} \qquad (5\text{–}17)$$

It is simple to relate the electric field to the contact potential V_0, since the $\mathcal{E}$ field at any x is the negative of the potential gradient at that point:

$$\mathcal{E}(x) = -\frac{d\mathcal{V}(x)}{dx} \quad \text{or} \quad -V_0 = \int_{-x_{p0}}^{x_{n0}} \mathcal{E}(x)dx \qquad (5\text{–}18)$$

Thus the negative of the contact potential is simply the area under the $\mathcal{E}(x)$ vs. x triangle. This relates the contact potential to the width of the depletion region:

$$V_0 = -\frac{1}{2}\mathcal{E}_0 W = \frac{1}{2}\frac{q}{\epsilon}N_d x_{n0} W \qquad (5\text{–}19)$$

Since the balance of charge requirement is $x_{n0}N_d = x_{p0}N_a$, and W is simply $x_{p0} + x_{n0}$, we can write $x_{n0} = WN_a/(N_a + N_d)$ in Eq. (5–19):

$$V_0 = \frac{1}{2}\frac{q}{\epsilon}\frac{N_a N_d}{N_a + N_d}W^2 \qquad (5\text{–}20)$$

By solving for W, we have an expression for the width of the transition region in terms of the contact potential, the doping concentrations, and known constants q and ϵ.

$$W = \left[\frac{2\epsilon V_0}{q}\left(\frac{N_a + N_d}{N_a N_d}\right)\right]^{1/2} = \left[\frac{2\epsilon V_0}{q}\left(\frac{1}{N_a} + \frac{1}{N_d}\right)\right]^{1/2} \quad (5\text{--}21)$$

There are several useful variations of Eq. (5–21); for example, V_0 can be written in terms of the doping concentrations with the aid of Eq. (5–8):

$$W = \left[\frac{2\epsilon kT}{q^2}\left(\ln\frac{N_a N_d}{n_i^2}\right)\left(\frac{1}{N_a} + \frac{1}{N_d}\right)\right]^{1/2} \quad (5\text{--}22)$$

We can also calculate the penetration of the transition region into the n and p materials:

$$x_{p0} = \frac{W N_d}{N_a + N_d} = \frac{W}{1 + N_a/N_d} = \left\{\frac{2\epsilon V_0}{q}\left[\frac{N_d}{N_a(N_a + N_d)}\right]\right\}^{1/2} \quad (5\text{--}23a)$$

$$x_{n0} = \frac{W N_a}{N_a + N_d} = \frac{W}{1 + N_d/N_a} = \left\{\frac{2\epsilon V_0}{q}\left[\frac{N_a}{N_d(N_a + N_d)}\right]\right\}^{1/2} \quad (5\text{--}23b)$$

As expected, Eqs. (5–23) predict that the transition region extends farther into the side with the lighter doping. For example, if $N_a \ll N_d$, x_{p0} is large compared with x_{n0}. This agrees with our qualitative argument that a deep penetration is necessary in lightly doped material to "uncover" the same amount of space charge as for a short penetration into heavily doped material.

Another important result of Eq. (5–21) is that the transition width W varies as the square root of the potential across the region. In the derivation to this point, we have considered only the equilibrium contact potential V_0. In Section 5.3 we shall see that an applied voltage can increase or decrease the potential across the transition region by aiding or opposing the equilibrium electric field. Therefore, Eq. (5–21) predicts that an applied voltage will increase or decrease the width of the transition region as well.

Boron is implanted into an n-type Si sample ($N_d = 10^{16}$ cm^{-3}), forming an abrupt junction of square cross section, with area = 2×10^{-3} cm^2. Assume that the acceptor concentration in the p-type region is $N_a = 4 \times 10^{18}$ cm^{-3}. Calculate V_0, x_{n0}, x_{p0}, Q_+, and $\mathscr{E}_0$ for this junction at equilibrium (300 K). Sketch $\mathscr{E}(x)$ and charge density to scale, as in Fig. 5–12.   **EXAMPLE 5–2**

From Eq. (5–8),   **SOLUTION**

$$V_0 = \frac{kT}{q}\ln\frac{N_a N_d}{n_i^2} = 0.0259\ln\frac{4 \times 10^{34}}{2.25 \times 10^{20}}$$

$$= 0.0259\ln(1.78 \times 10^{14}) = 0.85\ V$$

From Eq. (5–21),

$$W = \left[\frac{2\epsilon V_0}{q}\left(\frac{1}{N_a} + \frac{1}{N_d}\right)\right]^{1/2}$$

$$= \left[\frac{2(11.8 \times 8.85 \times 10^{-14})(0.85)}{1.6 \times 10^{-19}}(0.25 \times 10^{-18} + 10^{-16})\right]^{1/2}$$

$$= 3.34 \times 10^{-5} \text{ cm} = 0.334 \text{ μm}$$

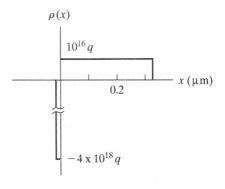

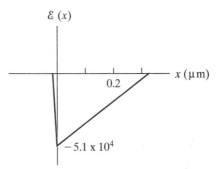

From Eq. (5–23),

$$x_{n0} = \frac{3.34 \times 10^{-5}}{1 + 0.0025} \simeq 0.333 \text{ μm}$$

$$x_{p0} = \frac{3.34 \times 10^{-5}}{1 + 400} \simeq 8.3 \times 10^{-8} \text{ cm} = 8.3 \text{ Å}$$

Note that $x_{n0} \simeq W$.

$$Q_+ = -Q_- = qAx_{n0}N_d = (1.6 \times 10^{-19})(2 \times 10^{-3})(3.33 \times 10^{-5})(10^{16})$$

$$= 1.07 \times 10^{-10} \, C$$

$$\mathscr{E}_0 = \frac{-qN_d x_{n0}}{\epsilon} = \frac{-(1.6 \times 10^{-19})(10^{16})(3.3 \times 10^{-5})}{(11.8)(8.85 \times 10^{-14})}$$

$$= -5.1 \times 10^4 \, V/cm$$

One useful feature of a p-n junction is that current flows quite freely in the p to n direction when the p region has a positive external voltage bias relative to n (forward bias and forward current), whereas virtually no current flows when p is made negative relative to n (reverse bias and reverse current). This asymmetry of the current flow makes the p-n junction diode very useful as a *rectifier*. While rectification is an important application, it is only the beginning of a host of uses for the biased junction. Biased p-n junctions can be used as voltage-variable capacitors, photocells, light emitters, and many more devices which are basic to modern electronics. Two or more junctions can be used to form transistors and controlled switches.

   In this section we begin with a qualitative description of current flow in a biased junction. With the background of the previous section, the basic features of current flow are relatively simple to understand, and these qualitative concepts form the basis for the analytical description of forward and reverse currents in a junction.

**5.3
FORWARD- AND
REVERSE-BIASED
JUNCTIONS;
STEADY STATE
CONDITIONS**

5.3.1 Qualitative Description of Current Flow at a Junction

We assume that an applied voltage bias V appears across the transition region of the junction rather than in the neutral n and p regions. Of course, there will be some voltage drop in the neutral material, if a current flows through it. But in most p-n junction devices, the length of each region is small compared with its area, and the doping is usually moderate to heavy; thus the resistance is small in each neutral region, and only a small voltage drop can be maintained outside the space charge (transition) region. For almost all calculations it is valid to assume that an applied voltage appears entirely across the transition region. We shall take V to be positive when the external bias is positive on the p side relative to the n side.

Since an applied voltage changes the electrostatic potential barrier and thus the electric field within the transition region, we would expect changes in the various components of current at the junction (Fig. 5–13). In addition, the separation of the energy bands is affected by the applied bias, along with the width of the depletion region. Let us begin by examining qualitatively the effects of bias on the important features of the junction.

The *electrostatic potential barrier* at the junction is lowered by a forward bias V_f from the equilibrium contact potential V_0 to the smaller value $V_0 - V_f$. This lowering of the potential barrier occurs because a forward bias (p positive with respect to n) raises the electrostatic potential on the p side relative to the n side. For a reverse bias ($V = -V_r$) the opposite occurs; the electrostatic potential of the p side is depressed relative to the n side, and the potential barrier at the junction becomes larger ($V_0 + V_r$).

The *electric field* within the transition region can be deduced from the potential barrier. We notice that the field decreases with forward bias, since the applied electric field opposes the built-in field. With reverse bias the field at the junction is increased by the applied field, which is in the same direction as the equilibrium field.

The change in electric field at the junction calls for a change in the *transition region width W*, since it is still necessary that a proper number of positive and negative charges (in the form of uncompensated donor and acceptor ions) be exposed for a given value of the $\mathscr{E}$ field. Thus we would expect the width W to decrease under forward bias (smaller $\mathscr{E}$, fewer uncompensated charges) and to increase under reverse bias. Equations (5–21) and (5–23) can be used to calculate W, x_{p0}, and x_{n0} if V_0 is replaced by the new barrier height[8] $V_0 - V$.

The *separation of the energy bands* is a direct function of the electrostatic potential barrier at the junction. The height of the electron energy barrier is simply the electronic charge q times the height of the electrostatic potential barrier. Thus the bands are separated less $[q(V_0 - V_f)]$ under forward bias than at equilibrium, and more $[q(V_0 + V_r)]$ under reverse bias. We assume the Fermi level deep inside each neutral region is essentially the equilibrium value (we shall return to this assumption later); therefore, the shifting of the energy bands under bias implies a separation of the Fermi levels on either side of the junction, as mentioned in Section 5.2.2. Under forward bias, the Fermi level on the n side E_{Fn} is above E_{Fp} by the energy qV_f; for reverse bias, E_{Fp} is qV_r joules higher than E_{Fn}. *In energy units of electron volts, the Fermi levels in the two neutral regions are separated by an energy (eV) numerically equal to the applied voltage (V).*

---

[8]With bias applied to the junction, the 0 in the subscripts of x_{n0} and x_{p0} does not imply equilibrium. Instead, it signifies the origin of a new set of coordinates, $x_n = 0$ and $x_p = 0$, as defined later in Fig. 5–15.

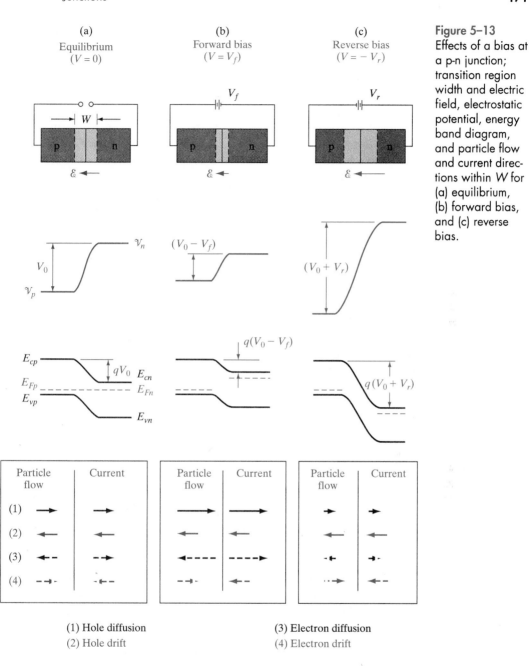

Figure 5–13
Effects of a bias at a p-n junction; transition region width and electric field, electrostatic potential, energy band diagram, and particle flow and current directions within W for (a) equilibrium, (b) forward bias, and (c) reverse bias.

(1) Hole diffusion   (3) Electron diffusion
(2) Hole drift   (4) Electron drift

The *diffusion current* is composed of majority carrier electrons on the n side surmounting the potential energy barrier to diffuse to the p side, and holes surmounting their barrier from p to n.[9] There is a distribution of energies for electrons in the n-side conduction band (Fig. 3–16), and some electrons in the high-energy "tail" of the distribution have enough energy to diffuse from n to p at equilibrium in spite of the barrier. With forward bias, however, the barrier is lowered (to $V_0 - V_f$), and many more electrons in the n-side conduction band have sufficient energy to diffuse from n to p over the smaller barrier. Therefore, the electron diffusion current can be quite large with forward bias. Similarly, more holes can diffuse from p to n under forward bias because of the lowered barrier. For reverse bias the barrier becomes so large ($V_0 + V_r$) that virtually no electrons in the n-side conduction band or holes in the p-side valence band have enough energy to surmount it. Therefore, the diffusion current is usually negligible for reverse bias.

The *drift current* is relatively insensitive to the height of the potential barrier. This sounds strange at first, since we normally think in terms of material with ample carriers, and therefore we expect drift current to be simply proportional to the applied field. The reason for this apparent anomaly is the fact that the drift current is limited *not* by *how fast* carriers are swept down the barrier, *but* rather *how often*. For example, minority carrier electrons on the p side which wander into the transition region will be swept down the barrier by the $\mathscr{E}$ field, giving rise to the electron component of drift current. However, this current is small not because of the size of the barrier, but because there are very few minority electrons in the p side to participate. Every electron on the p side which diffuses to the transition region will be swept down the potential energy hill, whether the hill is large or small. The electron drift current does not depend on how fast an individual electron is swept from p to n, but rather on how many electrons are swept down the barrier per second. Similar comments apply regarding the drift of minority holes from the n side to the p side of the junction. To a good approximation, therefore, the electron and hole drift currents at the junction are independent of the applied voltage.

The supply of minority carriers on each side of the junction required to participate in the drift component of current is generated by thermal excitation of electron–hole pairs. For example, an EHP created near the junction on the p side provides a minority electron in the p material. If the EHP is generated within a diffusion length L_n of the transition region, this electron can diffuse to the junction and be swept down the barrier to the n side. The resulting current due to drift of generated carriers across the junction is commonly called the *generation current* since its magnitude depends entirely on

[9]Remember that the potential energy barriers for electrons and holes are directed oppositely. The barrier for electrons is apparent from the energy band diagram, which is always drawn for electron energies. For holes, the potential energy barrier at the junction has the same shape as the electrostatic potential barrier (the conversion factor between electrostatic potential and hole energy is +q). A simple check of these two barrier directions can be made by asking the directions in which carriers are swept by the $\mathscr{E}$ field within the transition region—a hole is swept in the direction of $\mathscr{E}$, from n to p (swept down the potential "hill" for holes); an electron is swept opposite to $\mathscr{E}$, from p to n (swept down the potential energy "hill" for electrons).

the rate of generation of EHPs. As we shall discuss later, this generation current can be increased greatly by optical excitation of EHPs near the junction (the p-n junction *photodiode*).

The *total current* crossing the junction is composed of the sum of the diffusion and drift components. As Fig. 5–13 indicates, the electron and hole diffusion currents are both directed from p to n (although the particle flow directions are opposite to each other), and the drift currents are from n to p. The *net* current crossing the junction is zero at equilibrium, since the drift and diffusion components cancel for each type of carrier (the equilibrium electron and hole components need not be equal, as in Fig. 5–13, as long as the net hole current and the net electron current are each zero). Under reverse bias, both diffusion components are negligible because of the large barrier at the junction, and the only current is the relatively small (and essentially voltage-independent) generation current from n to p. This generation current is shown in Fig. 5–14, in a sketch of a typical *I–V* plot for a p-n junction. In this figure the positive direction for the current *I* is taken from p to n, and the applied voltage *V* is positive when the positive battery terminal is connected to p and the negative terminal to n. The only current flowing in this p-n junction diode for negative *V* is the small current *I*(gen.) due to carriers generated in the transition region or minority carriers which diffuse to the junction and are collected. The current at $V = 0$ (equilibrium) is zero since the generation and diffusion currents cancel:[10]

$$I = I(\text{diff.}) - |I(\text{gen.})| = 0 \quad \text{for } V = 0 \qquad (5\text{–}24)$$

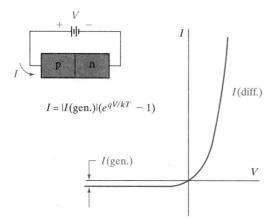

Figure 5–14
I–V characteristic of a p-n junction.

---

[10]The total current *I* is the sum of the generation and diffusion components. However, these components are oppositely directed, *I*(diff.) being positive and *I*(gen.) being negative for the chosen reference direction. To avoid confusion of signs, we use here the magnitude of the drift current |*I*(gen.)| and include its negative sign in Eq. (5–24). Thus when we write the term –|*I*(gen.)|, there is no doubt that the generation current is in the negative current direction. This approach emphasizes the fact that the two components of current add with opposite signs to give the total current.

As we shall see in the next section, an applied forward bias $V = V_f$ increases the probability that a carrier can diffuse across the junction, by the factor $\exp(qV_f/kT)$. Thus the diffusion current under forward bias is given by its equilibrium value multiplied by $\exp(qV/kT)$; similarly, for reverse bias the diffusion current is the equilibrium value reduced by the same factor, with $V = -V_r$. Since the equilibrium diffusion current is equal in magnitude to $|I(\text{gen.})|$, the diffusion current with applied bias is simply $|I(\text{gen.})|$ $\exp(qV/kT)$. The total current I is then the diffusion current minus the absolute value of the generation current, which we will now refer to as I_0:

$$I = I_0(e^{qV/kT} - 1) \tag{5-25}$$

In Eq. (5–25) the applied voltage V can be positive or negative, $V = V_f$ or $V = -V_r$. When V is positive and greater than a few kT/q ($kT/q = 0.0259\ \text{V}$ at room temperature), the exponential term is much greater than unity. The current thus increases exponentially with forward bias. When V is negative (reverse bias), the exponential term approaches zero and the current is $-I_0$, which is in the n to p (negative) direction. This negative generation current is also called the *reverse saturation current*. The striking feature of Fig. 5–14 is the nonlinearity of the *I–V* characteristic. Current flows relatively freely in the forward direction of the diode, but almost no current flows in the reverse direction.

5.3.2 Carrier Injection

From the discussion in the previous section, we expect the minority carrier concentration on each side of a p-n junction to vary with the applied bias because of variations in the diffusion of carriers across the junction. The equilibrium ratio of hole concentrations on each side

$$\frac{p_p}{p_n} = e^{qV_0/kT} \tag{5-26}$$

becomes with bias (Fig. 5–13)

$$\frac{p(-x_{p0})}{p(x_{n0})} = e^{q(V_0 - V)/kT} \tag{5-27}$$

This equation uses the altered barrier $V_0 - V$ to relate the steady state hole concentrations on the two sides of the transition region with either forward or reverse bias (V positive or negative). For low-level injection we can neglect changes in the majority carrier concentrations. Although the absolute increase of the majority carrier concentration is equal to the increase of the minority carrier concentration in order to maintain space charge neutrality, the relative change in majority carrier concentration can be assumed to vary only slightly with bias compared with equilibrium values. With this simplification we can write the ratio of Eq. (5–26) to (5–27) as

$$\frac{p(x_{n0})}{p_n} = e^{qV/kT} \quad \text{taking } p(-x_{p0}) = p_p \qquad (5\text{–}28)$$

With forward bias, Eq. (5–28) suggests a greatly increased minority carrier hole concentration at the edge of the transition region on the n side $p(x_{n0})$ than was the case at equilibrium. Conversely, the hole concentration $p(x_{n0})$ under reverse bias (V negative) is reduced below the equilibrium value p_n. The exponential increase of the hole concentration at x_{n0} with forward bias is an example of *minority carrier injection*. As Fig. 5–15 suggests, a forward bias V results in a steady state injection of excess holes into the n region and electrons into the p region. We can easily calculate the excess hole concentration Δp_n at the edge of the transition region x_{n0} by subtracting the equilibrium hole concentration from Eq. (5–28),

$$\Delta p_n = p(x_{n0}) - p_n = p_n(e^{qV/kT} - 1) \qquad (5\text{–}29)$$

and similarly for excess electrons on the p side,

$$\Delta n_p = n(-x_{p0}) - n_p = n_p(e^{qV/kT} - 1) \qquad (5\text{–}30)$$

From our study of diffusion of excess carriers in Section 4.4.4, we expect that injection leading to a steady concentration of Δp_n excess holes at x_{n0} will produce a *distribution* of excess holes in the n material. As the holes diffuse deeper into the n region, they recombine with electrons in the n material, and the resulting excess hole distribution is obtained as a solution of the diffusion equation, Eq. (4–34b). If the n region is long compared with the hole diffusion length L_p, the solution is exponential, as in Eq. (4–36). Similarly, the injected electrons in the p material diffuse and recombine, giving an exponential distribution of excess electrons. For convenience, let us define two new coordinates (Fig. 5–15): Distances measured in the x-direction in the n material from x_{n0} will be designated x_n; distances in the p material measured in the $-x$-direction with $-x_{p0}$ as the origin will be called x_p. This convention will simplify the mathematics considerably. We can write the diffusion equation as in Eq. (4–34) for each side of the junction and solve for the distributions of excess carriers (δn and δp) assuming long p and n regions:

$$\delta n(x_p) = \Delta n_p e^{-x_p/L_n} = n_p(e^{qV/kT} - 1)e^{-x_p/L_n} \qquad (5\text{–}31\text{a})$$

$$\delta p(x_n) = \Delta p_n e^{-x_n/L_p} = p_n(e^{qV/kT} - 1)e^{-x_n/L_p} \qquad (5\text{–}31\text{b})$$

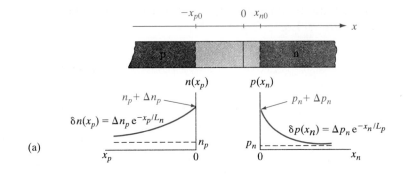

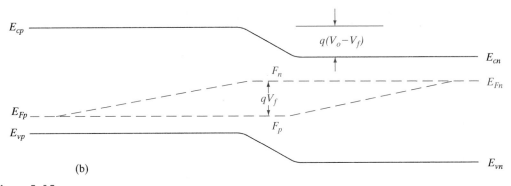

Figure 5–15
Forward-biased junction: (a) minority carrier distributions on the two sides of the transition region and definitions of distances x_n and x_p measured from the transition region edges; (b) variation of the quasi-Fermi levels with position.

The hole diffusion current at any point x_n, in the n material can be calculated from Eq. (4–40):

$$I_p(x_n) = -qAD_p \frac{d\delta p(x_n)}{dx_n} = qA\frac{D_p}{L_p}\Delta p_n e^{-x_n/L_p} = qA\frac{D_p}{L_p}\delta p(x_n) \quad (5\text{--}32)$$

where A is the cross-sectional area of the junction. Thus the hole diffusion current at each position x_n is proportional to the excess hole concentration at that point.[11] The total hole current injected into the n material at the junction can be obtained simply by evaluating Eq. (5–32) at x_{n0}:

$$I_p(x_n = 0) = \frac{qAD_p}{L_p}\Delta p_n = \frac{qAD_p}{L_p}p_n(e^{qV/kT} - 1) \quad (5\text{--}33)$$

[11]With carrier injection due to bias, it is clear that the equilibrium Fermi levels cannot be used to describe carrier concentrations in the device. It is necessary to use the concept of quasi-Fermi levels, taking into account the spatial variations of the carrier concentrations.

By a similar analysis, the injection of electrons into the p material leads to an electron current at the junction of

$$I_n(x_p = 0) = -\frac{qAD_n}{L_n}\Delta n_p = -\frac{qAD_n}{L_n}n_p(e^{qV/kT} - 1) \qquad (5\text{-}34)$$

The minus sign in Eq. (5–34) means that the electron current is opposite to the x_p-direction; that is, the true direction of I_n is in the +x-direction, adding to I_p in the total current (Fig. 5–16). If we neglect recombination in the transition region, which is known as the Shockley ideal diode approximation, we can consider that each injected electron reaching $-x_{p0}$ must pass through x_{n0}. Thus the total diode current I at x_{n0} can be calculated as the sum of $I_p(x_n = 0)$ and $-I_n(x_p = 0)$. If we take the +x-direction as the reference direction for the total current I, we must use a minus sign with $I_n(x_p)$ to account for the fact that x_p is defined in the $-x$-direction:

$$I = I_p(x_n = 0) - I_n(x_p = 0) = \frac{qAD_p}{L_p}\Delta p_n + \frac{qAD_n}{L_n}\Delta n_p \qquad (5\text{-}35)$$

$$I = qA\left(\frac{D_p}{L_p}p_n + \frac{D_n}{L_n}n_p\right)(e^{qV/kT} - 1) = I_0(e^{qV/kT} - 1) \qquad (5\text{-}36)$$

Equation (5–36) is the *diode equation*, having the same form as the qualitative relation Eq. (5–25). Nothing in the derivation excludes the possibility that the bias voltage V can be negative; thus the diode equation describes the total current through the diode for either forward or reverse bias. We can calculate the current for reverse bias by letting $V = -V_r$:

$$I = qA\left(\frac{D_p}{L_p}p_n + \frac{D_n}{L_n}n_p\right)(e^{-qV_r/kT} - 1) \qquad (5\text{-}37a)$$

If V_r is larger than a few kT/q, the total current is just the reverse saturation current

$$I = -qA\left(\frac{D_p}{L_p}p_n + \frac{D_n}{L_n}n_p\right) = -I_0 \qquad (5\text{-}37b)$$

One implication of Eq. (5–36) is that the total current at the junction is dominated by injection of carriers from the more heavily doped side into the side with lesser doping. For example, if the p material is very heavily doped and the n region is lightly doped, the minority carrier concentration on the p side (n_p) is negligible compared with the minority carrier concentration on the n side (p_n). Thus the diode equation can be approximated by injection of holes only, as in Eq. (5–33). This means that the charge stored in

Figure 5–16
Two methods for calculating junction current from the excess minority carrier distributions: (a) diffusion currents at the edges of the transition region; (b) charge in the distributions divided by the minority carrier lifetimes; (c) the diode equation.

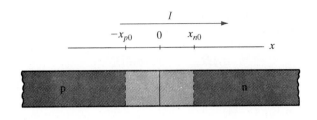

(a)

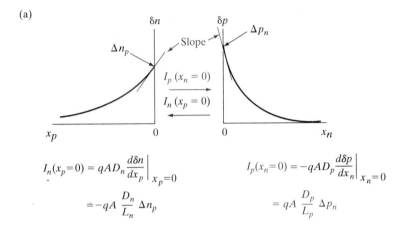

$$I_n(x_p=0) = qAD_n \frac{d\delta n}{dx_p}\bigg|_{x_p=0}$$

$$\dot{=} -qA\frac{D_n}{L_n}\Delta n_p$$

$$I_p(x_n=0) = -qAD_p \frac{d\delta p}{dx_n}\bigg|_{x_n=0}$$

$$= qA\frac{D_p}{L_p}\Delta p_n$$

(b)

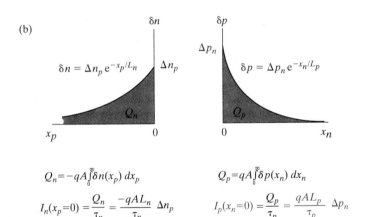

$$Q_n = -qA\int_0^\infty \delta n(x_p)\, dx_p$$

$$I_n(x_p=0) = \frac{Q_n}{\tau_n} = \frac{-qAL_n}{\tau_n}\Delta n_p$$

$$Q_p = qA\int_0^\infty \delta p(x_n)\, dx_n$$

$$I_p(x_n=0) = \frac{Q_p}{\tau_p} = \frac{qAL_p}{\tau_p}\Delta p_n$$

(c)

$$I = I_p(x_n=0) - I_n(x_p=0) = qA\left(\frac{D_p}{L_p}\Delta p_n + \frac{D_n}{L_n}\Delta n_p\right)$$

$$= qA\left(\frac{D_p p_n}{L_p} + \frac{D_n n_p}{L_n}\right)(e^{qV/kT} - 1)$$

the minority carrier distributions is due mostly to holes on the n side. For example, to double the hole current in this p$^+$-n junction one should not double the p$^+$ doping, but rather reduce the n-type doping by a factor of two. This structure is called a p$^+$-n junction, where the + superscript simply means heavy doping. Another characteristic of the p$^+$-n or n$^+$-p structure is that the transition region extends primarily into the lightly doped region, as we found in the discussion of Eq. (5–23). Having one side heavily doped is a useful arrangement for many practical devices, as we shall see in our discussions of switching diodes and transistors. This type of junction is common in devices which are fabricated by counterdoping. For example, an n-type Si sample with $N_d = 10^{14}$ cm^{-3} can be used as the substrate for an implanted or diffused junction. If the doping of the p region is greater than 10^{19} cm^{-3} (typical of diffused junctions), the structure is definitely p$^+$-n, with n_p more than five orders of magnitude smaller than p_n. Since this configuration is common in device technology, we shall return to it in much of the following discussion.

Figure 5–15b shows the quasi-Fermi levels as a function of position for a p-n junction in forward bias. The equilibrium E_F is split into the quasi-Fermi levels, F_n and F_p which are separated within W by an energy qV caused by the applied bias, V. This energy represents the deviation from equilibrium (see Section 4.3.3). In forward bias in the depletion region we thus get

$$pn = n_i^2 e^{(F_n - F_p)/kT} = n_i^2 e^{(qV/kT)} \qquad (5\text{–}38)$$

On either side of the junction, it is the minority carrier quasi-Fermi level that varies the most. The majority carrier concentration is not affected much, so the majority carrier quasi-Fermi level is close to the original E_F. We see that the quasi-Fermi levels are more or less flat within the depletion region, which appears to be inconsistent with what we learned in Section 4.4.6 about the current flow being proportional to the gradient of the quasi-Fermi levels. Keeping in mind that for an ideal diode, the electron (and hole) current is constant across the depletion region, we see that within the depletion region the product of the gradient of the quasi-Fermi level and the carrier concentration must be independent of position. For a given current, the gradient in the quasi-Fermi level must be significant for minority carriers, since the carrier concentration is small (see Eq. 4–52). On the other hand, for majority carriers, very little gradient is needed in the quasi-Fermi level. Within W there is an intermediate situation, where the carrier concentration is changing from majority on one side to minority on the other. Although there is some variation in F_n and F_p within W, it doesn't show up on the scale used in Fig. 5–15. A homework problem with typical values should help clarify the concept (Prob. 5.21). Outside of the depletion regions, the quasi-Fermi levels for the minority carriers vary linearly and eventually merge with the Fermi levels. In contrast, the minority carrier concentrations decay exponentially with distance. In fact it takes many diffusion lengths for the quasi-Fermi level to cross

E_i, where the minority carrier concentration is equal to the intrinsic carrier concentration, let alone approach E_F, where for example $\delta p(x_n) \simeq p_n$.

Another simple and instructive way of calculating the total current is to consider the injected current as supplying the carriers for the excess distributions (Fig. 5–16b). For example, $I_p(x_n = 0)$ must supply enough holes per second to maintain the steady state exponential distribution $\delta p(x_n)$ as the holes recombine. The total positive charge stored in the excess carrier distribution at any instant of time is

$$Q_p = qA \int_0^\infty \delta p(x_n) dx_n = qA\Delta p_n \int_0^\infty e^{-x_n/L_p} dx_n = qAL_p\Delta p_n \quad (5\text{–}39)$$

The average lifetime of a hole in the n-type material is τ_p. Thus, on the average, this entire charge distribution recombines and must be replenished every τ_p seconds. The injected hole current at $x_n = 0$ needed to maintain the distribution is simply the total charge divided by the average time of replacement:

$$I_p(x_n = 0) = \frac{Q_p}{\tau_p} = qA\frac{L_p}{\tau_p}\Delta p_n = qA\frac{D_p}{L_p}\Delta p_n \quad\quad (5\text{–}40)$$

using $D_p/L_p = L_p/\tau_p$.

This is the same result as Eq. (5–33), which was calculated from the diffusion currents. Similarly, we can calculate the negative charge stored in the distribution $\delta n(x_p)$ and divide by τ_n to obtain the injected electron current in the p material. This method, called the *charge control approximation*, illustrates the important fact that the minority carriers injected into either side of a p-n junction diffuse into the neutral material and recombine with the majority carriers. The minority carrier current [for example, $I_p(x_n)$] decreases exponentially with distance into the neutral region. Thus several diffusion lengths away from the junction, most of the total current is carried by the majority carriers. We shall discuss this point in more detail later in this section.

In summary, we can calculate the current at a p-n junction in two ways (Fig. 5–16): (a) from the slopes of the excess minority carrier distributions at the two edges of the transition regions and (b) from the steady state charge stored in each distribution. We add the hole current injected into the n material $I_p(x_n = 0)$ to the electron current injected into the p material $I_n(x_p = 0)$, after including a minus sign with $I_n(x_p)$ to conform with the conventional definition of positive current in the +x-direction. We are able to add these two currents because of the assumption that no recombination takes place within the transition region. Thus we effectively have the total electron and hole current at one point in the device (x_{n0}). Since the total current must be constant throughout the device (despite variations in the current components), I as described by Eq. (5–36) is the total current at every position x in the diode.

The drift of minority carriers can be neglected in the neutral regions *outside W*, because the minority carrier concentration is small compared with that of the majority carriers. If the minority carriers contribute to the total current at all, their contribution must be through diffusion (dependent on the *gradient* of the carrier concentration). Even a very small concentration of minority carriers can have an appreciable effect on the current if the spatial variation is large.

Calculation of the majority carrier currents in the two neutral regions is simple, once we have found the minority carrier current. Since the total current I must be constant throughout the device, the majority carrier component of current at any point is just the difference between I and the minority component (Fig. 5–17). For example, since $I_p(x_n)$ is proportional to the excess hole concentration at each position in the n material [Eq. (5–32)], it decreases exponentially in x_n with the decreasing $\delta p(x_n)$. Thus the electron component of current must increase appropriately with x_n to maintain the total current I. Far from the junction, the current in the n material is carried almost entirely by electrons. The physical explanation of this is that electrons must flow in from the n material (and ultimately from the negative terminal of the battery), to resupply electrons lost by recombination in the excess hole distribution near the junction. The electron current $I_n(x_n)$ includes sufficient

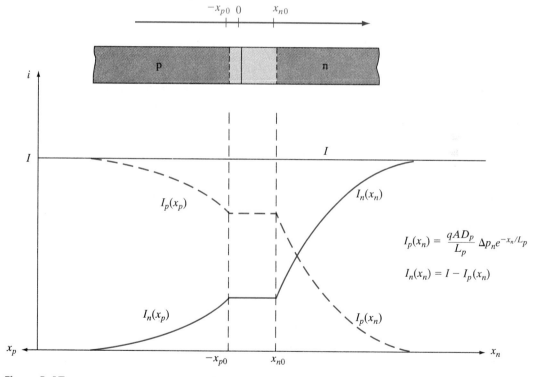

Figure 5–17
Electron and hole components of current in a forward-biased p-n junction. In this example, we have a higher injected minority hole current on the n-side than electron current on the p side because we have a lower n doping than p doping.

electron flow to supply not only recombination near x_{n0}, but also injection of electrons into the p region. Of course, the flow of electrons in the n material toward the junction constitutes a current in the +x-direction, contributing to the total current I.

One question that still remains to be answered is whether the majority carrier current is due to drift or diffusion or both, at different points in the diode. Near the junction (just outside of the depletion regions) the majority carrier concentration changes by exactly the same amount as minority carriers in order to maintain space charge neutrality. The majority carrier concentration can change rather fast, in a very short time scale known as the dielectric relaxation time, τ_D $(=\rho\epsilon)$, where ρ is the resistivity and ϵ is the dielectric constant. The relaxation time τ_D is the analog of the RC time constant in a circuit. Very far away from the junction (more than 3 to 5 diffusion lengths), the minority carrier concentration decays to a low, constant background value. Hence, the majority carrier concentration also becomes independent of position. Here, clearly the only possible current component is majority carrier drift current. When approaching the junction there is a spatially varying majority (and minority) carrier concentration and the majority carrier current changes from pure drift to drift and diffusion, although drift always dominates for majority carriers except in cases of very high levels of injection. Throughout the diode, the total current due to majority and minority carriers at any cross section is kept constant.

We thus note that the electric field in the neutral regions cannot be zero as we previously assumed; otherwise, there would be no drift currents. Thus our assumption that all of the applied voltage appears across the transition region is not completely accurate. On the other hand, the majority carrier concentrations are usually large in the neutral regions, so that only a small field is needed to drive the drift currents. Thus the assumption that junction voltage equals applied voltage is acceptable for most calculations.

EXAMPLE 5–3   Find an expression for the electron current in the n-type material of a forward-biased p-n junction.

SOLUTION   The total current is

$$I = qA\left(\frac{D_p}{L_p}p_n + \frac{D_n}{L_n}n_p\right)(e^{qV/kT} - 1)$$

The hole current on the n side is

$$I_p(x_n) = qA\frac{D_p}{L_p}p_n e^{-x_n/L_p}(e^{qV/kT} - 1)$$

Thus the electron current in the n material is

$$I_n(x_n) = I - I_p(x_n) = qA\left[\frac{D_p}{L_p}(1 - e^{-x_n/L_p})p_n + \frac{D_n}{L_n}n_p\right](e^{qV/kT} - 1)$$

This expression includes the supplying of electrons for recombination with the injected holes, and the injection of electrons across the junction into the p side.

5.3.3 Reverse Bias

In our discussion of carrier injection and minority carrier distributions, we have primarily assumed forward bias. The distributions for reverse bias can be obtained from the same equations (Fig. 5–18), if a negative value of V is introduced. For example, if $V = -V_r$ (p negatively biased with respect to n), we can approximate Eq. (5–29) as

$$\Delta p_n = p_n(e^{q(-V_r)/kT} - 1) \simeq -p_n \quad \text{for } V_r \gg kT/q \qquad (5\text{–}41)$$

and similarly $\Delta n_p \simeq -n_p$.

Thus for a reverse bias of more than a few tenths of a volt, the minority carrier concentration at each edge of the transition region becomes essentially zero as the excess concentration approaches the negative of the equilibrium concentration. The excess minority carrier concentrations in the neutral regions are still given by Eq. (5–31), so that depletion of carriers below the equilibrium values extends approximately a diffusion length beyond each side of the transition region. This reverse-bias depletion of minority carriers can be thought of as *minority carrier extraction*, analogous to the injection of forward bias. Physically, extraction occurs because minority carriers at the edges of the depletion region are swept down the barrier at the junction to the other side and are not replaced by an opposing diffusion of carriers. For example, when holes at x_{n0} are swept across the junction to the p side by the $\mathscr{E}$ field, a gradient in the hole distribution in the n material exists, and holes in the n region diffuse toward the junction. The steady state hole distribution in the n region has the inverted exponential shape of Fig. 5–18a. It is important to remember that although the reverse saturation current occurs at the junction by drift of carriers down the barrier, this current is fed from each side by diffusion toward the junction of minority carriers in the neutral regions. The rate of carrier drift across the junction (reverse saturation current) depends on the rate at which holes arrive at x_{n0} (and electrons at x_{p0}) by diffusion from the neutral material. These minority carriers are supplied by thermal generation, and we can show that the expression for the reverse saturation

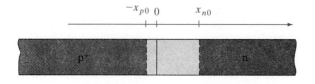

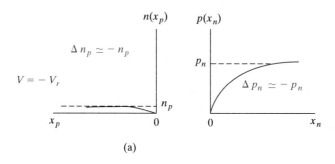

(a)

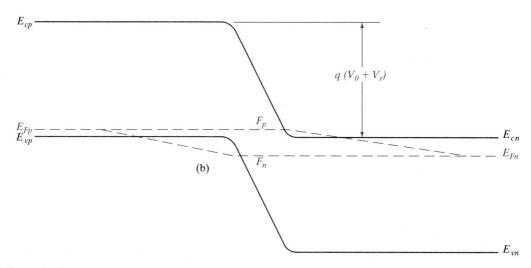

(b)

Figure 5–18
Reverse-biased p-n junction: (a) minority carrier distributions near the reverse-biased junction; (b) variation of the quasi-Fermi levels.

current, Eq. (5–38), represents the rate at which carriers are generated thermally within a diffusion length of each side of the transition region.

EXAMPLE 5–4

Consider a volume of n-type material of area A, with a length of one hole diffusion length L_p. The rate of thermal generation of holes within the volume is

$$AL_p \frac{p_n}{\tau_p} \quad \text{since } g_{\text{th}} = \alpha_r n_i^2 = \alpha_r n_n p_n = \frac{p_n}{\tau_p}$$

Assume that each thermally generated hole diffuses out of the volume before it can recombine. The resulting hole current is $I = qAL_p p_n/\tau_p$, which is the same as the saturation current for a p^+-n junction. We conclude that saturation current is due to the collection of minority carriers thermally generated within a diffusion length of the junction.

In reverse bias, the quasi-Fermi levels split in the opposite sense than in forward bias (Fig. 5–18b). The F_n moves farther away from E_c (close to E_v) and F_p moves farther away from E_v, reflecting the fact that in reverse bias we have fewer carriers than in equilibrium, unlike the forward bias case where we have an excess of carriers. In reverse bias, in the depletion region, we have

$$pn = n_i^2 e^{(F_n - F_p)/kT} \approx 0 \qquad (5\text{–}42)$$

It is interesting to note that the quasi-Fermi levels in reverse bias can go inside the bands. For example, F_p goes inside the conduction band on the n-side of the depletion region. However, we must remember that F_p is a measure of the hole concentration, and should be correlated with the valence band edge, E_v, and not with E_c. Hence, the band diagram simply reflects the fact that we have very few holes in this region, even fewer than the already small equilibrium minority carrier hole concentration (Fig. 5–18a). Similar observations can be made about the electrons.

We have found that a p-n junction biased in the reverse direction exhibits a small, essentially voltage-independent saturation current. This is true until a critical reverse bias is reached, for which *reverse breakdown* occurs (Fig. 5–19). At this critical voltage (V_{br}) the reverse current through the diode increases sharply, and relatively large currents can flow with little further increase in voltage. The existence of a critical breakdown voltage introduces almost a right-angle appearance to the reverse characteristic of most diodes.

There is nothing inherently destructive about reverse breakdown. If the current is limited to a reasonable value by the external circuit, the p-n junction

**5.4
REVERSE-BIAS
BREAKDOWN**

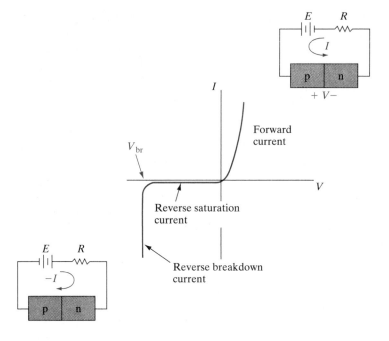

Figure 5–19
Reverse break-
down in a p-n
junction.

can be operated in reverse breakdown as safely as in the forward-bias condition. For example, the maximum reverse current which can flow in the device of Fig. 5–19 is $(E - V_{br})/R$; the series resistance R can be chosen to limit the current to a safe level for the particular diode used. If the current is not limited externally, the junction can be damaged by excessive reverse current, which overheats the device as the maximum power rating is exceeded. It is important to remember, however, that such destruction of the device is not necessarily due to mechanisms unique to reverse breakdown; similar results occur if the device passes excessive current in the forward direction.[12] As we shall see in Section 5.4.4, useful devices called *breakdown diodes* are designed to operate in the reverse breakdown regime of their characteristics.

Reverse breakdown can occur by two mechanisms, each of which requires a critical electric field in the junction transition region. The first mechanism, called the *Zener effect*, is operative at low voltages (up to a few volts reverse bias). If the breakdown occurs at higher voltages (from a few volts to thousands of volts), the mechanism is *avalanche breakdown*. We shall discuss these two mechanisms in this section.

5.4.1 Zener Breakdown

When a heavily doped junction is reverse biased, the energy bands become crossed at relatively low voltages (i.e., the n-side conduction band appears

---

[12]The dissipated power (IV) in the junction is of course greater for a given current in the breakdown regime than would be the case for forward bias, simply because V is greater.

opposite the p-side valence band). As Fig. 5–20 indicates, the crossing of the bands aligns the large number of empty states in the n-side conduction band opposite the many filled states of the p-side valence band. If the barrier separating these two bands is narrow, tunneling of electrons can occur, as discussed in Section 2.4.4. Tunneling of electrons from the p-side valence band to the n-side conduction band constitutes a reverse current from n to p; this is the *Zener effect.*

The basic requirements for tunneling current are a large number of electrons separated from a large number of empty states by a narrow barrier of finite height. Since the tunneling probability depends upon the width of the barrier (d in Fig. 5–20), it is important that the metallurgical junction be sharp and the doping high, so that the transition region W extends only a very short distance from each side of the junction. If the junction is not abrupt, or if either side of the junction is lightly doped, the transition region W will be too wide for tunneling.

As the bands are crossed (at a few tenths of a volt for a heavily doped junction), the tunneling distance d may be too large for appreciable tunneling. However, d becomes smaller as the reverse bias is increased, because the higher electric fields result in steeper slopes for the band edges. This assumes that the transition region width W does not increase appreciably with reverse bias. For low voltages and heavy doping on each side of the junction, this is a good assumption. However, if Zener breakdown does not occur with reverse bias of a few volts, avalanche breakdown will become dominant.

In the simple covalent bonding model (Fig. 3–1), the Zener effect can be thought of as *field ionization* of the host atoms at the junction. That is, the reverse bias of a heavily doped junction causes a large electric field within W; at a critical field strength, electrons participating in covalent bonds may be torn from the bonds by the field and accelerated to the n side of the junction. The electric field required for this type of ionization is on the order of 10^6 V/cm.

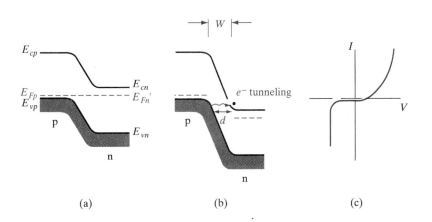

Figure 5–20
The Zener effect: (a) heavily doped junction at equilibrium; (b) reverse bias with electron tunneling from p to n; (c) I–V characteristic.

5.4.2 Avalanche Breakdown

For lightly doped junctions electron tunneling is negligible, and instead, the breakdown mechanism involves the *impact ionization* of host atoms by energetic carriers. Normal lattice-scattering events can result in the creation of EHPs if the carrier being scattered has sufficient energy. For example, if the electric field $\mathscr{E}$ in the transition region is large, an electron entering from the p side may be accelerated to high enough kinetic energy to cause an ionizing collision with the lattice (Fig. 5–21a). A single such interaction results in *carrier multiplication*; the original electron and the generated electron are both swept to the n side of the junction, and the generated hole is swept to

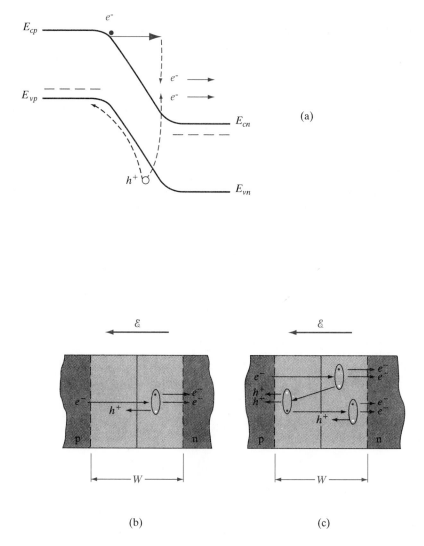

Figure 5–21
Electron-hole pairs created by impact ionization: (a) band diagram of a p-n junciton in reverse bias showing (primary) electron gaining kinetic energy in the field of the depletion region, and creating a (secondary) electron-hole pair by impact ionization, the primary electron losing most of its kinetic energy in the process; (b) a single ionizing collision by an incoming electron in the depletion region of the junction; (c) primary, secondary and tertiary collisions.

the p side (Fig. 5–21b). The degree of multiplication can become very high if carriers generated within the transition region also have ionizing collisions with the lattice. For example, an incoming electron may have a collision with the lattice and create an EHP; each of these carriers has a chance of creating a new EHP, and each of those can also create an EHP, and so forth (Fig. 5–21c). This is an *avalanche* process, since each incoming carrier can initiate the creation of a large number of new carriers.

We can make an approximate analysis of avalanche multiplication by assuming that a carrier of either type has a probability P of having an ionizing collision with the lattice while being accelerated a distance W through the transition region. Thus for n_{in} electrons entering from the p side, there will be Pn_{in} ionizing collisions and an EHP (secondary carriers) for each collision. After the Pn_{in} collisions by the primary electrons, we have the primary plus the secondary electrons, $n_{in}(1 + P)$. After a collision, each EHP moves effectively a distance of W within the transition region. For example, if an EHP is created at the center of the region, the electron drifts a distance $W/2$ to n and the hole $W/2$ to p. Thus the probability that an ionizing collision will occur due to the motion of the secondary carriers is still P in this simplified model. For $n_{in}P$ secondary pairs there will be $(n_{in}P)P$ ionizing collisions and $n_{in}P^2$ tertiary pairs. Summing up the total number of electrons out of the region at n after many collisions, we have

$$n_{out} = n_{in}(1 + P + P^2 + P^3 + \ldots) \qquad (5\text{–}43)$$

assuming no recombination. In a more comprehensive theory we would include recombination as well as different probabilities for ionizing collisions by electrons and holes. In our simple theory, the electron multiplication M_n is

$$M_n = \frac{n_{out}}{n_{in}} = 1 + P + P^2 + P^3 + \cdots = \frac{1}{1 - P} \qquad (5\text{–}44a)$$

as can be verified by direct division. As the probability of ionization P approaches unity, the carrier multiplication (and therefore the reverse current through the junction) increases without limit. Actually, the limit on the current will be dictated by the external circuit.

The relation between multiplication and P was easy to write in Eq. (5–44a); however, the relation of P to parameters of the junction is much more complicated. Physically, we expect the ionization probability to increase with increasing electric field, and therefore to depend on the reverse bias. Measurements of carrier multiplication M in junctions near breakdown lead to an empirical relation

$$M = \frac{1}{1 - (V/V_{br})^{\mathbf{n}}} \qquad (5\text{–}44b)$$

where the exponent $\mathbf{n}$ varies from about 3 to 6, depending on the type of material used for the junction.

Figure 5–22
Variation of
avalanche break-
down voltage in
abrupt p⁺-n junc-
tions, as a func-
tion of donor
concentration on
the n side, for sev-
eral semiconduc-
tors. [After S.M.
Sze and G. Gib-
bons, *Applied
Physics Letters*,
vol. 8, p. 111
(1966).]

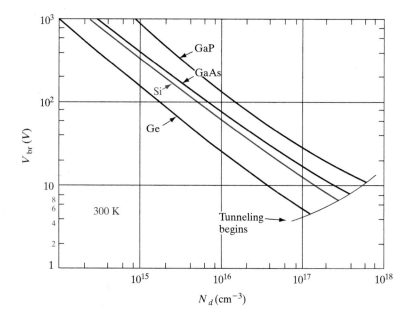

In general, the critical reverse voltage for breakdown increases with the band gap of the material, since more energy is required for an ionizing collision. Also, the peak electric field within W increases with increased doping on the more lightly doped side of the junction. Therefore, V_{br} decreases as the doping increases, as Fig. 5–22 indicates.

5.4.3 Rectifiers

The most obvious property of a p-n junction is its *unilateral* nature; that is, to a good approximation it conducts current in only one direction. We can think of an *ideal diode* as a short circuit when forward biased and as an open circuit when reverse biased (Fig. 5–23a). The p-n junction diode does not quite fit this description, but the I–V characteristics of many junctions can be approximated by the ideal diode in series with other circuit elements to form an equivalent circuit. For example, most forward-biased diodes exhibit an *offset voltage E_0* (see Fig. 5–33), which can be approximated in a circuit model by a battery in series with the ideal diode (Fig. 5–23b). The series battery in the model keeps the ideal diode turned off for applied voltages less than E_0. From Section 5.6.1 we expect E_0 to be approximately the contact potential of the junction. In some cases the approximation to the actual diode characteristic is improved by adding a series resistor R to the circuit equivalent (Fig. 5–23c). The equivalent circuit approximations illustrated in Fig. 5–23 are called *piecewise-linear equivalents*, since the approximate characteristics are linear over specific ranges of voltage and current.

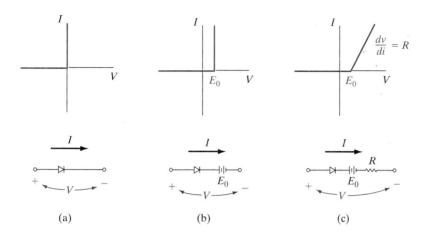

(a)   (b)   (c)

Figure 5–23
Piecewise-linear approximations of junction diode characteristics: (a) the ideal diode; (b) ideal diode with an offset voltage; (c) ideal diode with an offset voltage and a resistance to account for slope in the forward characteristic.

An ideal diode can be placed in series with an a-c voltage source to provide *rectification* of the signal. Since current can flow only in the forward direction through the diode, only the positive half-cycles of the input sine wave are passed. The output voltage is a *half-rectified sine wave*. Whereas the input sinusoid has zero average value, the rectified signal has a positive average value and therefore contains a d-c component. By appropriate filtering, this d-c level can be extracted from the rectified signal.

The unilateral nature of diodes is useful for many other circuit applications that require *waveshaping*. This involves alteration of a-c signals by passing only certain portions of the signal while blocking other portions.

Junction diodes designed for use as rectifiers should have *I–V* characteristics as close as possible to that of the ideal diode. The reverse current should be negligible, and the forward current should exhibit little voltage dependence (negligible *forward resistance R*). The reverse breakdown voltage should be large, and the offset voltage E_0 in the forward direction should be small. Unfortunately, not all of these requirements can be met by a single device; compromises must be made in the design of the junction to provide the best diode for the intended application.

From the theory derived in Section 5.3 we can easily list the various requirements for good rectifier junctions. *Band gap* is obviously an important consideration in choosing a material for rectifier diodes. Since n_i is small for large band gap materials, the reverse saturation current (which depends on thermally generated carriers) decreases with increasing E_g. A rectifier made with a wide band gap material can be operated at higher temperatures, because thermal excitation of EHPs is reduced by the increased band gap. Such temperature effects are critically important in rectifiers, which must carry large currents in the forward direction and are thereby subjected to appreciable heating. On the other hand, the contact potential and offset voltage E_0 generally increase with E_g. This drawback is usually outweighed by the advantages of low n_i; for example, Si is generally preferred over Ge for power rectifiers because of its wider band gap, lower leakage current, and higher breakdown voltage, as well as its more convenient fabrication properties.

The *doping concentration* on each side of the junction influences the avalanche breakdown voltage, the contact potential, and the series resistance of the diode. If the junction has one highly doped side and one lightly doped side (such as a p⁺-n junction), the lightly doped region determines many of the properties of the junction. From Fig. 5–22 we see that a high-resistivity region should be used for at least one side of the junction to increase the breakdown voltage V_{br}. However, this approach tends to increase the forward resistance R of Fig. 5–23c, and therefore contributes to the problems of thermal effects due to I^2R heating. To reduce the resistance of the lightly doped region, it is necessary to make its area large and reduce its length. Therefore, the physical *geometry* of the diode is another important design variable. Limitations on the practical area for a diode include problems of obtaining uniform starting material and junction processing over large areas. Localized flaws in junction uniformity can cause premature reverse breakdown in a small region of the device. Similarly, the lightly doped region of the junction cannot be made arbitrarily short. One of the primary problems with a short, lightly doped region is an effect called *punch-through*. Since the transition region width W increases with reverse bias and extends primarily into the lightly doped region, it is possible for W to increase until it fills the entire length of this region (Prob. 5.33). The result of punch-through is a breakdown below the value of V_{br} expected from Fig. 5–22.

In devices designed for use at high reverse bias, care must be taken to avoid premature breakdown across the edge of the sample. This effect can be reduced by *beveling* the edge or by diffusing a *guard ring* to isolate the junction from the edge of the sample (Fig. 5–24). The electric field is lower at the beveled edge of the sample in Fig. 5–24b than it is in the main body of the device. Similarly, the junction at the lightly doped p guard ring of Fig. 5–24c breaks down at higher voltage than the p⁺-n junction. Since the depletion region is wider in the p ring than in the p⁺ region, the average electric field is smaller at the ring for a given diode reverse voltage.

Figure 5–24 Beveled edge and guard ring to prevent edge breakdown under reverse bias: (a) diode with beveled edge; (b) closeup view of edge, showing reduction of depletion region near the bevel; (c) guard ring.

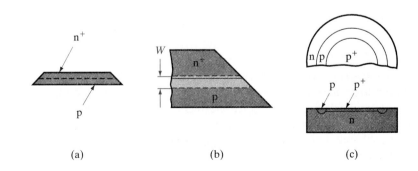

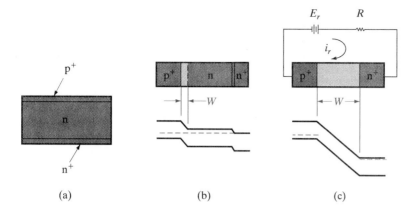

Figure 5–25
A p+-n-n+ junction
diode: (a) device
configuration;
(b) zero-bias
condition;
(c) reverse-
biased to
punch-through.

   In fabricating a p^+-n or a p-n^+ junction, it is common to terminate the lightly doped region with a heavily doped layer of the same type (Fig. 5–25a), to ease the problem of making ohmic contact to the device. The result is a p^+-n-n^+ structure with the p^+-n layer serving as the active junction, or a p^+-p-n^+ device with an active p-n^+ junction. The lightly doped center region determines the avalanche breakdown voltage. If this region is short compared with the minority carrier diffusion length, the excess carrier injection for large forward currents can increase the conductivity of the region significantly. This type of *conductivity modulation*, which reduces the forward resistance R, can be very useful for high-current devices. On the other hand, a short, lightly doped center region can also lead to punch-through under reverse bias, as in Fig. 5–25c.

   The mounting of a rectifier junction is critical to its ability to handle power. For diodes used in low-power circuits, glass or plastic encapsulation or a simple header mounting is adequate. However, high-current devices that must dissipate large amounts of heat require special mountings to transfer thermal energy away from the junction. A typical Si power rectifier is mounted on a molybdenum or tungsten disk to match the thermal expansion properties of the Si. This disk is fastened to a large stud of copper or other thermally conductive material that can be bolted to a heat sink with appropriate cooling.

5.4.4 The Breakdown Diode

As we discussed earlier in this section, the reverse-bias breakdown voltage of a junction can be varied by choice of junction doping concentrations. The breakdown mechanism is the Zener effect (tunneling) for abrupt junctions with extremely heavy doping; however, the more common breakdown is avalanche (impact ionization), typical of more lightly doped or graded junctions. By varying the doping we can fabricate diodes with specific breakdown

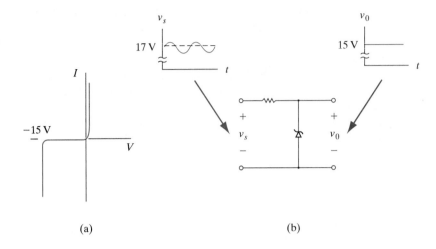

(a)   (b)

voltages ranging from less than one volt to several hundred volts. If the junction is well designed, the breakdown will be sharp and the current after breakdown will be essentially independent of voltage (Fig. 5–26a). When a diode is designed for a specific breakdown voltage, it is called a *breakdown diode*. Such diodes are also called *Zener diodes*, despite the fact that the actual breakdown mechanism is usually the avalanche effect. This error in terminology is due to an early mistake in identifying the first observations of breakdown in p-n junctions.

Breakdown diodes can be used as *voltage regulators* in circuits with varying inputs. The 15-V breakdown diode of Fig. 5–26 holds the circuit output voltage v_0 constant at 15 V, while the input varies at voltages greater than 15 V. For example, if v_s is a rectified and filtered signal composed of a 17-V d-c component and a 1-V ripple variation above and below 17 V, the output v_0 will remain constant at 15 V. More complicated voltage regulator circuits can be designed using breakdown diodes, depending on the type of signal being regulated and the nature of the output load. In a similar application, such a device can be used as a *reference diode*; since the breakdown voltage of a particular diode is known, the voltage across it during breakdown can be used as a reference in circuits that require a known value of voltage.

5.5
TRANSIENT AND
A-C CONDITIONS

We have considered the properties of p-n junctions under equilibrium conditions and with steady state current flow. Most of the basic concepts of junction devices can be obtained from these properties, except for the important behavior of junctions under transient or a-c conditions. Since most solid state devices are used for switching or for processing a-c signals, we cannot claim to understand p-n junctions without knowing at least the basics of time-dependent processes. Unfortunately, a complete analysis of these effects involves more mathematical manipulation than is appropriate for an introductory discussion. Basically, the problem involves solving the various current flow equations in two simultaneous variables, space and time. We can, however, obtain the basic results for several special cases which represent typical time-dependent applications of junction devices.

In this section we investigate the important influence of excess carriers in transient and a-c problems. The switching of a diode from its forward state to its reverse state is analyzed to illustrate a typical transient problem. Finally, these concepts are applied to the case of small a-c signals to determine the equivalent capacitance of a p-n junction.

5.5.1 Time Variation of Stored Charge

Another look at the excess carrier distributions of a p-n junction under bias (e.g., Fig. 5–15) tells us that any change in current must lead to a change of charge stored in the carrier distributions. Since time is required in building up or depleting a charge distribution, however, the stored charge must inevitably lag behind the current in a time-dependent problem. This is inherently a capacitive effect, as we shall see in Section 5.5.4.

For a proper solution of a transient problem, we must use the time-dependent continuity equations, Eqs. (4-31). We can obtain each component of the current at position x and time t from these equations; for example, from Eq. (4–31a) we can write

$$-\frac{\partial J_p(x,t)}{\partial x} = q\frac{\delta p(x,t)}{\tau_p} + q\frac{\partial p(x,t)}{\partial t} \tag{5-45}$$

To obtain the instantaneous current density, we can integrate both sides at time t to obtain

$$J_p(0) - J_p(x) = q\int_0^x\left[\frac{\delta p(x,t)}{\tau_p} + \frac{\partial p(x,t)}{\partial t}\right]dx \tag{5-46}$$

For injection into a long n region from a p^+ region, we can take the current at $x_n = 0$ to be all hole current, and J_p at $x_n = \infty$ to be zero. Then the total injected current, including time variations, is

$$i(t) = i_p(x_n = 0, t) = \frac{qA}{\tau_p}\int_0^\infty \delta p(x_n,t)dx_n + qA\frac{\partial}{\partial t}\int_0^\infty \delta p(x_n,t)dx_n$$

$$\boxed{i(t) = \frac{Q_p(t)}{\tau_p} + \frac{dQ_p(t)}{dt}} \tag{5-47}$$

This result indicates that the hole current injected across the p^+-n junction (and therefore approximately the total diode current) is determined by two charge storage effects: (1) the usual recombination term Q_p/τ_p in which the excess carrier distribution is replaced every τ_p seconds, and (2) a charge buildup (or depletion) term dQ_p/dt, which allows for the fact that the distribution of excess carriers can be increasing or decreasing in a time-dependent problem. For steady state the dQ_p/dt term is zero, and Eq. (5–47) reduces to Eq. (5–40), as expected. In fact, we could have written Eq. (5–47) intuitively rather than having obtained it from the continuity equation, since it is reasonable that the hole current injected at any given time must supply minority carriers for recombination and for whatever variations occur in the total stored charge.

We can solve for the stored charge as a function of time for a given current transient. For example, the step turn-off transient (Fig. 5–27a), in which a current I is suddenly removed at $t = 0$, leaves the diode with stored charge. Since the excess holes in the n region must die out by recombination with the matching excess electron population, some time is required for $Q_p(t)$ to reach zero. Solving Eq. (5–47) with Laplace transforms, with $i(t > 0) = 0$ and $Q_p(0) = I\tau_p$, we obtain

$$0 = \frac{1}{\tau_p} Q_p(s) + s Q_p(s) - I\tau_p$$

$$Q_p(s) = \frac{I\tau_p}{s + 1/\tau_p}$$

$$Q_p(t) = I\tau_p e^{-t/\tau_p} \tag{5–48}$$

As expected, the stored charge dies out exponentially from its initial value $I\tau_p$ with a time constant equal to the hole lifetime in the n material.

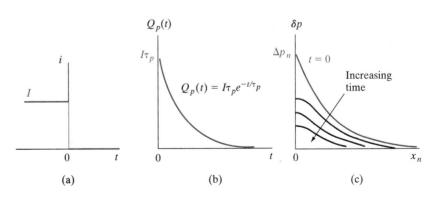

Figure 5–27
Effects of a step turn-off transient in a p^+-n diode: (a) current through the diode; (b) decay of stored charge in the n-region; (c) excess hole distribution in the n-region as a function of time during the transient.

An important implication of Fig. 5–27 is that even though the current is suddenly terminated, the voltage across the junction persists until Q_p disappears. Since the excess hole concentration can be related to junction voltage by formulas derived in Section 5.3.2, we can presumably solve for $v(t)$. We already know that at any time during the transient, the excess hole concentration at $x_n = 0$ is

$$\Delta p_n(t) = p_n(e^{qv(t)/kT} - 1) \qquad (5\text{--}49)$$

so that finding $\Delta p_n(t)$ will easily give us the transient voltage. Unfortunately, it is not simple to obtain $\Delta p_n(t)$ exactly from our expression for $Q_p(t)$. The problem is that the hole distribution does not remain in the convenient exponential form it has in steady state. As Fig. 5–27c suggests, the quantity $\delta p(x_n, t)$ becomes markedly nonexponential as the transient proceeds. For example, since the injected hole current is proportional to the gradient of the hole distribution at $x_n = 0$ (Fig. 5–16a), zero current implies zero gradient. Thus the slope of the distribution must be exactly zero at $x_n = 0$ throughout the transient.[13] This zero slope at the point of injection distorts the exponential distribution, particularly in the region near the junction. As time progresses in Fig. 5–27c, δp (and therefore δn) decreases as the excess electrons and holes recombine. To find the exact expression for $\delta p(x_n, t)$ during the transient would require a rather difficult solution of the time-dependent continuity equation.

An approximate solution for $v(t)$ can be obtained by assuming an exponential distribution for δp at every instant during the decay. This type of *quasi-steady state* approximation neglects distortion due to the slope requirement at $x_n = 0$ and the effects of diffusion during the transient. Thus we would expect the calculation to give rather crude results. On the other hand, such a solution can give us a feeling for the variation of junction voltage during the transient. If we take

$$\delta p(x_n, t) = \Delta p_n(t)e^{-x_n/L_p} \qquad (5\text{--}50)$$

we have for the stored charge at any instant

$$Q_p(t) = qA\int_0^\infty \Delta p_n(t)e^{-x_n/L_p}dx_n = qAL_p\Delta p_n(t) \qquad (5\text{--}51)$$

Relating $\Delta p_n(t)$ to $v(t)$ by Eq. (5–49) we have

$$\Delta p_n(t) = p_n(e^{qv(t)/kT} - 1) = \frac{Q_p(t)}{qAL_p} \qquad (5\text{--}52)$$

[13]We notice that, while the *magnitude* of δp cannot change instantaneously, the *slope* must go to zero immediately. This can occur in a small region near the junction with negligible redistribution of charge at $t = 0$.

Thus in the quasi-steady state approximation, the junction voltage varies according to

$$v(t) = \frac{kT}{q} \ln\left(\frac{I\tau_p}{qAL_pp_n}e^{-t/\tau_p} + 1\right) \tag{5-53}$$

during the turn-off transient of Fig. 5–27. This analysis, while not accurate in its details, does indicate clearly that the voltage across a p-n junction cannot be changed instantaneously, and that stored charge can present a problem in a diode intended for switching applications.

Many of the problems of stored charge can be reduced by designing a p^+-n diode (for example) with a very narrow n region. If the n region is shorter than a hole diffusion length, very little charge is stored. Thus, little time is required to switch the diode on and off. This type of structure, called the *narrow base diode*, is considered in Prob. 5.35. The switching process can be made still faster by purposely adding recombination centers, such as Au atoms in Si, to increase the recombination rate.

5.5.2   Reverse Recovery Transient

In most switching applications a diode is switched from forward conduction to a reverse-biased state, and vice versa. The resulting stored charge transient is somewhat more complicated than for a simple turn-off transient, and therefore it requires slightly more analysis. An important result of this example is that a reverse current much larger than the normal reverse saturation current can flow in a junction during the time required for readjustment of the stored charge.

Let us assume a p^+-n junction is driven by a square wave generator that periodically switches from $+E$ to $-E$ volts (Fig. 5–28a). While E is positive the diode is forward biased, and in steady state the current I_f flows through the junction. If E is much larger than the small forward voltage of the junction, the source voltage appears almost entirely across the resistor, and the current is approximately $i = I_f \simeq E/R$. After the generator voltage is reversed ($t > 0$), the current must initially reverse to $i = I_r \simeq -E/R$. The reason for this unusually large reverse current through the diode is that the stored charge (and hence the junction voltage) cannot be changed instantaneously. Therefore, just as the current is reversed, the junction voltage remains at the small forward-bias value it had before $t = 0$. A voltage loop equation then tells us that the large reverse current $-E/R$ must flow temporarily. While the current is negative through the junction, the slope of the $\delta p(x_n)$ distribution must be positive at $x_n = 0$.

As the stored charge is depleted from the neighborhood of the junction (Fig. 5–28b), we can find the junction voltage again from Eq. (5–49). As long as Δp_n is positive, the junction voltage $v(t)$ is positive and small; thus $i \simeq -E/R$ until Δp_n, goes to zero. When the stored charge is depleted and Δp_n be-

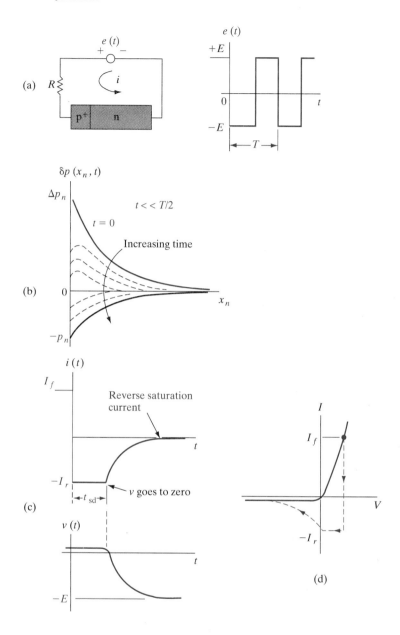

Figure 5–28
Storage delay time in a p⁺-n diode: (a) circuit and input square wave; (b) hole distribution in the n-region as a function of time during the transient; (c) variation of current and voltage with time; (d) sketch of transient current and voltage on the device I–V characteristic.

comes negative, the junction exhibits a negative voltage. Since the reverse-bias voltage of a junction can be large, the source voltage begins to divide between R and the junction. As time proceeds, the magnitude of the reverse current becomes smaller as more of $-E$ appears across the reverse-biased junction, until finally the only current is the small reverse saturation current which is characteristic of the diode. The time t_{sd} required for the stored charge

Figure 5–29
Effects of storage
delay time on
switching signal:
(a) switching volt-
age; (b) diode
current.

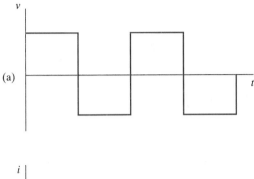

(a)

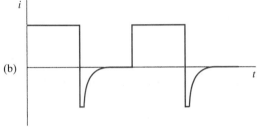

(b)

(and therefore the junction voltage) to become zero is called the *storage delay time.* This delay time is an important figure of merit in evaluating diodes for switching applications. It is usually desirable that t_{sd} be small compared with the switching times required (Fig. 5–29). The critical parameter determining t_{sd} is the carrier lifetime (τ_p for the example of the p$^+$-n junction). Since the recombination rate determines the speed with which excess holes can disappear from the n region, we would expect t_{sd} to be proportional to τ_p. In fact, an exact analysis of the problem of Fig. 5–28 leads to the result

$$t_{sd} = \tau_p \left[\text{erf}^{-1}\left(\frac{I_f}{I_f + I_r} \right) \right]^2 \tag{5-54}$$

where the error function (erf) is a tabulated function. Although the exact solution leading to Eq. (5–54) is too lengthy for us to consider here, an approximate result can be obtained from the quasi-steady state assumption.

EXAMPLE 5–5

Assume a p$^+$-n diode is biased in the forward direction, with a current I_f. At time $t = 0$ the current is switched to $- I_r$. Use the appropriate boundary conditions to solve Eq. (5–47) for $Q_p(t)$. Apply the quasi-steady state approximation to find the storage delay time t_{sd}.

From Eq. (5–47),

$$i(t) = \frac{Q_p(t)}{\tau_p} + \frac{dQ_p(t)}{dt} \quad \text{for } t < 0, Q_p = I_f\tau_p$$

Using Laplace transforms,

$$-\frac{I_r}{s} = \frac{Q_p(s)}{\tau_p} + sQ_p(s) - I_f\tau_p$$

$$Q_p(s) = \frac{I_f\tau_p}{s + 1/\tau_p} - \frac{I_r}{s(s + 1/\tau_p)}$$

$$Q_p(t) = I_f\tau_p e^{-t/\tau_p} + I_r\tau_p(e^{-t/\tau_p} - 1) = \tau_p[-I_r + (I_f + I_r)e^{-t/\tau_p}]$$

Assuming that $Q_p(t) = qAL_p\Delta p_n(t)$ as in Eq. (5–52),

$$\Delta p_n(t) = \frac{\tau_p}{qAL_p}[-I_r + (I_f + I_r)e^{-t/\tau_p}]$$

This is set to equal zero when $t = t_{sd}$, and we obtain:

$$t_{sd} = -\tau_p \ln\left[\frac{I_r}{I_f + I_r}\right] = \tau_p \ln\left(1 + \frac{I_f}{I_r}\right)$$

An important result of Eq. (5–54) is that τ_p can be calculated in a straight-forward way from a measurement of storage delay time. In fact, measurement of t_{sd} from an experimental arrangement such as Fig. 5–28a is a common method of measuring lifetimes. In some cases this is a more convenient technique than the photoconductive decay measurement discussed in Section 4.3.2.

As in the case of the turn-off transient of the previous section, the storage delay time can be reduced by introducing recombination centers into the diode material, thus reducing the carrier lifetimes, or by utilizing the narrow base diode configuration.

5.5.3 Switching Diodes

In discussing rectifiers we emphasized the importance of minimizing the reverse-bias current and the power losses under forward bias. In many applications, time response can be important as well. If a junction diode is to be used to switch rapidly from the conducting to the nonconducting state and back again, special consideration must be given to its charge control properties. We have discussed the equations governing the turn-on time and the reverse recovery time of a junction. From Eqs. (5–47) and (5–54) it is clear that a diode with fast switching properties must either store very little charge in the neutral regions for steady forward currents, or have a very short carrier lifetime, or both.

As mentioned above, we can improve the switching speed of a diode by adding efficient recombination centers to the bulk material. For Si diodes, Au doping is useful for this purpose. To a good approximation the carrier lifetime varies with the reciprocal of the recombination center concentration. Thus, for example, a p^+-n Si diode may have $\tau_p = 1$ µs and a reverse recovery time of 0.1 µs before Au doping. If the addition of 10^{14} Au atoms/cm^3 reduces the lifetime to 0.1 µs and t_{sd} to 0.01 µs, 10^{15} cm^{-3} Au atoms could reduce τ_p to 0.01 µs and t_{sd} to 1 ns (10^{-9} s). This process cannot be continued indefinitely, however. The reverse current due to generation of carriers from the Au centers in the depletion region becomes appreciable with large Au concentration (Section 5.6.2). In addition, as the Au concentration approaches the lightest doping of the junction, the equilibrium carrier concentration of that region can be affected.

A second approach to improving the diode switching time is to make the lightly doped neutral region shorter than a minority carrier diffusion length. This is the *narrow base diode* (Prob. 5.35). In this case the stored charge for forward conduction is very small, since most of the injected carriers diffuse through the lightly doped region to the end contact. When such a diode is switched to reverse conduction, very little time is required to eliminate the stored charge in the narrow neutral region. The mathematics involved in Prob. 5.35 is particularly interesting, because it closely resembles the calculations we shall make in analyzing the bipolar junction transistor in Chapter 7.

5.5.4 Capacitance of p-n Junctions

There are basically two types of capacitance associated with a junction: (1) the *junction capacitance* due to the dipole in the transition region and (2) the *charge storage capacitance* arising from the lagging behind of voltage as current changes, due to charge storage effects.[14] Both of these capacitances are important, and they must be considered in designing p-n junction devices for use with time-varying signals. The junction capacitance (1) is dominant under reverse-bias conditions, and the charge storage capacitance (2) is dominant when the junction is forward biased. In many applications of p-n junctions, the capacitance is a limiting factor in the usefulness of the device; on the other hand, there are important applications in which the capacitance discussed here can be useful in circuit applications and in providing important information about the structure of the p-n junction.

---

[14]The capacitance (1) above is also referred to as *transition region capacitance* or *depletion layer capacitance*; (2) is often called the *diffusion capacitance*.

The junction capacitance of a diode is easy to visualize from the charge distribution in the transition region (Fig. 5–12). The uncompensated acceptor ions on the p side provide a negative charge, and an equal positive charge results from the ionized donors on the n side of the transition region. The capacitance of the resulting dipole is slightly more difficult to calculate than is the usual parallel plate capacitance, but we can obtain it in a few steps.

Instead of the common expression $C = |Q/V|$, which applies to capacitors in which charge is a linear function of voltage, we must use the more general definition

$$C = \left|\frac{dQ}{dV}\right| \tag{5-55}$$

since the charge Q on each side of the transition region varies nonlinearly with the applied voltage (Fig. 5–30a). We can demonstrate this nonlinear dependence by reviewing the equations for the width of the transition region (W) and the resulting charge. The equilibrium value of W was found in Eq. (5–21) to be

$$W = \left[\frac{2\epsilon V_0}{q}\left(\frac{N_a + N_d}{N_a N_d}\right)\right]^{1/2} \quad (equilibrium) \tag{5-56}$$

Since we are dealing with the nonequilibrium case with voltage V applied, we must use the altered value of the electrostatic potential barrier ($V_0 - V$), as discussed in relation to Fig. 5–13. The proper expression for the width of the transition region is then

$$W = \left[\frac{2\epsilon(V_0 - V)}{q}\left(\frac{N_a + N_d}{N_a N_d}\right)\right]^{1/2} \quad (with\ bias) \tag{5-57}$$

In this expression the applied voltage V can be either positive or negative to account for forward or reverse bias. As expected, the width of the transition region is increased for reverse bias and is decreased under forward bias. Since the uncompensated charge Q on each side of the junction varies with the transition region width, variations in the applied voltage result in corresponding variations in the charge, as required for a capacitor. The value of Q can be written in terms of the doping concentration and transition region width on each side of the junction (Fig. 5–12):

$$|Q| = qAx_{n0}N_d = qAx_{p0}N_a \tag{5-58}$$

Relating the total width of the transition region W to the individual widths x_{n0} and x_{p0} from Eqs. (5–23) we have

$$x_{n0} = \frac{N_a}{N_a + N_d}W, \quad x_{p0} = \frac{N_d}{N_a + N_d}W \tag{5-59}$$

and therefore the charge on each side of the dipole is

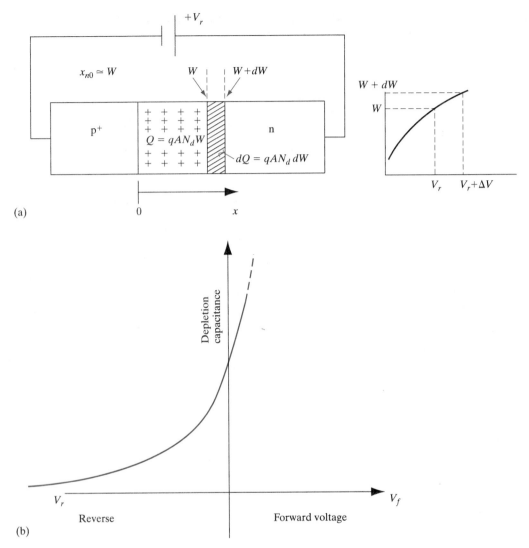

(a)

(b)

Figure 5–30
Depletion capacitance of a junction: (a) p⁺-n junction showing variation of depletion edge on n side with reverse bias. Electrically, the structure looks like a parallel plate capacitor whose dielectric is the depletion region, and the plates are the space charge neutral regions; (b) variation of depletion capacitance with reverse bias [Eq. (5–63)]. We neglect x_{p0} in the heavily-doped p⁺ material.

$$|Q| = qA\frac{N_dN_a}{N_d + N_a}W = A\left[2q\epsilon(V_0 - V)\frac{N_dN_a}{N_d + N_a}\right]^{1/2} \qquad (5\text{–}60)$$

Thus the charge is indeed a nonlinear function of applied voltage. From this expression and the definition of capacitance in Eq. (5–55), we can calculate the junction capacitance C_j. Since the voltage that varies the charge in the transition region is the barrier height $(V_0 - V)$, we must take the derivative with respect to this potential difference:

$$C_j = \left| \frac{dQ}{d(V_0 - V)} \right| = \frac{A}{2} \left[\frac{2q\epsilon}{(V_0 - V)} \frac{N_d N_a}{N_d + N_a} \right]^{1/2} \tag{5-61}$$

The quantity C_j is a *voltage-variable capacitance*, since C_j is proportional to $(V_0 - V)^{-1/2}$. There are several important applications for variable capacitors, including use in tuned circuits. The p-n junction device which makes use of the voltage-variable properties of C_j is called a *varactor*. We shall discuss this device further in Section 5.5.5.

Although the dipole charge is distributed in the transition region of the junction, the form of the parallel plate capacitor formula is obtained from the expressions for C_j and W (Fig. 5–30a):

$$C_j = \epsilon A \left[\frac{q}{2\epsilon(V_0 - V)} \frac{N_d N_a}{N_d + N_a} \right]^{1/2} = \frac{\epsilon A}{W} \tag{5-62}$$

In analogy with the parallel plate capacitor, the transition region width W corresponds with the plate separation of the conventional capacitor.

In the case of an asymmetrically doped junction, the transition region extends primarily into the less heavily doped side, and the capacitance is determined by only one of the doping concentrations (Fig. 5–30a). For a p^+-n junction, $N_a \gg N_d$ and $x_{n0} \simeq W$, while x_{p0} is negligible. The capacitance is then (Fig. 5–30b)

$$C_j = \frac{A}{2} \left[\frac{2q\epsilon}{V_0 - V} N_d \right]^{1/2} \quad \text{for } p^+\text{-}n \tag{5-63}$$

It is therefore possible to obtain the doping concentration of the lightly doped n region from a measurement of capacitance. For example, in a reverse-biased junction the applied voltage $V = -V_r$ can be made much larger than the contact potential V_0, so that the latter becomes negligible. If the area of the junction can be measured, a reliable value of N_d results from a measurement of C_j. However, these equations were obtained by assuming a sharp step junction. Certain modifications must be made in the case of a graded junction (Section 5.6.4 and Prob. 5.38).

The junction capacitance dominates the reactance of a p-n junction under reverse bias; for forward bias, however, the charge storage, or diffusion capacitance C_s becomes dominant. It has been recently shown[15] that the various time-dependent current components as well as the boundary conditions affect the diffusion capacitance in forward bias. We need to specify where the stored charges are extracted, and where the relevant voltage drops occur.

[15]S. Laux and K. Hess, "Revisiting the Analytic Theory of P-N Junction Impedance: Improvements Guided by Computer Simulation Leading to a New Equivalent Circuit," *IEEE Trans. Elec. Dev.*, 46(2), p. 396 (Feb. 1999).

To illustrate the calculation let us look at the simplified case of a symmetric, abrupt p-n junction where the doping levels N_a and N_d are equal.

We will consider two cases. For the long diode, which we have been dealing with so far, the diffusion lengths are assumed to be small compared to the lengths of the p and n regions. In this case, the injected minority carriers on either side of the depletion region decay exponentially to their equilibrium value long before they reach the ohmic contacts (Fig. 5–31a). On the other hand, the diffusion lengths in a short diode are assumed to be long compared to the length of the p and n regions (Fig. 5–31b). In the short diode, the injected excess minority carrier concentrations decrease almost linearly to zero at the ohmic contacts designated $x = -a$ and $x = c$ in Fig. 5–31b. Minority carrier distributions are discussed in Probs. 5.34-5.36 and Section 5.3.2. We will discuss the almost linear excess carrier distribution in a narrow region in Section 7.4.1.

The total current in the diode is the sum of the particle currents and the displacement current evaluated at any suitable location (chosen here at $x = 0$).

$$J_t = J_n(0) + J_p(0) + J_d(0) \qquad (5\text{–}64)$$

For the general case of time-dependent voltage and currents, we need to solve the hole and electron current continuity equations (Eq. 4–31a and 4-31b) for J_n and J_p and also take the time-derivative of Poisson's equation (Eq. 5–14) to obtain the displacement current:

$$J_d(x) = \frac{\partial}{\partial t}[\epsilon \mathscr{E}(x)] \qquad (5\text{–}65a)$$

We can integrate Poisson's equation between 0 and c, and take the derivative with respect to time to get

$$J_d(0) = q\frac{\partial}{\partial t}\int_0^c (n - p)dx + J_d(c) \qquad (5\text{–}65b)$$

We notice that the dopant charges do not appear here because they are time independent. Laux and Hess show in their paper that for most practical cases $J_d(c) = 0$, and that the displacement current $J_d(0)$ originates from a time-varying voltage across the depletion capacitance that was discussed earlier in this section.

Integrating the electron continuity equation (Eq. 4–31b) from $-a$ to 0 in the p-region, and the hole continuity equation (Eq. 4–31a) from 0 to c in the n-region, we can get the sum of the electron and hole particle current densities, at $x = 0$.

$$J_n(0) + J_p(0) = q\int_{-a}^c R\,dx + q\int_{-a}^0 \frac{\partial n}{\partial t}dx + q\int_0^c \frac{\partial p}{\partial t}dx + J_n(-a) + J_p(c) \qquad (5\text{–}65c)$$

where R is the net recombination rate at each point x.

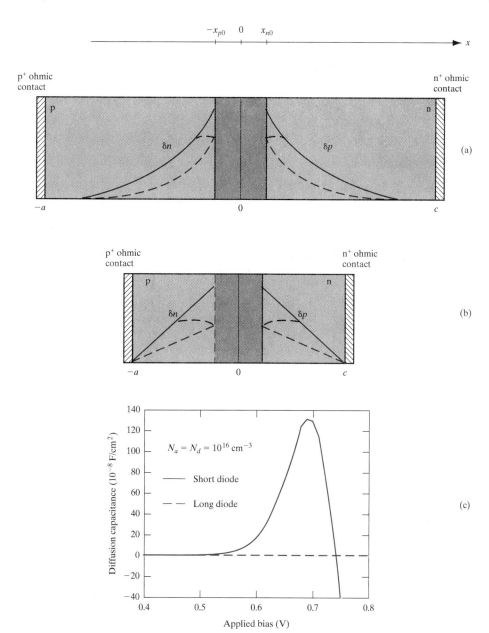

Figure 5–31

Diffusion capacitance in p-n junctions. (a) Steady-state minority carrier distribution for a forward bias, V (colored lines), and reduced forward bias, V-ΔV (dashed colored lines) in a long diode. The transient case when the current is reduced suddenly is shown by the black, dashed lines. Although the carrier distributions can change quickly near the junctions, they stay close to the original steady-state distributions far from the junctions at first. Gradually, the carrier distributions approach the new steady-state distributions for V-ΔV (dashed colored lines); (b) minority carrier distributions in a short diode; (c) diffusion capacitance as a function of forward bias in long and short diodes.

Let us see what the physical interpretation of each term in Eq (5–65c) is. For a long base diode, since the minority carrier concentrations reduce to zero before we reach the ohmic contacts at $-a$ and c, the terms $J_n(-a)$ and $J_p(c)$ are zero. Furthermore, for the steady state case, the terms

$$q \int_{-a}^{0} \frac{\partial n}{\partial t} dx + q \int_{0}^{c} \frac{\partial p}{\partial t} dx$$

are also zero because the time derivatives are zero. We see then that the dc current is given by the first term, which is the integrated carrier recombination rate. This is exactly the charge control model of the diode that we discussed in regard to Fig. 5–16.

For the time-dependent case, the integrands in the second and third terms in Eq. (5–65c) can be expressed, for example for the holes, by the chain rule as

$$q \int_{0}^{c} \frac{\partial p}{\partial t} dx = q \int_{0}^{c} \frac{\partial p}{\partial V} \cdot \frac{\partial V}{\partial t} dx = q \frac{\partial}{\partial V} \left\{ \int_{0}^{c} p\,dx \right\} \frac{\partial V}{\partial t} \qquad (5\text{--}66)$$

We have interchanged the order of integration with respect to x and differentiation with respect to t. Equation (5–66) is in the form of a current, with a (diffusion) capacitance times the voltage ramp rate. There is a similar contribution from the electrons.

In conventional theories of the diffusion capacitance due to stored minority carriers, the second and third terms in Eq. (5–65c) are erroneously considered to be the only contributors. Furthermore, the stored charge, for example for holes, is approximately set equal to

$$q \int_{x_{n0}}^{c} p\,dx$$

in the neutral region, rather than the correct expression in Eq. (5–65c) which considers both the neutral and the depletion regions. Also, in conventional theories, we assume that all the applied voltage is dropped across the depletion region. In reality, there can be a significant fraction of the applied bias dropped across the neutral region from x_{n0} to c and from $-a$ to $-x_{p0}$.

More importantly, Laux and Hess have shown that the first term in Eq. (5–65c) due to carrier recombination cancels most of the diffusion capacitance in long base diodes. Physically, the reason for this cancellation of the capacitance effect is that if the injected minority carriers (holes) recombine on the n side between 0 and c, they cannot be fully "*reclaimed*" at the injecting ohmic contact at $-a$ where the external voltage is changed, and similarly for electron injection.

For steady state, holes lost due to EHP recombination in the diode must be replenished at $-a$ (and electrons at c). For capacitive effects to be manifested, however, we must consider the transient case. We must determine the transfer of charge through the external terminals, as a function of the applied voltage variation at those terminals. To understand why the reclaimable charge is less than the total stored minority carrier charge, let us consider the transient conditions in a p$^+$-n diode, as discussed in Section 5.5.2. As shown in Fig. 5–28, when the forward bias is reduced, the minority carrier hole concentration at the edge of the depletion region is reduced, and therefore the slope of the hole distribution changes near the junction. This reduction occurs by some holes near $x_n = 0$ moving to the left towards the p$^+$ ohmic contact. The arrival of holes at the p$^+$ contact is referred to as reclaimed charge. Not all of the reduction in the hole distribution (shown in color in Fig. 5–31a) occurs by reclaiming holes at the p$^+$ contact, however. From the shape of the hole distribution within the n region, there obviously continues to be a diffusion of holes to the right also, toward the n$^+$ ohmic contact. In a long diode, these holes do not make it all the way to the n$^+$ ohmic contact because they recombine with electrons on the way. These recombined electrons have to be replenished by the n$^+$ ohmic contact. The key point is that because some of the holes are diffusing to the right, not all the holes in the stored distribution can be extracted (reclaimed) at the p$^+$ ohmic contact at the left, when the forward bias is reduced by a small amount in the transient case. The resulting capacitance–voltage behavior (Fig. 5–31c) for long base diodes shows almost zero diffusion capacitance.

The situation is somewhat different for narrow or short-base diodes. Since the minority carrier diffusion lengths are much longer than the length of the diode, there is negligible carrier recombination within the charge distribution, and the term

$$q\int_{-a}^{c} R\,dx$$

in Eq. (5–65c) is small. On the other hand, since most of the injected minority carriers now reach the ohmic contacts, the fourth and fifth terms in Eq. (5–65c) are large, unlike for the long base case. To understand physically why the reclaimed hole charge at $-a$ is less than the total stored charge in the short diode, we must recognize once again that for capacitive effects to be manifested, we need to consider the transient case. When the current is reduced suddenly, the slope of the hole distribution at $x_n = 0$ reduces, but the slope at $x = c$ does not (Fig. 5–31b). Because most holes reach the n$^+$ contact (at c), there is a reduction in the "reclaimable" hole charge at the p$^+$-ohmic contact (at $-a$). Hence, the net charge that is driven through the external circuit is reduced, and the diffusion capacitance due to minority carrier storage is reduced for the short diode, although not as drastically as for the long diode case. An exact solution of the continuity equation in this case shows that the reclaimable charge is 2/3 of the total stored charge.

There is an exponentially increasing diffusion capacitance with applied forward bias for the short diode (Fig. 5–31c). For a triangular minority carrier charge distribution (Fig. 5–31b), the stored hole charge on the n side is given by half the product of the height times the base of the triangle (Prob. 5.34).

$$Q_p = \frac{1}{2}qA(c - x_{no})(\Delta p_n) = \frac{1}{2}qA(c - x_{no})p_n\left(e^{\frac{qV}{kT}} - 1\right) \quad (5\text{–}67a)$$

Since the reclaimable charge is 2/3 of this, the diffusion capacitance is:

$$C_s = \frac{dQ_p}{dV} = \frac{1}{3}\frac{q^2}{kT}A(c - x_{no})p_n e^{\frac{qV}{kT}}$$ (5–67b)

There is a similar contribution from the stored electrons on the p-side. Laux and Hess show in their paper that because of the voltage drop in the neutral regions, and the possibility of conductivity modulation occurring there due to high carrier concentration at large forward biases, the diffusion capacitance becomes negative around the built-in voltage, V_0. Most Si p-n junctions in practice behave like short-base diodes, while laser diodes made in direct bandgap (short lifetime) semiconductors often correspond to the long base case.

Similarly, we can determine the *a-c conductance* by allowing small changes in the current. For example, for a long diode, we get:

$$G_s = \frac{dI}{dV} = \frac{qAL_p p_n}{\tau_p}\frac{d}{dV}(e^{qV/kT}) = \frac{q}{kT}I$$ (5–67c)

5.5.5 The Varactor Diode

The term *varactor* is a shortened form of *variable reactor*, referring to the voltage-variable capacitance of a reverse-biased p-n junction. The equations derived in Section 5.5.4 indicate that junction capacitance depends on the applied voltage and the design of the junction. In some cases a junction with fixed reverse bias may be used as a capacitance of a set value. More commonly the varactor diode is designed to exploit the voltage-variable properties of the junction capacitance. For example, a varactor (or a set of varactors) may be used in the tuning stage of a radio receiver to replace the bulky variable plate capacitor. The size of the resulting circuit can be greatly reduced, and its dependability is improved. Other applications of varactors include use in harmonic generation, microwave frequency multiplication, and active filters.

If the p-n junction is abrupt, the capacitance varies as the square root of the reverse bias V_r [Eq. (5–61)]. In a graded junction, however, the capacitance can usually be written in the form

$$C_j \propto V_r^{-n} \quad \text{for } V_r \gg V_0$$ (5–68a)

For example, in a linearly graded junction the exponent **n** is one-third (Prob. 5.38). Thus the voltage sensitivity of C_j is greater for an abrupt junction than for a linearly graded junction. For this reason, varactor diodes are often made by epitaxial growth techniques, or by ion implantation. The epitaxial layer and the substrate doping profile can be designed to obtain junctions for which the exponent **n** in Eq. (5–68a) is greater than one-half. Such junctions are called *hyperabrupt junctions*.

In the set of doping profiles shown in Fig. 5–32, the junction is assumed p$^+$-n so that the depletion layer width W extends primarily into the n side.

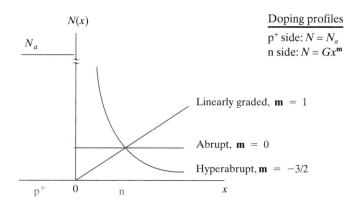

$N(x)$

N_a

p^+ 0 n x

Doping profiles
p^+ side: $N = N_a$
n side: $N = Gx^m$

Linearly graded, **m** $= 1$

Abrupt, **m** $= 0$

Hyperabrupt, **m** $= -3/2$

Figure 5–32
Graded junction
profiles: linearly
graded, abrupt,
hyperabrupt.

Three types of doping profiles on the n side are illustrated, with the donor distribution $N_d(x)$ given by Gx^m, where G is a constant and the exponent **m** is 0, 1, or $-\frac{3}{2}$. We can show (Prob. 5.37) that the exponent **n** in Eq. (5–68a) is $1/(\mathbf{m} + 2)$ for the p^+-n junction. Thus for the profiles of Fig. 5–32, **n** is $\frac{1}{2}$ for the abrupt junction and $\frac{1}{3}$ for the linearly graded junction. The hyperabrupt junction[16] with **m** $= -\frac{3}{2}$ is particularly interesting for certain varactor applications, since for this case **n** $= 2$ and the capacitance is proportional to V_r^{-2}. When such a capacitor is used with an inductor L in a resonant circuit, the resonant frequency varies linearly with the voltage applied to the varactor.

$$\omega_r = \frac{1}{\sqrt{LC}} \propto \frac{1}{\sqrt{V_r^{-n}}} \propto V_r, \quad \text{for } \mathbf{n} = 2 \qquad (5\text{--}68b)$$

Because of the wide variety of C_j vs. V_r dependencies available by choosing doping profiles, varactor diodes can be designed for specific applications. For some high-frequency applications, varactors can be designed to exploit the forward-bias charge storage capacitance in short diodes.

The approach we have taken in studying p-n junctions has focused on the basic principles of operation, neglecting secondary effects. This allows for a relatively uncluttered view of carrier injection and other junction properties, and illuminates the essential features of diode operation. To complete the description, however, we must now fill in a few details which can affect the operation of junction devices under special circumstances.

Most of the deviations from the simple theory can be treated by fairly straightforward modifications of the basic equations. In this section we shall investigate the most important deviations and alter the theory wherever possible. In a few cases, we shall simply indicate the approach to be taken and

**5.6
DEVIATIONS
FROM THE SIMPLE
THEORY**

[16]It is clear that $N_d(x)$ cannot become arbitrarily large at $x = 0$. However, the **m** $= -\frac{3}{2}$ profile can be approximated a short distance away from the junction.

the result. The most important alterations to the simple diode theory are the effects of contact potential and changes in majority carrier concentration on carrier injection, recombination and generation within the transition region, ohmic effects, and the effects of graded junctions.

5.6.1 Effects of Contact Potential on Carrier Injection

If the forward-bias $I–V$ characteristics of various semiconductor diodes are compared, it becomes clear that the band gap has an important influence on carrier injection. For example, Fig. 5–33 compares the low-temperature characteristics of heavily doped diodes having various band gaps. One obvious feature of this figure is that the $I–V$ characteristics appear "square"; that is, the current is very small until a critical forward bias is reached, and then the current increases rapidly. This is typical of exponentials plotted on such a scale. However, it is significant that the limiting voltage is slightly less than the value of the band gap in electron volts.

The reason for the small current at low voltages for these devices can be understood from a simple rearrangement of the diode equation. If we rewrite Eq. (5–36) for a forward-biased p^+-n diode (with $V \gg kT/q$) and include the exponential form for the minority carrier concentration p_n, we obtain

$$I = \frac{qAD_p}{L_p}p_n e^{qV/kT} = \frac{qAD_p}{L_p}N_v e^{[qV-(E_{Fn}-E_{vn})]/kT} \tag{5–69}$$

Hole injection into the n material is small if the forward bias V is much less than $(E_{Fn} - E_{vn})/q$. For a p^+-n diode, this quantity is essentially the contact potential, since the Fermi level is near the valence band on the p side. If the n region is also heavily doped, the contact potential is almost equal to the

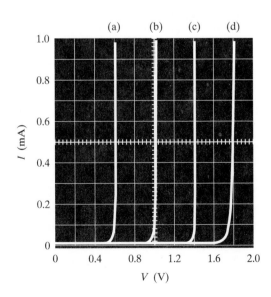

Figure 5–33
I–V characteristics of heavily doped p-n junction diodes at 77 K, illustrating the effects of contact potential on the forward current:
(a) Ge, $E_g \simeq$ 0.7 eV;
(b) Si, $E_g \simeq$ 1.4 eV;
(c) GaAs, $E_g \simeq$ 1.4 eV;
(d) GaAsP, $E_g \simeq$ 1.9 eV.

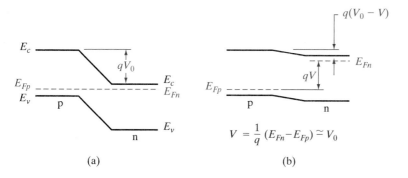

Figure 5–34
Examples of contact potential for a heavily doped p-n junction: (a) at equilibrium; (b) approaching the maximum forward bias $V = V_0$.

band gap (Fig. 5–34). This accounts for the dramatic increase in diode current near the band gap voltage in Fig. 5–33. Contributing to the small current at lower voltages is the fact that the minority carrier concentration $p_n = n_i^2/N_d$ is very small at low temperature (n_i small) and with heavy doping (N_d large).

The limiting forward bias across a p-n junction is equal to the contact potential, as in Fig. 5–34(b). This effect is not predicted by the simple diode equation, for which the current increases exponentially with applied voltage. The reason this important result is excluded in the simple theory is that in Eq. (5–28) we neglect changes in the majority carrier concentrations on either side of the junction. This assumption is valid only for low injection levels; for large injected carrier concentrations, the excess majority carriers become important compared with the majority doping. For example, at low injection $\Delta n_p = \Delta p_p$ is important compared with the equilibrium minority electron concentration n_p, but is negligible compared with the majority hole concentration p_p; this was the basis for neglecting Δp_p in Eq. (5–28). For high injection levels, however, Δp_p can be comparable to p_p and we must write Eq. (5–27) in the form

$$\frac{p(-x_{p0})}{p(x_{n0})} = \frac{p_p + \Delta p_p}{p_n + \Delta p_n} = e^{q(V_0 - V)/kT} = \frac{n_n + \Delta n_n}{n_p + \Delta n_p} \qquad (5\text{--}70)$$

From Eq. (5–38), we get at either edge of the depletion region,

$$pn = p(-x_{p0})n(-x_{p0}) = p(x_{n0})n(x_{n0}) = n_i^2 e^{\frac{F_n - F_p}{kT}} = n_i^2 e^{qV/kT} \qquad (5\text{--}71a)$$

For example at $-x_{po}$ we then get

$$(p_p + \Delta p_p)(n_p + \Delta n_p) = n_i^2 e^{qV/kT} \qquad (5\text{--}71b)$$

Keeping in mind that $\Delta p_p = \Delta n_p$, $n_p \ll \Delta n_p$, and in high level injection $p_p < \Delta p_p$, we approximately get

$$\Delta n_p = n_i e^{qV/2kT} \qquad (5\text{--}72)$$

The rest of the derivation is very similar to that in Section 5.3.2. Hence, the diode current in high level injection scales as

$$I \propto e^{qV/2kT} \qquad (5\text{--}73)$$

5.6.2 Recombination and Generation in the Transition Region

In analyzing the p-n junction, we have assumed that recombination and thermal generation of carriers occur primarily in the neutral p and n regions, outside the transition region. In this model, forward current in the diode is carried by recombination of excess minority carriers injected into each neutral region by the junction. Similarly, the reverse saturation current is due to the thermal generation of EHPs in the neutral regions and the subsequent diffusion of the generated minority carriers to the transition region, where they are swept to the other side by the field. In many devices this model is adequate; however, a more complete description of junction operation should include recombination and generation within the transition region itself.

When a junction is forward biased, the transition region contains excess carriers of both types, which are in transit from one side of the junction to the other. Unless the width of the transition region W is very small compared with the carrier diffusion lengths L_n and L_p, significant recombination can take place within W. An accurate calculation of this recombination current is complicated by the fact that the recombination rate, which depends on the carrier concentrations [Eq. (4–5)], varies with position within the transition region. Analysis of the recombination kinetics shows that the current due to recombination within W is proportional to n_i and increases with forward bias according to approximately $\exp(qV/2kT)$. On the other hand, current due to recombination in the neutral regions is proportional to p_n and n_p [Eq. (5–36)] and therefore to n_i^2/N_d and n_i^2/N_a, and increases according to $\exp(qV/kT)$. The diode equation can be modified to include this effect by including the parameter **n**:

$$\boxed{I = I_0{'}(e^{qV/\mathbf{n}kT} - 1)} \qquad (5\text{--}74)$$

where **n** varies between 1 and 2, depending on the material and temperature. Since **n** determines the departure from the ideal diode characteristic, it is often called the *ideality factor*.

The ratio of the two currents

$$\frac{I(\text{recombination in neutral regions})}{I(\text{recombination in transition region})} \propto \frac{n_i^2 e^{qV/kT}}{n_i e^{qV/2kT}} \propto n_i e^{qV/2kT} \qquad (5\text{--}75)$$

becomes small for wide band gap materials, low temperatures (small n_i), and for low voltage. Thus the forward current for low injection in a Si diode is likely to be dominated by recombination in the transition region, while a Ge diode may follow the usual diode equation. In either case, injection through W into the neutral regions becomes more important with increased voltage. Therefore, **n** in Eq. (5–74) may vary from ~2 at low voltage to ~1 at higher voltage.

Just as recombination within W can affect the forward characteristics, the reverse current through a junction can be influenced by carrier *generation* in the transition region. We found in Section 5.3.3 that the reverse saturation current can be accounted for by the thermal generation of EHPs within a diffusion length of either side of the transition region. The generated minority carriers diffuse to the transition region, where they are swept to the other side of the junction by the electric field (Fig. 5–35). However, carrier generation can take place within the transition region itself. If W is small compared with L_n or L_p, band-to-band generation of EHPs within the transition region is not important compared with generation in the neutral regions. However, the lack of free carriers within the space charge of the transition region can create a current due to the net generation of carriers by *emission from recombination centers*. Of the four generation-recombination processes depicted in Fig. 5–36, the two capture rates R_n and R_p are negligible

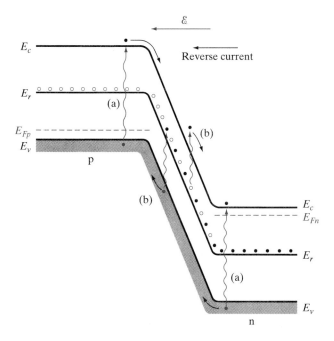

Figure 5–35
Current in a reverse-biased p-n junction due to thermal generation of carriers by (a) band-to-band EHP generation, and (b) generation from a recombination level.

Figure 5–36
Capture and gen-
eration of carriers
at a recombina-
tion center: (a)
capture and gen-
eration of elec-
trons and holes;
(b) hole capture
and generation
processes re-
drawn in terms of
valence band
electron excitation
to E_r (hole genera-
tion) and electron
deexcitation from
E_r to E_v (hole cap-
ture by E_r).

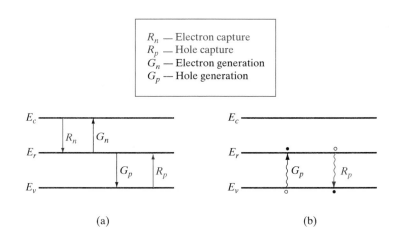

R_n — Electron capture
R_p — Hole capture
G_n — Electron generation
G_p — Hole generation

(a)   (b)

within W because of the very small carrier concentrations in the reverse-bias space charge region. Therefore, a recombination level E_r near the center of the band gap can provide carriers through the thermal generation rates G_n and G_p. Each recombination center alternately emits an electron and a hole; physically, this means that an electron at E_r is thermally excited to the conduction band (G_n) and a valence band electron is subsequently excited thermally to the empty state on the recombination level, leaving a hole behind in the valence band (G_p). The process can then be repeated over and over, providing electrons for the conduction band and holes for the valence band. Normally, these emission processes are exactly balanced by the corresponding capture processes R_n and R_p. However, in the reverse-bias transition region, generated carriers are swept out before recombination can occur, and net generation results.

Of course, the importance of thermal generation within W depends on the temperature and the nature of the recombination centers. A level near the middle of the band gap is most effective, since for such centers neither G_n nor G_p requires thermal excitation of an electron over more than about half the band gap. If no recombination level is available, this type of generation is negligible. However, in most materials recombination centers exist near the middle of the gap due to trace impurities or lattice defects. Generation from centers within W is most important in materials with large band gaps, for which band-to-band generation in the neutral regions is small. Thus for Si, generation within W is generally more important than for a narrower band gap material such as Ge.

The saturation current due to generation in the neutral regions was found to be essentially independent of reverse bias. However, generation within W naturally increases as W increases with reverse bias. As a result, the reverse current can increase almost linearly with W, or with the square root of reverse-bias voltage.

5.6.3 Ohmic Losses

In deriving the diode equation we assumed that the voltage applied to the device appears entirely across the junction. Thus we neglected any voltage drop in the neutral regions or at the external contacts. For most devices this is a valid assumption; the doping is usually fairly high, so that the resistivity of each neutral region is low, and the area of a typical diode is large compared with its length. However, some devices do exhibit ohmic effects, which cause significant deviation from the expected *I–V* characteristic.

We can seldom represent ohmic losses in a diode accurately by including a simple resistance in series with the junction. The effects of voltage drops outside the transition region are complicated by the fact that the voltage drop depends on the current, which in turn is dictated by the voltage across the junction. For example, if we represent the series resistance of the p and n regions by R_p and R_n, respectively, we can write the junction voltage V as

$$V = V_a - I[R_p(I) + R_n(I)] \tag{5–76}$$

where V_a is the external voltage applied to the device. As the current increases, there is an increasing voltage drop in R_p and R_n, and the junction voltage V decreases. This reduction in V lowers the level of injection so that the current increases more slowly with increased bias. A further complication in calculating the ohmic loss is that the conductivity of each neutral region increases with increasing carrier injection. Since the effects of Eq. (5–76) are most pronounced at high injection levels, this *conductivity modulation* by the injected excess carriers can reduce R_p and R_n significantly.

Ohmic losses are purposely avoided in properly designed devices by appropriate choices of doping and geometry. Therefore, deviations of the current generally appear only for very high currents, outside the normal operating range of the device.

Figure 5–37 shows the forward and reverse current–voltage characteristics of a p-n junction on a semi-log scale, both for an ideal Shockley diode as well as for non-ideal devices. For an ideal forward-based diode, we get a straight line on a semi-log plot reflecting the exponential dependence of current on voltage. On the other hand, taking into account all the second order effects discussed in Section 5.6, we see various regions of operation. At low current levels, we see the enhanced generation–recombination current, leading to a higher diode ideality factor (**n** = 2). For moderate currents, we get ideal low-level injection and diffusion-limited current (**n** = 1). At higher currents, we get high level injection and **n** = 2, while at even higher currents, the ohmic drops in the space charge neutral regions become important.

Similarly, in reverse bias, in an ideal diode, we have a constant, voltage-independent reverse saturation current. However, in actuality, we get an enhanced, voltage-dependent generation–recombination leakage current. At very high reverse biases, the diode breaks down reversibly due to avalanche effects.

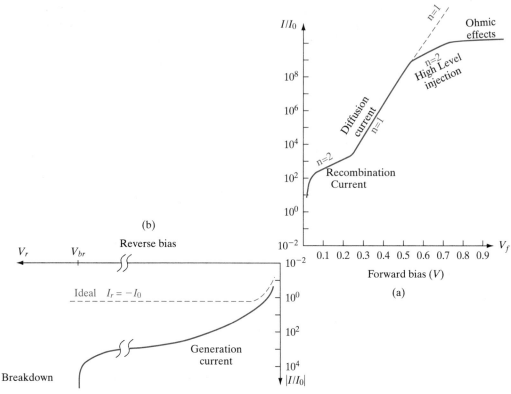

Figure 5-37
Forward and reverse current-voltage characteristics plotted on semi-log scales, with current normalized with respect to saturation current, I_0; (a) the ideal forward characteristic is an exponential with an ideality factor, **n** = 1 (dashed straight line on log-linear plot). The actual forward characteristics of a typical diode (colored line) have four regimes of operation; (b) ideal reverse characteristic (dashed line) is a voltage-independent current = $-I_0$. Actual leakage characteristics (colored line) are higher due to generation in the depletion region, and also show breakdown at high voltages.

5.6.4 Graded Junctions

While the abrupt junction approximation accurately describes the properties of many epitaxially grown junctions, it is often inadequate in analyzing diffused or implanted junction devices. For shallow diffusions, in which the diffused impurity profile is very steep (Fig. 5-38a), the abrupt approximation is usually acceptable. If the impurity profile is spread out into the sample, however, a graded junction can result (Fig. 5-38b). Several of the expressions we have derived for the abrupt junction must be modified for this case (see Section 5.5.5).

The graded junction problem can be solved analytically if, for example, we make a linear approximation of the net impurity distribution near the

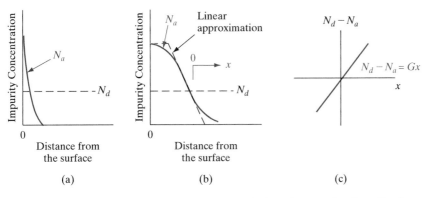

Figure 5–38
Approximations to diffused junctions: (a) shallow diffusion (abrupt); (b) deep drive-in diffusion with source removed (graded); (c) linear approximation to the graded junction.

junction (Fig. 5–38c). We assume that the graded region can be described approximately by

$$N_d - N_a = Gx \qquad (5\text{–}77)$$

where G is a grade constant giving the slope of the net impurity distribution. In Poisson's equation [Eq. (5–14)], the linear approximation becomes

$$\frac{d\mathscr{E}}{dx} = \frac{q}{\epsilon}(p - n + N_d^+ - N_a^-) \simeq \frac{q}{\epsilon}Gx \qquad (5\text{–}78)$$

within the transition region. In this approximation we assume complete ionization of the impurities and neglect the carrier concentrations in the transition region, as before. The net space charge varies linearly over W, and the electric field distribution is therefore parabolic. The expressions for contact potential and junction capacitance are different from the abrupt junction case (Fig. 5–39 and Prob. 5.38), since the electric field is no longer linear on each side of the junction.

In a graded junction the usual depletion approximation is often inaccurate. If the grade constant G is small, the carrier concentrations $(p - n)$ can be important in Eq. (5–78). Similarly, the usual assumption of negligible space charge outside the transition region is questionable for small G. It would be more accurate to refer to the regions just outside the transition region as quasi-neutral rather than neutral. Thus the edges of the transition region are not sharp as Fig. 5–39 implies but are spread out in x. These effects complicate calculations of junction properties, and a computer must be used in solving the problem accurately.

Most of the conclusions we have made regarding carrier injection, recombination and generation currents, and other properties are qualitatively applicable to graded junctions, with some alterations in the functional form of the resulting equations. Therefore, we can apply most of our basic concepts of junction theory to reasonably graded junctions as long as we remember that certain modifications should be made in accurate computations.

Figure 5–39
Properties of the
graded junction
transition region:
(a) net impurity
profile; (b) net
charge distribu-
tion; (c) electric
field; (d) electro-
static potential.

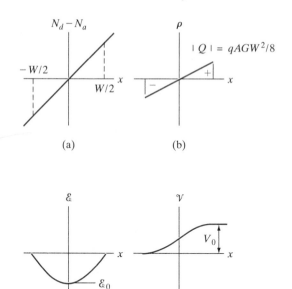

$N_d - N_a$

$-W/2$

$W/2$

x

(a)

ρ

$|Q| = qAGW^2/8$

$+$

$-$

x

(b)

$\mathcal{E}$

x

$\mathcal{E}_0$

(c)

$\mathcal{V}$

V_0

x

(d)

**5.7
METAL–
SEMICONDUCTOR
JUNCTIONS**

Many of the useful properties of a p-n junction can be achieved by simply forming an appropriate metal–semiconductor contact. This approach is obviously attractive because of its simplicity of fabrication; also, as we shall see in this section, metal–semiconductor junctions are particularly useful when high-speed rectification is required. On the other hand, we also must be able to form nonrectifying (ohmic) contacts to semiconductors. Therefore, this section deals with both rectifying and ohmic contacts.

5.7.1 Schottky Barriers

In Section 2.2.1 we discussed the work function $q\Phi_m$ of a metal in a vacuum. An energy of $q\Phi_m$ is required to remove an electron at the Fermi level to the vacuum outside the metal. Typical values of Φ_m for very clean surfaces are 4.3V for Al and 4.8V for Au. When negative charges are brought near the metal surface, positive (image) charges are induced in the metal. When this image force is combined with an applied electric field, the effective work function is somewhat reduced. Such barrier lowering is called the *Schottky effect*, and this terminology is carried over to the discussion of potential barriers arising in metal–semiconductor contacts. Although the Schottky effect is only a part of the explanation of metal–semiconductor contacts, rectifying contacts are generally referred to as *Schottky barrier diodes*. In this section we shall see how such barriers arise in metal–semiconductor contacts. First we consider barriers in ideal metal–semiconductor junctions, and then in Section 5.7.4 we will include effects which alter the barrier height.

When a metal with work function $q\Phi_m$ is brought in contact with a semiconductor having a work function $q\Phi_s$, charge transfer occurs until the Fermi levels align at equilibrium (Fig. 5–40). For example, when $\Phi_m > \Phi_s$, the semiconductor Fermi level is initially higher than that of the metal before contact is made. To align the two Fermi levels, the electrostatic potential of the semiconductor must be raised (i.e., the electron energies must be lowered) relative to that of the metal. In the n-type semiconductor of Fig. 5–40 a depletion region W is formed near the junction. The positive charge due to uncompensated donor ions within W matches the negative charge on the metal. The electric field and the bending of the bands within W are similar to effects already discussed for p-n junctions. For example, the depletion width W in the semiconductor can be calculated from Eq. (5–21) by using the p^+-n approximation (i.e., by assuming the negative charge in the dipole is a thin sheet of charge to the left of the junction). Similarly, the junction capacitance is $A\epsilon_s/W$, as in the p^+-n junction.[17]

The equilibrium contact potential V_0, which prevents further net electron diffusion from the semiconductor conduction band into the metal, is the difference in work function potentials $\Phi_m - \Phi_s$. The potential barrier height Φ_B for electron injection from the metal into the semiconductor conduction

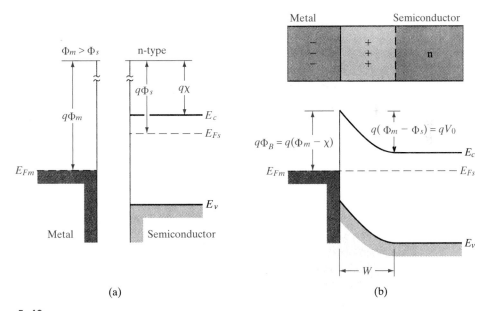

(a)   (b)

Figure 5–40
A Schottky barrier formed by contacting an n-type semiconductor with a metal having a larger work function: (a) band diagrams for the metal and the semiconductor before joining; (b) equilibrium band diagram for the junction.

---

[17]While the properties of the Schottky barrier depletion region are similar to the p^+-n, it is clear that the analogy does not include forward-bias hole injection, which is dominant for the p^+-n but not for the contact of Fig. 5–40.

band is $\Phi_m - \chi$, where $q\chi$ (called the *electron affinity*) is measured from the vacuum level to the semiconductor conduction band edge. The equilibrium potential difference V_0 can be decreased or increased by the application of either forward- or reverse-bias voltage, as in the p-n junction.

Figure 5–41 illustrates a Schottky barrier on a p-type semiconductor, with $\Phi_m < \Phi_s$. In this case aligning the Fermi levels at equilibrium requires a positive charge on the metal side and a negative charge on the semiconductor side of the junction. The negative charge is accommodated by a depletion region W in which ionized acceptors (N_a^-) are left uncompensated by holes. The potential barrier V_0 retarding hole diffusion from the semiconductor to the metal is $\Phi_s - \Phi_m$, and as before this barrier can be raised or lowered by the application of voltage across the junction. In visualizing the barrier for holes, we recall from Fig. 5–11 that the electrostatic potential barrier for positive charge is opposite to the barrier on the electron energy diagram.

The two other cases of ideal metal–semiconductor contacts ($\Phi_m < \Phi_s$ for n-type semiconductors, and $\Phi_m > \Phi_s$ for p-type) result in nonrectifying contacts. We will save treatment of these cases for Section 5.7.3, where ohmic contacts are discussed.

5.7.2 Rectifying Contacts

When a forward-bias voltage V is applied to the Schottky barrier of Fig. 5–40b, the contact potential is reduced from V_0 to $V_0 - V$ (Fig. 5–42a). As a result, electrons in the semiconductor conduction band can diffuse across the depletion

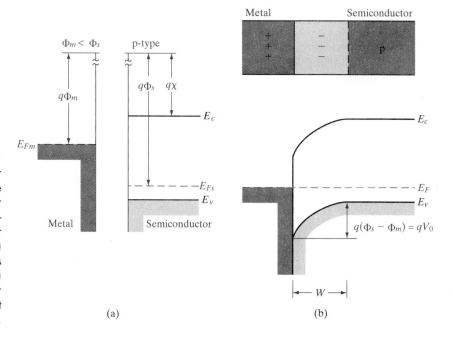

Figure 5–41
Schottky barrier between a p-type semiconductor and a metal having a smaller work function: (a) band diagrams before joining; (b) band diagram for the junction at equilibrium.

(a)   (b)

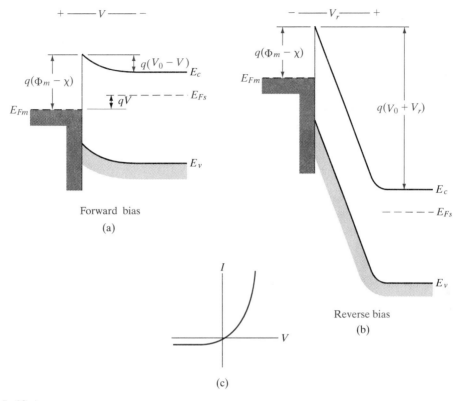

Figure 5–42
Effects of forward and reverse bias on the junction of Fig. 5–40: (a) forward bias; (b) reverse bias; (c) typical current-voltage characteristic.

region to the metal. This gives rise to a forward current (metal to semiconductor) through the junction. Conversely, a reverse bias increases the barrier to $V_0 + V_r$, and electron flow from semiconductor to metal becomes negligible. In either case flow of electrons from the metal to the semiconductor is retarded by the barrier $\Phi_m - \chi$. The resulting diode equation is similar in form to that of the p-n junction

$$I = I_0(e^{qV/kT} - 1) \qquad (5\text{–}79)$$

as Fig. 5–42c suggests. In this case the reverse saturation current I_0 is not simply derived as it was for the p-n junction. One important feature we can predict intuitively, however, is that the saturation current should depend upon the size of the barrier Φ_B for electron injection from the metal into the semiconductor. This barrier (which is $\Phi_m - \chi$ for the ideal case shown in Fig. 5–42) is unaffected by the bias voltage. We expect the probability of an electron in the metal surmounting this barrier to be given by a Boltzmann factor. Thus

$$I_0 \propto e^{-q\Phi_B/kT} \qquad\qquad (5\text{--}80)$$

The diode equation (5–79) applies also to the metal–p-type semiconductor junction of Fig. 5–41. In this case forward voltage is defined with the semiconductor biased positively with respect to the metal. Forward current increases as this voltage lowers the potential barrier to $V_0 - V$ and holes flow from the semiconductor to the metal. Of course, a reverse voltage increases the barrier for hole flow and the current becomes negligible.

In both of these cases the Schottky barrier diode is rectifying, with easy current flow in the forward direction and little current in the reverse direction. We also note that the forward current in each case is due to the injection of *majority* carriers from the semiconductor into the metal. The absence of minority carrier injection and the associated storage delay time is an important feature of Schottky barrier diodes. Although some minority carrier injection occurs at high current levels, these are essentially majority carrier devices. Their high-frequency properties and switching speed are therefore generally better than typical p-n junctions.

In the early days of semiconductor technology, rectifying contacts were made simply by pressing a wire against the surface of the semiconductor. In modern devices, however, the metal—semiconductor contact is made by depositing an appropriate metal film on a clean semiconductor surface and defining the contact pattern photolithographically. Schottky barrier devices are particularly well suited for use in densely packed integrated circuits, because fewer photolithographic masking steps are required compared to p-n junction devices.

5.7.3 Ohmic Contacts

In many cases we wish to have an *ohmic* metal–semiconductor contact, having a linear *I–V* characteristic in both biasing directions. For example, the surface of a typical integrated circuit is a maze of p and n regions, which must be contacted and interconnected. It is important that such contacts be ohmic, with minimal resistance and no tendency to rectify signals.

Ideal metal–semiconductor contacts are ohmic when the charge induced in the semiconductor in aligning the Fermi levels is provided by majority carriers (Fig. 5–43). For example, in the $\Phi_m < \Phi_s$ (n-type) case of Fig. 5–43a, the Fermi levels are aligned at equilibrium by transferring electrons from the metal to the semiconductor. This raises the semiconductor electron energies (lowers the electrostatic potential) relative to the metal at equilibrium (Fig. 5–43b). In this case the barrier to electron flow between the metal and the semiconductor is small and easily overcome by a small voltage. Similarly, the case $\Phi_m > \Phi_s$ (p-type) results in easy hole flow across the junction (Fig. 5–43d). Unlike the rectifying contacts dis-

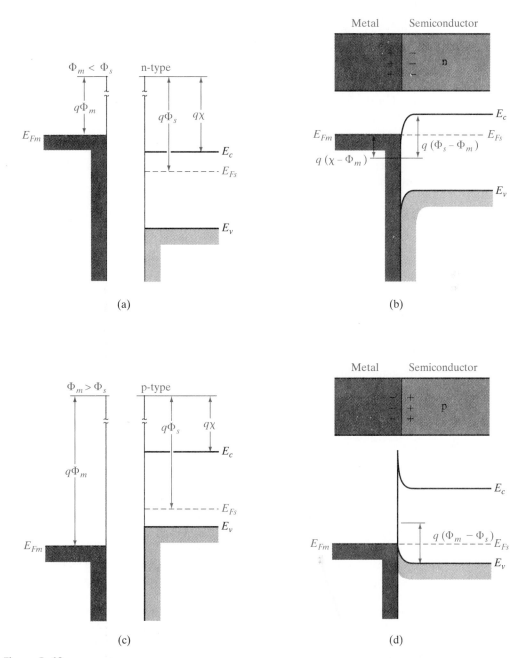

Figure 5–43
Ohmic metal–semiconductor contacts: (a) $\Phi_m < \Phi_s$ for an n-type semiconductor, and (b) the equilibrium band diagram for the junction; (c) $\Phi_m > \Phi_s$ for a p-type semiconductor, and (d) the junction at equilibrium.

cussed previously, no depletion region occurs in the semiconductor in these cases since the electrostatic potential difference required to align the Fermi levels at equilibrium calls for accumulation of majority carriers in the semiconductor.

A practical method for forming ohmic contacts is by doping the semiconductor heavily in the contact region. Thus if a barrier exists at the interface, the depletion width is small enough to allow carriers to tunnel through the barrier. For example, Au containing a small percentage of Sb can be alloyed to n-type Si, forming an n^+ layer at the semiconductor surface and an excellent ohmic contact. Similarly, p-type material requires a p^+ surface layer in contact with the metal. In the case of Al on p-type Si, the metal contact also provides the acceptor dopant. Thus the required p^+ surface layer is formed during a brief heat treatment of the contact after the Al is deposited.

5.7.4 Typical Schottky Barriers

The discussion of ideal metal–semiconductor contacts does not include certain effects of the junction between the two dissimilar materials. Unlike a p-n junction, which occurs within a single crystal, a Schottky barrier junction includes a termination of the semiconductor crystal. The semiconductor surface contains *surface states* due to incomplete covalent bonds and other effects,which can lead to charges at the metal–semiconductor interface. Furthermore, the contact is seldom an atomically sharp discontinuity between the semiconductor crystal and the metal. There is typically a thin interfacial layer, which is neither semiconductor nor metal. For example, silicon crystals are covered by a thin (10–20 Å) oxide layer even after etching or cleaving in atmospheric conditions. Therefore, deposition of a metal on such a Si surface leaves a glassy interfacial layer at the junction. Although electrons can tunnel through this thin layer, it does affect the barrier to current transport through the junction.

Because of surface states, the interfacial layer, microscopic clusters of metal–semiconductor phases, and other effects, it is difficult to fabricate junctions with barriers near the ideal values predicted from the work functions of the two isolated materials. Therefore, measured barrier heights are used in device design. In compound semiconductors the interfacial layer introduces states in the semiconductor band gap that pin the Fermi level at a fixed position, regardless of the metal used (Fig. 5–44). For example, a collection of interface states located $0.7 \sim 0.9$ eV below the conduction band pins E_F at the surface of n-type GaAs, and the Schottky barrier height is determined from this pinning effect rather than by the work function of the metal. An interesting case is n-type InAs (Fig. 5–44b), in which E_F at the interface is pinned *above* the conduction band edge. As a result, ohmic contact to n-type InAs can be made by depositing virtually any metal on the surface. For Si, good Schottky barriers are formed by various metals, such as Au or Pt. In

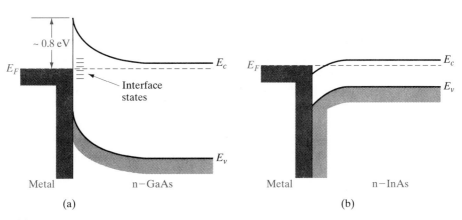

Figure 5–44
Fermi level pinning by interface states in compound semiconductors: (a) E_F is pinned near $E_C - 0.8$ eV
in n-type GaAs, regardless of the choice of metal; (b) E_F is pinned above E_C in n-type InAs, providing an
excellent ohmic contact.

the case of Pt, heat treatment results in a platinum silicide layer, which provides a reliable Schottky barrier with $\Phi_B \simeq 0.85$ V on n-type Si.

A full treatment of Schottky barrier diodes results in a forward current equation of the form

$$I = ABT^2 e^{-q\Phi_B/kT} e^{qV/\mathbf{n}kT} \qquad (5\text{--}81)$$

where B is a constant containing parameters of the junction properties and
$\mathbf{n}$ is a number between 1 and 2, similar to the ideality factor in Eq. (5–74)
but arising from different reasons. The mathematics of this derivation is similar to that of *thermionic emission*, and the factor B corresponds to an effective Richardson constant in the thermionic problem.

Thus far we have discussed p-n junctions formed within a single semiconductor (*homojunctions*) and junctions between a metal and a semiconductor.
The third important class of junctions consist of those between two lattice-
matched semiconductors with different band gaps (*heterojunctions*). We discussed lattice-matching in Section 1.4.1. The interface between two such
semiconductors may be virtually free of defects, and continuous crystals containing single or multiple heterojunctions can be formed. The availability of
heterojunctions and multilayer structures in compound semiconductors opens
a broad range of possibilities for device development. We will discuss many
of these applications in later chapters, including heterojunction bipolar transistors, field-effect transistors, and semiconductor lasers.

**5.8
HETERO-
JUNCTIONS**

When semiconductors of different band gaps, work functions, and electron affinities are brought together to form a junction, we expect discontinuities in the energy bands as the Fermi levels line up at equilibrium (Fig. 5–45). The discontinuities in the conduction band ΔE_c and the valence band ΔE_v accommodate the difference in band gap between the two semiconductors ΔE_g. In an ideal case, ΔE_c would be the difference in electron affinities $q(\chi_2 - \chi_1)$, and ΔE_v would be found from $\Delta E_g - \Delta E_c$. This is known as the Anderson affinity rule. In practice, the band discontinuities are found experimentally for particular semiconductor pairs. For example, in the commonly used system GaAs–AlGaAs (see Figs. 3-6 and 3-13), the direct band gap difference ΔE_g^{Γ} between the wider band gap AlGaAs and the narrower band gap GaAs is apportioned approximately $\frac{2}{3}$ in the conduction band and $\frac{1}{3}$ in the valence band for the heterojunction. The built-in contact potential is divided between the two semiconductors as required to align the Fermi levels at equilibrium. The resulting depletion region on each side of the heterojunction and the amount of built-in potential on each side (making up the contact potential V_0) are found by solving Poisson's equation with the boundary condition of continuous electric flux density, $\epsilon_1 \mathscr{E}_1 = \epsilon_2 \mathscr{E}_2$ at the junction. The barrier that electrons must overcome in moving from the n side to the p side may be quite different from the barrier for holes moving from p to n. The depletion region on each side is analogous to that described in Eq. (5–23), except that we must account for the different dielectric constants in the two semiconductors.

To draw the band diagram for any semiconductor device involving homojunctions or heterojunctions, we need material parameters such as the bandgap and the electron affinity which depend on the semiconductor material but not on the doping, and the workfunction which depends on the semiconductor as well as the doping. The electron affinity and workfunction are referenced to the vacuum level. The true vacuum level (or global vacuum level), E_{vac}, is the potential energy reference when an electron is taken out of the semiconductor to infinity, where it sees no forces. Hence, the true vacuum level is a constant (Fig. 5–45). That introduces an apparent contradiction, however, because looking at the band bending in a semiconductor device, it seems to imply that the electron affinity in the semiconductor changes as a function of position, which is impossible because the electron affinity is a material parameter. Therefore, we need to introduce the new concept of the local vacuum level, E_{vac} (loc), which varies along with and parallel to the conduction band edge, thereby keeping the electron affinity constant. The local vacuum level tracks the potential energy of an electron if it is moved just outside of the semiconductor, but not far away. The difference between the local and global vacuum levels is due to the electrical work done against the fringing electric fields of the depletion region, and is equal to the potential energy qV_0 due to the built-in contact potential V_0 in equilibrium. This potential energy can, of course, be modified by an applied bias.

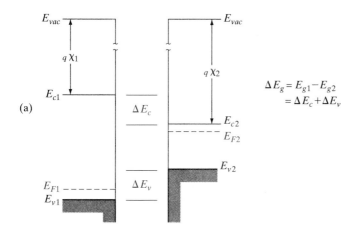

$$\Delta E_g = E_{g1} - E_{g2}$$
$$= \Delta E_c + \Delta E_v$$

Figure 5–45
An ideal hetero-
junction between
a p-type, wide
band gap semi-
conductor an n-
type narrower
band gap semi-
conductor: (a)
band diagrams
before joining; (b)
band discontinu-
ities and band
bending at
equilibrium.

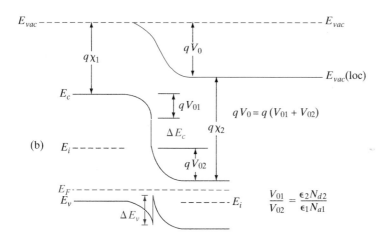

$$qV_0 = q(V_{01} + V_{02})$$

$$\frac{V_{01}}{V_{02}} = \frac{\epsilon_2 N_{d2}}{\epsilon_1 N_{a1}}$$

   To draw the band diagram for a heterojunction accurately, we must not only use the proper values for the band discontinuities but also account for the band bending in the junction. To do this, we must solve Poisson's equation across the heterojunction, taking into account the details of doping and space charge, which generally requires a computer solution. We can, however, sketch an approximate diagram without a detailed calculation. Given the experimental band offsets ΔE_v and ΔE_c, we can proceed as follows:

1.   Align the Fermi level with the two semiconductor bands separated. Leave space for the transition region.

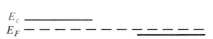

E_c ————
E_F — — — — — — — — — ————

E_v ————

2. The metallurgical junction ($x = 0$) is located near the more heavily doped side. At $x = 0$ put ΔE_v and ΔE_c, separated by the appropriate band gaps.

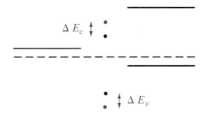

ΔE_c ↕

$↕ \Delta E_v$

3. Connect the conduction band and valence band regions, keeping the band gap constant in each material.

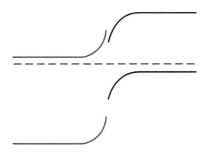

Steps 2 and 3 of this procedure are where the exact band bending is important and must be obtained by solving Poisson's equation. In step 2 we must use the band offset values ΔE_c and ΔE_v for the specific pair of semiconductors in the heterojunction.

EXAMPLE 5–6   For heterojunctions in the GaAs–AlGaAs system, the direct (Γ) band gap difference ΔE_g^Γ is accommodated approximately $\frac{2}{3}$ in the conduction band and $\frac{1}{3}$ in the valence band. For an Al composition of 0.3, the AlGaAs is direct (see Fig. 3–6) with $\Delta E_g^{\Gamma} = 1.85$ eV. Sketch the band diagrams for two heterojunction cases: N^+-$Al_{0.3}Ga_{0.7}As$ on n-type GaAs, and N^+-$Al_{0.3}Ga_{0.7}As$ on p^+-GaAs.[18]

Taking $\Delta E_g = 1.85 - 1.43 = 0.42$ eV, the band offsets are $\Delta E_c = 0.28$ eV and **SOLUTION** $\Delta E_v = 0.14$ eV. In each case we draw the equilibrium Fermi level, add the appropriate bands far from the junction, add the band offsets while estimating the relative amounts of band bending and position of $x = 0$ for the particular doping on the two sides, and finally sketch the band edges so that E_g is maintained in each separate semiconductor right up to the heterojunction at $x = 0$.

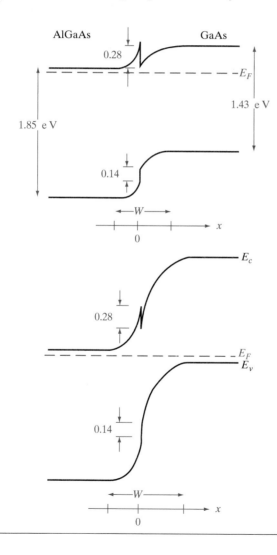

[18]In discussing heterojunctions, it is common to use capital N or P to designate the wide band gap material.

A particularly important example of a heterojunction is shown in Figure 5–46, in which heavily n-type AlGaAs is grown on lightly doped GaAs. In this example the discontinuity in the conduction band allows electrons to spill over from the N^+-AlGaAs into the GaAs, where they become trapped in the potential well. As a result, electrons collect on the GaAs side of the heterojunction and move the Fermi level above the conduction band in the GaAs near the interface. These electrons are confined in a narrow potential well in the GaAs conduction band. If we construct a device in which conduction occurs parallel to the interface, the electrons in such a potential well form a *two-dimensional electron gas* with very interesting device properties. As we shall see in Chapter 6, electron conduction in such a potential well can result in very high mobility electrons. This high mobility is due to the fact that the electrons in this well come from the AlGaAs, and not from doping in the GaAs. As a result, there is negligible impurity scattering in the GaAs well, and the mobility is controlled almost entirely by lattice scattering (phonons). At low temperatures, where phonon scattering is low, the mobility in this region can be very high. If the band-bending in the GaAs conduction band is strong enough, the potential well may be extremely narrow, so that discrete states such as E_1 and E_2 in Fig. 5–46 are formed. We will return to this example in Chapter 6.

Another obvious feature of Fig. 5–46 is that the concept of a contact potential barrier qV_0 for both electrons and holes in a homojunction is no longer valid for the heterojunction. In Fig. 5–46 the barrier for electrons qV_n is smaller than the barrier for holes qV_p. This property of a heterojunction can be used to alter the relative injection of electrons and holes, as we shall see in Section 7.9.

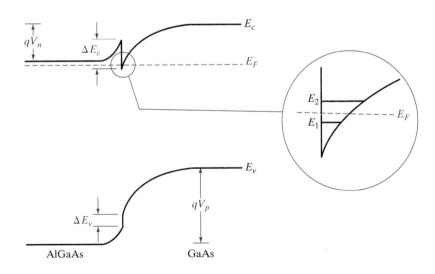

Figure 5–46
A heterojunction between N^+-AlGaAs and lightly doped GaAs, illustrating the potential well for electrons formed in the GaAs conduction band. If this well is sufficiently thin, discrete states (such as E_1 and E_2) are formed, as discussed in Section 2.4.3.

5.1 Design an oxide mask to block P diffusion in Si at 1000°C for 30 minutes using a design criterion that the mask thickness should be eight times the diffusion length. If we grow this oxide using a wet oxidation process at 1100°C, how long must we do the oxidation? Calculate the total number of Si atoms that are consumed from the wafer in the process, for a 200 mm diameter wafer.

5.2 When impurities are diffused into a sample from an unlimited source such that the surface concentration N_0 is held constant, the impurity distribution (profile) is given by

$$N(x, t) = N_0 \operatorname{erfc}\left(\frac{x}{2\sqrt{Dt}}\right)$$

where D is the diffusion coefficient for the impurity, t is the diffusion time, and erfc is the complementary error function.

If a certain number of impurities are placed in a thin layer on the surface before diffusion, and if no impurities are added and none escape during diffusion, a gaussian distribution is obtained:

$$N(x, t) = \frac{N_s}{\sqrt{\pi Dt}} e^{-(x/2\sqrt{Dt})^2}$$

where N_s is the quantity of impurity placed on the surface (atoms/cm^2) prior to $t = 0$. Notice that this expression differs from Eq. (4–44) by a factor of two. Why?

Figure P5–2 gives curves of the complementary error function and gaussian factors for the variable u, which in our case is $x/2\sqrt{Dt}$. Assume that boron is diffused into n-type Si (uniform $N_d = 5 \times 10^{16}$cm^{-3}) at 1000°C for 30 min. The diffusion coefficient for B in Si at this temperature is $D = 3 \times 10^{-14}$ cm^2/s.

(a) Plot $N_a(x)$ after the diffusion, assuming that the surface concentration is held constant at $N_0 = 5 \times 10^{20}$ cm^{-3}. Locate the position of the junction below the surface.

(b) Plot $N_a(x)$ after the diffusion, assuming that B is deposited in a thin layer on the surface prior to diffusion ($N_s = 5 \times 10^{13}$ cm^{-2}), and no additional B atoms are available during the diffusion. Locate the junction for this case.

Hint: Plot the curves on five-cycle semilog paper, with an abscissa varying from zero to $\frac{1}{2}$ μm. In plotting $N_a(x)$, choose values of x that are simple multiples of $2\sqrt{Dt}$.

5.3 A 900 nm oxide is grown on (100) Si in wet oxygen at 1100°C. How long does it take to grow the first 200 nm, the next 300 nm and the final 400 nm?

A square window (1 mm × 1 mm) is etched in this oxide and the wafer is re-oxidized at 1150°C in wet oxygen such that the oxide thickness *outside* of the window region increases to 2000 nm. Draw a cross section of the wafer and mark off all the thicknesses, dimensions and oxide–Si interfaces relative to the original Si surface. Calculate the step heights in Si and in the oxide at the edge of the window.

Figure P5–2

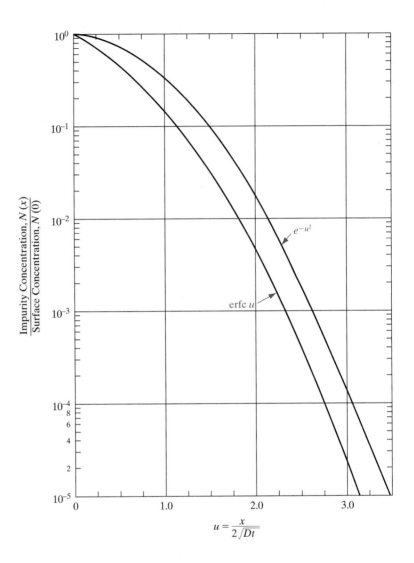

$$u = \frac{x}{2\sqrt{Dt}}$$

5.4 We wish to do an As implant into a Si wafer with a 0.1 μm oxide such that the peak lies at the oxide–silicon interface, with a peak value of 5×10^{19} cm^{-3}. What implant parameters (energy, dose and beam current) would you choose? The scan area is 200 cm^2, and the desired implant time is 20 s. Assume similar range statistics in oxide and Si.

5.5 We want to implant 5×10^{14} cm^{-2} B into Si at an average depth of 0.5 μm. We have an implanter which has a *maximum* acceleration voltage of 150 kV. How can we achieve this profile if we have singly and doubly charged B in the machine? Suppose the doubly ionized beam current is 0.1 mA, how long will the implant take if the scan area is 100 cm^2? By doing clever ion implanter source design, Dr. Boron Maximus has increased the beam current by a factor of 1000. From a dose uniformity point of view is this good or bad?

5.6 Assuming a constant (unlimited) source diffusion of P at 1000°C into p-type Si ($N_a = 2 \times 10^{16}$ cm^{-3}), calculate the time required to achieve a junction depth of 1 micron. See equations in Prob. 5.2.

5.7 We are interested in patterning the structure shown in Fig. P5-7. Design the mask aligner optics in terms of numerical aperture of the lens and the wavelength of the source.

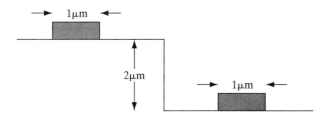

Figure P5–7

5.8 In a p$^+$-n junction the hole diffusion current in the neutral n material is given by Eq. (5–32). What are the electron diffusion and electron drift components of current at point x_n in the neutral n region?

5.9 An abrupt Si p-n junction has $N_a = 10^{18}$ cm^{-3} on one side and $N_d = 5 \times 10^{15}$ cm^{-3} on the other.

(a) Calculate the Fermi level positions at 300 K in the p and n regions.

(b) Draw an equilibrium band diagram for the junction and determine the contact potential V_0 from the diagram.

(c) Compare the results of part (b) with V_0 as calculated from Eq. (5–8).

5.10 The junction described in Prob. 5.9 has a circular cross section with diameter of 10 μm. Calculate x_{n0}, x_{p0}, Q_+, and $\mathscr{E}_0$ for this junction at equilibrium (300 K). Sketch $\mathscr{E}(x)$ and charge density to scale, as in Fig. 5–12.

5.11 The electron injection efficiency of a junction is I_n/I at $x_p = 0$.

(a) Assuming the junction follows the simple diode equation, express I_n/I in terms of the diffusion constants, diffusion lengths, and equilibrium minority carrier concentrations.

(b) Show that I_n/I can be written as $[1 + L_n^p p_p \mu_p^n / L_p^n n_n \mu_n^p]^{-1}$, where the superscripts refer to the n and p regions. What should be done to increase the electron injection efficiency of a junction?

5.12 A Si p$^+$-n junction has a donor doping of 5×10^{16} cm^{-3} on the n side and a cross-sectional area of 10^{-3} cm^2. If $\tau_p = 1$ μs and $D_p = 10$ cm^2/s, calculate the current with a forward bias of 0.5 V at 300 K.

5.13 (a) Explain physically why the charge storage capacitance is unimportant for reverse-biased junctions.

(b) Assuming that a GaAs junction is doped to equal concentrations on the n and p sides, would you expect electron or hole injection to dominate in forward bias? Explain.

5.14 (a) A Si p^+-n junction 10^{-2} cm^2 in area has $N_d = 10^{15}$ cm^{-3} doping on the n side. Calculate the junction capacitance with a reverse bias of 10 V.

(b) An abrupt p^+-n junction is formed in Si with a donor doping of $N_d = 10^{15}$ cm^{-3}. What is the depletion region thickness W just prior to avalanche breakdown?

5.15 Using Eqs. (5–17) and (5–23), show that the peak electric field in the transition region is controlled by the doping on the more lightly doped side of the junction.

5.16 An abrupt Si p-n junction has the following properties at 300 K:

p side	n side	$A = 10^{-4}$ cm^2
$N_a = 10^{17}$ cm^{-3}	$N_d = 10^{15}$	
$\tau_n = 0.1$ μs	$\tau_p = 10$ μs	
$\mu_p = 200$ cm^2/V-s	$\mu_n = 1300$	
$\mu_n = 700$	$\mu_p = 450$	

Draw an equilibrium band diagram for this junction, including numerical values for the Fermi level position relative to the intrinsic level on each side. Find the contact potential from the diagram and check your answer with the analytical expression for V_0.

5.17 A long p^+-n diode is forward biased with current I flowing. The current is suddenly tripled at $t = 0$.

(a) What is the slope of the hole distribution at $x_n = 0$ just after the current is tripled?

(b) Assuming the voltage is always $\gg kT/q$, relate the final junction voltage (at $t = \infty$) to the initial voltage (before $t = 0$).

5.18 Assume that the doping concentration N_a on the p side of an abrupt junction is the same as N_d on the n side. Each side is many diffusion lengths long. Find the expression for the hole current I_p in the p-type material.

5.19 A Si p-n junction with cross-sectional area, $A = 0.001$ cm^2 is formed with $N_a = 10^{15}$ cm^{-3}, $N_d = 10^{17}$ cm^{-3}. Calculate:

(a) Contact potential, V_0.

(b) Space-charge width at equilibrium (zero bias).

(c) Current with a forward bias of 0.5 V. Assume that the current is diffusion dominated. Assume $\mu_n = 1500$ cm^2/V-s, $\mu_p = 450$ cm^2/V-s, $\tau_n = \tau_p = 2.5$ μs. Which carries most of the current, electrons or holes and why? If you wanted to double the electron current, what should you do?

5.20 An n^+-p junction with a long p-region has the following properies: $N_a = 10^{16}$ cm^{-3}; $D_p = 13$ cm^2/s; $\mu_n = 1000$ cm^2/V-s; $\tau_n = 2$μs; $n_i = 10^{10}$ cm^{-3}. If we apply 0.7 V forward bias to the junction at 300 K, what is the electric field in the p-region far from the junction?

5.21 For the diode in Problem 5.16, draw the band diagram qualitatively under forward and reverse bias showing the quasi-Fermi levels.

5.22 In a p$^+$-n junction, the n-doping N_d is doubled. How do the following change if everything else remains unchanged? Indicate only increase or decrease.

(a) Junction capacitance

(b) Built-in potential

(c) Breakdown voltage

(d) Ohmic losses

5.23 The junction of problem (5.16) is forward biased by 0.5 V. What is the forward current? What is the current at a reverse bias of –0.5 V?

5.24 In the junction of problem (5.16), what is the total depletion capacitance at –4 V?

5.25 A p$^+$-n Si diode ($V_0 = 0.956$ V) has a donor doping of 10^{17} cm^{-3} and an n-region width = 1 μm. Does it break down by avalanche or punchthrough?

5.26 Calculate the capacitance for the following Si n$^+$-p junction.

$N_a = 10^{15}$ cm^{-3}

Area = 0.001 cm^2

Reverse bias = 1, 5 and 10 V

Plot $1/C^2$ vs. V_R

Demonstrate that the slope yields N_a. Repeat calculations for $N_a = 10^{17}$ cm^{-3}.

Since the doping is not specified on the n$^+$ side, use a suitable approximation.

5.27 We assumed in Section 5.2.3 that carriers are excluded within W and that the semiconductor is neutral outside W. This is known as the *depletion approximation*. Obviously, such a sharp transition is unrealistic. In fact, the space charge varies over a distance of several *Debye lengths*, given by

$$L_D = \left[\frac{\epsilon_s kT}{q^2 N_d} \right]^{1/2} \quad \text{on the n side.}$$

Calculate the Debye length on the n side for Si junctions having $N_a = 10^{18}$ cm^{-3} on the p-side and $N_d = 10^{14}, 10^{16}$, and 10^{18} cm^{-3} on the n-side and compare with the size of W in each case.

5.28 We have a symmetric p-n silicon junction ($N_a = N_d = 10^{17}$ cm^{-3}). If the peak electric field in the junction at breakdown is 5×10^5 V/cm, what is the reverse breakdown voltage in this junction?

5.29 We wish to design a p$^+$-n diode such that the avalanche breakdown and punchthrough both occur at 15 V. Assume the relative dielectric constant of the semiconductor is 10, V_0 is 0.5 V, and the breakdown field is 1 MV/cm. Determine the width and doping of the n-region.

5.30 A long p$^+$-n junction has its forward bias current switched from I_{F1} to I_{F2} at $t = 0$. Find an expression for the stored charge Q_p as a function of time in the n-region.

5.31 A long p^+-n diode is forward biased with current I flowing. The current is suddenly doubled at $t = 0$.

Assume that the stored charge in the n region can be represented by an exponential at each instant, for simplicity. Write the expression for the instantaneous current as a sum of recombination current and current due to changes in the stored charge. Using proper boundary conditions, solve this equation for the instantaneous hole distribution and find the expression for the instantaneous junction voltage.

5.32 The diode of Fig. 5–23c is used in a simple half-wave rectifier circuit in which the diode is placed in series with a load resistor. Assume that the diode offset voltage E_0 is 0.4 V and that $R = dv/di = 400\ \Omega$. For a load resistor of 1 kΩ and a sinusoidal input of $2 \sin \omega t$, sketch the output voltage (across the load resistor) over two cycles.

5.33 An abrupt p^+-n junction is formed in Si with a donor doping of $N_d = 10^{15}\ \text{cm}^{-3}$. What is the minimum thickness of the n region to ensure avalanche breakdown rather than punchthrough?

5.34 Assume holes are injected from a p^+-n junction into a short n region of length l. If $\delta p(x_n)$ varies linearly from Δp_n at $x_n = 0$ to zero at the ohmic contact $(x_n = l)$, find the steady state charge in the excess hole distribution Q_p and the current I.

5.35 Assume that a p^+-n diode is built with an n region width l smaller than a hole diffusion length $(l < L_p)$. This is the so-called *narrow base diode*. Since for this case holes are injected into a short n region under forward bias, we cannot use the assumption $\delta p(x_n = \infty) = 0$ in Eq. (4–35). Instead, we must use as a boundary condition the fact that $\delta p = 0$ at $x_n = l$.

(a) Solve the diffusion equation to obtain

$$\delta p(x_n) = \frac{\Delta p_n \left[e^{(l-x_n)/L_p} - e^{(x_n-l)/L_p} \right]}{e^{l/L_p} - e^{-l/L_p}}$$

(b) Show that the current in the diode is

$$I = \left(\frac{qAD_p p_n}{L_p} \operatorname{ctnh} \frac{l}{L_p} \right) (e^{qV/kT} - 1)$$

5.36 Given the narrow base diode result (Prob. 5.35), (a) calculate the current due to recombination in the n region, and (b) show that the current due to recombination at the ohmic contact is

$$I(\text{ohmic contact}) = \left(\frac{qAD_p p_n}{L_p} \operatorname{csch} \frac{l}{L_p} \right) (e^{qV/kT} - 1)$$

5.37 Assume that a p^+-n junction is built with a graded n region in which the doping is described by $N_d(x) = Gx^m$. The depletion region $(W \cong x_{n0})$ extends from essentially the junction at $x = 0$ to a point W within the n region. The singularity at $x = 0$ for negative **m** can be neglected.

(a) Integrate Gauss's law across the depletion region to obtain the maximum value of the electric field $\mathscr{E}_0 = -qGW^{(m+1)}/\epsilon(m+1)$.

(b) Find the expression for $\mathscr{E}(x)$, and use the result to obtain $V_0 - V = qGW^{(m+2)}/\epsilon(m+2)$.

(c) Find the charge Q due to ionized donors in the depletion region; write Q explicitly in terms of $(V_0 - V)$.

(d) Using the results of (c), take the derivative $dQ/d(V_0 - V)$ to show that the capacitance is

$$C_j = A\left[\frac{qG\epsilon^{(m+1)}}{(m+2)(V_0 - V)}\right]^{1/(m+2)}$$

5.38 Assume a linearly graded junction as in Fig. 5–39, with a doping distribution described by Eq. (5–77). The doping is symmetrical, so that $x_{p0} = x_{n0} = W/2$.

(a) Integrate Eq. (5–78) to show that

$$\mathscr{E}(x) = \frac{q}{2\epsilon}G\left[x^2 - \left(\frac{W}{2}\right)^2\right]$$

(b) Show that the width of the depletion region is

$$W = \left[\frac{12\epsilon(V_0 - V)}{qG}\right]^{1/3}$$

(c) Show that the junction capacitance is

$$C_j = A\left[\frac{qG\epsilon^2}{12(V_0 - V)}\right]^{1/3}$$

5.39 Design an ohmic contact for n-type GaAs using InAs, with an intervening graded InGaAs region (see Fig. 5–44).

5.40 (a) Using Eq. (5–8), calculate the contact potential V_0 of a Si p-n junction operating at 300 K for $N_a = 10^{14}$ and 10^{19} cm^{-3}, with $N_d = 10^{14}, 10^{15}, 10^{16}, 10^{17}, 10^{18}$, and 10^{19} cm^{-3} in each case and plot vs. N_d.

(b) Plot the maximum electric field $\mathscr{E}_0$ vs. N_d for the junctions described in (a).

5.41 A Schottky barrier is formed between a metal having a work function of 4.3 eV and p-type Si (electron affinity = 4 eV). The acceptor doping in the Si is 10^{17} cm^{-3}.

(a) Draw the equilibrium band diagram, showing a numerical value for qV_0.

(b) Draw the band diagram with 0.3 V forward bias. Repeat for 2 V reverse bias.

5.42 What is the conductivity of a piece of Ge ($n_i = 2.5 \times 10^{13}$ cm^{-3}) doped with 5×10^{13} cm^{-3} donors and 2.5×10^{13} cm^{-3} acceptors? ($D_n = 100$ cm^2/s, $D_p = 50$ cm^2/s). If the electron affinity of Ge = 4.0 eV, and we put down a metal electrode with work function = 4.5 eV, what is the work function difference? Do you expect this to be a Schottky barrier or an ohmic contact?

READING LIST   **Arienzo, M., S. S. Iyer, B. S. Meyerson, G. L. Patton, and J. M. C. Stork.** "Si-Ge Alloys: Growth, Properties, and Applications." *Solid Surface Science* 48/49 (May 1991): 377– 86.

Chang, L. L., and L. Esaki. "Semiconductor Quantum Heterostructures." *Physics Today* 45 (October 1992): 36– 43.

Ghandhi, S. K. *VLSI Fabrication Principles*, 2nd ed. New York: Wiley, 1994.

Hummel, R. E. *Electronic Properties of Materials,* 2nd ed. Berlin: Springer-Verlag, 1993.

Jaeger, R. C. *Modular Series on Solid State Devices: Vol V. Introduction to Microelectronic Fabrication.* Reading, MA: Addison-Wesley, 1988.

Karunasiri, R. P. G., and K. L. Wang. "Quantum Devices Using SiGe/Si Heterostructures." *Journal of Vacuum Science and Technology* B 9 (July– August 1992): 2064– 71.

Li, S. S. *Semiconductor Physical Electronics.* New York: Plenum Press, 1993.

Muller, R. S., and T. I. Kamins. *Device Electronics for Integrated Circuits.* New York: Wiley, 1986.

Neamen, D. A. *Semiconductor Physics and Devices: Basic Principles.* Homewood, IL: Irwin, 1992.

Neudeck, G. W. *Modular Series on Solid State Devices: Vol II. The PN Junction Diode.* Reading, MA: Addison-Wesley, 1983.

Shockley, W. "The Theory of P-N Junctions in Semiconductors and P-N Junction Transistors." *Bell Syst. Tech. J.* 28 (1949), 435.

Ryssell, H., and I. Ruge. *Ion-Implantation.* Chichester: Wiley, 1986.

Tove, P. A. "Formation and Characterization of Metal–Semiconductor Junctions." *Vacuum* 36 (October 1986): 659–67.

Wang, S. *Fundamentals of Semiconductor Theory and Device Physics.* Englewood Cliffs, NJ: Prentice Hall, 1989.

Wolf, S., and R. N. Tauber. *Silicon Processing for the VLSI Era.* Sunset Beach, CA: Lattice Press, 1986.

Wolfe, C. M., G. E. Stillman, and N. Holonyak, Jr. *Physical Properties of Semiconductors.* Englewood Cliffs, NJ: Prentice Hall, 1989.

Chapter 6
Field-Effect Transistors

The modern era of semiconductor electronics was ushered in by the invention of the bipolar transistor in 1948 by Bardeen, Brattain, and Shockley at the Bell Telephone Laboratories. This device, along with its field-effect counterpart, has had an enormous impact on virtually every area of modern life. In this chapter we will learn about the operation, applications, and fabrication of the field-effect transistor (FET).

The field-effect transistor comes in several forms. In a *junction* FET (called a *JFET*) the control (*gate*) voltage varies the depletion width of a reverse-biased p-n junction. A similar device results if the junction is replaced by a Schottky barrier (*metal–semiconductor* FET, called a *MESFET*). Alternatively, the metal gate electrode may be separated from the semiconductor by an insulator (*metal-insulator-semiconductor* FET, called a *MISFET*). A common special case of this type uses an oxide layer as the insulator (*MOSFET*).

In Chapter 5 we found that two dominant features of p-n junctions are the injection of minority carriers with forward bias and a variation of the depletion width W with reverse bias. These two p-n junction properties are used in two important types of transistors. The *bipolar junction transistor (BJT)* discussed in Chapter 7 uses the injection of minority carriers across a forward-biased junction, and the junction field effect transistor discussed in this chapter depends on control of a junction depletion width under reverse bias. The FET is a majority carrier device, and is therefore often called a *unipolar* transistor. The BJT, on the other hand, operates by the injection and collection of *minority* carriers. Since the action of both electrons and holes is important in this device, it is called a *bipolar* transistor. Like its bipolar counterpart, the FET is a three-terminal device in which the current through two terminals is controlled at the third. Unlike the BJT, however, field-effect devices are controlled by a voltage at the third terminal rather than by a current.

The history of BJTs and FETs is rather interesting. It was the FET that was proposed first in 1930 by Lilienfeld, but he never got it to work because he did not fully appreciate the role of surface defects or surface states. In the process of trying to demonstrate experimentally such a field effect transistor, Bardeen and Brattain somewhat serendipitously invented the first bipolar transistor, the Ge point contact transistor. This major breakthrough was rapidly followed by Shockley's extension of the concept to the BJT. It was only much later, after the problem of surface states was resolved by growing an oxide insulator on Si, that the first MOSFET was demonstrated in 1960 by Kahng and Atalla. Although the BJT reigned supreme in the early days of semiconductor integrated electronics, it has been gradually supplanted in most applications by the Si MOSFET. The main reason is, unlike BJTs, the various types of FET are characterized by a high *input impedance*, since the control voltage is applied to a reverse-biased junction or Schottky barrier, or across an insulator. These devices are particularly well suited for controlled switching between a conducting state and a nonconducting state, and are therefore useful in digital circuits. They are also suitable for integration of many devices on a single chip, as we shall see in Chapter 9. In fact, millions of MOS transistors are commonly used together in semiconductor memory devices and microprocessors.

**6.1
TRANSISTOR
OPERATION**

We begin this section with a general discussion of amplification and switching, the basic circuit functions performed by transistors. The transistor is a three-terminal device with the important feature that the current through two terminals can be controlled by small changes we make in the current or voltage at the third terminal. This control feature allows us to amplify small a-c signals or to switch the device from an *on* state to an *off* state and back. These two operations, *amplification* and *switching*, are the basis of a host of electronic functions. This section provides a brief introduction to these operations, as a foundation for understanding both bipolar and field-effect transistors.

6.1.1 The Load Line

Consider a two-terminal device that has a nonlinear *I–V* characteristic, as in Fig. 6–1. We might determine this curve experimentally by measuring the current for various applied voltages, or by using an oscilloscope called a *curve tracer*, which varies *I* and *V* repetitively and displays the resulting curve. When such a device is biased with the simple battery–resistor combination

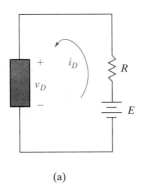

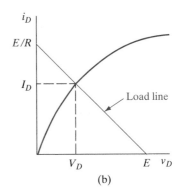

Figure 6–1
A two-terminal nonlinear device: (a) biasing circuit; (b) *I–V* characteristic and load line.

(a)

(b)

shown in the figure, steady state values of I_D and V_D are attained. To find these values we begin by writing a loop equation around the circuit:[1]

$$E = i_D R + v_D \qquad (6–1)$$

This gives us one equation describing the circuit, but it contains two unknowns (i_D and v_D). Fortunately, we have another equation of the form $i_D = f(v_D)$ in the curve of Fig. 6–1b, giving us two equations with two unknowns. The steady state current and voltage are found by a simultaneous solution of these two equations. However, since one equation is analytical and the other is graphical, we must first put them into the same form. It is easy to make the linear equation (6–1) graphical, so we plot it on Fig. 6–1b to find the simultaneous solution. The end points of the line described by Eq. (6–1) are at E when $i_D = 0$ and at E/R when $v_D = 0$. The two graphs cross at $v_D = V_D$ and $i_D = I_D$, the steady state values of current and voltage for the device with this biasing circuit.

Now let's add a third terminal which somehow controls the I–V characteristic of the device. For example, assume that the device current–voltage curve can be moved up the current axis by increasing the control voltage as in Fig. 6–2b. This results in a family of i_D–v_D curves, depending upon the choice of v_G. We can still write the loop equation (6–1) and draw it on the set of curves, but now the simultaneous solution depends on the value of v_G. In the example of Fig. 6–2, V_G is 0.5 V and the d-c values of I_D and V_D are found at the intersection to be 10 mA and 5 V, respectively. Whatever the value of the control voltage v_G at the third terminal, values of I_D and V_D are obtained from points along the line representing Eq. (6–1). This is called the *load line*.

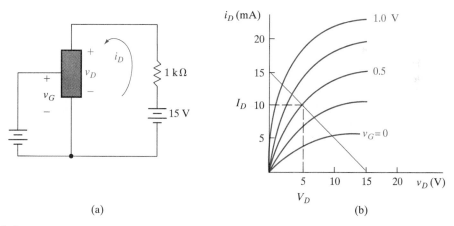

(a)   (b)

Figure 6–2
A three-terminal nonlinear device that can be controlled by the voltage at the third terminal v_G: (a) biasing circuit; (b) I–V characteristic and load line. If V_G = 0.5 V, the d-c values of I_D and V_D are as shown by the dashed lines.

---

[1] We use i_D to symbolize the total current, I_D for the d-c value, and i_d for the a-c component. A similar scheme is used for other currents and voltages.

6.1.2 Amplification and Switching

If an a-c source is added to the control voltage, we can achieve large variations in i_D by making small changes in v_G. For example, as v_G varies about its d-c value by 0.25 V in Fig. 6–2, v_d varies about its d-c value V_D by 2 V. Thus the amplification of the a-c signal is $2/0.25 = 8$. If the curves for equal changes in v_G are equally spaced on the i_D axis, a faithful amplified version of the small control signal can be obtained. This type of voltage-controlled amplification is typical of field-effect transistors For bipolar transistors, a small control current is used to achieve large changes in the device current, achieving current amplification.

Another important circuit function of transistors is the controlled switching of the device off and on. In the example of Fig. 6–2, we can switch from the bottom of the load line ($i_D = 0$) to almost the top ($i_D \simeq E/R$) by appropriate changes in v_G. This type of switching with control at a third terminal is particularly useful in digital circuits.

6.2
THE JUNCTION
FET

In a *junction FET (JFET)* the voltage-variable depletion region width of a junction is used to control the effective cross-sectional area of a conducting *channel*. In the device of Fig. 6–3, the current I_D flows through an n-type channel between two p$^+$ regions. A reverse bias between these p$^+$ regions

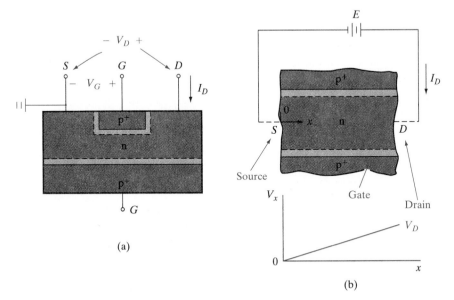

(a)

(b)

Figure 6–3
Simplified cross-sectional view of a junction FET: (a) transistor geometry; (b) detail of the channel and voltage variation along the channel with $V_G = 0$ and small I_D.

2

2

and the channel causes the depletion regions to intrude into the n material, and therefore the effective width of the channel can be restricted. Since the resistivity of the channel region is fixed by its doping, the channel resistance varies with changes in the effective cross-sectional area. By analogy, the variable depletion regions serve as the two doors of a gate, which open and close on the conducting channel.

In Fig. 6–3 electrons in the n-type channel drift from left to right, opposite to current flow. The end of the channel from which electrons flow is called the *source*, and the end toward which they flow is called the *drain*. The p$^+$ regions are called *gates*. If the channel were p-type, holes would flow from the source to the drain, in the same direction as the current flow, and the gate regions would be n$^+$. It is common practice to connect the two gate regions electrically; therefore, the voltage V_G refers to the potential from each gate region G to the source S. Since the conductivity of the heavily doped p$^+$ regions is high, we can assume that the potential is uniform throughout each gate. In the lightly doped channel material, however, the potential varies with position (Fig. 6–3b). If the channel of Fig. 6–3 is considered as a distributed resistor carrying a current I_D it is clear that the voltage from the drain end of the channel D to the source electrode S must be greater than the voltage from a point near the source end to S. For low values of current we can assume a linear variation of voltage V_x in the channel, varying from V_D at the drain end to zero at the source end (Fig. 6–3b).

6.2.1 Pinch-off and Saturation

In Figure 6–4 we consider the channel in a simplified way by neglecting voltage drops between the source and drain electrodes and the respective ends of the channel. For example, we assume that the potential at the drain end of the channel is the same as the potential at the electrode D. This is a good approximation if the source and drain regions are relatively large, so that there is little resistance between the ends of the channel and the electrodes. In Fig. 6–4 the gates are short circuited to the source ($V_G = 0$), such that the potential at $x = 0$ is the same as the potential everywhere in the gate regions. For very small currents, the widths of the depletion regions are close to the equilibrium values (Fig. 6–4a). As the current I_D is increased, however, it becomes important that V_x is large near the drain end and small near the source end of the channel. Since the reverse bias across each point in the gate-to-channel junction is simply V_x when V_G is zero, we can estimate the shape of the depletion regions as in Fig. 6–4b. The reverse bias is relatively large near the drain ($V_{GD} = -V_D$) and decreases toward zero near the source. As a result, the depletion region intrudes into the channel near the drain, and the effective channel area is constricted.

Figure 6–4
Depletion regions
in the channel of
a JFET with zero
gate bias for sev-
eral values of V_D:
(a) linear range;
(b) near pinch-off;
(c) beyond pinch-
off.

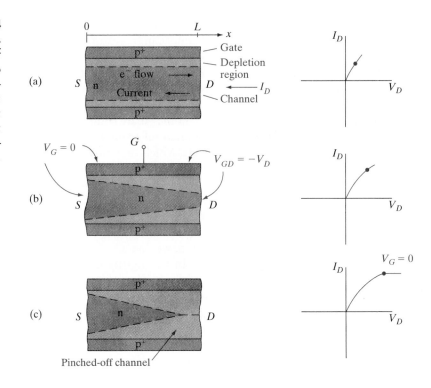

Since the resistance of the constricted channel is higher, the I–V plot for the channel begins to depart from the straight line that was valid at low current levels. As the voltage V_D and current I_D are increased still further, the channel region near the drain becomes more constricted by the depletion regions and the channel resistance continues to increase. As V_D is increased, there must be some bias voltage at which the depletion regions meet near the drain and essentially *pinch off* the channel (Fig. 6–4c). When this happens, the current I_D cannot increase significantly with further increase in V_D. Beyond pinch-off the current is *saturated* approximately at its value at pinch-off.[2] Once electrons from the channel enter the electric field of the depletion region, they are swept through and ultimately flow to the positive drain contact. After the current saturates beyond pinch-off, the differential channel resistance dV_D/dI_D becomes very high. To a good approximation, we can calculate the current at the critical pinch-off voltage and assume there is no further increase in current as V_D is increased.

[2] *Saturation* is used by device engineers in more different contexts than any other word. We have discussed velocity saturation, reverse saturation current of a junction, and now the saturation of FET characteristics. In Chapter 7, we will discuss saturation of a BJT. The student has probably also reached saturation by now in trying to absorb these various meanings.

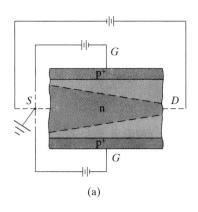

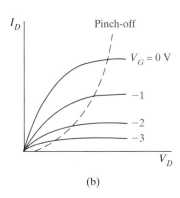

Figure 6–5
Effects of a negative gate bias: (a) increase of depletion region widths with V_G negative; (b) family of current–voltage curves for the channels as V_G is varied.

(a)   (b)

6.2.2 Gate Control

The effect of a negative gate bias $-V_G$ is to increase the resistance of the channel and induce pinch-off at a lower value of current (Fig. 6–5). Since the depletion regions are larger with V_G negative, the effective channel width is smaller and its resistance is higher in the low-current range of the characteristic. Therefore, the slopes of the I_D vs. V_D curves below pinch-off become smaller as the gate voltage is made more negative (Fig. 6–5b). The pinch-off condition is reached at a lower drain-to-source voltage, and the saturation current is lower than for the case of zero gate bias. As V_G is varied, a family of curves is obtained for the I–V characteristic of the channel, as in Fig. 6–5b.

Beyond the pinch-off voltage the drain current I_D is controlled by V_G. By varying the gate bias we can obtain amplification of an a-c signal. Since the input control voltage V_G appears across the reverse-biased gate junctions, the input impedance of the device is high.

We can calculate the pinch-off voltage rather simply by representing the channel in the approximate form of Fig. 6–6. If the channel is symmetrical and the effects of the gates are the same in each half of the channel region, we can restrict our attention to the channel half-width $h(x)$, measured from the center line ($y = L$). The metallurgical half-width of the channel (i.e., neglecting the depletion region) is a. We can find the pinch-off voltage by calculating the reverse bias between the n channel and the p$^+$ gate at the drain end of the channel ($x = L$). For simplicity we shall assume that the channel width at the drain decreases uniformly as the reverse bias increases to pinch-off. If the reverse bias between the gate and the drain is $-V_{GD}$, the width of the depletion region at $x = L$ can be found from Eq. (5–57):

$$W(x = L) = \left[\frac{2\epsilon(-V_{GD})}{qN_d}\right]^{1/2} \quad (V_{GD} \text{ negative}) \qquad (6\text{--}2)$$

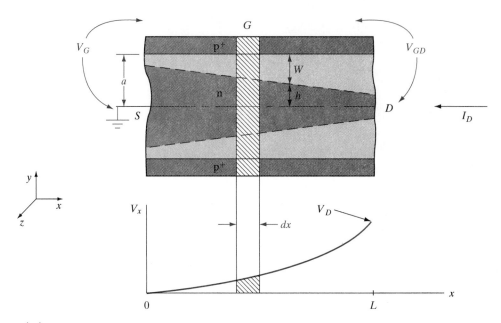

Figure 6–6
Simplified diagram of the channel with definitions of dimensions and differential volume for calculations.

In this expression we assume the equilibrium contact potential V_0 is negligible compared with V_{GD} and the depletion region extends primarily into the channel for the p$^+$-n junction. Including V_0 is left for Prob. 6.2.

Pinch-off occurs at the drain end of the channel when

$$h(x = L) = a - W(x = L) = 0 \tag{6-3}$$

that is, when $W(x = L) = a$. If we define the value of $-V_{GD}$ at pinch-off as V_p, we have

$$\left[\frac{2\epsilon V_P}{qN_d} \right]^{1/2} = a$$

$$\boxed{V_P = \frac{qa^2 N_d}{2\epsilon}} \tag{6-4}$$

The pinch-off voltage V_p is a positive number; its relation to V_D and V_G is

$$V_P = -V_{GD}(\text{pinch-off}) = -V_G + V_D \tag{6-5}$$

where V_G is zero or negative for proper device operation. A forward bias on the gate would cause hole injection from the p$^+$ regions into the channel, eliminating the field-effect control of the device. From Eq. (6–5) it is clear that pinch-off results from a combination of gate-to-source voltage and drain-to-source voltage. Pinch-off is reached at a lower value of V_D (and therefore a lower I_D) when a negative gate bias is applied, in agreement with Fig. 6–5b.

6.2.3   Current–Voltage Characteristics

Calculation of the exact channel current is complicated, although the mathematics is relatively straightforward below pinch-off. The approach we shall take is to find the expression for I_D just at pinch-off, and then assume the saturation current beyond pinch-off remains fairly constant at this value.

In the coordinate system defined in Fig. 6–6, the center of the channel at the source end is taken as the origin. The length of the channel in the x-direction is L, and the depth of channel in the z-direction is Z. We shall call the resistivity of the n-type channel material ρ (valid only in the neutral n material, outside the depletion regions). If we consider the differential volume of neutral channel material $Z2h(x)dx$, the resistance of the volume element is $\rho dx/Z2h(x)$ [see Eq. (3–44)]. Since the current does not change with distance along the channel, I_D is related to the differential voltage change in the element dV_x by the conductance of the element:

$$I_D = \frac{Z2h(x)}{\rho}\frac{dV_x}{dx} \qquad (6\text{--}6)$$

The term $2h(x)$ is the channel width at x.

The half-width of the channel at point x depends on the local reverse bias between gate and channel $-V_{Gx}$:

$$h(x) = a - W(x) = a - \left[\frac{2\epsilon(-V_{Gx})}{qN_d}\right]^{1/2} = a\left[1 - \left(\frac{V_x - V_G}{V_P}\right)^{1/2}\right] \qquad (6\text{--}7)$$

since $V_{Gx} = V_G - V_x$ and $V_p = qa^2N_d/2\epsilon$. Implicit in Eq. (6–7) is the assumption that the expression for $W(x)$ can be obtained by a simple extension of Eq. (6–2) to point x in the channel. This is called the *gradual channel approximation*; it is valid if $h(x)$ does not vary abruptly in any element dx.

The voltage V_{Gx} will be negative since the gate voltage V_G is chosen zero or negative for proper operation. Substituting Eq. (6–7) into Eq. (6–6), we have

$$\frac{2Za}{\rho}\left[1 - \left(\frac{V_x - V_G}{V_P}\right)^{1/2}\right]dV_x = I_D dx \qquad (6\text{--}8)$$

We can solve this equation to obtain

$$\boxed{I_D = G_0 V_P\left[\frac{V_D}{V_P} + \frac{2}{3}\left(-\frac{V_G}{V_P}\right)^{3/2} - \frac{2}{3}\left(\frac{V_D - V_G}{V_P}\right)^{3/2}\right]} \qquad (6\text{--}9)$$

where V_G is negative and $G_0 \equiv 2aZ/\rho L$ is the conductance of the channel for negligible $W(x)$, i.e., with no gate voltage and low values of I_D. This equation is valid only up to pinch-off, where $V_D - V_G = V_p$. If we assume the saturation current remains essentially constant at its value at pinch-off, we have

$$I_D(\text{sat.}) = G_0 V_P \left[\frac{V_D}{V_P} + \frac{2}{3}\left(-\frac{V_G}{V_P} \right)^{3/2} - \frac{2}{3} \right]$$

$$= G_0 V_P \left[\frac{V_G}{V_P} + \frac{2}{3}\left(+\frac{V_G}{V_P} \right)^{3/2} + \frac{1}{3} \right] \qquad (6\text{--}10)$$

where

$$\frac{V_D}{V_P} = 1 + \frac{V_G}{V_P}$$

The resulting family of *I–V* curves for the channel agrees with the results we predicted qualitatively (Fig. 6–5b). The saturation current is greatest when V_G is zero and becomes smaller as V_G is made negative.

We can represent the device biased in the saturation region by an equivalent circuit where changes in drain current are related to gate voltage changes by

$$g_m(\text{sat.}) = \frac{\partial I_D(\text{sat.})}{\partial V_G} = G_0 \left[1 - \left(-\frac{V_G}{V_P} \right)^{1/2} \right] \qquad (6\text{--}11)$$

The quantity g_m is the *mutual transconductance*, with units (A/V) called siemens (S), sometimes called mhos. As a figure of merit for FET devices it is common to describe the transconductance per unit channel width Z. This quantity g_m/Z is usually given in units of millisiemens per millimeter.

It is found experimentally that a square-law characteristic closely approximates the drain current in saturation:

$$I_D(\text{sat.}) \simeq I_{DSS}\left(1 + \frac{V_G}{V_P} \right)^2, \quad (V_G \text{ negative}) \qquad (6\text{--}12)$$

where I_{DSS} is the saturated drain current with $V_G = 0$.

The appearance of a constant value of channel resistivity (in the G_0 term) in Eqs. (6–9)–(6–11) implies that the electron mobility is constant. As mentioned in Sec. 3.4.4, electron velocity saturation at high fields may make this assumption invalid. This is particularly likely for very short channels, where even moderate drain voltage can result in a high field along the channel. Another departure from the ideal model results from the fact that the effective channel length decreases as the drain voltage is increased beyond pinch-off, as Fig. 6–4(c) suggests. In short-channel devices this effect can cause I_D to increase beyond pinch-off, since L appears in the denominator of Eq.

(6–10), in G_0. Therefore, the assumption of constant saturation current is not valid for very short-channel devices.

The depletion of the channel discussed above for a JFET can be accomplished by the use of a reverse-biased Schottky barrier instead of a p-n junction. The resulting device is called a MESFET, indicating that a metal–semiconductor junction is used. This device is useful in high-speed digital or microwave circuits, where the simplicity of Schottky barriers allows fabrication to close geometrical tolerances. There are particular speed advantages for MESFET devices in III–V compounds such as GaAs or InP, which have higher mobilities and carrier drift velocities than Si.

6.3 THE METAL–SEMICONDUCTOR FET

6.3.1 The GaAs MESFET

Figure 6–7 shows schematically a simple MESFET in GaAs. The substrate is undoped or doped with chromium, which has an energy level near the center of the GaAs band gap. In either case the Fermi level is near the center of the gap, resulting in very high resistivity material ($\sim 10^8$ Ω-cm), generally called *semi-insulating* GaAs. On this nonconducting substrate a thin layer of lightly-doped n-type GaAs is grown epitaxially, to form the channel region of the FET.[3] The photolithographic processing consists of defining patterns in the metal layers for source and drain ohmic contacts (e.g., Au–Ge) and for the Schottky barrier gate (e.g., Al). By reverse biasing the Schottky gate, the channel can be depleted to the semi-insulating substrate, and the resulting *I–V* characteristics are similar to the JFET device.

By using GaAs instead of Si, a higher electron mobility is available (see Appendix III), and furthermore GaAs can be operated at higher temperatures (and therefore higher power levels). Since no diffusions are involved in Fig. 6–7, close geometrical tolerances can be achieved and the MESFET can be made very small. Gate lengths $L \lesssim 0.25$ μm are common in these devices. This is important at high frequencies, since drift time and capacitances must be kept to a minimum.

It is possible to avoid the epitaxial growth of the n-type layer and the etched isolation in Fig. 6–7 by using ion implantation. Starting with a semi-insulating GaAs substrate, a thin n-type layer at the surface of each transistor region can be formed by implanting Si or a column VI donor impurity such as Se. This implantation requires an anneal to remove the radiation damage, but the epitaxial growth step is eliminated. In either the fully implanted device or the epitaxial device of Fig. 6–7, the source and drain contacts may be improved by further n[+] implantation in these regions. Because

---

[3]In many cases a high resistivity GaAs epitaxial layer (called a *buffer layer*) is grown between the two layers shown in Fig. 6–7.

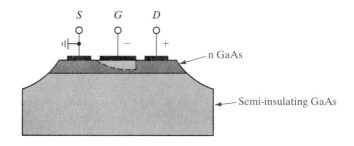

of the relative simplicity of implanted GaAs MESFETs and the isolation be-tween devices provided by the semi-insulating substrate, these structures are commonly used in GaAs integrated circuits.

6.3.2 The High Electron Mobility Transistor (HEMT)

Since the metal–semiconductor field effect transistor (MESFET) is compat-ible with the use of III–V compounds, it is possible to exploit the band gap en-gineering available with heterojunctions in these materials. In order to maintain high transconductance in a MESFET, the channel conductivity must be as high as possible. Obviously, the conductivity can be increased by in-creasing the doping in the channel and thus the carrier concentration. How-ever, increased doping also causes increased scattering by the ionized impurities, which leads to a degradation of mobility (see Fig. 3–23). What is needed is a way of creating a high electron concentration in the channel of a MESFET by some means other than doping. A clever approach to this re-quirement is to grow a thin undoped well (e.g., GaAs) bounded by wider band gap, doped barriers (e.g., AlGaAs). This configuration, called *modulation dop-ing*, results in conductive GaAs when electrons from the doped AlGaAs bar-riers fall into the well and become trapped there, as shown in Fig. 6–8a. Since the donors are in the AlGaAs rather than the GaAs, there is no impurity scat-tering of electrons in the well. If a MESFET is constructed with the channel along the GaAs well (perpendicular to the page in Fig. 6–8), we can take ad-vantage of this reduced scattering and resulting higher mobility. The effect is especially strong at low temperature where lattice (phonon) scattering is also low. This device is called a *modulation doped field-effect transistor (MODFET)* and is also called a *high electron mobility transistor (HEMT)*.

In Fig. 6–8a we have left out the band-bending expected at the AlGaAs/ GaAs interfaces. Based upon the discussion in Section 5.8, we expect the electrons to accumulate at the corners of the well due to band-bending at the heterojunction. In fact, only one heterojunction is required to trap elec-trons, as shown in Fig. 6–8b. Generally, the donors in the AlGaAs layer are purposely separated from the interface by ~100 Å. Using this configuration, we can achieve a high electron concentration in the channel while retaining high mobility, since the GaAs channel region is spatially separated from the ionized impurities which provide the free carriers.

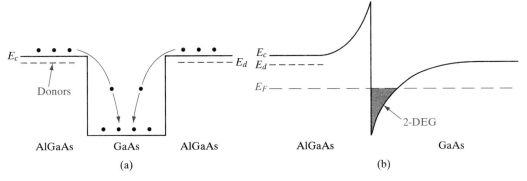

Figure 6–8
(a) Simplified view of modulation doping, showing only the conduction band. Electrons in the donor-doped AlGaAs fall into the GaAs potential well and become trapped. As a result, the undoped GaAs becomes n-type, without the scattering by ionized donors which is typical of bulk n-type material. (b) Use of a single AlGaAs/GaAs heterojunction to trap electrons in the undoped GaAs. The thin sheet of charge due to free electrons at the interface forms a two-dimensional electron gas (2-DEG), which can be exploited in HEMT devices.

In Fig. 6–8b, mobile electrons generated by the donors in the AlGaAs diffuse into the small band gap GaAs layer, and they are prevented from returning to the AlGaAs by the potential barrier at the AlGaAs/GaAs interface. The electrons in the (almost) triangular well form a two-dimensional electron gas (sometimes abbreviated 2-DEG). Sheet carrier densities as high as $10^{12} cm^{-2}$ can be obtained at a single interface such as that shown in Fig. 6–8b. Ionized impurity scattering is greatly reduced simply by separating the electrons from the donors. Also, screening effects due to the extremely high density of the two-dimensional electron gas can reduce ionized impurity scattering further. In properly designed structures, the electron transport approaches that of bulk GaAs with no impurities, so that mobility is limited by lattice scattering. As a result, mobilities above 250,000 cm^2/V-s at 77 K and 2,000,000 cm^2/V-s at 4 K can be achieved.

The advantages of a HEMT are its ability to locate a large electron density ($\sim 10^{12} cm^{-2}$) in a very thin layer (< 100 Å thick) very close to the gate while simultaneously eliminating ionized impurity scattering. The AlGaAs layer in a HEMT is fully depleted under normal operating conditions, and since the electrons are confined to the heterojunction, device behavior closely resembles that of a MOSFET. The advantages of the HEMT over the Si MOSFET are the higher mobility and maximum electron velocity in GaAs compared with Si, and the smoother interfaces possible with an AlGaAs/GaAs heterojunction compared with the Si/SiO$_2$ interface. The high performance of the HEMT translates into an extremely high cutoff frequency, and devices with fast access times.

Although we have discussed the HEMT in terms of the AlGaAs/GaAs heterojunction, other materials are also promising, such as the InGaAsP/InP system. A motivation for avoiding Al$_x$Ga$_{1-x}$As is the presence of a deep-level defect called the DX center for $x > 0.2$, which traps electrons and impairs the HEMT operation. Since very thin layers are involved, materials with slight lattice mismatch can be grown to form *pseudomorphic* HEMTs. An example of such a system is the use of a thin layer of InGaAs grown pseudomorphically on GaAs, followed by AlGaAs. An advantage of this system is that a useful band discontinuity can be achieved using AlGaAs of low enough Al composition to avoid the DX center problem.

The HEMT, or MODFET, is also referred to as a *two-dimensional electron gas FET (2-DEG FET, or TEGFET)* to emphasize the fact that conduction along the channel occurs in a thin sheet of charge. The device has also been called a *separately doped FET (SEDFET)*, to emphasize the fact that the doping occurs in a separate region from the channel.

6.3.3 Short Channel Effects

As mentioned in Section 6.2.3, a variety of modifications to the simple theory of JFET and MESFET operation must be made when the channel length is small (typically < 1 μm). In the past, these short-channel effects would be considered unusual, but now it is common to encounter FET devices in which these effects dominate the *I–V* characteristics. For example, high-field effects occur when 1 V appears across a channel length of 1 μm (10^{-4} cm), giving an electric field of 10 kV/cm.

A simple piecewise-linear approximation to the velocity-field curve assumes a constant mobility (linear) dependence up to some critical field $\mathscr{E}_c$ and a constant saturation velocity v_s for higher fields. For Si a better approximation is

$$v_d = \frac{\mu\mathscr{E}}{1 + \mu\mathscr{E}/v_s} \tag{6–13}$$

where μ is the low-field mobility. These two approximations are shown in Fig. 6–9 (a). If we assume that the electrons passing through the channel drift with a constant saturation velocity v_s, the current takes a simple form

$$I_D = qnv_sA = qN_dv_sZh \tag{6–14}$$

where h is a slow function of V_G. In this case the saturated current follows the velocity saturation, and does not require a true pinch-off in the sense of depletion regions meeting at some point in the channel. In the saturated velocity case, the transconductance g_m is essentially constant, in contrast with the constant mobility case described by Eq. (6–11). As shown in Fig. 6–9(b), the $I_D - V_D$ curves are more evenly spaced if constant saturation velocity dominates, compared with the V_G-dependent spacing shown in Fig. 6–5(b) for the long-channel constant-mobility case.

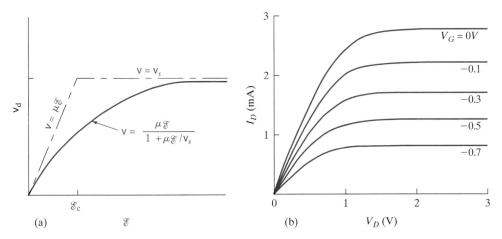

Figure 6–9
Effects of electron velocity saturation at high electric fields: (a) approximations to the saturation of drift velocity with increasing field; (b) drain current–voltage characteristics for the saturated velocity case, showing almost equally spaced curves with increasing gate voltage.

Most devices operate with characteristics intermediate between the constant mobility and the constant velocity regimes. Depending on the details of the field distribution, it is possible to divide up the channel into regions dominated by the two extreme cases, or to use an approximation such as Eq. (6–13).

Another important short-channel effect, described in Section 6.2.3, is the reduction in effective channel length after pinch-off as the drain voltage is increased. This effect is not significant in long-channel devices, since the change in L due to intrusion of the depletion region is a minor fraction of the total channel length. In short-channel devices, however, the effective channel length can be substantially shortened, leading to a slope in the saturated I–V characteristic that is analogous to the Early (base-width narrowing) effect in bipolar transistors discussed in Section 7.7.2.

One of the most widely used electronic devices, particularly in digital integrated circuits, is the *metal–insulator–semiconductor (MIS) transistor*. In this device the channel current is controlled by a voltage applied at a gate electrode that is isolated from the channel by an insulator. The resulting device may be referred to generically as an insulated-gate field effect transistor (IGFET). However, since most such devices are made using silicon for the semiconductor, SiO_2 for the insulator, and metal or heavily doped polysilicon for the gate electrode, the term *MOS field-effect transistor* (MOSFET) is commonly used.

6.4
THE METAL–
INSULATOR–
SEMICONDUCTOR
FET

Figure 6–10
An enhancement-
type n-channel
MOSFET: (a) iso-
metric view of
device and equi-
librium band dia-
gram along
channel; (b) drain
current–voltage
output characteris-
tics as a function
of gate voltage.

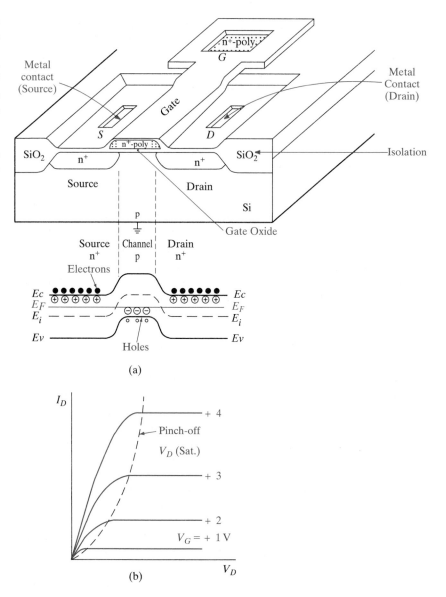

(a)

(b)

6.4.1 Basic Operation and Fabrication

The basic MOS transistor is illustrated in Fig. 6–10a for the case of an enhancement-mode n-channel device formed on a p-type Si substrate. The n^+ source and drain regions are diffused or implanted into a relatively light-ly doped p-type substrate, and a thin oxide layer separates the conducting gate from the Si surface. No current flows from drain to source without a conducting n channel between them. This can be understood clearly by looking at the band diagram of the MOSFET in equilibrium along the channel (Fig. 6–10a). The Fermi level is flat in equilibrium. The conduction band is close to the Fermi level in the n$^+$ source/drain, while the valence band is closer to the Fermi level in the p-type material. Hence, there is a potential barrier for an electron to go from the source to the drain, corresponding to the built-in potential of the back-to-back p-n junctions between the source and drain.

When a positive voltage is applied to the gate relative to the substrate (which is connected to the source in this case), positive charges are in effect deposited on the gate metal. In response, negative charges are induced in the underlying Si, by the formation of a depletion region and a thin surface region containing mobile electrons. These induced electrons form the channel of the FET, and allow current to flow from drain to source. As Fig. 6–10b suggests, the effect of the gate voltage is to vary the conductance of this induced channel for low drain-to-source voltage, analogous to the JFET case. Since electrons are electrostatically induced in the p-type channel region, the channel becomes less p-type, and therefore the valence band moves down, farther away from the Fermi level. This obviously reduces the barrier for electrons between the source, channel and the drain. If the barrier is reduced sufficiently by applying a gate voltage in excess of what is known as the *threshold voltage*, V_T, there is significant current flow from the source to the drain. Thus, one view of a MOSFET is that it is a gate-controlled potential barrier. It is very important to have high-quality, low-leakage p-n junctions in order to ensure a low off-state leakage in the MOSFET. For a given value of V_G there will be some drain voltage V_D for which the current becomes saturated, after which it remains essentially constant.

The threshold voltage V_T is the minimum gate voltage required to induce the channel. In general, the positive gate voltage of an n-channel device (such as that shown in Fig. 6–11) must be larger than some value V_T before a conducting channel is induced. Similarly, a p-channel device (made on an n-type substrate with p-type source and drain implants or diffusions) requires a gate voltage more negative than some threshold value to induce the required positive charge (mobile holes) in the channel. There are exceptions to this general rule, however, as we shall see. For example, some n-channel devices have a channel already with zero gate voltage, and in fact a negative gate voltage is required to turn the device off. Such a "normally on" device is called a *depletion-mode* transistor, since gate voltage is used to deplete a channel which exists at equilibrium. The more common MOS transistor is "normally off" with zero gate voltage, and operates in the *enhancement mode* by applying gate voltage large enough to induce a conducting channel.

An alternative view of a MOSFET is that it is a gate-controlled resistor. If the (positive) gate voltage exceeds the threshold voltage in an n-channel device, electrons are induced in the p-type substrate. Since this channel is connected to the n$^+$ source and drain regions, the structure looks electrically like an induced n-type resistor. As the gate voltage increases, more electron

Figure 6–11
n-channel
MOSFET cross-
sections under dif-
ferent operating
conditions: (a) lin-
ear region for
$V_G > V_T$ and
$V_D < (V_G - V_T)$;
(b) onset of
saturation
at pinch-off,
$V_G > V_T$ and
$V_D = (V_G - V_T)$;
(c) strong satura-
tion, $V_G > V_T$ and
$V_D > (V_G - V_T)$.

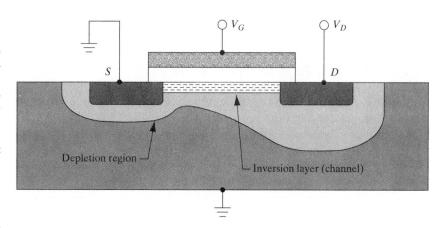

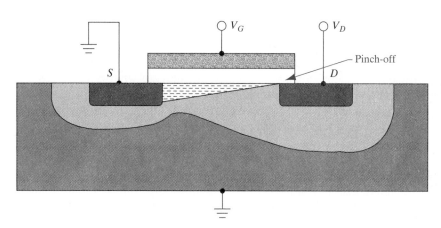

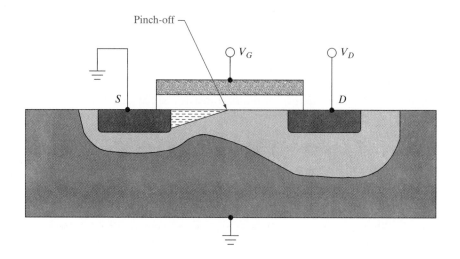

charge is induced in the channel and, therefore, the channel becomes more conducting. The drain current initially increases linearly with the drain bias (the *linear* regime) (Fig. 6–10b). As more drain current flows in the channel, however, there is more ohmic voltage drop along the channel such that the channel potential varies from zero near the grounded source to whatever the applied drain potential is near the drain end of the channel. Hence, the voltage difference between the gate and the channel reduces from V_G near the source to $(V_G - V_D)$ near the drain end. Once the drain bias is increased to the point that $(V_G - V_D) = V_T$, threshold is barely maintained near the drain end, and the channel is said to be pinched off. Increasing the drain bias beyond this point (V_{DSAT}) causes the point at which the channel gets pinched off to move more and more into the channel, closer to the source end (Fig. 6–11c). Electrons in the channel are pulled into the pinch-off region and travel at the saturation drift velocity because of the very high longitudinal electric field along the channel. Now, the drain current is said to be in the *saturation* region because it does not increase with drain bias significantly (Fig. 6–10b). Actually, there is a slight increase of drain current with drain bias due to various effects such as channel length modulation and drain-induced barrier lowering (DIBL) that will be discussed in Section 6.5.10.

The MOS transistor is particularly useful in digital circuits, in which it is switched from the "off" state (no conducting channel) to the "on" state. The control of drain current is obtained at a gate electrode which is insulated from the source and drain by the oxide. Thus the d-c input impedance of an MOS circuit can be very large.

Both n-channel and p-channel MOS transistors are in common usage. The n-channel type illustrated in Fig. 6–10 is generally preferred because it takes advantage of the fact that the electron mobility in Si is larger than the mobility of holes. In much of the discussion to follow we will use the n-channel (p-type substrate) example, although the p-channel case will be kept in mind also.

Let us give a very simplified description of how such an n-channel MOSFET can be fabricated. A much more detailed discussion is given in Section 9.3.1. An ultra-thin (~5–10 nm) dry thermal silicon dioxide is grown on the p-type substrate. This serves as the gate insulator between the conducting gate and the channel. We immediately cover it with LPCVD of polysilicon, which is doped very heavily n^+ using P diffusion in order to make it behave electrically like a metal electrode. The doped polysilicon layer is then patterned to form the gates, and etched anisotropically by RIE to achieve vertical walls (Section 5.1.7). The gate itself is used as an implant mask for an n^+ implant which forms the source/drain junctions abutted to the gate edges, but is blocked from the channel region. Such a scheme is called a *self-aligned process* because we did not have to use a separate lithography step for the source/drain formation. Self-alignment is simple and is very useful because we thereby guarantee that there will be some overlap of the gate with the source/drain but not too much overlap. The advantages of this are discussed in Section 6.5.8. The implanted dopants must be annealed for reasons discussed in Section 5.1.4. Finally, the MOSFETs have to be properly interconnected according to the circuit layout, using metallization. This involves LPCVD of an oxide dielectric, etching contact holes by RIE, sputter depositing a suitable metal such as Al, patterning and etching it.

As shown in Fig. 6–10a, the MOSFET is surrounded on all sides by a thick SiO$_2$ layer. This layer provides critical electrical isolation between adjacent transistors on an integrated circuit. We shall see in Section 9.3.1 that such *isolation* or *field* regions can be formed in several ways, such as *LOCal Oxidation of Silicon (LOCOS)*. Briefly, it involves depositing a LPCVD Si$_3$N$_4$ layer over the entire substrate before the fabrication of the MOSFETs, patterning and etching it so that it is removed only in the isolation regions, but not in the *active* regions where the MOSFETs will be formed subsequently. A boron *channel stop* implant is then done in the isolation regions. Exploiting the useful property of Si$_3$N$_4$ that it blocks thermial oxidation, a thick LOCOS oxide is selectively grown by wet oxidation only in the isolation regions. The reason why a thick field oxide layer and a boron channel stop implant leads to electrical isolation is discussed in Section 6.5.5.

6.4.2 The Ideal MOS Capacitor

The surface effects that arise in an apparently simple MOS structure are actually quite complicated. Although many of these effects are beyond the scope of this discussion, we will be able to identify those which control typical MOS transistor operation. We begin by considering an uncomplicated idealized case, and then include effects encountered in real surfaces in the next section.

Some important definitions are made in the energy band diagram of Fig. 6–12a. The work function characteristic of the metal (see Section 2.2.1) can be defined in terms of the energy required to move an electron from the Fermi level to outside the metal. In MOS work it is more convenient to use a *modified work function* $q\Phi_m$ for the metal–oxide interface. The energy $q\Phi_m$ is measured from the metal Fermi level to the conduction band of the oxide.[4] Similarly, $q\Phi_s$ is the modified work function at the semiconductor–oxide interface. In this idealized case we assume that $\Phi_m = \Phi_s$, so there is no difference in the two work functions. Another quantity that will be useful in later discussions is $q\phi_F$, which measures the position of the Fermi level below the intrinsic level E_i for the semiconductor. This quantity indicates how strongly p-type the semiconductor is [see Eq. (3–25)].

The MOS structure of Fig. 6–12a is essentially a capacitor in which one plate is a semiconductor. If we apply a negative voltage between the metal and the semiconductor (Fig. 6–12b), we effectively deposit a negative charge on the metal. In response, we expect an equal net positive charge to accumulate

[4]On the MOS band diagrams of this section we show a break in the electron energy scale leading to the insulator conduction band, since the band gap of SiO$_2$ (or other typical insulators) is much greater than that of the Si.

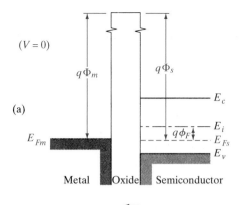

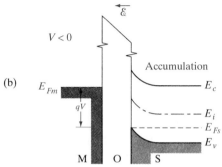

(a)

(b)

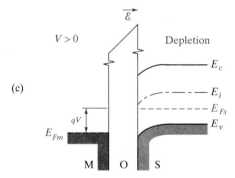

(c)

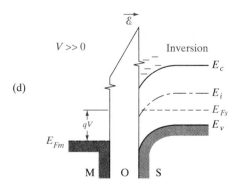

(d)

Figure 6–12
Band diagram for the ideal MOS structure at: (a) equilibrium; (b) negative voltage causes hole accumulation in the p-type semiconductor; (c) positive voltage depletes holes from the semiconductor surface: (d) a larger positive voltage causes inversion— a "n-type" layer at the semiconductor surface.

at the surface of the semiconductor. In the case of a p-type substrate this occurs by *hole accumulation* at the semiconductor–oxide interface.

Since the applied negative voltage *depresses* the electrostatic potential of the metal relative to the semiconductor, the electron energies are *raised* in the metal relative to the semiconductor.[5] As a result, the Fermi level for the metal E_{Fm} lies above its equilibrium position by qV, where V is the applied voltage.

Since Φ_m and Φ_s do not change with applied voltage, moving E_{Fm} up in energy relative to E_{Fs} causes a tilt in the oxide conduction band. We expect such a tilt since an electric field causes a gradient in E_i (and similarly in E_v and E_c) as described in Section 4.4.2:

$$\mathscr{E}(x) = \frac{1}{q}\frac{dE_i}{dx} \qquad \text{(see 4–26)}$$

The energy bands of the semiconductor bend near the interface to accommodate the accumulation of holes. Since

$$p = n_i e^{(E_i - E_F)/kT} \qquad \text{(see 3–25)}$$

it is clear that an increase in hole concentration implies an increase in $E_i - E_F$ at the semiconductor surface.

Since no current passes through the MOS structure, there can be no variation in the Fermi level within the semiconductor. Therefore, if $E_i - E_F$ is to increase, it must occur by E_i moving up in energy near the surface. The result is a bending of the semiconductor bands near the interface. We notice in Fig. 6–12b that the Fermi level near the interface lies closer to the valence band, indicating a larger hole concentration than that arising from the doping of the p-type semiconductor.

In Fig. 6–12c we apply a positive voltage from the metal to the semiconductor. This raises the potential of the metal, lowering the metal Fermi level by qV relative to its equilibrium position. As a result, the oxide conduction band is again tilted. We notice that the slope of this band, obtained by simply moving the metal side down relative to the semiconductor side, is in the proper direction for the applied field, according to Eq. (4–26).

The positive voltage deposits positive charge on the metal and calls for a corresponding net negative charge at the surface of the semiconductor. Such a negative charge in p-type material arises from *depletion* of holes from the region near the surface, leaving behind uncompensated ionized acceptors. This is analogous to the depletion region at a p-n junction discussed in Section 5.2.3. In the depleted region the hole concentration decreases, moving E_i closer to E_F, and bending the bands down near the semiconductor surface.

If we continue to increase the positive voltage, the bands at the semiconductor surface bend down more strongly. In fact, a sufficiently large volt-

[5]Recall that an electrostatic potential diagram is drawn for positive test charges, in contrast with an electron energy diagram which is drawn for negative charges.

age can bend E_i *below* E_F (Fig. 6–12d). This is a particularly interesting case, since $E_F \gg E_i$ implies a large electron concentration in the conduction band.

The region near the semiconductor surface in this case has conduction properties typical of n-type material, with an electron concentration given by Eq. (3–25a). This n-type surface layer is formed not by doping, but instead by *inversion* of the originally p-type semiconductor due to the applied voltage. This inverted layer, separated from the underlying p-type material by a depletion region, is the key to MOS transistor operation.

We should take a closer look at the inversion region, since it becomes the conducting channel in the FET. In Fig. 6–13 we define a potential ϕ at any point x, measured relative to the equilibrium position of E_i. The energy $q\phi$ tells us the extent of band bending at x, and $q\phi_s$ represents the band bending at the surface. We notice that $\phi_s = 0$ is the *flat band* condition for this ideal MOS case (i.e., the bands look like Fig. 6–12a). When $\phi_s < 0$, the bands bend up at the surface, and we have hole accumulation (Fig. 6–12b). Similarly, when $\phi_s > 0$, we have depletion (Fig. 6–12c). Finally, when ϕ_s is positive and larger than ϕ_F, the bands at the surface are bent down such that $E_i(x = 0)$ lies below E_F, and inversion is obtained (Fig. 6–12d).

While it is true that the surface is inverted whenever ϕ_s is larger than ϕ_F, a practical criterion is needed to tell us whether a true n-type conducting channel exists at the surface. The best criterion for *strong inversion* is that the surface should be as strongly n-type as the substrate is p-type. That is, E_i should lie as far below E_F at the surface as it is above E_F far from the surface. This condition occurs when

$$\phi_s(\text{inv.}) = 2\phi_F = 2\frac{kT}{q}\ln\frac{N_a}{n_i} \qquad (6\text{–}15)$$

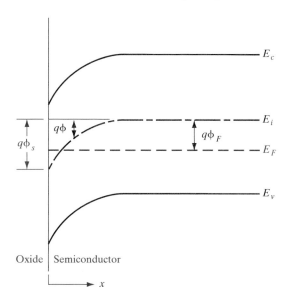

Oxide | Semiconductor

x

Figure 6–13
Bending of the semiconductor bands at the onset of strong inversion: the surface potential ϕ_s is twice the value of ϕ_F in the neutral p material.

A surface potential of ϕ_F is required to bend the bands down to the intrinsic condition at the surface ($E_i = E_F$), and E_i must then be depressed another $q\phi_F$ at the surface to obtain the condition we call strong inversion.

The electron and hole concentrations are related to the potential $\phi(x)$ defined in Fig. 6–13. Since the equilibrium electron concentration is

$$n_0 = n_i e^{(E_F - E_i)/kT} = n_i e^{-q\phi_F/kT} \tag{6–16}$$

we can easily relate the electron concentration at any x to this value:

$$n = n_i e^{-q(\phi_F - \phi)/kT} = n_0 e^{q\phi/kT} \tag{6–17}$$

and similarly for holes:

$$p_0 = n_i e^{q\phi_F/kT} = N_a^- \tag{6–18a}$$

$$p = p_0 e^{-q\phi/kT} \tag{6–18b}$$

at any x. We could combine these equations with Poisson's equation (6–19) and the usual charge density expression (6–20) to solve for $\phi(x)$:

$$\frac{\partial^2 \phi}{\partial x^2} = -\frac{\rho(x)}{\epsilon_s} \tag{6–19}$$

$$\rho(x) = q(N_d^+ - N_a^- + p - n) \tag{6–20}$$

Let us solve this equation to determine the total integrated charge per unit area, Q_s, as a function of the surface potential, ϕ_s. Substituting Eqs. (6–16), (6–17) and (6–18) for the electron and hole concentrations in Eqs. (6–19) and (6–20), we get

$$\frac{\partial^2 \phi}{\partial x^2} = \frac{\partial}{\partial x}\left(\frac{\partial \phi}{\partial x}\right) = -\frac{q}{\epsilon_s}\left[p_0\left(e^{-\frac{q\phi}{kT}} - 1\right) - n_0\left(e^{\frac{q\phi}{kT}} - 1\right)\right] \tag{6–21}$$

It should be kept in mind that

$$\frac{-\partial \phi}{\partial x}$$

is the electric field, $\mathscr{E}$, at a depth x.

Integrating Eq. (6–21) from the bulk (where the bands are flat, the electric fields are zero and the carrier concentrations are determined solely by the doping), towards the surface, we get

$$\int_0^{\frac{\partial \phi}{\partial x}}\left(\frac{\partial \phi}{\partial x}\right)d\left(\frac{\partial \phi}{\partial x}\right) = -\frac{q}{\epsilon_s}\int_0^{\phi}\left[p_0\left(e^{\frac{-q\phi}{kT}} - 1\right) - n_0\left(e^{\frac{q\phi}{kT}} - 1\right)\right]d\phi \tag{6–22}$$

After integration, we then get

$$\mathscr{E}^2 = \left(\frac{2kT\,p_0}{\epsilon_s}\right)\left[\left(e^{-\frac{q\phi}{kT}} + \frac{q\phi}{kT} - 1\right) + \frac{n_0}{p_0}\left(e^{\frac{q\phi}{kT}} - \frac{q\phi}{kT} - 1\right)\right] \tag{6–23}$$

A particularly important case is at the surface ($x = 0$) where the surface perpendicular electric field, $\mathscr{E}_s$, becomes

$$\mathscr{E}_s = \frac{\sqrt{2}kT}{qL_D}\left[\left(e^{-\frac{q\phi_s}{kT}} + \frac{q\phi_s}{kT} - 1\right) + \frac{n_0}{p_0}\left(e^{\frac{q\phi_s}{kT}} - \frac{q\phi_s}{kT} - 1\right)\right]^{\frac{1}{2}} \quad (6\text{–}24)$$

where we have introduced a new term, the *Debye screening length*,

$$L_D = \sqrt{\frac{\epsilon_s kT}{q^2 p_0}} \quad (6\text{–}25)$$

The Debye length is a very important concept in semiconductors. It gives us an idea of the distance scale in which charge imbalances are screened or smeared out. For example, if we think of inserting a positively charged sphere in an n-type semiconductor we know that the mobile electrons will crowd around the sphere. If we move away from the sphere by several Debye lengths, the positively charged sphere and the negative electron cloud will look like a neutral entity. Not surprisingly, L_D depends inversely on doping because the higher the carrier concentration, the more easily screening takes place. For n-type material we should use n_0 in Eq. (6–25).

By using Gauss' law at the surface, we can relate the integrated space charge per unit area to the electric displacement, keeping in mind that the electric field or displacement deep in the substrate is zero.

$$Q_s = -\epsilon_s \mathscr{E}_s \quad (6\text{–}26)$$

The space charge density per unit area Q_s in Eq. (6–26) is plotted as a function of the surface potential ϕ_s in Fig. 6–14. We see from Eq. (6–24) and Fig. 6–14 that when the surface potential is zero (flatband conditions), the net space charge is zero. This is because the fixed dopant charges are cancelled by the mobile carrier charges at flatband. When the surface potential is negative, it attracts and forms an accumulation layer of the majority carrier holes at the surface. The first term in Eq. (6–24) is the dominant one, and the accumulation space charge increases very strongly (exponentially) with negative surface potential. It is easy to see why by looking at Eq. (6–18), which gives the surface hole concentration in a p-type semiconductor as a function of surface potential. Since the bandbending decreases as a function of depth, the integrated accumulation charge involves averaging over depth and introduces a factor of 2 in the exponent. Mathematically, this is due to the square root in Eq. (6–24). It must be noted that since this charge is due to the mobile majority carriers (holes in this case), the charge piles up near the oxide–silicon interface, since typical accumulation layer thicknesses are ~20 nm. Also, because of the exponential dependence of accumulation charge on surface potential, the bandbending is generally small or is said to be *pinned* to nearly zero.

On the other hand, for a positive surface potential, we see from Eq. (6–24) that initially the second (linear) term is the dominant one. Although

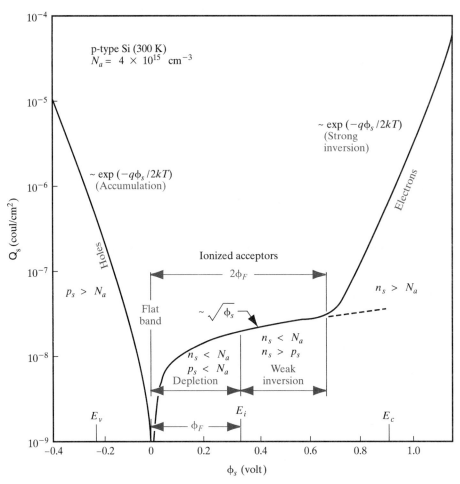

Figure 6–14
Variation of space-charge density in the semiconductor as a function of the surface potential ϕ_s for p-type silicon with $N_a = 4 \times 10^{15}$ cm^{-3} at room temperature. p_s and n_s are the hole and electron concentrations at the surface, ϕ_F is the potential difference between the Fermi level and the intrinsic level of the bulk. (Garrett and Brattain, Phys. Rev., 99, 376 (1995).)

the exponential term, $\exp(q\phi_s/kT)$ is very large, it is multiplied by the ratio of the minority to majority carrier concentration which is very small, and is initially negligible. Hence, the space charge for small positive surface potentials increases as $\sim \sqrt{\phi_s}$ as shown in Fig. 6–14. As discussed in detail later in this section, this corresponds to the depletion region charge due to the exposed, fixed immobile dopants (acceptors in this case). The depletion width typically extends over several hundred nm. At some point, the band bending is twice the Fermi potential ϕ_F, which is enough for the onset of strong inversion. Now, the exponential term $\exp(q\phi_s(\text{inv.})/kT)$ multiplied by the minority carrier concentration n_0 is equal to the majority carrier concentration

p_0. Hence, for band bending beyond this point, it becomes the dominant term. As in the case of accumulation, the mobile inversion charge now increases very strongly with bias, as indicated by Eq. (6–17), and shown in Fig. 6–14. The typical inversion layer thicknesses are ~5 nm, and the surface potential now is essentially pinned at $2\phi_F$.

It may be pointed out that in accumulation, and especially in inversion, the carriers are confined in the x-direction in narrow, essentially triangular potential wells, causing quantum mechanical particle-in-a-box states or sub-bands, similar to those discussed in Chapter 2. However, the carriers are free in the other directions (parallel to the oxide–silicon interface). This leads to a 2-dimensional electron gas (2DEG) or hole gas, with a "staircase" constant density of states, as discussed in Appendix IV. The detailed analysis of these effects is, unfortunately, beyond the scope of our discussion here.

The charge distribution, electric field, and electrostatic potential for the inverted surface are sketched in Fig. 6–15. For simplicity we use the de-pletion approximation of Chapter 5 in this figure, assuming complete deple-tion for $0 < x < W$, and neutral material for $x > W$. In this approximation the charge per unit area[6] due to uncompensated acceptors in the depletion re-gion is $-qN_aW$. The positive charge Q_m on the metal is balanced by the neg-ative charge Q_s in the semiconductor, which is the depletion layer charge plus the charge due to the inversion region Q_n:

$$Q_m = -Q_s = qN_aW - Q_n \qquad (6\text{–}27)$$

The width of the inversion region is exaggerated in Fig. 6–15 for illus-trative purposes. Actually, the width of this region is generally less than 100 Å. Thus we have neglected it in sketching the electric field and potential dis-tribution. In the potential distribution diagram we see that an applied volt-age V appears partially across the insulator (V_i) and partially across the depletion region of the semiconductor (ϕ_s):

$$V = V_i + \phi_s \qquad (6\text{–}28)$$

The voltage across the insulator is obviously related to the charge on either side, divided by the capacitance:

$$V_i = \frac{-Q_s d}{\epsilon_i} = \frac{-Q_s}{C_i} \qquad (6\text{–}29)$$

where ϵ_i is permittivity of the insulator and C_i is the insulator capacitance per unit area. The charge Q_s will be negative for n-channel, giving a positive V_i.

Using the depletion approximation, we can solve for W as a function of ϕ_s (Prob. 6.8). The result is the same as would be obtained for an n⁺-p junc-tion in Chapter 5, for which the depletion region extends almost entirely into the p region:

[6]In this chapter we will use charge per unit area Q and capacitance per unit area C to avoid carrying A throughout the discussion.

Figure 6–15
Approximate distributions of charge, electric field, and electrostatic potential in the ideal MOS capacitor in inversion. The relative width of the inverted region is exaggerated for illustrative purposes, but is neglected in the field and potential diagrams.

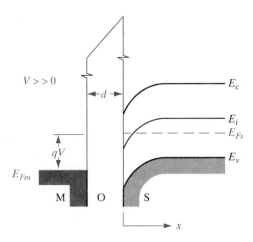

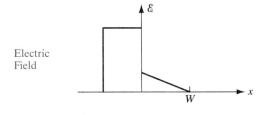

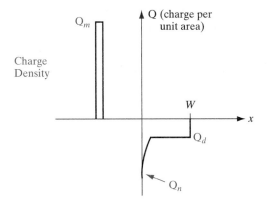

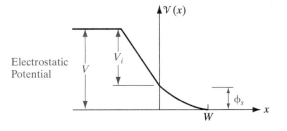

$$W = \left[\frac{2\epsilon_s\phi_s}{qN_a}\right]^{1/2} \tag{6-30}$$

This depletion region grows with increased voltage across the capacitor until strong inversion is reached. After that, further increases in voltage result in stronger inversion rather than in more depletion. Thus the maximum value of the depletion width is

$$W_m = \left[\frac{2\epsilon_s\phi_s(\text{inv.})}{qN_a}\right]^{1/2} = 2\left[\frac{\epsilon_s kT \ln(N_a/n_i)}{q^2 N_a}\right]^{1/2} \tag{6-31}$$

using Eq. (6–15). We know the quantities in this expression, so W_m can be calculated.

Find the maximum width of the depletion region for an ideal MOS capacitor on p-type Si with $N_a = 10^{16}$ cm^{-3}.   **EXAMPLE 6–1**

The relative dielectric constant of Si is 11.8 from Appendix III. We get ϕ_F **SOLUTION** from Eq. (6–15):

$$\phi_F = \frac{kT}{q}\ln\frac{N_a}{n_i} = 0.0259\ln\frac{10^{16}}{1.5 \times 10^{10}} = 0.347 \text{ V}$$

Thus

$$W_m = 2\sqrt{\frac{\epsilon_s\phi_F}{qN_a}} = 2\left[\frac{(11.8)(8.85 \times 10^{-14})(0.347)}{(1.6 \times 10^{-19})(10^{16})}\right]^{1/2}$$

$$= 3.01 \times 10^{-5} \text{ cm} = 0.301 \text{ μm}$$

The charge per unit area in the depletion region Q_d at strong inversion is[7]

$$Q_d = -qN_aW_m = -2(\epsilon_s qN_a\phi_F)^{1/2} \tag{6-32}$$

The applied voltage must be large enough to create this depletion charge plus the surface potential $\phi_s(\text{inv.})$. The *threshold* voltage required for strong inversion, using Eqs. (6–15), (6–28), and (6–29), is

$$V_T = -\frac{Q_d}{C_i} + 2\phi_F \quad (\textit{ideal case}) \tag{6-33}$$

---

[7]In the p-channel (n-type substrate) case, for which ϕ_F is negative, we use $Q_d = +qN_dW_m = 2(\epsilon_s qN_d|\phi_F|)^{1/2}$.

This assumes the negative charge at the semiconductor surface Q_s at inversion is mostly due to the depletion charge Q_d. The threshold voltage represents the minimum voltage required to achieve strong inversion, and is an extremely important quantity for MOS transistors. We will see in the next section that other terms must be added to this expression for real MOS structures.

The capacitance–voltage characteristics of this ideal MOS structure (Fig. 6–16) vary depending on whether the semiconductor surface is in accumulation, depletion, or inversion.

Since the capacitance for MOSFETs is voltage dependent, we must use the more general expression in Eq. (5–55) for the voltage-dependent semiconductor capacitance,

$$C_s = \frac{dQ}{dV} = \frac{dQ_s}{d\phi_s} \qquad (6\text{–}34)$$

Actually, if one looks at the electrical equivalent circuit of a MOS capacitor or MOSFET, it is the series combination of a fixed, voltage-independent gate oxide (insulator) capacitance, and a voltage-dependent semiconductor capacitance (defined according to Eq. (6–34)), such that the overall MOS capacitance becomes voltage dependent. The semiconductor capacitance itself can be determined from the slope of the Q_s versus ϕ_s plot (Fig. 6–14). It is clear that the semiconductor capacitance in accumulation is very high because the slope is so steep; i.e., the accumulation charge changes a lot with surface potential. Hence, the series capacitance in accumulation is basically the insulator capacitance, C_i. Since, for negative voltage, holes are accumulated at the surface (Fig. 6–12b), the MOS structure appears almost like a

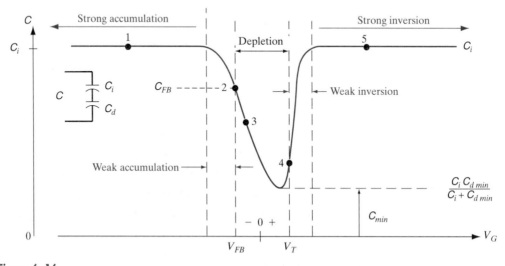

Figure 6–16

Capacitance–voltage relation for an n-channel (p-substrate) MOS capacitor. The dashed curve for $V > V_T$ is observed at high measurement frequencies. The flat band voltage V_{FB} will be discussed in Section 6.4.3. When the semiconductor is in depletion, the semiconductor capacitance C_s is denoted as C_d.

parallel-plate capacitor, dominated by the insulator properties $C_i = \epsilon_i/d$ (point 1 in Fig. 6–16). As the voltage becomes less negative, the semiconductor surface is depleted. Thus a depletion-layer capacitance C_d is added in series with C_i:

$$C_d = \frac{\epsilon_s}{W} \qquad (6\text{--}35)$$

where ϵ_s is the semiconductor permittivity and W is the width of the depletion layer from Eq. (6–30). The total capacitance is

$$C = \frac{C_i C_d}{C_i + C_d} \qquad (6\text{--}36)$$

The capacitance decreases as W grows (point 3) until finally inversion is reached at V_T (point 4). In the depletion region, the small signal semiconductor capacitance is given by the same formula (Eq. 6-34) which gives the variation of the (depletion) space charge with surface potential. Since the charge increases as $\sim\sqrt{\phi_s}$, the depletion capacitance will obviously decrease as $1/\sqrt{\phi_s}$, exactly as for the depletion capacitance of a p-n junction (See Eq. 5–63).

After inversion is reached, the small signal capacitance depends on whether the measurements are made at high (typically ~1 MHz) or low (typically ~1–100 Hz) frequency, where "high" and "low" are with respect to the generation–recombination rate of the minority carriers in the inversion layer. If the gate voltage is varied rapidly, the charge in the inversion layer cannot change in response, and thus does not contribute to the small signal a-c capacitance. Hence, the semiconductor capacitance is at a minimum, corresponding to a maximum depletion width.

On the other hand, if the gate bias is changed slowly, there is time for minority carriers to be generated in the bulk, drift across the depletion region to the inversion layer, or go back to the substrate and recombine. Now, the semiconductor capacitance, using the same Eq. (6–34), is very large because we saw in Fig. 6–14 that the inversion charge increases exponentially with ϕ_s. Hence, the low frequency MOS series capacitance in strong inversion is basically C_i once again.

What is the frequency dependence of the capacitance in accumulation (Fig. 6–12a)? We get a very high capacitance both at low and high frequencies because the majority carriers in the accumulation layer can respond much faster than minority carriers. While minority carriers respond on the time scale of generation–recombination times (typically hundreds of microseconds in Si), majority carriers respond on the time scale of the dielectric relaxation time, $\tau_D = \rho\epsilon$, where ρ is the resistivity and ϵ is the permittivity. τ_D is analogous to the RC time constant of a system, and is small for the majority carriers ($\sim10^{-13}$s). As an interesting aside, it may be pointed out that in inversion, although the high frequency capacitance for MOS capacitors is low, it is high ($= C_i$) for MOSFETs because now the inversion charge can flow in readily and very fast $(\sim\tau_D)$ from the source/drain regions rather than having to be created by generation–recombination in bulk.

EXAMPLE 6–2

Using the conditions of Example 6–1 and a 100-Å-thick SiO$_2$ layer, we can calculate major points on the C–V curve of Fig. 6–16. The relative dielectric constant of SiO$_2$ is 3.9.

$$C_i = \frac{\epsilon_i}{d} = \frac{(3.9)(8.85 \times 10^{-14})}{10^{-6}} = 3.45 \times 10^{-7} \text{F/cm}^2$$

$$Q_d = -qN_aW_m = -(1.6 \times 10^{-19})(10^{16})(0.301 \times 10^{-4})$$
$$= -4.82 \times 10^{-8} \text{C/cm}^2$$

$$V_T = -\frac{Q_d}{C_i} + 2\phi_F = \frac{4.82 \times 10^{-8}}{34.5 \times 10^{-8}} + 2(0.347) = 0.834 \text{ V}$$

At maximum depletion

$$C_d = \frac{\epsilon_s}{W_m} = \frac{(11.8)(8.85 \times 10^{-14})}{0.301 \times 10^{-4}} = 3.47 \times 10^{-8} \text{ F/cm}^2$$

$$C_{\min} = \frac{C_iC_d}{C_i + C_d} = \frac{34.5 \times 3.47}{34.5 + 3.47}10^{-8} = 3.15 \times 10^{-8} \text{ F/cm}^2$$

6.4.3 Effects of Real Surfaces

When MOS devices are made using typical materials (e.g., n$^+$ polysilicon-SiO$_2$–Si), departures from the ideal case described in the previous section can strongly affect V_T and other properties. First, there is a work function difference between the doped polysilicon gate and substrate, which depends on the substrate doping. Here, the heavily doped polysilicon acts as a metal electrode. Second, there are inevitably charges at the Si–SiO$_2$ interface and within the oxide which must be taken into account.

Work Function Difference. We expect Φ_s to vary depending on the doping of the semiconductor. Figure 6–17 illustrates the work function potential difference $\Phi_{ms} = \Phi_m - \Phi_s$ for n$^+$ polysilicon on Si as the doping is varied. We note that Φ_{ms} is always negative for this case, and is most negative for heavily doped p-type Si (i.e., for E_F close to the valence band).

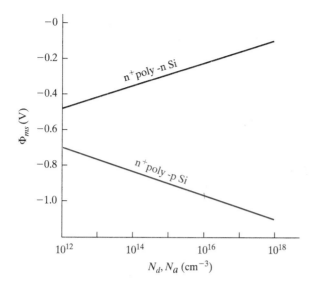

Figure 6–17
Variation of the metal–semiconductor work function potential differ-ence Φ_{ms} with substrate doping concentration, for n^+ poly-Si.

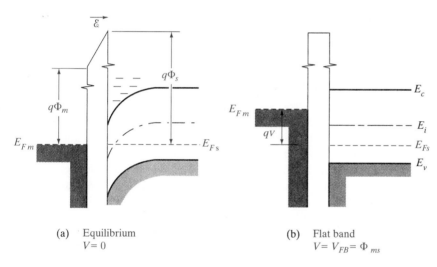

Figure 6–18
Effect of a nega-tive work function difference ($\Phi_{ms} < 0$): (a) band bending and formation of negative charge at the semicon-ductor surface; (b) achievement of the flat band con-dition by applica-tion of a negative voltage.

If we try to construct an equilibrium diagram with Φ_{ms} negative (Fig. 6–18a), we find that in aligning E_F we must include a tilt in the oxide conduction band (implying an electric field). Thus the metal is positively charged and the semi-conductor surface is negatively charged at equilibrium, to accommodate the work function difference. As a result, the bands bend down near the semicon-ductor surface. In fact, if Φ_{ms} is sufficiently negative, an inversion region can exist with no external voltage applied. To obtain the *flat band* condition pictured in Fig. 6–18b, we must apply a negative voltage to the metal ($V_{FB} = \Phi_{ms}$).

Figure 6–19
Effects of charges
in the oxide
and at the inter-
face: (a) defini-
tions of charge
densities (C/cm²)
due to various
sources; (b) repre-
senting these
charges as an
equivalent sheet
of positive charge
Q_i at the oxide-
semiconductor
interface. This
positive charge
induces an equiv-
alent negative
charge in the
semiconductor,
which requires
a negative
gate voltage to
achieve the flat
band condition.

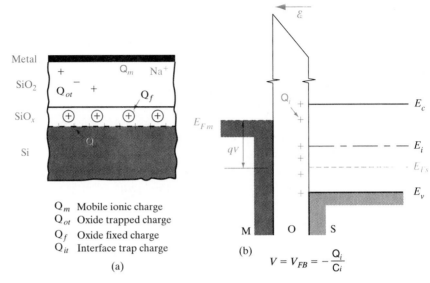

Q_m Mobile ionic charge
Q_{ot} Oxide trapped charge
Q_f Oxide fixed charge
Q_{it} Interface trap charge

(a)

(b)

$$V = V_{FB} = -\frac{Q_i}{C_i}$$

Interface Charge. In addition to the work function difference, the equilibrium MOS structure is affected by charges in the insulator and at the semiconductor–oxide interface (Fig. 6–19). For example, alkali metal ions (particularly Na⁺) can be incorporated inadvertently in the oxide during growth or subsequent processing steps. Since sodium is a common contaminant, it is necessary to use extremely clean chemicals, water, gases, and processing environment to minimize its effect on dielectric layers. Sodium ions introduce positive charges (Q_m) in the oxide, which in turn induce negative charges in the semiconductor. The effect of such positive ionic charges in the oxide depends upon the number of ions involved and their distance from the semiconductor surface (Prob. 6.13). The negative charge induced in the semiconductor is greater if the Na⁺ ions are near the interface than if they are farther away. The effect of this ionic charge on threshold voltage is complicated by the fact that Na⁺ ions are relatively mobile in SiO_2, particularly at elevated temperatures, and can thus drift in an applied electric field. Obviously, a device with V_T dependent on its past history of voltage bias is unacceptable. Fortunately, Na contamination of the oxide can be reduced to tolerable levels by proper care in processing. The oxide also contains trapped charges (Q_{ot}) due to imperfections in the SiO_2.

In addition to oxide charges, a set of positive charges arises from *interface states* at the Si–SiO₂ interface. These charges, which we will call Q_{it}, result from the sudden termination of the semiconductor crystal lattice at the oxide interface. Near the interface is a transition layer (SiO_x) containing fixed charges (Q_f). As oxidation takes place in forming the SiO_2 layer, Si is removed from the surface and reacts with the oxygen. When the oxidation is stopped, some ionic Si is left near the interface. These ions, along with un-completed Si bonds at the surface, result in a sheet of positive charge Q_f near the interface. This charge depends on oxidation rate and subsequent heat treatment, and also on crystal orientation. For carefully treated Si–SiO$_2$ interfaces, typical charge densities due to Q_{it} and Q_f are about 10^{10} charges/cm^2 for samples with {100} surfaces. The interface charge density is about a factor of ten higher on {111} surfaces. That is why MOS devices are generally made on {100} Si.

For simplicity we will include the various oxide and interface charges in an *effective* positive charge at the interface Q_i (C/cm^2). The effect of this charge is to induce an equivalent negative charge in the semiconductor. Thus an additional component must be added to the flat band voltage:

$$V_{FB} = \Phi_{ms} - \frac{Q_i}{C_i} \qquad (6\text{--}37)$$

Since the difference in work function and the positive interface charge both tend to bend the bands down at the semiconductor surface, a negative voltage must be applied to the metal relative to the semiconductor to achieve the flat band condition of Fig. 6–19b.

6.4.4 Threshold Voltage

The voltage required to achieve flat band should be added to the threshold voltage equation (6–33) obtained for the ideal MOS structure (for which we assumed a zero flat band voltage)

$$\boxed{V_T = \Phi_{ms} - \frac{Q_i}{C_i} - \frac{Q_d}{C_i} + 2\phi_F} \qquad (6\text{--}38)$$

Thus the voltage required to create strong inversion must be large enough to first achieve the flat band condition (Φ_{ms} and Q_i/C_i terms), then accommodate the charge in the depletion region (Q_d/C_i), and finally to induce the inverted region ($2\phi_F$). This equation accounts for the dominant threshold voltage effects in typical MOS devices. It can be used for both n-type and p-type substrates[8] if appropriate signs are included for each term (Fig. 6–20). Typically Φ_{ms} is negative, although its value varies as in Fig. 6–17. The interface charge is positive, so the contribution of the $-Q_i/C_i$ term is negative for either substrate type. On the other hand, the charge in the depletion region is negative for ionized acceptors (p-type substrate, n-channel device) and is positive for ionized donors (n-type substrate, p channel). Also, the term ϕ_F,

---

[8]It is important to remember that n-channel devices are made on p-type substrates, and p-channel devices have n-type substrates.

Figure 6–20
Influence of materials parameters on threshold voltage: (a) the threshold voltage equation indicating signs of the various contributions; (b) variation of V_T with substrate doping for n-channel and p-channel n^+ poly-SiO_2–Si devices.

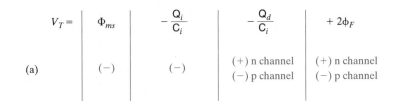

$V_T =$	Φ_{ms}	$-\dfrac{Q_i}{C_i}$	$-\dfrac{Q_d}{C_i}$	$+2\phi_F$
(a)	$(-)$	$(-)$	$(+)$ n channel $(-)$ p channel	$(+)$ n channel $(-)$ p channel

(b)

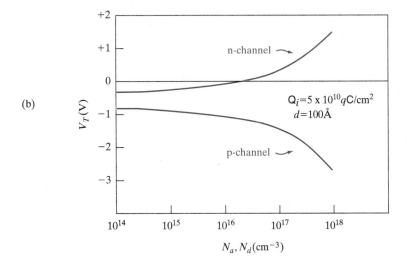

which is defined as $(E_i - E_F)/q$ in the neutral substrate, can be positive or negative, depending on the conductivity type of the substrate. Considering the signs in Fig. 6–20, we see that all four terms give negative contributions in the p-channel case. Thus we expect negative threshold voltages for typical p-channel devices. On the other hand, n-channel devices may have either positive or negative threshold voltages, depending on the relative values of terms in Eq. (6–38).

All terms in Eq. (6–38) except Q_i/C_i depend on the doping in the substrate. The terms Φ_{ms} and ϕ_F have relatively small variations as E_F is moved up or down by the doping. Large changes can occur in Q_d, which varies with the square root of the doping impurity concentration as in Eq. (6–32). We illustrate the variation of threshold voltage with substrate doping in Fig. 6–20. As expected from Eq. (6–38), V_T is always negative for the p-channel case. In the n-channel case, the negative flat band voltage terms can dominate for lightly doped p-type substrates, resulting in a negative threshold voltage. However, for more heavily doped substrates, the increasing contribution of N_a to the Q_d term dominates, and V_T becomes positive.

We should pause here and consider what positive or negative V_T means for the two cases. In a p-channel device we expect to apply a negative voltage from metal to semiconductor in order to induce the positive charges in the channel. In this case a negative threshold voltage means that the negative voltage we apply must be larger than V_T in order to achieve strong inversion. In the n-channel case we expect to apply a positive voltage to the metal to induce the channel. Thus a positive value for V_T means the applied voltage must be larger than this threshold value to obtain strong inversion and a conducting n channel. On the other hand, a negative V_T in this case means that a channel exists at $V = 0$ due to the Φ_{ms} and Q_i effects (Figs. 6-18 and 6-19), and we must apply a negative voltage V_T to turn the device off. Since lightly doped substrates are desirable to maintain a high breakdown voltage for the drain junction, Fig. 6-20 suggests that V_T will be negative for n-channel devices made by standard processing. This tendency for the formation of depletion mode (normally on) n-channel transistors is a problem which must be dealt with by special fabrication methods to be described in Section 6.5.5.

We can calculate V_T for the MOS structure described in Examples 6–1 and 6–2, including the effects of flat band voltage. If n⁺ polysilicon is used for the gate, Fig. 6–17 indicates $\Phi_{ms} = -0.95$ V for $N_a = 10^{16}$ cm^{-3}. Assuming an interface charge of $5 \times 10^{10}\, q$(C/cm²), we obtain **EXAMPLE 6–3**

$$V_T = \Phi_{ms} + 2\phi_F - \frac{1}{C_i}(Q_i + Q_d)$$

$$= -0.95 + 0.694 - \frac{(5 \times 10^{10} \times 1.6 \times 10^{-19}) - 4.82 \times 10^{-8}}{34.5 \times 10^{-8}}$$

$$= -0.14\ V$$

This value corresponds to the $N_a = 10^{16}$ cm^{-3} point in Fig. 6–20 for the n-channel case.

6.4.5 MOS Capacitance–Voltage Analysis

Let us see how the various parameters of a MOS device such as insulator thickness, substrate doping, and V_T can be determined from the C–V characteristics (Fig. 6–21). First, the shape of the C–V curve depends upon the type of substrate doping. If the high frequency capacitance is large for negative gate biases and small for positive biases, it is a p-type substrate, and vice versa. From the low frequency C–V curve for p-type material, as the gate bias is made more positive (or less negative), the capacitance goes down slowly in depletion and then rises rapidly in inversion. As a result, the low frequency C–V is not quite symmetric in shape. For n-type substrates, the C–V curves would be the mirror image of Fig. 6–21.

Figure 6–21
Fast interface
state determina-
tion: (a) High fre-
quency and low
frequency C–V
curves showing
impact of fast in-
terface states; (b)
energy levels in
the bandgap due
to fast interface
states; (c) equiva-
lent circuit of
MOS structure
showing capaci-
tance components
due to gate oxide
(C_i), depletion
layer in the chan-
nel (C_d) and fast
interface states
(C_{it}).

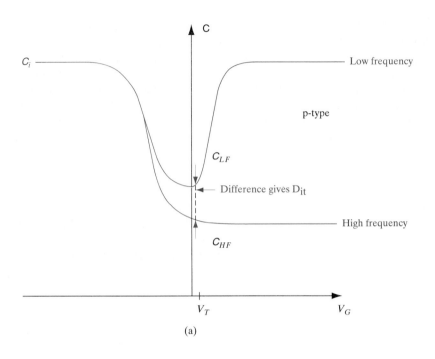

(a)

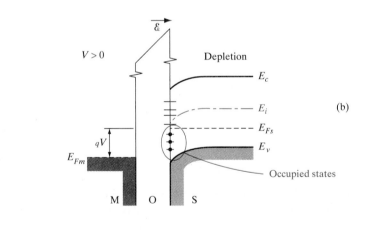

(b)

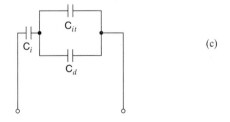

(c)

The capacitance $C_i = \epsilon_i/d$ in accumulation or strong inversion (at low frequencies) gives us the insulator thickness, d. The minimum MOS capacitance, C_{min}, is the series combination of C_i and the minimum depletion capacitance, $C_{dmin} = \epsilon_s/W_m$, corresponding to the maximum depletion width. We can in principle use the measurement of C_{min} to determine the substrate doping. However, from Eq. (6–31) we see that the dependence of W_m on N_a is complicated, and we get a transcendental equation which can only be solved numerically. Actually, an approximate, iterative solution exists which gives us N_a in terms of the minimum depletion capacitance, C_{dmin}.

$$N_a = 10^{\,[30.388 + 1.683\,\log C_{dmin} - 0.03177(\log C_{d\,min})^2]} \tag{6–39}$$

where C_{dmin} is in F/cm^2.

Once the substrate doping is obtained, we can determine the flatband capacitance from it. It can be shown that the semiconductor capacitance at flatband C_{FB} (point 2 in Fig. 6–16) is determined from the Debye length capacitance

$$C_{debye} = \frac{\sqrt{2}\epsilon_s}{L_D} \tag{6–40}$$

where the Debye length is dependent on doping as described in Eq. (6–25). The overall MOS flatband capacitance, C_{FB}, is the series combination of C_{debye} and C_i. We can thus determine V_{FB} corresponding to the C_{FB}. Once C_i, V_{FB} and substrate doping are obtained, all terms in the V_T expression (Eq. 6-38) are known. Interestingly, the threshold voltage V_T does not correspond to exactly the minimum of the C–V characteristics, C_{min}, but a slightly higher capacitance marked as point 4 in Fig. 6–16. In fact, it corresponds to the series combination of C_i and $2C_{dmin}$, rather than the series combination of C_i and C_{dmin}. The reason for this is that when we change the gate bias around strong inversion, the change of charge in the semiconductor is the sum of the change in depletion charge and the mobile inversion charge, where the two are equal in magnitude at the onset of strong inversion.

We can also determine MOS parameters such as the *fast interface state* density, D_{it}, and mobile ion charges, Q_m, from C–V measurements (Figs. 6-21 and 6-22). The term fast interface state refers to the fact that these defects can change their charge state relatively fast in response to changes of the gate bias. As the surface potential in a MOS device is varied, the fast interface states or traps in the bandgap can move above or below the Fermi level in response to the bias, because their positions relative to the band-edges are fixed (Fig. 6–21b). Keeping in mind the property of the Fermi–Dirac distribution that energy levels below the Fermi level have a high probability of occupancy by electrons, while levels above the Fermi level tend to be empty, we see that a fast interface state moving above the Fermi level would tend to give up its trapped electron to the semiconductor (or equivalently capture a hole). Conversely, the same fast interface state below the Fermi level captures an electron (or gives up a hole). It obviously makes sense to talk in terms of electrons or holes, depending on which is the majority carrier in the semiconductor. Since charge storage results in capacitance, the fast interface states give rise to a capacitance which is in parallel with the depletion capacitance in the channel (and hence is additive), and this combination is in series with the insulator capcitance C_i. The fast interface states can keep pace with low frequency variations of the gate bias (~1–1000 Hz), but not at extremely high frequencies (~1 MHz). So the fast interface states contribute to the low frequency capacitance C_{LF}, but not the high frequency capacitance C_{HF}. Clearly, from the difference between the two, we ought to be able to compute the fast interface state density. Although we will not do the detailed derivation here, it can be shown that

$$D_{it} = \frac{1}{q}\left(\frac{C_i C_{LF}}{C_i - C_{LF}} - \frac{C_i C_{HF}}{C_i - C_{HF}}\right) cm^{-2}eV^{-1} \qquad (6\text{–}41)$$

While the fast interface states can respond quickly to voltage changes, the fixed oxide charges Q_f, as the name implies, do not change their charge state regardless of the gate bias or surface potential. As mentioned above, the effect of these charges on the flatband and threshold voltage depends not only on the number of charges but also their location relative to the oxide–silicon interface (Fig. 6–22). Hence, we must take a weighted sum of these charges, counting charges closer to the oxide–silicon interface more heavily than those that are farther away. This position dependence is the basis of what is called the *bias-temperature stress test* for measuring the mobile ion content, Q_m. We heat up the MOS device to ~200–300°C (to make the ions more mobile) and apply a positive gate bias to generate a field of ~1 MV/cm within the oxide. After cooling the capacitor to room temperature, the C–V characteristics are measured. We have seen how V_{FB} can be determined from the C–V curve, using Eq. (6–40) and $C_i.V_{FB}$ is also given by Eq. (6–37). The positive bias repels positive mobile ions such as Na$^+$ to the oxide–silicon interface so that they contribute fully to a flatband voltage we can call V_{FB}^+. Next, the capacitor is heated up again, a negative bias is applied so that the ions drift to the gate electrode, and another C–V measurement is made. Now, the mobile ions are too far away to affect the semiconductor bandbending, but induce an equal and opposite charge on the gate electrode. From the resulting C–V, the new flatband, V_{FB}^-, is determined. From the difference of the two flatband voltages, we can determine the mobile ion content using

$$Q_m = C_i(V_{FB}^+ - V_{FB}^-) \qquad (6\text{–}42)$$

6.4.6 Time-dependent Capacitance Measurements

During C–V measurements, if the gate bias is varied rapidly from accumulation to inversion, the depletion width can momentarily become greater

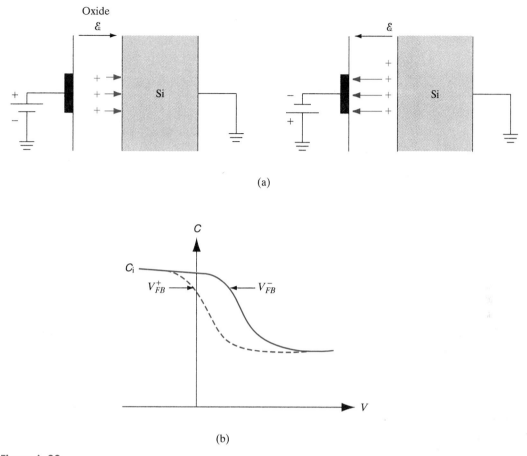

Oxide

(a)

(b)

Figure 6–22
Mobile ion determination: (a) Movement of mobile ions due to positive and negative bias-temperature stress; (b) C–V characteristics under positive (dashed line) and negative (solid line) bias-temperature.

than the theoretical maximum for gate biases beyond V_T. This phenomenon is known as *deep depletion*, and causes the MOS capacitance to drop below the theoretical minimum, C_{min}, for a transient period. After a time period characteristic of the minority carrier lifetime, which determines the rate of generation of the minority carriers in the MOS device, the depletion width collapses back to the theoretical maximum (and the capacitance recovers to C_{min}). This capacitance transient, C–t, forms the basis of a powerful technique to measure the lifetime, known as the Zerbst technique. It was shown by Zerbst that by plotting the C–t data as in Fig. 6–23, the slope is inversely proportional to the lifetime (τ).

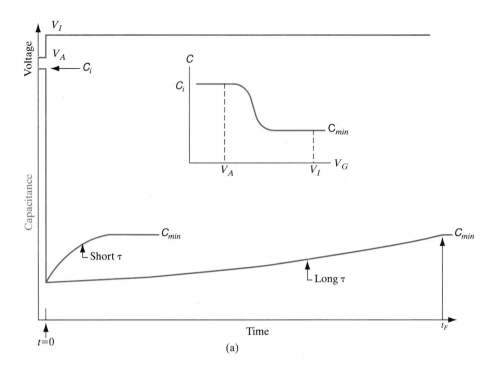

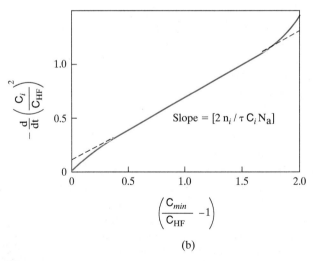

Figure 6–23
Zerbst plots: (a) time-dependent MOS capacitance (C_{HF}) due to the application of a step voltage V_A (which puts the capacitor in accumulation) to V_I (which puts the capacitor in inversion); (b) extraction of minority carrier lifetimes from MOS capacitance-time data.

6.4.7 Current—voltage Characteristics of MOS Gate Oxides

An ideal gate insulator does not conduct any current, but for real insulators there can be some leakage current which varies with the voltage or electric field across the gate oxide. By looking at the band diagram of the MOS system perpendicular to the oxide–silicon interface (Fig. 6–24), we see that for electrons in the conduction band, there is a barrier, ΔE_c (= 3.1 eV). Although electrons with energy less than this barrier cannot go through the oxide classically, it was discussed in Chapter 2 that quantum mechanically electrons can tunnel through a barrier, especially if the barrier thickness is sufficiently small. The detailed calculation of the *Fowler–Nordheim tunneling* curent for electrons going from the Si conduction band to the conduction band of SiO_2, and then having the electrons "hop" along in the oxide to the gate electrode involves solving the Schrödinger equation for the electron wave function. The Fowler–Nordheim tunneling current I_{FN} can be expressed as a function of the electric field in the gate oxide:

$$I_{FN} \propto \mathscr{E}_{ox}^2 \exp\left(\frac{-B}{\mathscr{E}_{ox}}\right) \qquad (6\text{–}43)$$

where B is a constant depending on m_n^* and the barrier height.

As gate oxides are made thinner in successive generations of MOSFETs, the tunneling barrier in the gate oxide becomes so thin that the electrons in the conduction band of Si can tunnel through the gate oxide and emerge in the gate, without having to go via the conduction band of the gate oxide. This is known as *direct tunneling* rather than Fowler–Nordheim tunneling. The overall physics is similar, but some of the details are different. For instance, Fowler–Nordheim tunneling involves a triangular barrier, while direct tunneling is through a trapezoidal barrier (Fig. 6–24a). Such tunneling currents are becoming a major problem in modern devices because the useful feature of high input impedance for MOS devices is degraded.

Prolonged charge transport through gate oxides can ultimately cause catastrophic electrical breakdown of the oxides. This is known as *Time-Dependent Dielectric Breakdown (TDDB)*. One of the popular models that explains this degradation involves electrons tunneling into the conduction band of the gate oxide from the negative electrode (cathode), then gaining energy from the electric field, thus becoming "hot" electrons in the gate oxide. If they gain sufficient energy, they can cause impact ionization within the oxide and create electron–hole pairs. The electrons are accelerated toward the (positive) Si substrate, while the holes travel toward the gate. However, electron and hole mobilities are extremely small in SiO_2. Hole mobilities are particularly low (~0.01 cm^2/Vs). Hence, there is a great propensity for these impact-generated holes to be trapped at defect sites within the oxide, near the cathode. The resulting band diagram (Fig. 6–25) is altered by this sheet of

Figure 6–24
Current–voltage
characteristics of
gate oxides: (a)
Fowler–Nordheim
and direct tunnel-
ing through thin
gate oxides;
(b) plot of
Fowler–Nordheim
tunneling leakage
current as a func-
tion of electric
field across
the oxide.

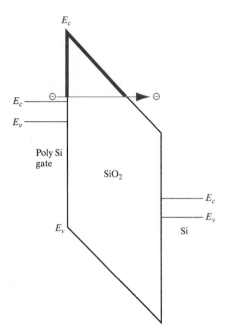

Fowler–Nordheim tunneling

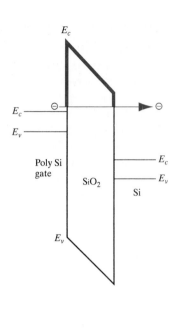

Direct tunneling

(a)

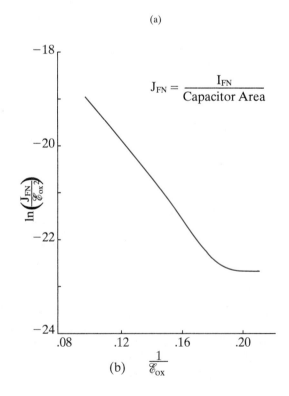

$$J_{FN} = \frac{I_{FN}}{\text{Capacitor Area}}$$

(b)

trapped positive charge, which causes the internal electric field between this point and the gate to increase. A similar distortion of the electric field near the Si anode is created by the trapped impact-generated electrons. However, the steepest slope in Fig. 6–25, and therefore the highest field, is near the gate. As a result, the barrier for electron tunneling from the gate into the oxide is reduced. More electrons can tunnel into the oxide, and cause more impact ionization. We get a positive feedback effect that can lead to a runaway TDDB process.

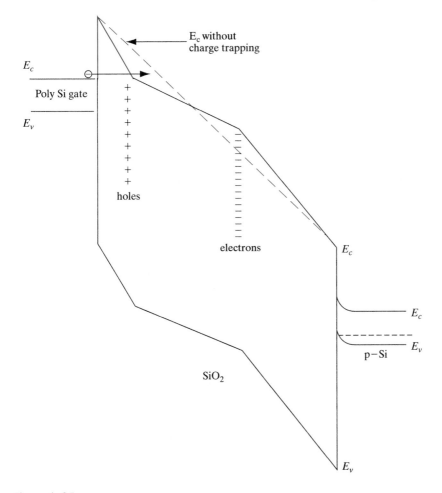

Figure 6–25
Time-dependent dielectric breakdown of oxides: Band diagram of a MOS device showing the band edges in the polysilicon gate, oxide and Si substrate. Trapped holes and electrons in the oxide distort the band edges, and increase the electric field in the oxide near the gate. The tunneling barrier width is seen to be less than if there were no charge trapping (dashed line).

The MOS transistor is also called a surface field-effect transistor, since it depends on control of current through a thin channel at the surface of the semiconductor (Fig. 6–10). When an inversion region is formed under the gate, current can flow from drain to source (for an n-channel device). In this section we analyze the conductance of this channel and find the $I_D - V_D$ characteristics as a function of gate voltage V_G. As in the JFET case, we will find these characteristics below saturation and then assume I_D remains essentially constant above saturation.

6.5.1 Output Characteristics

The applied gate voltage V_G is accounted for by Eq. (6–28) plus the voltage required to achieve flat band:

$$V_G = V_{FB} - \frac{Q_s}{C_i} + \phi_s \qquad (6\text{–}44)$$

The induced charge Q_s in the semiconductor is composed of mobile charge Q_n and fixed charge in the depletion region Q_d. Substituting $Q_n + Q_d$ for Q_s, we can solve for the mobile charge:

$$Q_n = -C_i\left[V_G - \left(V_{FB} + \phi_s - \frac{Q_d}{C_i}\right)\right] \qquad (6\text{–}45)$$

At threshold the term in brackets can be written $V_G - V_T$ from Eq. (6–38).

With a voltage V_D applied, there is a voltage rise V_x from the source to each point x in the channel. Thus the potential $\phi_s(x)$ is that required to achieve strong inversion $(2\phi_F)$ plus the voltage V_x:

$$Q_n = -C_i\left[V_G - V_{FB} - 2\phi_F - V_x - \frac{1}{C_i}\sqrt{2q\epsilon_s N_a(2\phi_F + V_x)}\right] \qquad (6\text{–}46)$$

If we neglect the variation of $Q_d(x)$ with bias V_x, Eq. (6–46) can be simplified to

$$Q_n(x) = -C_i(V_G - V_T - V_x) \qquad (6\text{–}47)$$

This equation describes the mobile charge in the channel at point x (Fig. 6–26). The conductance of the differential element dx is $\bar{\mu}_n Q_n(x)Z/dx$, where Z is the width of the channel and $\bar{\mu}_n$ is a *surface* electron mobility (indicating the mobility in a thin region near the surface is not the same as in the bulk material). At point x we have

$$I_D dx = \bar{\mu}_n Z |Q_n(x)| dV_x \qquad (6\text{–}48)$$

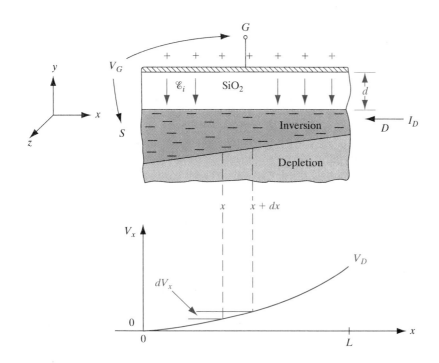

Figure 6–26
Schematic view of the n-channel region of a MOS transistor under bias below pinch-off, and the variation of voltage V_x along the conducting channel.

Integrating from source to drain,

$$\int_0^L I_D dx = \bar{\mu}_n ZC_i \int_0^{V_D} (V_G - V_T - V_x)dV_x$$

$$I_D = \frac{\bar{\mu}_n ZC_i}{L}[(V_G - V_T)V_D - \tfrac{1}{2}V_D^2] \tag{6–49}$$

where

$$\frac{\bar{\mu}_n ZC_i}{L} = k_N$$

determines the conductance and transconductance of the *n*-channel MOSFET (see Eqs. (6–51) and (6–54)).

In this analysis the depletion charge Q_d in the threshold voltage V_T is simply the value with no drain current. This is an approximation, since $Q_d(x)$ varies considerably when V_D is applied, to reflect the variation in V_x (see Fig. 6–26b). However, Eq. (6–49) is a fairly accurate description of drain current for low values of V_D, and is often used in approximate design calculations because of its simplicity. A more accurate and general expression is obtained by including the variation of $Q_d(x)$. Performing the integration of Eq. (6–48) using Eq (6–46) for $Q_n(x)$, one obtains

$$I_D = \frac{\bar{\mu}_n Z C_i}{L}$$

$$\times \left\{ (V_G - V_{FB} - 2\phi_F - \tfrac{1}{2}V_D)V_D - \frac{2}{3}\frac{\sqrt{2\epsilon_s q N_a}}{C_i}[(V_D + 2\phi_F)^{3/2} - (2\phi_F)^{3/2}] \right\} \quad (6\text{–}50)$$

The drain characteristics that result from these questions are shown in Fig. 6–10c. If the gate voltage is above threshold ($V_G > V_T$), the drain current is described by Eq. (6–50) or approximately by Eq. (6–49) for low V_D. Initially the channel appears as an essentially linear resistor, dependent on V_G. The conductance of the channel in this linear region can be obtained from Eq. (6–49) with $V_D \ll (V_G - V_T)$:

$$g = \frac{\partial I_D}{\partial V_D} \simeq \frac{Z}{L}\bar{\mu}_n C_i(V_G - V_T) \quad (6\text{–}51)$$

where $V_G > V_T$ for a channel to exist.

As the drain voltage is increased, the voltage across the oxide decreases near the drain, and Q_n becomes smaller there. As a result the channel becomes pinched off at the drain end, and the current saturates. The saturation condition is approximately given by

$$V_D(\text{sat.}) \simeq V_G - V_T \quad (6\text{–}52)$$

The drain current at saturation remains essentially constant for larger values of drain voltage. Substituting Eq. (6–52) into Eq. (6–49), we obtain

$$I_D(\text{sat.}) \simeq \tfrac{1}{2}\bar{\mu}_n C_i\frac{Z}{L}(V_G - V_T)^2 = \frac{Z}{2L}\bar{\mu}_n C_i V_D^2(\text{sat.}) \quad (6\text{–}53)$$

for the approximate value of drain current at saturation.

The transconductance in the saturation range can be obtained approximately by differentiating Eq. (6–53) with respect to the gate voltage:

$$g_m(\text{sat.}) = \frac{\partial I_D(\text{sat.})}{\partial V_G} \simeq \frac{Z}{L}\bar{\mu}_n C_i(V_G - V_T) \quad (6\text{–}54)$$

The derivations presented here are based on the n-channel device. For the p-channel enhancement transistor the voltages V_D, V_G, and V_T are negative, and current flows from source to drain (Fig. 6–27).

6.5.2   Transfer Characteristics

The output characteristics plot the drain current as a function of the drain bias, with gate bias as a parameter (Fig. 6–27). On the other hand, the *transfer* characteristics plot the output drain current as a function of the input gate bias, for fixed drain bias (Fig. 6–28a). Clearly, in the linear region, I_D versus V_G should be a straight line from Eq. (6–49). The intercept of this line on the V_G axis is the linear region threshold voltage, V_T (lin.) and the slope

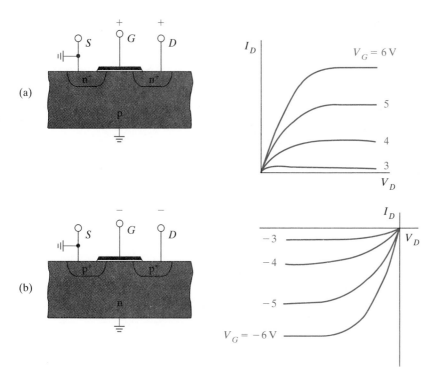

Figure 6–27
Drain current–
voltage character-
istics for enhance-
ment transistors:
(a) for n-channel
V_D, V_G, V_T, and I_D
are positive; (b)
for p-channel all
these quantities
are negative.

(divided by the applied V_D) gives us the linear value of k_N, k_N(lin.), of the n-channel MOSFET. If we look at actual data, however, we see that while the characteristics are approximately linear at low gate bias, at high gate biases the drain current increases sub-linearly. The transconductance, g_m (lin.), in the linear region can be obtained by differentiating the right hand side of Eq. (6–49) with respect to gate bias. The g_m (lin.) is plotted as a function of V_G in Fig. 6–28b. It may be noted that the transconductance is zero below V_T because there is little drain current. It goes through a maximum at the point of inflection of the I_D–V_G curve, and then decreases. This decrease is due to two factors that will be discussed in Sections 6.5.3 and 6.5.8: degradation of the effective channel mobility as a function of increasing transverse electric field across the gate oxide, and source/drain series resistance.

For the transfer characteristics in the saturation region, since Eq. (6–53) shows a quadratic dependence of I_D on V_G, we get a linear behavior by plotting not the drain current, but rather the square root of I_D, as a function of V_G (Fig. 6–29). In this case the intercept gives us the threshold voltage in the saturation region, V_T(sat.). We shall see in Section 6.5.10 that due to effects such as drain induced barrier lowing (DIBL), for short channel length MOS-FETs the V_T(sat.) can be lower than V_T(lin.), while the long channel values are similar. Similarly, the slope of the transfer characteristics can be used to determine the value of k_N in the saturation region, k_N(sat.) for the n-channel MOSFET, which can be different from k_N(lin.) for short channel devices.

Figure 6–28
Linear region transfer characteristics: (a) plot of drain current versus gate voltage for MOSFETs in the linear region; (b) transconductance as a function of gate bias.

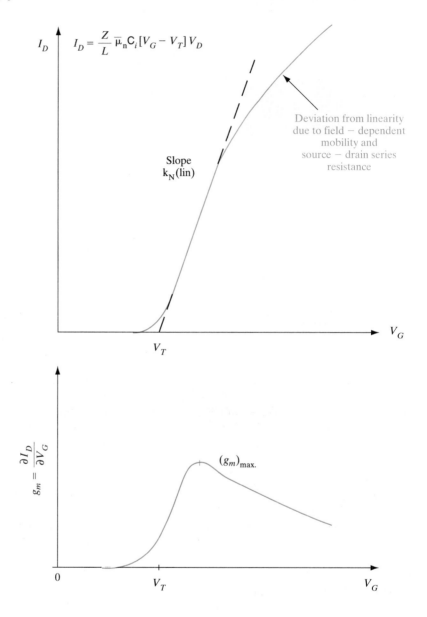

6.5.3 Mobility Models

The mobility of carriers in the channel of a MOSFET is lower than in bulk semiconductors because there are additional scattering mechanisms. Since carriers in the channel are very close to the semiconductor–oxide interface, they are scattered by surface roughness and by coulombic interaction with fixed charges in the gate oxide. When the carriers travel in the inversion layer from the source to the drain, they encounter microscopic roughness on an

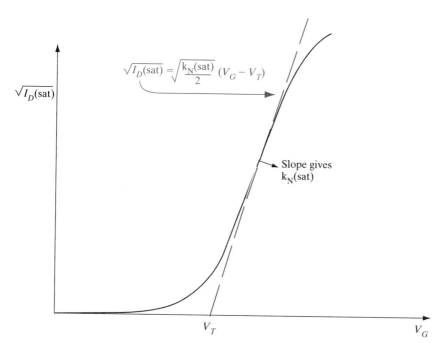

Figure 6–29
Saturation region
transfer character-
istics: plot of
square root of the
drain current ver-
sus gate voltage
for MOSFETs.

In the figure:

$$\sqrt{I_D(\text{sat})} = \sqrt{\frac{k_N(\text{sat})}{2}}\,(V_G - V_T)$$

Slope gives $k_N(\text{sat})$

$\sqrt{I_D(\text{sat})}$ (vertical axis)

V_T, V_G (horizontal axis)

atomistic scale at the oxide–silicon interface and undergo scattering because, as discussed in Section 3.4.1, any deviation from a perfectly periodic crystal potential results in scattering. This mobility degradation increases with the gate bias because a higher gate bias draws the carriers closer to the oxide–silicon interface, where they are more influenced by the interfacial roughness.

It is very interesting to note that if we plot the effective carrier mobility in the MOSFET as a function of the average transverse electric field in the middle of the inversion layer, we get what is known as a "universal" mobility degradation curve for any MOSFET, which is independent of the technology or device structural parameters such as oxide thickness and channel doping (Fig. 6–30). We can apply Gauss's law to the region marked by the colored box in Fig. 6–31, which encloses all the depletion charge and half of the inversion charge in the channel. We see that the average transverse field in the middle of the inversion region is given by

$$\mathscr{E}_{eff} = \frac{1}{\epsilon_s}\left(Q_d + \frac{1}{2}Q_n\right) \qquad (6\text{–}55a)$$

While this model works quite well for electrons, for reasons that are not clearly understood at present, it has to be modified slightly for holes in the sense that the average transverse field must now be defined as

$$\mathscr{E}_{eff} = \frac{1}{\epsilon_s}\left(Q_d + \frac{1}{3}Q_n\right) \qquad (6\text{–}55b)$$

Figure 6–30
Inversion layer
electron mobility
versus effective
transverse field, at
various tempera-
tures. The trian-
gles, circles and
squares refer to
different MOSFETs
with different gate
oxide thicknesses
and channel dop-
ings. (After Sabnis
and Clemens,
IEEE IEDM,
1979).

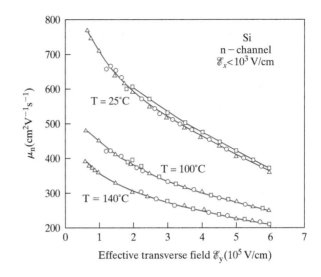

Figure 6–30
Inversion layer electron mobility versus effective transverse field, at various temperatures. The triangles, circles and squares refer to different MOSFETs with different gate oxide thicknesses and channel dopings. (After Sabnis and Clemens, *IEEE IEDM,* 1979).

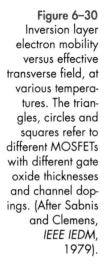

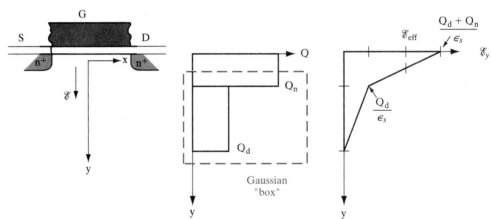

Figure 6–31
Determination of effective transverse field. Idealized charge distribution and transverse electric field in the inversion layer and depletion layer, as a function of depth in the channel of a MOSFET. The region to which we apply Gauss's law is shown in color.

This degradation of mobility with gate bias is often compactly described by writing the drain current expression as

$$I_D = \frac{\bar{\mu}_n Z C_i}{L\{1 + \theta(V_G - V_T)\}}\left[(V_G - V_T)V_D - \frac{1}{2}V_D^2\right] \qquad (6\text{–}56)$$

where θ is called the *mobility degradation parameter*. Because of the additional $(V_G - V_T)$ term in the denominator, the drain current increases sub-linearly with gate bias for high gate voltages.

In addition to this dependence of the channel mobility on gate bias or transverse electric field, there is also a strong dependence on drain bias or the longitudinal electric field. As shown in Fig. 3–24, the carrier drift velocity increases linearly with electric field (ohmic behavior) until the field reaches $\mathscr{E}_{sat}$; in other words, the mobility is constant up to $\mathscr{E}_{sat}$. After this, the velocity saturates at v_s, and it can no longer be described in terms of mobility. These effects can be described as:

$$v = \mu\mathscr{E} \text{ for } \mathscr{E} < \mathscr{E}_{sat} \tag{6-57}$$

$$\text{and } v = v_s \text{ for } \mathscr{E} > \mathscr{E}_{sat} \tag{6-58}$$

The maximum longitudinal electric field near the drain end of the channel is approximately given by the voltage drop along the pinch-off region, $(V_D-V_D(\text{sat.}))$, divided by the length of the pinch-off region, ΔL.

$$\mathscr{E}_{max} = \left(\frac{V_D - V_D(\text{sat.})}{\Delta L}\right) \tag{6-59}$$

From a two-dimensional solution of the Poisson equation near the drain end, one can show that the pinch-off region ΔL shown in Fig. 6–11c is approximately equal to $\sqrt{(3dx_j)}$, where d is the gate oxide thickness and x_j is the source/drain junction depth. The factor of 3 is due to the ratio of the dielectric constant for Si to that of SiO_2.

6.5.4 Short Channel MOSFET I-V Characteristics

In short channel devices, the analysis has to be somewhat modified. As mentioned in the previous section, the effective channel mobility decreases with increasing transverse electric field perpendicular to the gate oxide (i.e., the gate bias). Furthermore, for very high longitudinal electric fields in the pinch-off region, the carrier velocity saturates (Fig. 3–24). For short channel lengths, the carriers travel at the saturation velocity over most of the channel. In that case, the drain current is given by the width times the channel charge per unit area times the saturation velocity.

$$I_D \approx ZC_i(V_G - V_T)v_s \tag{6-60}$$

As a result, the saturation drain current does not increase quadratically with $(V_G - V_T)$ as shown in Eq. (6–53), but rather shows a linear dependence (note the equal spacing of curves in Fig. 6–32). Due to the advances in Si device processing, particularly photolithography, MOSFETs used in modern integrated circuits tend to have short channels, and are commonly described by Eq. (6–60) rather than Eq. (6–53).

6.5.5 Control of Threshold Voltage

Since the threshold voltage determines the requirements for turning the MOS transistor on or off, it is very important to be able to adjust V_T in designing the device. For example, if the transistor is to be used in a circuit driven by a 3-V

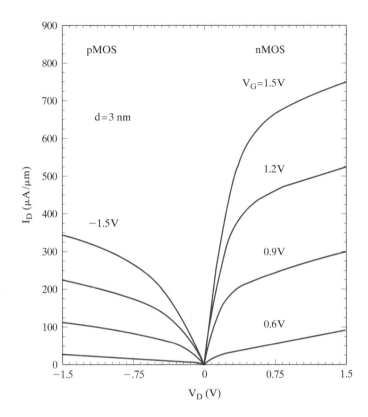

Figure 6–32
Experimental output characteristics of n-channel and p-channel MOSFETs with 0.1 μm channel lengths. The curves exhibit almost equal spacing, indicating a linear dependence of I_D on V_G, rather than a quadratic dependence. We also see that I_D is not constant but increases somewhat with V_D in the saturation region. The p-channel devices have lower currents because hole mobilities are lower than electron mobilities.

battery, it is clear that a 4-V threshold voltage is unacceptable. Some applications require not only a low value of V_T, but also a precisely controlled value to match other devices in the circuit.

All of the terms in Eq. (6–38) can be controlled to some extent. The work function potential difference Φ_{ms} is determined by choice of the gate conductor material; ϕ_F depends on the substrate doping; Q_i can be reduced by proper oxidation methods and by using Si grown in the (100) orientation; Q_d can be adjusted by doping of the substrate; and C_i depends on the thickness and dielectric constant of the insulator. We shall discuss here several methods of controlling these quantities in device fabrication.

Choice of Gate Electrode. Since V_T depends on Φ_{ms}, the choice of the gate electrode material (i.e., the gate electrode work function) has an impact on the threshold voltage. When MOSFETs were first made in the 1960's, they used Al gates. However, since Al has a low melting point, it precluded the use of a self-aligned source/drain technology because that required a high temperature source/drain implant anneal after the gate formation. Hence, Al was supplanted by n⁺ doped LPCVD polysilicon *refractory* (high melting point) gates, where the Fermi level lines up with the conduction band edge in Si. While this works quite well for n-channel MOSFETs, we shall see in Section 9.3.1 that it can create problems for p-channel MOSFETs. Therefore, sometimes, a p^+ doped polysilicon gate is used for p-channel devices. Refractory metal gates with suitable work functions are also being researched as possible replacements for doped polysilicon. One attractive candidate is tungsten, whose work function is such that the Fermi level happens to lie near the mid-gap of Si.

Control of C_i.   Since a low value of V_T and a high drive current is usually desired, a thin oxide layer is used in the gate region to increase $C_i = \epsilon_i / d$ in Eq. (6–38). From Fig. 6–20 we see that increasing C_i makes V_T less negative for p-channel devices and less positive for n-channel with $-Q_d > Q_i$. For practical considerations, the gate oxide thickness is generally $20 - 100$ Å $(2 - 10$ nm) in modern devices having submicron gate length. An example of such a device is shown in Figure 6–33. The gate oxide, easily observable in

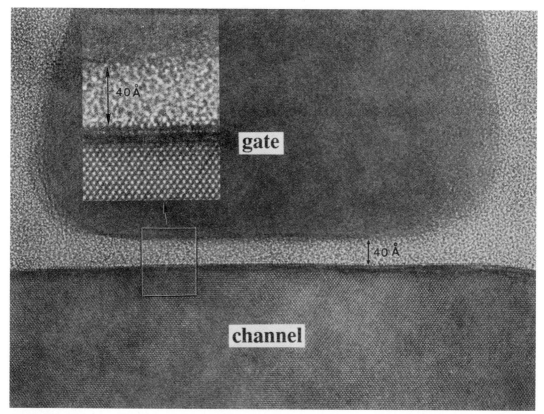

Figure 6–33

Cross section of a MOSFET. This high resolution transmission electron micrograph of a silicon Metal–Oxide Semiconductor Field Effect Transistor shows the silicon channel and metal gate separated by a thin (40Å, 4nm) silicon–dioxide insulator. The inset shows a magnified view of the three regions, in which individual rows of atoms in the crystalline silicon can be distinguished. (Photograph courtesy of AT&T Bell Laboratories.)

this micrograph, is 40Å thick. The interfacial layer between the crystalline silicon and the amorphous SiO_2 is also observable.

Although a low threshold voltage is desirable in the gate region of a transistor, a large value of V_T is needed between devices. For example, if a number of transistors are interconnected on a single Si chip, we do not want inversion layers to be formed inadvertently between devices (generally called the *field*). One way to avoid such parasitic channels is to increase V_T in the field by using a very thick oxide. Figure 6–34 illustrates a transistor with a gate oxide 10 nm thick and a field oxide of 0.5 μm.

EXAMPLE 6–4

Consider an n^+ polysilicon-SiO_2–Si p-channel device with $N_d = 10^{16}$ cm^{-3} and $Q_i = 5 \times 10^{10}q$ C/cm^2. Calculate V_T for a gate oxide thickness of 0.01 μm and repeat for a field oxide thickness of 0.5 μm.

SOLUTION

Values of ϕ_F, Q_i, and Q_d can be obtained from Examples 6–2 and 6–3 if we use appropriate signs as in Fig. 6–20a. The value of C_i for the thin oxide case is the same as in Example 6–2. From Fig. 6–17, $\Phi_{ms} = -0.25$ V.

$$V_T = -0.25 - 0.694 - \frac{8 \times 10^{-9} + 4.82 \times 10^{-8}}{34.5 \times 10^{-8}} = -1.1 \text{ V}$$

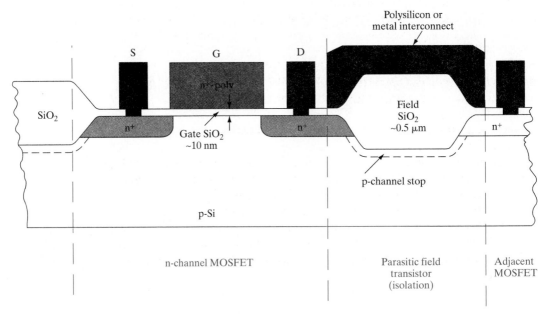

Figure 6–34
Thin oxide in the gate region and thick oxide in the field between transistors for V_T control (not to scale).

This value corresponds to that expected from Fig. 6–20b. In the field region where $d = 0.5$ μm,

$$V_T = -0.944 - \frac{5.62 \times 10^{-8}}{6.9 \times 10^{-9}} = -9.1 \text{ V}$$

The value of C_i can also be controlled by varying ϵ_i. A SiO_2 layer which has some N incorporated in it, leading to the formation of a silicon oxynitride, is often used. Such silicon oxynitrides have slightly higher ϵ_i and C_i than SiO_2, with excellent interface properties. Other high dielectric constant materials such as Ta_2O_5, ZrO_2 and ferroelectrics (e.g., barium–strontium–titanate) are also being investigated as replacements for SiO_2 as the gate dielectric in MOSFETs in order to increase $C_i = \epsilon_i/d$ and, therefore, the drive current of the MOSFET. Generally speaking, we cannot use these high dielectric constant materials directly on the Si substrate; a very thin (~0.5 nm) interfacial SiO_2 layer is needed to achieve a low fast interface state density. It is clear from the expression for C_i that for these high dielectric constant materials, a physically thicker layer, d, can be used than for SiO_2 and still achieve a certain C_i. This is very useful for reducing the tunneling leakage current through the gate dielectric, discussed in Section 6.4.7. A physically thicker layer implies a wider tunneling barrier with a reduced tunneling probability.

Threshold Adjustment by Ion Implantation. The most valuable tool for controlling threshold voltage is ion implantation (Section 5.1.4). Since very precise quantities of impurity can be introduced by this method, it is possible to maintain close control of V_T. For example, Fig. 6–35 illustrates a boron implantation through the gate oxide of a p-channel device such that the implanted peak occurs just below the Si surface. The negatively charged boron acceptors serve to reduce the effects of the positive depletion charge Q_d. As a result, V_T becomes less negative. Similarly, a shallow boron implant into the p-type substrate of an n-channel transistor can make V_T positive, as required for an enhancement device.

If the implantation is performed at higher energy, or into the bare Si instead of through an oxide layer, the impurity distribution lies deeper below the surface. In such cases the essentially gaussian impurity concentration profile cannot be approximated by a spike at the Si surface. Therefore, effects of distributed charge on the Q_d term of Eq. (6–38) must be considered. Calculations of the effects on V_T in this case are more complicated, and the shift of threshold voltage with implantation dose is often obtained empirically instead.

The implantation energy required for shallow V_T adjustment implants is low (50–100 keV), and relatively low doses are needed. A typical V_T adjustment requires only about 10 s of implantation for each wafer, and therefore this procedure is compatible with large-scale production requirements.

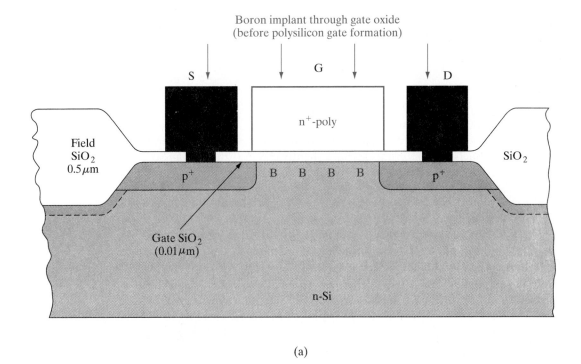

(a)

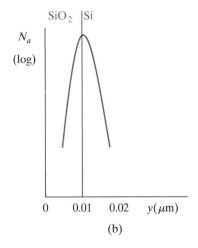

(b)

Figure 6–35

Adjustment of V_T in a p-channel transistor by boron implantation: (a) boron ions are implanted through the thin gate oxide but are absorbed within the thick oxide regions; (b) variation of implanted boron concentration in the gate region—here the peak of the boron distribution lies just below the Si surface.

For the p-channel transistor of Example 6–4, calculate the boron ion dose F_B (B$^+$ ions/cm^2) required to reduce V_T from -1.1 V to -0.5 V. Assume that the implanted acceptors form a sheet of negative charge just below the Si surface.

EXAMPLE 6–5

$$-0.5 = -1.1 + \frac{qF_B}{C_i}$$

SOLUTION

$$F_B = \frac{3.45 \times 10^{-7}}{1.6 \times 10^{-19}}(0.6) = 1.3 \times 10^{12} \text{ cm}^{-2}$$

For a beam current of 10 μA scanned over a 650-cm^2 target area,

$$\frac{10^{-5}(\text{C/s})}{650 \text{ cm}^2}t(s) = 1.3 \times 10^{12} \text{ (ions/cm}^2) \times 1.6 \times 10^{-19}(\text{C/ion})$$

The implant time is $t = 13.5$ s.

If the implantation is continued to higher doses, V_T can be moved past zero to the *depletion-mode* condition (Fig. 6–36). This capability provides considerable flexibility to the integrated-circuit designer, by allowing enhancement- and depletion-mode devices to be incorporated on the same chip. For example, a depletion-mode transistor can be used instead of a resistor as a load element for the enhancement device. Thus an array of MOS transistors can be fabricated in an IC layout, with some adjusted by implantation to have the desired enhancement mode V_T and others implanted to become depletion loads.

As mentioned above, V_T control is important not only in the MOSFETs but also in the isolation or field regions. In addition to using a thick field oxide, we can do a *channel stop implant* (so called because it stops turning on

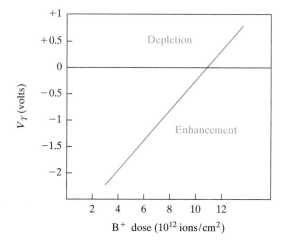

Figure 6–36
Typical variation of V_T for a p-channel device with increased implanted boron dose. The originally enhancement p-channel transistor becomes a depletion-mode device ($V_T > 0$) by sufficient B implantation.

a parasitic channel in the isolation regions) selectively in the isolation regions under the field oxide (Fig. 6–34). Generally, a B channel stop implant is used for n-channel devices. (It must be noted that such an acceptor implant will raise the field thresholds for n-channel MOSFETs made in a p-substrate, but will *decrease* the field thresholds for p-MOSFETs made in an n-substrate).

6.5.6 Substrate Bias Effects

In the derivation of Eq. (6–49) for current along the channel, we assumed that the source S was connected to the substrate B (Fig. 6–27). In fact, it is possible to apply a voltage between S and B (Fig. 6–37). With a reverse bias between the substrate and the source (V_B negative for an n-channel device), the depletion region is widened and the threshold gate voltage required to achieve inversion must be increased to accommodate the larger Q_d. A simplified view of the result is that W is widened uniformly along the channel, so that Eq. (6–32) should be changed to

$$Q_d' = -[2\epsilon_s qN_a(2\phi_F - V_B)]^{1/2} \qquad (6\text{–}61)$$

The change in threshold voltage due to the substrate bias is

$$\Delta V_T = \frac{\sqrt{2\epsilon_s qN_a}}{C_i}[(2\phi_F - V_B)^{1/2} - (2\phi_F)^{1/2}] \qquad (6\text{–}62)$$

If the substrate bias V_B is much larger than $2\phi_F$ (typically ~0.6 V), the threshold voltage is dominated by V_B and

$$\Delta V_T \simeq \frac{\sqrt{2\epsilon_s qN_a}}{C_i}(-V_B)^{1/2} \quad \text{(n channel)} \qquad (6\text{–}63)$$

where V_B will be negative for the n-channel case. As the substrate bias is increased, the threshold voltage becomes more positive. The effect of this bias becomes more dramatic as the substrate doping is increased, since ΔV_T is also proportional to $\sqrt{N_a}$. For a p-channel device the bulk-to-source voltage V_B is positive to achieve a reverse bias, and the approximate change ΔV_T for $V_B \gg 2\phi_F$ is

$$\Delta V_T \simeq \frac{\sqrt{2\epsilon_s qN_d}}{C_i}V_B^{1/2} \quad \text{(p channel)} \qquad (6\text{–}64)$$

Thus the p-channel threshold voltage becomes more negative with substrate bias.

The substrate bias effect (also called the *body effect*) increases V_T for either type of device. This effect can be used to raise the threshold voltage of a marginally enhancement device ($V_T \approx 0$) to a somewhat larger and more manageable value. This can be an asset for n-channel devices particularly (see Fig. 6–20). The effect can present problems, however, in MOS integrated circuits for which it is impractical to connect each source region to the

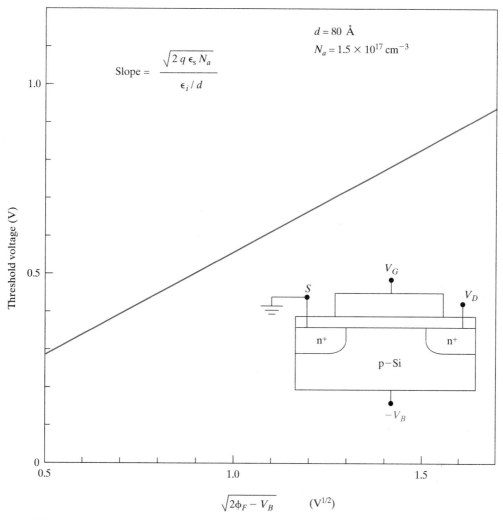

$$\text{Slope} = \frac{\sqrt{2\,q\,\epsilon_s\,N_a}}{\epsilon_i\,/\,d}$$

$d = 80 \text{ Å}$

$N_a = 1.5 \times 10^{17} \text{ cm}^{-3}$

Threshold voltage (V)

1.0

0.5

0

V_G

S

V_D

n^+ n^+

$p-\text{Si}$

$-V_B$

0.5   1.0   1.5

$$\sqrt{2\phi_F - V_B} \qquad (\text{V}^{1/2})$$

Figure 6–37

Threshold voltage dependence on substrate bias resulting from application of a voltage V_B from the substrate (i.e., bulk) to the source. For n channel, V_B must be zero or negative to avoid forward bias of the source junction. For p channel, V_B must be zero or positive.

substrate. In these cases, possible V_T shifts due to the body effect must be taken into account in the circuit design.

6.5.7 Subthreshold Characteristics

If we look at the drain current expression (Eq. 6-53), it appears that the current abruptly goes to zero as soon as V_G is reduced to V_T. In reality, there is still some drain conduction below threshold, and this is known as *subthreshold conduction.* This current is due to weak inversion in the channel between flat-band and threshold (for bandbending between zero and $2\phi_F$), which leads to a diffusion current from source to drain. The drain current in the subthreshold region is equal to

$$I_D = \mu(C_d + C_{it})\frac{Z}{L}\left(\frac{kT}{q}\right)^2\left(1 - e^{\frac{-qV_D}{kT}}\right)\left(e^{\frac{q(V_G - V_T)}{c_r kT}}\right) \qquad (6\text{--}65)$$

where

$$c_r = \left[1 + \frac{C_d + C_{it}}{C_i}\right]$$

It can be seen that I_D depends exponentially on gate bias, V_G. However, V_D has little influence once V_D exceeds a few kT/q. Obviously, if we plot $\ln I_D$ as a function of gate bias V_G, we should get a linear behavior in the subthreshold regime, as shown in Fig. 6–38a. The slope of this line (or more precisely the reciprocal of the slope) is known as the subthreshold slope, S, which has typical values of ~70 mV/decade at room temperature for state-of-the-art MOSFETs. This means that a change in the input V_G of 70 mV will change the output I_D by an order of magnitude. Clearly, the smaller the value of S, the better the transistor is as a switch. A small value of S means a small change in the input bias can modulate the output current considerably.

It can be shown that the expression for S is given by

$$S = \frac{dV_G}{d(\log I_D)} = \ln 10 \frac{dV_G}{d(\ln I_D)} = 2.3\frac{kT}{q}\left[1 + \frac{C_d + C_{it}}{C_i}\right] \qquad (6\text{--}66)$$

Here, the factor ln 10 (= 2.3) is introduced to change from $\log_{10}$ to natural logarithm, ln. This equation can be understood by looking at the electrical equivalent circuit of the MOSFET in terms of the capacitors (Fig. 6–38b). Between the gate and the substrate, we find the gate capacitance, C_i, in series with the parallel combination of the depletion capacitance in the channel, C_d, and the fast interface state capacitance, $C_{it} = qD_{it}$. The expression in brackets in Eq. (6–66) is simply the capacitor divider ratio which tells us what fraction of the applied gate bias, V_G, appears at the Si–SiO$_2$ interface as the surface potential. Ultimately, it is the surface potential that is responsible for modulating the barrier between source and drain, and therefore the drain current. Hence, S is a measure of the efficacy of the gate potential in modulating I_D. From Eq. (6–66), we observe that S is improved by reducing the gate oxide thickness, which is reasonable because if the gate electrode is closer to the channel, the gate control is obviously better. The value of S is higher for heavy channel doping (which increases the depletion capacitance) or if the silicon–oxide interface has many fast interface states.

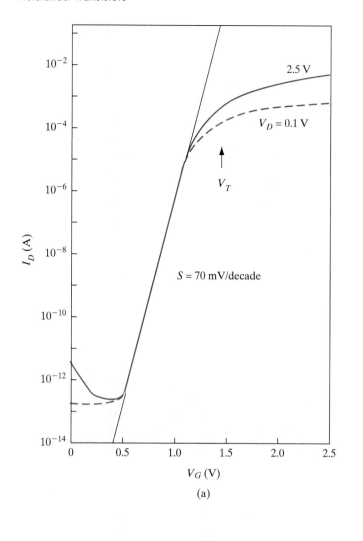

(a)

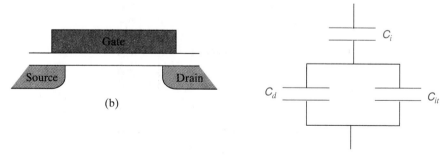

(b)

Figure 6-38
Subthreshold conduction in MOSFETs: (a) Semi-log plot of I_D versus V_G; (b) equivalent circuit showing capacitor divider which determines subthreshold slope.

For a very small gate voltage, the subthreshold current is reduced to the leakage current of the source/drain junctions. This determines the off-state leakage current, and therefore the standby power dissipation in many complementary MOS (CMOS) circuits involving both n-channel and p-channel MOSFETs. It also underlines the importance of having high quality source/drain junctions. From the subthreshold characteristics, it can be seen that if the V_T of a MOSFET is too low, it cannot be turned off fully at $V_G = 0$. Also, unavoidable statistical variations of V_T cause drastic variations of the subthreshold leakage current. On the other hand, if V_T is too high, one sacrifices drive current, which depends on the difference between the power supply voltage and V_T. For these reasons, the V_T of MOSFETs has historically been designed to be ~0.7 V. However, with the recent advent of various types of low voltage, low power portable electronics, there are new challenges in device and circuit design to optimize speed and power dissipation.

6.5.8 Equivalent Circuit for the MOSFET

When we attempt to draw an equivalent circuit of a MOSFET, we find that in addition to the intrinsic MOSFET itself, there are a variety of parasitic elements associated with it. An important addition to the gate capacitance is the so-called *Miller overlap capacitance* due to the overlap between the gate and the drain region (Fig. 6–39). This capacitance is particularly problematic because it represents a feedback path between the output drain terminal and the input gate terminal. One can measure the Miller capacitance at high frequency by holding the gate at ground ($V_G = 0$) so that an inversion layer is not formed in the channel. Thereby, most of the measured capacitance between gate and drain is due to the Miller capacitance, rather than the gate capacitance C_i. It is possible to minimize this capacitance by using a so-called *self-aligned gate*. In this process, the gate itself is used to mask the source/drain implants, thereby achieving alignment. Even in this design, however, there is still a certain amount of overlap because of the lateral straggle or spread of the implanted dopants underneath the gate, further exacerbated by the lateral diffusion which occurs during high temperature annealing. This spread of the source/drain junctions under the gate edge determines what is called the channel length reduction, ΔL_R (Fig. 6–40). Hence, we get the electrical or "effective" channel length, L_{eff}, in terms of the physical gate length, L as

$$L_{eff} = L - \Delta L_R \tag{6–67}$$

There can also be a width reduction, ΔZ, which changes the effective width, Z_{eff}, from the physical width Z of the MOSFET. The width reduction results from the electrical isolation regions that are formed around all transistors, generally by LOCOS. The LOCOS isolation technique is discussed in Section 9.3.1.

Another very important parameter in the equivalent circuit is the source/drain series resistance, $R_{SD} = (R_S + R_D)$, because it degrades the drain current and transconductance. For a certain applied drain bias to the source/

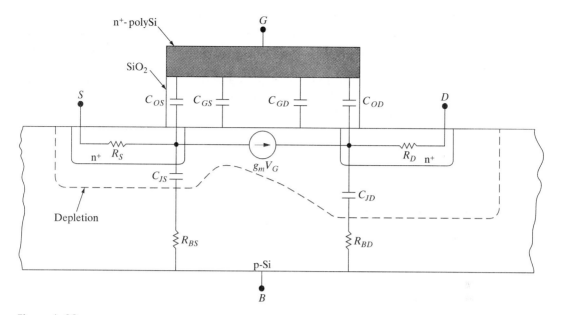

Figure 6–39

Equivalent circuit of a MOSFET, showing the passive capacitive and resistive components. The gate capacitance C_i is the sum of the distributed capacitances from the gate to the source-end of the channel (C_{GS}) and the drain-end (C_{GD}). In addition, we have an overlap capacitance (where the gate electrode overlaps the source/drain junctions) from the gate-to-source (C_{OS}) and gate-to-drain (C_{OD}). C_{OD} is also known as the Miller overlap capacitance. We also have p-n junction depletion capacitances associated with the source (C_{JS}) and drain (C_{JD}). The parasitic resistances include the source/drain series resistances (R_S and R_D), and the resistances in the substrate between the bulk contact and the source and drain (R_{BS} and R_{BD}). The drain current can be modeled as a (gate) voltage-controlled constant-current source.

drain terminals, part of the applied voltage is "wasted" as an ohmic voltage drop across these resistances, depending on the drain current (or gate bias). Hence, the actual drain voltage applied to the intrinsic MOSFET itself is less; this causes I_D to increase sub-linearly with V_G.

We can determine R_{SD}, along with ΔL_R, from the overall resistance of the MOSFET in the linear region,

$$\left(\frac{V_D}{I_D} \right)$$

This corresponds to the intrinsic channel impedance R_{Ch}, plus the source-drain resistance R_{SD}. Modifying Eq. (6–51) we get

$$\frac{V_D}{I_D} = R_{Ch} + R_{SD} = \frac{L - \Delta L_R}{Z - \Delta Z} \frac{1}{\mu_n C_i (V_G - V_T)} + R_{SD} \qquad (6\text{–}68)$$

We can measure V_D/I_D in the linear range for various MOSFETs having the same width, but different channel lengths, as a function of substrate bias. Varying the substrate bias changes the V_T through the body effect, and therefore the slope of the straight lines that result from plotting the overall resistance as a function of L. The lines pass through a point, having values which correspond to ΔL_R and R_{SD}, as shown in Fig. 6–40.

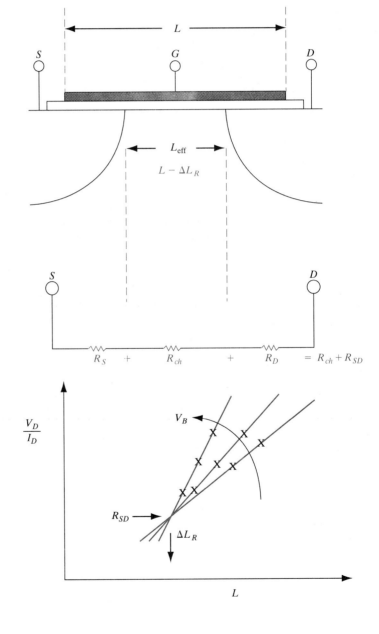

Figure 6–40 Determination of length reduction and source/drain series resistance in a MOSFET. The overall resistance of a MOSFET in the linear region is plotted as a function of channel length, for various substrate biases. The x mark data points for three different physical gate lengths L.

6.5.9 MOSFET Scaling and Hot Electron Effects

Much of the progress in semiconductor integrated circuit technology can be attributed to the ability to shrink or *scale* the devices. Scaling down MOSFETs has a multitude of benefits. From Table 6–1, we see the benefits of scaling in terms of the improvement of packing density, speed and power dissipation. A key concept in scaling, first due to Dennard at IBM, is that the various structural parameters of the MOSFET should be scaled in concert if the device is to keep functioning properly. In other words, if lateral dimensions such as the channel length and width are reduced by a factor of K, so should the vertical dimensions such as source/drain junction depths (x_j) and gate insulator thickness (Table 6–1). Scaling of depletion widths is achieved indirectly by scaling up doping concentrations. However, if we simply reduced the dimensions of the device and kept the power supply voltages the same, the internal electric fields in the device would increase. For ideal scaling, power supply voltages should also be reduced to keep the internal electric fields reasonably constant from one technology generation to the next. Unfortunately, in practice, power supply voltages are not scaled hand-in-hand with the device dimensions, partly because of other system-related constraints. The longitudinal electric fields in the pinch-off region, and the transverse electric fields across the gate oxide, increase with MOSFET scaling. A variety of problems then arise which are generically known as hot electron effects and short channel effects (Fig. 6–41).

When an electron travels from the source to the drain along the channel, it gains kinetic energy at the expense of electrostatic potential energy in the pinch-off region, and becomes a "hot" electron. At the conduction band edge, the electron only has potential energy; as it gains more kinetic energy,

Table 6–1 Scaling rules for MOSFETs according to a constant factor K. The horizontal and vertical dimensions are scaled by the same factor. The voltages are also scaled to keep the internal electric fields more or less constant, and the hot carrier effects manageable.

	Scaling factor
Surface dimensions (L,Z)	1/K
Vertical dimensions (d, x_j)	1/K
Impurity Concentrations	K
Current, Voltages	1/K
Current Density	K
Capacitance (per unit area)	K
Transconductance	1
Circuit Delay Time	1/K
Power Dissipation	$1/K^2$
Power Density	1
Power-Delay Product	$1/K^3$

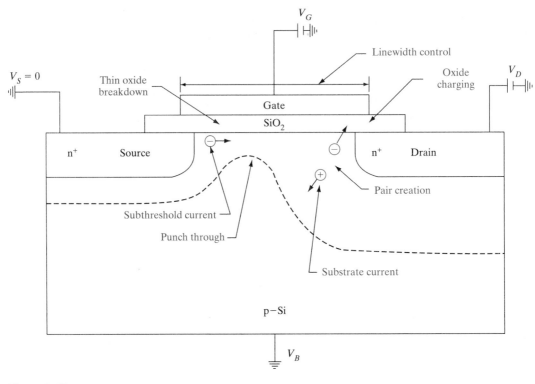

Figure 6–41
Short channel effects in MOSFETs. As MOSFETs are scaled down, potential problems due to short channel effects include hot carrier generation (electron-hole pair creation) in the pinch-off region,
punchthrough breakdown between source and drain, and thin gate oxide breakdown.

it moves higher up in the conduction band. A few of the electrons can become
energetic enough to surmount the 3.1 eV potential barrier between the Si
channel and the gate oxide (Fig. 6–25). Some of these injected hot electrons
can go through the gate oxide and be collected as gate current, thereby reducing the input impedance. More importantly, some of these electrons can
be trapped in the gate oxide as fixed oxide charges. According to Eq. (6–37),
this increases the flatband voltage, and therefore the V_T. In addition, these
energetic hot carriers can rupture Si–H bonds that exist at the Si–SiO$_2$ interface, creating fast interface states that degrade MOSFET parameters such
as transconductance and subthreshold slope, with stress. The results of such
hot carrier degradation are shown in Fig. 6–42, where we see the increase of
V_T and decrease of slope, and therefore transconductance, with stress. The solution to this problem is to use what is known as a *lightly doped drain* (LDD).

As discussed in more detail in Section 9.3.1, by reducing the doping concentration in the source/drain, the depletion width at the reverse-biased drain-channel junction is increased and the electric field is reduced.

   Hot carrier effects are less problematic for holes in p-channel MOSFETs than for electrons in n-channel devices for two reasons. The channel mobility of holes is approximately half that of electrons; hence, for the same electric field, there are fewer hot holes than hot electrons. Unfortunately, the lower hole mobility is also responsible for lower drive currents in p-channel than

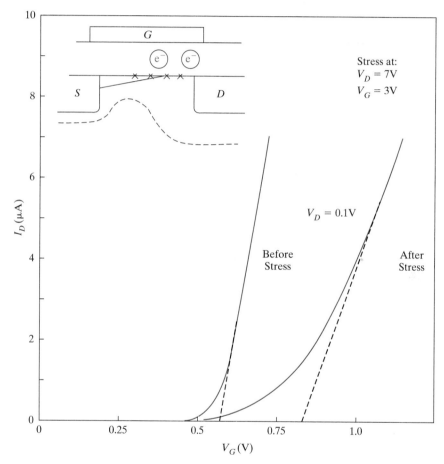

Figure 6–42
Hot carrier degradation in MOSFETs. The linear region transfer characteristics before and after hot carrier stress indicate an increase of V_T and decrease of transconductance (or channel mobility) due to hot electron damage. The damage can be due to hot electron injection into the gate oxide which increases the fixed oxide charge, and increasing fast interface state densities at the oxide-silicon interface (indicated by x).

in n-channel. Also, the barrier for hole injection in the valence band between Si and SiO_2 is higher (5 eV) than for electrons in the conduction band (3.1 eV), as shown in Fig. 6–25. Hence, while LDD is mandatory for n-channel, it is often not used for p-channel devices.

One "signature" for hot electron effects is substrate current (Fig. 6–43). As the electrons travel towards the drain and become hot, they can create secondary electron–hole pairs by impact ionization (Fig. 6–41). The secondary electrons are collected at the drain, and cause the drain current in saturation to increase with drain bias at high voltages, thereby leading to a decrease of the output impedance. The secondary holes are collected at the substrate as substrate current. This current can create circuit problems such as noise or

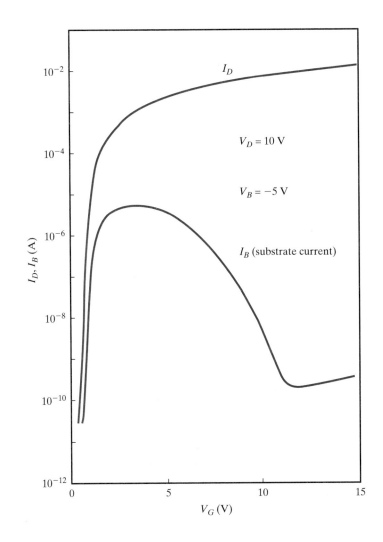

Figure 6–43
Substrate current in a MOSFET. The substrate current in n-channel MOSFETs due to impact-generated holes in the pinch-off region, as a function of gate bias. The substrate current initially increases with V_G because of the corresponding increase of I_D. However, for even higher V_G, the MOSFET goes from the saturation to the linear region, and the high electric fields in the pinch-off region decrease, causing less impact ionization. (After Kamata, et. al., Jpn. J. Appl. Phys., 15 (1976), 1127.)

latchup in CMOS circuits (Section 9.3.1). It can also be used as a monitor for hot electron effects. As shown in Fig. 6–43, substrate current initially increases with gate bias (for a fixed, high drain bias), goes through a peak and then decreases. The reason for this behavior is that initially, as the gate bias increases, the drain current increases and thereby provides more primary carriers into the pinch-off region for impact ionization. However, for even higher gate bias, the MOSFET goes from the saturation region into the linear region when the fixed V_D drops below $V_D(\text{sat.}) = (V_G - V_T)$. The longitudinal electric field in the pinch-off region drops, thereby reducing the impact ionization rates. Hot electron reliability studies are done under "worst case" conditions of peak substrate current. These are generally done under accelerated conditions of higher-than-normal operating voltages so that if there are any potential problems, they show up in a reasonable time period. The degradation data is then extrapolated to the actual operating conditions.

6.5.10   Drain-Induced Barrier Lowering

If small channel length MOSFETs are not scaled properly, and the source/drain junctions are too deep or the channel doping is too low, there can be unintended electrostatic interactions between the source and the drain known as *Drain-Induced Barrier Lowering (DIBL)*. This leads to punchthrough leakage or breakdown between the source and the drain, and loss of gate control. The phenomenon can be understood from Fig. 6–44, where we have schematically plotted the surface potential along the channel for a long channel device and a short device. We see that as the drain bias is increased, the conduction band edge (which reflects the electron energies) in the drain is pulled down, and the drain-channel depletion width expands. For a long channel MOSFET, the drain bias does not affect the source-to-channel potential barrier, which corresponds to the built-in potential of the source-channel p-n junction. Hence, unless the gate bias is increased to lower this potential barrier, there is little drain current. On the other hand, for a short channel MOSFET, as the drain bias is raised and the conduction band edge in the drain is pulled down (with a concomitant increase of the drain depletion width), the source-channel potential barrier is lowered due to DIBL. This can be shown numerically by a solution of the two-dimensional Poisson equation in the channel region. Simplistically, the onset of DIBL is sometimes considered to correspond to the drain depletion region expanding and merging with the source depletion region, and causing punchthrough breakdown between source and drain. However, it must be kept in mind that DIBL is ultimately caused by the lowering of the source-junction potential barrier below the built-in potential. Hence, if we get DIBL in a MOSFET for a grounded substrate, the problem can be mitigated by applying a substrate reverse bias, because that raises the potential barrier at the source end. This works in spite of the fact that the drain depletion region interacts even more with

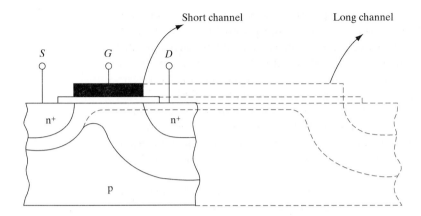

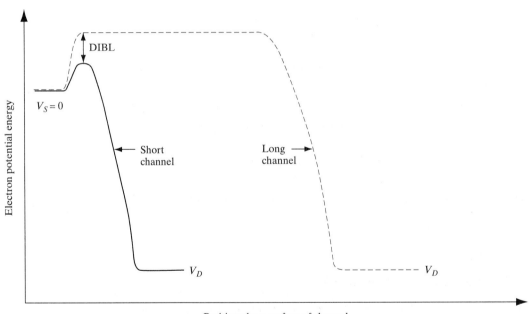

Figure 6–44
Drain-induced barrier lowering in MOSFETs. Cross-sections and potential distribution along the channel for a long channel and short channel MOSFET.

the source depletion region under such back bias. Once the source-channel barrier is lowered by DIBL, there can be significant drain leakage current, with the gate being unable to shut it off.

What are the solutions to this problem? The source/drain junctions must be made sufficiently shallow (i.e., scaled properly) as the channel lengths are reduced, to prevent DIBL. Secondly, the channel doping must be made sufficiently high to prevent the drain from being able to control the source junction. This is achieved by performing what is known as an *anti-punchthrough* implant in the channel. Sometimes, instead of such an implant throughout the channel (which can have undesirable consequences such as raising the V_T or the body effect), a localized implant is done only near the source/drains. These are known as *halo* or *pocket* implants. The higher doping reduces the source/drain depletion widths and prevents their interaction.

For short-channel MOSFETs, DIBL is related to the electrical modulation of the channel length in the pinch-off region, ΔL. Since the drain current is inversely proportional to the electrical channel length, we get

$$I_D \propto \frac{1}{L - \Delta L} = \frac{1}{L}\left(1 + \frac{\Delta L}{L}\right) \qquad (6\text{–}69)$$

for small pinch-off regions, ΔL. We assume that the fractional change in the channel length is proportional to the drain bias,

$$\frac{\Delta L}{L} = \lambda V_D \qquad (6\text{–}70)$$

where λ is the *channel length modulation parameter*. Hence, in the saturation region, the expression for the drain current becomes

$$I_D = \frac{Z}{2L}\bar{\mu}_n C_i (V_G - V_T)^2 (1 + \lambda V_D) \qquad (6\text{–}71)$$

This leads to a slope in the output characteristics, or a lowering of the output impedance (Fig. 6–32).

6.5.11   Short Channel and Narrow Width Effect

If we plot the threshold voltage as a function of channel length in MOSFETs, we find that V_T decreases with L for very small geometries. This effect is called the *short channel effect (SCE)*, and is somewhat similar to DIBL. The mechanism is due to something called *charge sharing* between the source/drain and the gate (Fig. 6–45)[9]. From the equation for the threshold voltage (6–38), we notice that one of the terms is the depletion charge under the gate.

---

[9] L. Yau, "A simple theory to predict the threshold voltage of short-channel IGFETs," Solid-State Electronics, 17 (1974): 1059.

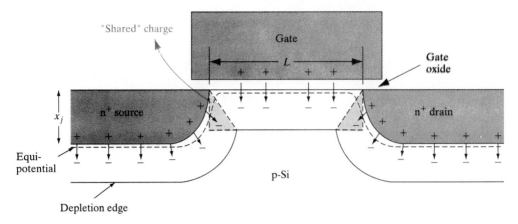

Figure 6–45
Short channel effect in a MOSFET. Cross-sectional view of MOSFET along the length showing depletion charge sharing (colored regions) between the gate, source and drain.

The equipotential lines in Fig. 6–45 designating the depletion regions curve around the contours of the source/drain junctions. Keeping in mind that the electric field lines are perpendicular to the equipotential contours, we see that the depletion charges that are physically underneath the gate in the approximately triangular regions near the source/drains have their field lines terminate not on the gate, but instead on the source/drains. Hence, electrically these depletion charges are "shared" with the source and drain regions and should not be counted in the V_T expression, Eq. (6–38). We can deal with this effect by replacing the orginal Q_d in the rectangular region underneath the gate by a lower Q_d in the trapezoidal region in Fig. 6–45. Clearly, for a long channel device, the triangular depletion charge regions near the source and drain are a very small fraction of the total depletion charge underneath the gate. However, as the channel lengths are reduced, the shared charge becomes a larger fraction of the total, and this results in a V_T *roll-off* as a function of L (Fig. 6–46). This is important because it is hard to control the channel lengths precisely in manufacturing. The channel length variations then lead to problems with V_T control.

In the last several years, another effect has been observed in n-channel MOSFETs with decreasing L. The V_T initially goes up before it goes down due to the short channel effect. This phenomenon has been dubbed the *reverse short channel effect (RSCE)*, and is due to interactions between Si point defects that are created during the source/drain implant and the B doping in the channel, causing the B to pile up near the source and drains, and thus raise the V_T.

Another related effect in MOSFETs is the *narrow width effect*, where the V_T goes up as the channel width Z is reduced for very narrow devices (Fig. 6–46). This can be understood from Fig. 6–47, where some of the depletion charges under the LOCOS isolation regions have field lines electri-

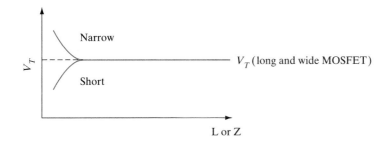

Figure 6–46
Roll-off of V_T with decreasing channel length, and increase of V_T with decreasing width.

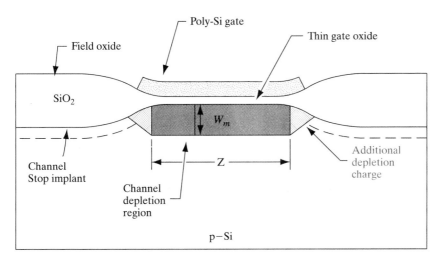

Figure 6–47
Narrow width effect in a MOSFET. Cross-sectional view of MOSFET along the width, showing additional depletion charge (colored regions) underneath the field or the LOCOS isolation regions.

cally terminating on the gate. Unlike the SCE, where the effective depletion charge is reduced due to charge sharing with the source/drain, here the depletion charge belonging to the gate is increased. The effect is not important for very wide devices, but becomes quite important as the widths are reduced below 1μm.

6.5.12   Gate-Induced Drain Leakage

If we examine the subthreshold characteristics shown in Fig. 6–38, we find that as the gate voltage is reduced below V_T, the subthreshold current drops and then bottoms out a level determined by the source/drain diode leakage. However, for even more negative gate biases we find that the off-state leakage current actually goes up as we try to turn off the MOSFET more for high V_D; this is known as *gate-induced drain leakage (GIDL)*. The same effect is seen at a fixed gate bias of near zero, for increasing drain bias. The reason for GIDL can be understood from Fig. 6–48, where we show the band diagram as a function of depth in the region where the gate overlaps the drain junction.

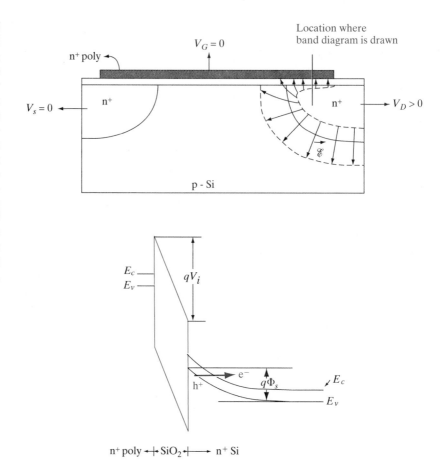

Figure 6–48
Gate-induced drain leakage in MOSFETs. The band diagram for the location shown in color is plotted as a function of depth in the gate–drain overlap region, indicating band-to-band tunneling and creation of electron–hole pairs in the drain region in the Si substrate.

As the gate is made more negative (or alternatively, for a fixed gate bias, the drain is made more positive), a depletion region forms in the n-type drain. Since the drain doping is high, the depletion widths tend to be narrow. If the bandbending is more than the bandgap E_g across a narrow depletion region, the conditions are conducive to band-to-band tunneling in this region, thereby creating electron–hole pairs. The electrons then go to the drain as GIDL. It must be emphasized that this tunneling is not through the gate oxide (Section 6.4.7), but entirely in the Si drain region. For GIDL to occur, the drain doping level should be moderate ($\sim10^{18}$ cm^{-3}). If it is much lower than this, the depletion widths and tunneling barriers are too wide. On the other hand, if the doping in the drain is very high, most of the voltage drops in the gate oxide, and the bandbending in the Si drain region drops below the value E_g. GIDL is an important factor in limiting the off-state leakage current in state-of-the-art MOSFETs.

6.1 Modify Eqs. (6–2) through (6–5) to include effects of the contact potential V_0. Define a true pinch-off voltage V_T to distinguish this case from V_P defined in Eq. (6–4).

6.2 Modify Eqs. (6–7) through (6–10) to include V_0. Let V_P be defined as in Eq. (6–4), and call the true pinch-off voltage V_T.

6.3 Assume the JFET shown in Fig. 6–6 is Si and has p^+ regions doped with 10^{18} acceptors/cm^3 and a channel with 10^{16} donors/cm^3. If the channel half-width a is 1 μm, compare V_P with V_0. What voltage V_{GD} is required to cause pinch-off when V_0 is included? With $V_G = -3$ V, at what value of V_D does the current saturate?

6.4 If the ratio $Z/L = 10$ for the JFET of Prob. 6.3, and $\mu_n = 1000$ cm^2/V-s, calculate $I_D(\text{sat})$ for $V_G = 0, -2, -4$, and -6 V. Plot $I_D(\text{sat})$ vs. $V_D(\text{sat})$.

6.5 For the JFET of Prob. 6.4, plot I_D vs. V_D for the same three values of V_G. Terminate each plot at the point of saturation.

6.6 The current I_D varies almost linearly with V_D in a JFET for low values of V_D.

   (a) Use the binomial expansion with $V_D/(-V_G) < 1$ to rewrite Eq. (6–9) as an approximation to this case.

   (b) Show that the expression for the channel conductance I_D/V_D in the linear range is the same as $g_m(\text{sat})$ given by Eq. (6–11).

   (c) What value of gate voltage V_G turns the device off such that the channel conductance goes to zero?

6.7 Use Eqs. (6–9) and (6–10) to calculate and plot $I_D(V_D, V_G)$ at 300 K for a Si JFET with $a = 1000$Å, $N_d = 7 \times 10^{17}$ cm^{-3}, $Z = 100$ μm, and $L = 5$ μm. Allow V_D to range from 0 to 5 V and allow V_G to take on values of $0, -1, -2, -3, -4$, and -5 V.

6.8 Show that the width of the depletion region in Fig. 6–15 is given by Eq. (6–30). Assume the carriers are completely swept out within W, as was done in Section 5.2.3.

6.9 An n^+-polysilicon-gate n-channel MOS transistor is made on a p-type Si substrate with $N_a = 5 \times 10^{15}$ cm^{-3}. The SiO$_2$ thickness is 100 Å in the gate region, and the effective interface charge Q_i is 4×10^{10} qC/cm^2.

   Find W_m, V_{FB}, and V_T.

6.10 An n^+ polysilicon-gate p-channel MOS transistor is made on an n-type Si substrate with $N_d = 5 \times 10^{16}$ cm^{-3}. The SiO$_2$ thickness is 100 Å in the gate region, and the effective interface charge Q_i is 2×10^{11} q C/cm^2. Sketch the C–V curve for this device and give important numbers for the scale.

6.11 Use Eq. (6–50) to calculate and plot $I_D(V_D, V_G)$ at 300 K for an n-channel Si MOSFET with an oxide thickness $d = 200$ Å, a channel mobility $\bar{\mu}_n = 1000$ cm^2/V-s, $Z = 100$ μm, $L = 5$ μm, and N_a of $10^{14}, 10^{15}, 10^{16}$, and 10^{17} cm^{-3}. Allow V_D to range from 0 to 5 V and allow V_G to take on values of $0, 1, 2, 3, 4$, and 5 V. Assume that $Q_i = 5 \times 10^{11}$ q C/cm^2.

6.12 Calculate the V_T of a Si-MOS transistor for a n^+-polysilicon gate with silicon oxide thickness = 50 Å, $N_a = 1 \times 10^{18}$ cm^{-3} and a fixed charge of $2 \times 10^{10} q$ C/cm^2. Is it an enhancement or depletion mode device? What B dose is required to change the V_T to 0 V? Assume a shallow B implant.

6.13 (a) Find the voltage V_{FB} required to reduce to zero the negative charge induced at the semiconductor surface by a sheet of positive charge Q_{ox} located x' below the metal.

(b) In the case of an arbitrary distribution of charge $\rho(x')$ in the oxide, show that

$$V_{FB} = -\frac{1}{C_i} \int_0^d \frac{x'}{d}\rho(x')dx'$$

6.14 The bias on a Si MOS capacitor is changed from inversion to accumulation mode. If the substrate doping is 10^{16} cm^{-3} donors, what is the change in the surface bandbending at 100°C?

6.15 A Si MOS capacitor has the high frequency C–V curve shown in Fig. P6-15 normalized to the capacitance in strong accumulation. Determine the oxide thickness and substrate doping assuming a gate-to-substrate work function difference of −0.35V

Figure P6–15

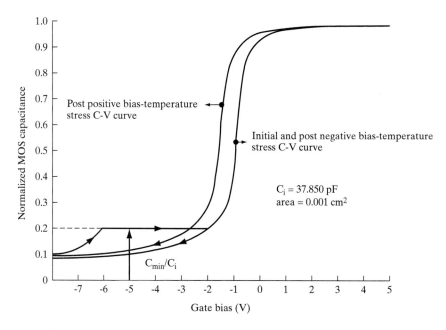

6.16 For the capacitor in Prob. 6.15, determine the initial flatband voltage.

6.17 For the capacitor in Prob. 6.15, determine the fixed oxide charge, Q_f, and the mobile ion content.

6.18 When an MOS transistor is biased with $V_D > V_D(\text{sat})$, the effective channel length is reduced by ΔL and the current I'_D is larger than $I_D(\text{sat})$, as shown in Fig. 6–32. Assuming that the depleted region ΔL is described by an expression similar to Eq. (6–30) with $V_D - V_D(\text{sat.})$ for the voltage across ΔL, show that the conductance beyond saturation is

$$g'_D = \frac{\partial I'_D}{\partial V_D} = I_D(\text{sat.})\frac{\partial}{\partial V_D}\left(\frac{L}{L - \Delta L}\right)$$

and find the expression for g'_D in terms of V_D.

6.19 Calculate the V_T of a Si n-channel MOSFET for a n$^+$-polysilicon gate with gate oxide thickness = 100 Å, $N_a = 10^{18}$ cm^{-3} and a fixed oxide charge of 5×10^{10} q C/cm^2. Repeat for a substrate bias of -2.5V.

6.20 For the MOSFET in Prob. 6.19, and $W = 50$ μm, $L = 2$ μm, calculate the drain current at $V_G = 5$V, $V_D = 0.1$V. Repeat for $V_G = 3$V, $V_D = 5$V. Assume an electron channel mobility $\bar{\mu}_n = 200$ cm^2/V-s, and the substrate is connected to the source.

6.21 An n-channel MOSFET with a 400 Å gate oxide requires its V_T to be lowered by 2 V. Using a 50 keV implant of singly-charged species and assuming the implant distribution is peaked at the oxide–Si interface and can be regarded as a sheet charge at the interface, what implant parameters (species, energy, dose and beam current) would you choose? The scan area is 200 cm^2, and the desired implant time is 20 s. Assume similar range statistics in oxide and Si.

6.22 For the MOSFET characteristics shown in Fig. P6-22, calculate:

1. Linear V_T and k_N

2. Saturation V_T and k_N

Assume channel mobility, $\bar{\mu}_n = 500$ cm^2/V-s and $V_{FB} = 0$.

Figure P6–22

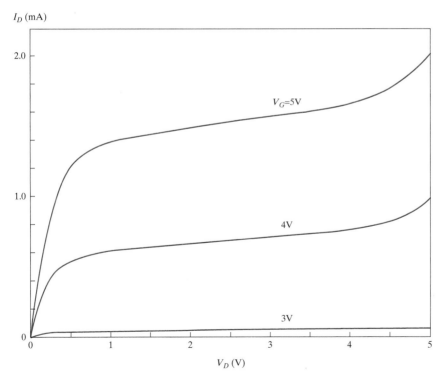

6.23 For Prob. 6.22, calculate the gate oxide thickness and substrate doping, either graphically or iteratively.

6.24 Assume that the inversion layer in a Si MOSFET can be treated as a 2-D electron gas trapped in an infinite rectangular potential well of width 100 Å. (In reality, it looks more like a triangular well.) Calculate the inversion charge per unit area assuming that the Fermi level lies midway between the second and third subbands. Assume $T = 77$ K, and effective mass $= 0.2\, m_0$. Assume also that the Fermi function can be treated as a rectangular function. Also sketch (E,k) for the first three subbands. Refer to Appendix IV.

6.25 The flat band voltage is shifted to -2V for an n^+-polysilicon-SiO$_2$-Si capacitor with parameters discussed in Example 6–2. Redraw Fig. 6–16 for this case and find the value of interface charge Q_i required to cause this shift in V_{FB}, with Φ_{ms} given by Fig. 6–17.

6.26 Plot I_D vs. V_D with several values of V_G for the thin-oxide p-channel transistor described in Example 6–4. Use the p-channel version of Eq. (6–49), and assume that $I_D(\text{sat})$ remains constant beyond pinch-off. Assume that $\overline{\mu}_p = 200$ cm^2/V-s, and $Z = 10L$.

6.27 A typical figure of merit for high-frequency operation of MOS transistors is the cutoff frequency $f_c = g_m/2\pi C_G LZ$, where the gate capacitance C_G is essentially C_i over most of the voltage range. Express f_c above pinch-off in terms of materials parameters and device dimensions, and calculate f_c for the transistor of Prob. 6.26, with $L = 1$ μm.

6.28 From Fig. 6–44 it is clear that the depletion regions of the source and drain junctions can meet for short channels, a condition called *punch-through*. Assume the source and drain regions of an n-channel Si MOSFET are doped with 10^{20} donors/cm^3 and the 1-μm-long channel is doped with 10^{16} acceptors/cm^3. If the source and substrate are grounded, what drain voltage will cause punch-through?

6.29 Calculate the substrate bias required to achieve enhancement-mode operation with $V_T = +0.5$ V for the n-channel device of Example 6–3. Comment on the practicality of this method of threshold control for thin-oxide transistors.

READING LIST   Dambkes, H. "Gallium Arsenide HEMTs for Low-Noise GHz Communications Engineering." *Microelectronics Journal* 20 (September–October 1989): 1–6.

Drummond, T. J., W. T. Masselink, and H. Morkoc. "Modulation-Doped GaAs/(Al,Ga)As Heterojunction Field-Effect Transistors: MODFETs." *Proceedings of the IEEE* 74 (June 1986): 773–822.

Frensley, W. R. "Gallium Arsenide Transistors." *Scientific American* 257 (August 1987): 80–87.

Inoue, K. "Recent Advances in InP-Based HEMT/HBT Device Technology." *Fourth International Conference on Indium Phosphide and Related Materials* (April 1992): 10–13.

Kahng, D. "A Historical Perspective on the Development of MOS Transistors and Related Devices," *IEEE Trans. Elec. Dev.*, ED-23 (1976): 655.

Morgan, D. V., and R. H. Williams, eds. *Physics and Technology of Heterojunction Devices.* London: P. Peregrinus, 1991.

Morkoc, H. "The HEMT: A Superfast Transistor." *IEEE Spectrum* 21 (February 1984): 28–35.

Muller, R. S., and T. I. Kamins. *Device Electronics for Integrated Circuits.* New York: Wiley, 1986.

Neamen, D. A. *Semiconductor Physics and Devices: Basic Principles.* Homewood, IL: Irwin, 1992.

Nguyen, L. D., L. E. Larson, and U. K. Mishra. "Ultra-High Speed Modulation-Doped Field-Effect Transistors: A Tutorial Review." *Proceedings of the IEEE* 80 (April 1992): 492–518.

Pavlidis, D. "Current Status of Heterojunction Bipolar and High-Electron Mobility Transistor Technologies." *Microelectronic Engineering* 19 (September 1992): 305–12.

Pierret, R. F. *Field Effect Devices.* Reading, MA: Addison-Wesley, 1990.

Sah, C. T. "Characteristics of the Metal–Oxide Semiconductor Transistors." *IEEE Trans. Elec. Dev.* ED-11 (1964): 324.

Sah, C. T. "Evolution of the MOS Transistor—From Conception to VLSI." *Proceedings of the IEEE* 76 (October 1988): 1280–1326.

Schroder, D. K. *Modular Series on Solid State Devices: Advanced MOS Devices.* Reading, MA: Addison-Wesley, 1987.

Shockley, W. and G. Pearson. "Modulation of Conductance of Thin Films of Semiconductors by Surface Charges." *Phys. Rev.* 74 (1948): 232.

Shur, M. *GaAs Devices and Circuits.* New York: Plenum Press, 1987.

Singh, J. *Semiconductor Devices.* New York: McGraw-Hill, 1994.

Smith, R. S. and I. G. Eddison. "Advanced Materials for GaAs Microwave Devices." *Advanced Materials* 4 (December 1992): 786–91.

Sze, S. M. *High-Speed Semiconductor Devices.* New York: Wiley, 1990.

Sze, S. M. *Physics of Semiconductor Devices.* New York: Wiley, 1981.

Uyemura, J. P. *Fundamentals of MOS Digital Integrated Circuits.* Reading, MA: Addison-Wesley, 1988.

Wang, S. *Fundamentals of Semiconductor Theory and Device Physics.* Englewood Cliffs, NJ: Prentice Hall, 1989.

Weisbuch, C., and B. Vinter. *Quantum Semiconductor Structures.* Boston: Academic Press, 1991.

Chapter 7
Bipolar Junction Transistors

We begin this chapter with a qualitative discussion of charge transport in a *bipolar junction transistor (BJT)*, to establish a sound physical understanding of its operation. Then we shall investigate carefully the charge distributions in the transistor and relate the three terminal currents to the physical characteristics of the device. Our aim is to gain a solid understanding of the current flow and control of the transistor and to discover the most important secondary effects that influence its operation. We shall discuss the properties of the transistor with proper biasing for amplification and then consider the effects of more general biasing, as encountered in switching circuits.

In this chapter we shall use the p-n-p transistor for most illustrations. The main advantage of the p-n-p for discussing transistor action is that hole flow and current are in the same direction. This makes the various mechanisms of charge transport somewhat easier to visualize in a preliminary explanation. Once these basic ideas are established for the p-n-p device, it is simple to relate them to the more widely used transistor, the n-p-n.

7.1 FUNDAMENTALS OF BJT OPERATION

The bipolar transistor is basically a simple device, and this section is devoted to a simple and largely qualitative view of BJT operation. We will deal with the details of these transistors in following sections, but first we must define some terms and gain physical understanding of how carriers are transported through the device. Then we can discuss how the current through two terminals can be controlled by small changes in the current at a third terminal.

Let us begin the discussion of bipolar transistors by considering the reverse-biased p-n junction diode of Fig. 7–1. According to the theory of Chapter 5, the reverse saturation current through this diode depends on the rate at which minority carriers are generated in the neighborhood of the junction. We found, for example, that the reverse current due to holes being swept from n to p is essentially independent of the size of the junction $\mathscr{E}$ field and hence is independent of the reverse bias. The reason given was that the hole current depends on how often minority holes are generated by EHP creation within a diffusion length of the junction—not upon how fast a particular hole is swept across the depletion layer by the field. As a result, it is

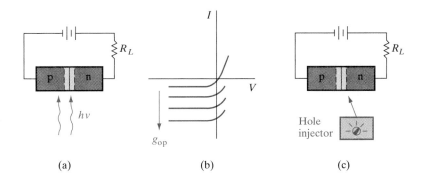

Figure 7–1
External control of
the current in a
reverse-biased p-n
junction: (a) opti-
cal generation;
(b) junction I–V
characteristics as
a function of EHP
generation; (c) mi-
nority carrier in-
jection by a
hypothetical
device.

possible to increase the reverse current through the diode by increasing the rate of EHP generation (Fig. 7–1b). One convenient method for accomplishing this is optical excitation of EHPs with light ($hv > E_g$), as in Section 4.3. With steady photoexcitation the reverse current will still be essentially independent of bias voltage, and if the dark saturation current is negligible, the reverse current is directly proportional to the optical generation rate g_{op}.

The example of external control of current through a junction by optical generation raises an interesting question: Is it possible to inject minority carriers in to the neighborhood of the junction *electrically* instead of optically? If so, we could control the junction reverse current simply by varying the rate of minority carrier injection. For example, let us consider a hypothetical *hole injection device* as in Fig. 7–1c. If we can inject holes at a predetermined rate into the n side of the junction, the effect on the junction current will resemble the effects of optical generation. The current from n to p will depend on the hole injection rate and will be essentially independent of the bias voltage. There are several obvious advantages to such external control of a current; for example, the current through the reverse-biased junction would vary very little if the load resistor R_L were changed, since the magnitude of the junction voltage is relatively unimportant. Therefore, such an arrangement should be a good approximation to a controllable constant current source.

A convenient hole injection device is a forward-biased p^+-n junction. According to Section 5.3.2, the current in such a junction is due primarily to holes injected from the p^+ region into the n material. If we make the n side of the forward-biased junction the same as the n side of the reverse-biased junction, the p^+-n-p structure of Fig. 7–2 results. With this configuration, injection of holes from the p^+-n junction into the center n region supplies the minority carrier holes to participate in the reverse current through the n-p junction. Of course, it is important that the injected holes do not recombine in the n region before they can diffuse to the depletion layer of the reverse-biased junction. Thus we must make the n region narrow compared with a hole diffusion length.

Figure 7–2
A p-n-p transistor:
(a) schematic rep-
resentation of a
p-n-p device with
a forward-biased
emitter junction
and a reverse-
biased collector
junction; (b) I–V
characteristics of
the reverse-biased
n-p junction as a
function of emitter
current.

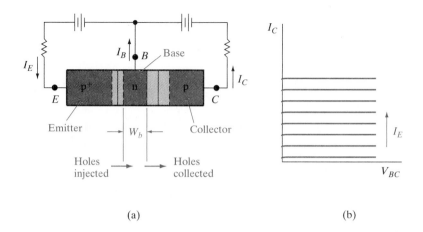

(a)   (b)

The structure we have described is a p-n-p bipolar junction transistor. The forward-biased junction which injects holes into the center n region is called the *emitter junction*, and the reverse-biased junction which collects the injected holes is called the *collector junction*. The p⁺ region, which serves as the source of injected holes, is called the *emitter*, and the p region into which the holes are swept by the reverse-biased junction is called the *collector*. The center n region is called the *base*, for reasons which will become clear in Section 7.3, when we discuss the historical development of transistor fabrication. The biasing arrangement in Fig. 7–2 is called the *common base* configuration, since the base electrode B is common to the emitter and collector circuits.

To have a good p-n-p transistor, we would prefer that almost all the holes injected by the emitter into the base be collected. Thus the n-type base region should be narrow, and the hole lifetime τ_p should be long. This requirement is summed up by specifying $W_b \ll L_p$, where W_b is the length of the *neutral* n material of the base (measured between the depletion regions of the emitter and collector junctions), and L_p is the diffusion length for holes in the base $(D_p\tau_p)^{1/2}$. With this requirement satisfied, an average hole injected at the emitter junction will diffuse to the depletion region of the collector junction without recombination in the base. A second requirement is that the current I_E crossing the emitter junction should be composed almost entirely of holes injected into the base, rather than electrons crossing from base to emitter. This requirement is satisfied by doping the base region lightly compared with the emitter, so that the p⁺-n emitter junction of Fig. 7–2 results.

It is clear that current I_E flows into the emitter of a properly biased p-n-p transistor and that I_C flows out at the collector, since the direction of hole flow is from emitter to collector. However, the base current I_B requires a bit more thought. In a good transistor the base current will be very small since I_E is essentially hole current, and the collected hole current I_C is almost equal to I_E. There must be some base current, however, due to requirements

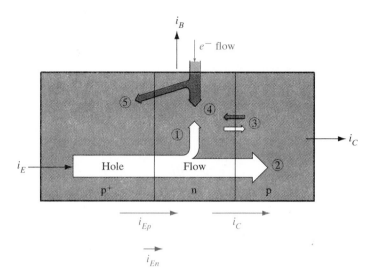

i_B

e^- flow

⑤ ④ ③

Hole Flow ①

i_E

i_C

p⁺ n p

②

i_{Ep} i_C

i_{En}

Figure 7–3
Summary of hole and electron flow in a p-n-p transistor with proper biasing: (1) injected holes lost to recombination in the base; (2) holes reaching the reverse-biased collector junction; (3) thermally generated electrons and holes making up the reverse saturation current of the collector junction; (4) electrons supplied by the base contact for recombination with holes; (5) electrons injected across the forward-biased emitter junction.

of electron flow into the n-type base region (Fig. 7–3). We can account for I_B physically by three dominant mechanisms:

(a)   There must be some recombination of injected holes with electrons in the base, even with $W_b \ll L_p$. The electrons lost to recombination must be resupplied through the base contact.

(b)   Some electrons will be injected from n to p in the forward biased emitter junction, even if the emitter is heavily doped compared to the base. These electrons must also be supplied by I_B.

(c)   Some electrons are swept into the base at the reverse-biased collector junction due to thermal generation in the collector. This small current reduces I_B by supplying electrons to the base.

The dominant sources of base current are (a) recombination in the base and (b) injection into the emitter region. Both of these effects can be greatly reduced by device design, as we shall see. In a well-designed transistor, I_B will be a very small fraction (perhaps one-hundredth) of I_E.

In an n-p-n transistor the three current directions are reversed, since electrons flow from emitter to collector and holes must be supplied to the base. The physical mechanisms for operation of the n-p-n can be understood simply by reversing the roles of electrons and holes in the p-n-p discussion.

In this section we shall discuss rather simply the various factors involved in transistor amplification. Basically, the transistor is useful in amplifiers because the currents at the emitter and collector are controllable by the relatively small base current. The essential mechanisms are easy to understand if various secondary effects are neglected. We shall use total current (d-c

**7.2
AMPLIFICATION
WITH BJTS**

plus a-c) in this discussion, with the understanding that the simple analysis applies only to d-c and to small-signal a-c at low frequencies. We can relate the terminal currents of the transistor i_E, i_B, and i_C by several important factors. In this introduction we shall neglect the saturation current at the collector (Fig. 7–3, component 3) and such effects as recombination in the transition regions. Under these assumptions, the collector current is made up entirely of those holes injected at the emitter which are not lost to recombination in the base. Thus i_C is proportional to the hole component of the emitter current i_{Ep}:

$$i_C = B i_{Ep} \tag{7–1}$$

The proportionality factor B is simply the fraction of injected holes which make it across the base to the collector; B is called the *base transport factor*. The total emitter current i_E is made up of the hole component i_{Ep} and the electron component i_{En}, due to electrons injected from base to emitter (component 5 in Fig. 7–3). The *emitter injection efficiency* γ is

$$\boxed{\gamma = \frac{i_{Ep}}{i_{En} + i_{Ep}}} \tag{7–2}$$

For an efficient transistor we would like B and γ to be very near unity; that is, the emitter current should be due mostly to holes ($\gamma \simeq 1$), and most of the injected holes should eventually participate in the collector current ($B \simeq 1$). The relation between the collector and emitter currents is

$$\frac{i_C}{i_E} = \frac{B i_{Ep}}{i_{En} + i_{Ep}} = B\gamma \equiv \alpha \tag{7–3}$$

The product $B\gamma$ is defined as the factor α, called the *current transfer ratio*, which represents the emitter-to-collector current amplification. There is no real amplification between these currents, since α is smaller than unity. On the other hand, the relation between i_C and i_B is more promising for amplification.

In accounting for the base current, we must include the rates at which electrons are lost from the base by injection across the emitter junction (i_{En}) and the rate of electron recombination with holes in the base. In each case, the lost electrons must be resupplied through the base current i_B. If the fraction of injected holes making it across the base *without* recombination is B, then it follows that $(1 - B)$ is the fraction *recombining* in the base. Thus the base current is

$$i_B = i_{En} + (1 - B)i_{Ep} \tag{7–4}$$

neglecting the collector saturation current. The relation between the collector and base currents is found from Eqs. (7–1) and (7–4):

$$\frac{i_C}{i_B} = \frac{B i_{Ep}}{i_{En} + (1 - B)i_{Ep}} = \frac{B[i_{Ep}/(i_{En} + i_{Ep})]}{1 - B[i_{Ep}/(i_{En} + i_{Ep})]} \tag{7–5}$$

$$\frac{i_C}{i_B} = \frac{B\gamma}{1 - B\gamma} = \frac{\alpha}{1 - \alpha} \equiv \beta \qquad (7\text{–}6)$$

The factor β relating the collector current to the base current is the *base-to-collector current amplification factor.*[1] Since α is near unity, it is clear that β can be large for a good transistor, and the collector current is large compared with the base current.

It remains to be shown that the collector current i_C can be controlled by variations in the small current i_B. In the discussion to this point, we have indicated control of i_C by the emitter current i_E, with the base current characterized as a small side effect. In fact, we can show from space charge neutrality arguments that i_B can indeed be used to determine the magnitude of i_C. Let us consider the transistor of Fig. 7–4, in which i_B is determined by a biasing circuit. For simplicity, we shall assume unity emitter injection efficiency and negligible collector saturation current. Since the n-type base region is electrostatically neutral between the two transition regions, the presence of excess holes in transit from emitter to collector calls for compensating excess electrons from the base contact. However, there is an important difference in the times which electrons and holes spend in the base. The average excess

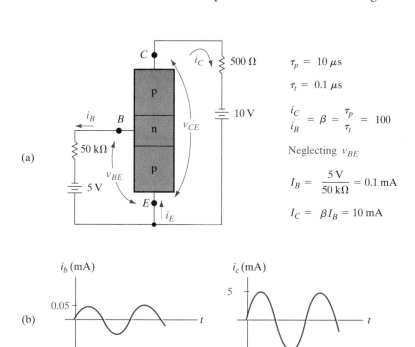

$\tau_p = 10\ \mu s$

$\tau_t = 0.1\ \mu s$

$$\frac{i_C}{i_B} = \beta = \frac{\tau_p}{\tau_t} = 100$$

Neglecting v_{BE}

$$I_B = \frac{5\ V}{50\ k\Omega} = 0.1\ mA$$

$$I_C = \beta I_B = 10\ mA$$

Figure 7–4
Example of amplification in a common-emitter transistor circuit: (a) biasing circuit; (b) addition of an a-c variation of base current i_b to the d-c value of I_B, resulting in an a-c component i_c.

[1] α is also called the *common-base current gain*; β is also called the *common-emitter current gain*.

hole spends a time τ_t, defined as the *transit time* from emitter to collector. Since the base width W_b is made small compared with L_p, this transit time is much less than the average hole lifetime τ_p in the base.[2] On the other hand, an average excess electron supplied from the base contact spends τ_p seconds in the base supplying space charge neutrality during the lifetime of an average excess hole. While the average electron waits τ_p seconds for recombination, many individual holes can enter and leave the base region, each with an average transit time τ_t. In particular, for each electron entering from the base contact, τ_p/τ_t holes can pass from emitter to collector while maintaining space charge neutrality. Thus the ratio of collector current to base current is simply

$$\frac{i_C}{i_B} = \beta = \frac{\tau_p}{\tau_t} \tag{7-7}$$

for $\gamma = 1$ and negligible collector saturation current.

If the electron supply to the base (i_B) is restricted, the traffic of holes from emitter to base is correspondingly reduced. This can be argued simply by supposing that the hole injection does continue despite the restriction on electrons from the base contact. The result would be a net buildup of positive charge in the base and a loss of forward bias (and therefore a loss of hole injection) at the emitter junction. Clearly, the supply of electrons through i_B can be used to raise or lower the hole flow from emitter to collector.

The base current is controlled independently in Fig. 7–4. This is called a *common-emitter* circuit, since the emitter electrode is common to the base and collector circuits. The emitter junction is clearly forward biased by the battery in the base circuit. The voltage drop in the forward-biased emitter junction is small, however, so that almost all of the voltage from collector to emitter appears across the reverse-biased collector junction. Since v_{BE} is small for the forward-biased junction, we can neglect it and approximate the base current as 5 V/50 kΩ = 0.1 mA. If $\tau_p = 10$ μs and $\tau_t = 0.1$ μs, β for the transistor is 100 and the collector current I_C is 10 mA. It is important to note that i_c is determined by β and the base current, rather than by the battery and resistor in the collector circuit (as long as these are of reasonable values to maintain a reverse-biased collector junction). In this example 5 V of the collector circuit battery voltage appears across the 500 Ω resistor, and 5 V serves to reverse bias the collector junction.

---

[2]This difference between average hole lifetime before recombination (τ_p) and the average time a hole spends in transit across the base (τ_t), may be confusing at first. How can the lifetime be longer than the time a hole actually spends in transit? The answer depends on the fact that holes are indistinguishable in the recombination kinetics. Think of an analogy with a shooting gallery, in which a good marksman fires slowly at a line of quickly moving ducks. Although many individual ducks make it across the firing line without being hit, the lifetime of an *average* duck within the firing line is determined by the time between shots. We can speak of the lifetime of an average duck because they are essentially indistinguishable. Similarly, the rate of recombination in the base (and therefore i_B) depends on the average lifetime τ_p and the distribution of the indistinguishable holes in the base region.

If a small a-c current i_b is superimposed on the steady state base current of Fig. 7–4a, a corresponding a-c current i_c appears in the collector circuit. The time-varying portion of the collector current will be i_b multiplied by the factor β, and current gain results.

We have neglected a number of important properties of the transistor in this introductory discussion, and many of these properties will be treated in detail below. We have established, however, the fundamental basis of operation for the bipolar transistor and have indicated in a simplified way how it can be used to produce current gain in an electronic circuit.

(a) Show that Eq. (7–7) is valid from arguments of the steady state replacement of stored charge. Assume that $\tau_n = \tau_p$.

EXAMPLE 7–1

(b) What is the steady state charge $Q_n = Q_p$ due to excess electrons and holes in the neutral base region for the transistor of Fig. 7–4?

(a) In steady state there are excess electrons and holes in the base. The charge in the electron distribution Q_n is replaced every τ_p seconds. Thus $i_B = Q_n/\tau_p$. The charge in the hole distribution Q_p is collected every τ_t seconds, and $i_C = Q_p/\tau_t$. For space charge neutrality, $Q_n = Q_p$, and

SOLUTION

$$\frac{i_C}{i_B} = \frac{Q_n/\tau_t}{Q_n/\tau_p} = \frac{\tau_p}{\tau_t}$$

(b) $Q_n = Q_p = i_C\tau_t = i_B\tau_p = 10^{-9}$ C.

The first transistor invented by Bardeen and Brattain in 1947 was the *point contact* transistor. In this device two sharp metal wires, or "cat's whiskers," formed an "emitter" of carriers and a "collector" of carriers. These wires were simply pressed onto a slab of Ge which provided a "base" or mechanical support, through which the injected carriers flowed. This basic invention rapidly led to the BJT, in which charge injection and collection was achieved using two p-n junctions in close proximity to each other. The p-n junctions in BJTs can be formed in a variety of ways using thermal diffusion, but modern devices are generally made using ion implantation (Section 5.1.4).

**7.3
BJT FABRICATION**

Let us review a simplified version of how to make a double polysilicon, self-aligned n-p-n Si BJT. This is the most commonly used, state-of-the-art technique for making BJTs for use in an IC. Use of n-p-n transistors is more popular than p-n-p devices because of the higher mobility of electrons compared to holes. The process steps are shown in cross-sectional view in Fig. 7–5. A p-type Si substrate is oxidized, windows are defined using photolithography and etched in the oxide. Using the photoresist and oxide as an implant mask, a donor with very small diffusivity in Si, such as As or Sb, is implanted into the open window to form a highly conductive n^+ layer (Fig. 7–5a). Subsequently,

Figure 7–5
Process flow for double polysilicon, self-aligned npn BJT: (a) n$^+$ buried layer formation; (b) n epitaxy followed by LOCOS isolation; (c) base/emitter window definition and (optional) masked "sinker" implant (P) into collector contact region; (d) intrinsic base implant using self-aligned oxide sidewall spacers; (e) self-aligned formation of n$^+$ emitter, as well as n$^+$ collector contact.

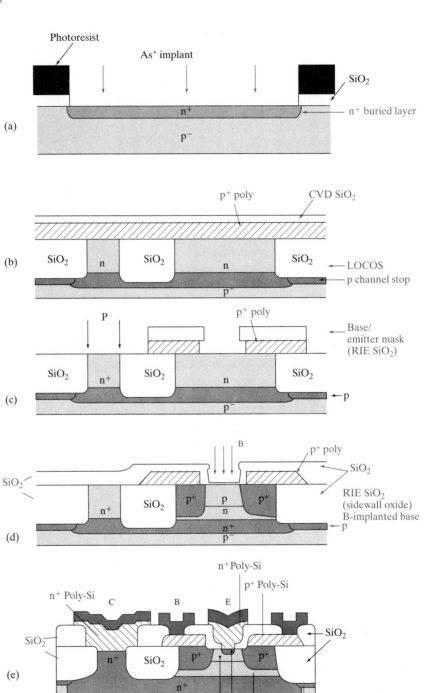

the photoresist and the oxide are removed, and a lightly doped n-type epitaxial layer is grown. During this high temperature growth, the implanted n^+
layer diffuses only slightly towards the surface and becomes a conductive
buried collector (also called a *sub-collector*). The n^+ sub-collector layer guarantees a low collector series resistance when it is connected subsequently to
the collector ohmic contact, sometimes through the use of an optional,
masked deep n^+ "sinker" implant or diffusion only in the collector contact region (Fig. 7–5c). The lightly doped n-type collector region above the n^+ sub-
collector in the part of the BJT where the base and emitter are formed
ensures a high base-collector reverse breakdown voltage. (It turns out that
wherever the sub-collector is formed, and subsequently the epitaxial layer is
grown on top, there is a notch or step in the substrate surface. This notch is
not explicitly shown in Fig. 7–5a. This notch is very useful as a marker of the
location of the sub-collectors because subsequently, we have to align the
LOCOS isolation mask with respect to the sub-collector.)

For integrated circuits involving not just discrete BJTs, but many interconnected transistors, there are issues involving electrical isolation of adjacent BJTs in order to ensure that there is no electrical cross-talk between
them. As described in Section 6.4.1, such isolation can be achieved by LOCOS
to form field or isolation oxides after a B channel stop implant (Fig. 7–5b). Another isolation scheme that is particularly well suited for high density bipolar
circuits involves the formation of shallow trenches by RIE, backfilled with
oxide and polysilicon (Section 9.3.1). In this process a nitride layer is patterned and used as an etch mask for an anisotropic etch of the silicon to form
the trench. Using reactive ion etching, a narrow trench about 1 μm deep can
be formed with very straight sidewalls. Oxidation inside the trench forms an
insulating layer, and the trench is then filled with oxide by Low-Pressure
Chemical Vapor Deposition (LPCVD).

A polysilicon layer is deposited by LPCVD, and doped heavily p^+ with B
either during deposition or subsequently by ion implantation. An oxide layer is
deposited next by LPCVD. Using photolithography with the base/emitter mask,
a window is etched in the polysilicon/oxide stack by RIE (Fig. 7–5c). A heavily
doped "extrinsic" p^+ base is formed by diffusion of B from the doped polysilicon layer into the substrate in order to provide a low resistance, high speed base
ohmic contact. An oxide layer is then deposited by LPCVD, which has the effect of closing up the base window that was etched previously, and B is implanted
into this window (Fig. 7–5d). This base implant forms a more lightly p doped
"intrinsic" base through which most of the current flows from the emitter to the
collector. The more heavily doped extrinsic base forms a collar around the intrinsic base, and serves to reduce the base series resistance. It is critical that the
base be enclosed well within the collector because otherwise it would be shorted to the p^- substrate. Finally, another LPCVD oxide layer is deposited to close
up the base window further, and the oxide is etched all the way to the Si substrate
by RIE, leaving oxide *spacers* on the sidewalls. Heavily n^+ doped (typically with
As) polysilicon is then deposited on the substrate, patterned and etched

forming *polysilicon emitter (polyemitter)* and collector contacts, as shown in Fig. 7–5e. (The use of two LPCVD polysilicon layers explains why this process is referred to as the double-polysilicon process.) Arsenic from the polysilicon is diffused into the substrate to form the n^+ emitter region nested within the base in a *self-aligned* manner, as well as the n^+ collector contact. Self alignment refers to the fact that a separate lithography step is not required to form the n^+ emitter region. We cleverly made use of the oxide sidewall spacers to ensure that the n^+ emitter region lies within the intrinsic p-type base. This is critical because otherwise the emitter gets shorted to the collector; we also want a gap between the n^+ emitter and the p^+ extrinsic base, because otherwise the emitter-base junction capacitance becomes too high. In the vertical direction, the difference between the emitter-base junction and the base-collector junction determines the base width. This is made very narrow in high gain, high speed BJTs.

Finally, an oxide layer is deposited by CVD, windows are etched in it corresponding to the emitter (E), base (B) and collector (C) contacts, and a suitable contact metal such as Al is sputter deposited to form the ohmic contacts. The Al is patterned photolithographically using the interconnect mask, and etched using RIE. The many ICs that are made simultaneously on the wafer are then separated into individual dies by sawing, mounted on suitable packages, and the various contacts are wire bonded to the external leads of the package.

7.4
MINORITY
CARRIER
DISTRIBUTIONS
AND TERMINAL
CURRENTS

In this section we examine the operation of a BJT in more detail. We begin our analysis by applying the techniques of previous chapters to the problem of hole injection into a narrow n-type base region. The mathematics is very similar to that used in the problem of the narrow base diode (Prob. 5.35). Basically, we assume holes are injected into the base at the forward-biased emitter, and these holes diffuse to the collector junction. The first step is to solve for the excess hole distribution in the base, and the second step is to evaluate the emitter and collector currents (I_E, I_C) from the gradient of the hole distribution on each side of the base. Then the base current (I_B) can be found from a current summation or from a charge control analysis of recombination in the base.

We shall at first simplify the calculations by making several assumptions:

1. Holes diffuse from emitter to collector; drift is negligible in the base region.

2. The emitter current is made up entirely of holes; the emitter injection efficiency is $\gamma = 1$.

3. The collector saturation current is negligible.

4. The active part of the base and the two junctions are of uniform cross-sectional area A; current flow in the base is essentially one-dimensional from emitter to collector.

5. All currents and voltages are steady state.

In later sections we shall consider the implications of imperfect injection efficiency, drift due to nonuniform doping in the base, structural effects such as different areas for the emitter and collector junctions, and capacitance and transit time effects in a-c operation.

7.4.1 Solution of the Diffusion Equation in the Base Region

Since the injected holes are assumed to flow from emitter to collector by diffusion, we can evaluate the currents crossing the two junctions by techniques used in Chapter 5. Neglecting recombination in the two depletion regions, the hole current entering the base at the emitter junction is the current I_E, and the hole current leaving the base at the collector is I_C. If we can solve for the distribution of excess holes in the base region, it is simple to evaluate the gradient of the distribution at the two ends of the base to find the currents. We shall consider the simplified geometry of Fig. 7–6, in which the base width is W_b between the two depletion regions, and the uniform cross-sectional area is A. The excess hole concentration at the edge of the emitter depletion region Δp_E and the corresponding concentration on the collector side of the base Δp_C are found from Eq. (5-29):

$$\Delta p_E = p_n(e^{qV_{EB}/kT} - 1) \tag{7–8a}$$

$$\Delta p_C = p_n(e^{qV_{CB}/kT} - 1) \tag{7–8b}$$

If the emitter junction is strongly forward biased ($V_{EB} \gg kT/q$) and the collector junction is strongly reverse biased ($V_{CB} \ll 0$), these excess concentrations simplify to

$$\Delta p_E \simeq p_n e^{qV_{EB}/kT} \tag{7–9a}$$

$$\Delta p_C \simeq -p_n \tag{7–9b}$$

We can solve for the excess hole concentration as a function of distance in the base $\delta p(x_n)$ by using the proper boundary conditions in the diffusion equation, Eq. (4–34b):

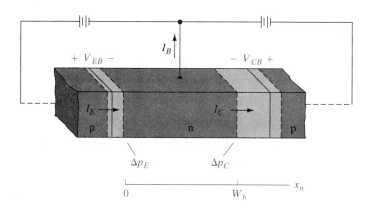

Figure 7–6
Simplified p-n-p transistor geometry used in the calculations.

$$\frac{d^2 \delta p(x_n)}{dx_n^2} = \frac{\delta p(x_n)}{L_p^2} \tag{7-10}$$

The solution of this equation is

$$\delta p(x_n) = C_1 e^{x_n/L_p} + C_2 e^{-x_n/L_p} \tag{7-11}$$

where L_p is the diffusion length of holes in the base region. Unlike the simple problem of injection into a long n region, we cannot eliminate one of the constants by assuming the excess holes disappear for large x_n. In fact, since $W_b \ll L_p$ in a properly designed transistor, most of the injected holes reach the collector at W_b. The solution is very similar to that of the narrow base diode problem. In this case the appropriate boundary conditions are

$$\delta p(x_n = 0) = C_1 + C_2 = \Delta p_E \tag{7-12a}$$

$$\delta p(x_n = W_b) = C_1 e^{W_b/L_p} + C_2 e^{-W_b/L_p} = \Delta p_C \tag{7-12b}$$

Solving for the parameters C_1 and C_2 we obtain

$$C_1 = \frac{\Delta p_C - \Delta p_E e^{-W_b/L_p}}{e^{W_b/L_p} - e^{-W_b/L_p}} \tag{7-13a}$$

$$C_2 = \frac{\Delta p_E e^{W_b/L_p} - \Delta p_C}{e^{W_b/L_p} - e^{-W_b/L_p}} \tag{7-13b}$$

These parameters applied to Eq. (7–11) give the full expression for the excess hole distribution in the base region. For example, if we assume that the collector junction is strongly reverse biased [Eq. (7–9b)] and the equilibrium hole concentration p_n is negligible compared with the injected concentration Δp_E, the excess hole distribution simplifies to

$$\delta p(x_n) = \Delta p_E \frac{e^{W_b/L_p} e^{-x_n/L_p} - e^{-W_b/L_p} e^{x_n/L_p}}{e^{W_b/L_p} - e^{-W_b/L_p}} \quad \text{(for } \Delta p_C \approx 0) \tag{7-14}$$

The various terms in Eq. (7–14) are sketched in Fig. 7–7, and the corresponding excess hole distribution in the base region is demonstrated for a moderate value of W_b/L_p. Note that $\delta p(x_n)$ varies almost linearly between the emitter and collector junction depletion regions. As we shall see, the slight deviation from linearity of the distribution indicates the small value of I_B caused by recombination in the base region.

7.4.2 Evaluation of the Terminal Currents

Having solved for the excess hole distribution in the base region, we can evaluate the emitter and collector currents from the gradient of the hole concentration at each depletion region edge. From Eq. (4–22b) we have

$$I_p(x_n) = -qAD_p \frac{d\delta p(x_n)}{dx_n} \tag{7-15}$$

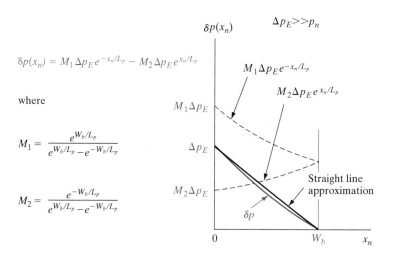

$$\delta p(x_n) = M_1 \Delta p_E e^{-x_n/L_p} - M_2 \Delta p_E e^{x_n/L_p}$$

where

$$M_1 = \frac{e^{W_b/L_p}}{e^{W_b/L_p} - e^{-W_b/L_p}}$$

$$M_2 = \frac{e^{-W_b/L_p}}{e^{W_b/L_p} - e^{-W_b/L_p}}$$

Figure 7–7
Sketch of the terms in Eq. (7–14), illustrating the linearity of the hole distribution in the base region. In this example, $W_b/L_p = \frac{1}{2}$.

This expression evaluated at $x_n = 0$ gives the hole component of the emitter current,

$$I_{Ep} = I_p(x_n = 0) = qA\frac{D_p}{L_p}(C_2 - C_1) \tag{7–16}$$

Similarly, if we neglect the electrons crossing from collector to base in the collector reverse saturation current, I_C is made up entirely of holes entering the collector depletion region from the base. Evaluating Eq. (7–15) at $x_n = W_b$ we have the collector current

$$I_C = I_p(x_n = W_b) = qA\frac{D_p}{L_p}(C_2 e^{-W_b/L_p} - C_1 e^{W_b/L_p}) \tag{7–17}$$

When the parameters C_1 and C_2 are substituted from Eqs. (7–13), the emitter and collector currents take a form that is most easily written in terms of hyperbolic functions:

$$I_{Ep} = qA\frac{D_p}{L_p}\left[\frac{\Delta p_E(e^{W_b/L_p} + e^{-W_b/L_p}) - 2\Delta p_C}{e^{W_b/L_p} - e^{-W_b/L_p}}\right]$$

$$I_{Ep} = qA\frac{D_p}{L_p}\left(\Delta p_E \text{ ctnh }\frac{W_b}{L_p} - \Delta p_C \text{ csch }\frac{W_b}{L_p}\right) \tag{7–18a}$$

$$I_C = qA\frac{D_p}{L_p}\left(\Delta p_E \text{ csch }\frac{W_b}{L_p} - \Delta p_C \text{ ctnh }\frac{W_b}{L_p}\right) \tag{7–18b}$$

Now we can obtain the value of I_B by a current summation, noting that the sum of the base and collector currents leaving the device must equal the emitter current entering. If $I_E \simeq I_{Ep}$ for $\gamma \simeq 1$,

$$I_B = I_E - I_C = qA\frac{D_p}{L_p}\left[(\Delta p_E + \Delta p_C)\left(\text{ctnh}\frac{W_b}{L_p} - \text{csch}\frac{W_b}{L_p}\right)\right]$$

$$\boxed{I_B = qA\frac{D_p}{L_p}\left[(\Delta p_E + \Delta p_C)\tanh\frac{W_b}{2L_p}\right]} \qquad (7\text{--}19)$$

By using the techniques of Chapter 5 we have evaluated the three terminal currents of the transistor in terms of the material parameters, the base width, and the excess concentrations Δp_E and Δp_C. Furthermore, since these excess concentrations are related in a straightforward way to the emitter and collector junction bias voltages by Eq. (7–8), it should be simple to evaluate the transistor performance under various biasing conditions. It is important to note here that Eqs. (7–18) and (7–19) are not restricted to the case of the usual transistor biasing. For example, Δp_c may be $-p_n$ for a strongly reverse-biased collector, or it may be a significant positive number if the collector is positively biased. The generality of these equations will be used in Section 7.5 in considering the application of transistors to switching circuits.

EXAMPLE 7–2

(a) Find the expression for the current I for the transistor connection shown if $\gamma = 1$.

(b) How does the current I divide between the base lead and the collector lead?

SOLUTION

(a) Since $V_{CB} = 0$, Eq. (7–8b) gives $\Delta p_C = 0$. Thus from Eq. (7–18a),

$$I_E = I = \frac{qAD_p}{L_p}\Delta p_E \,\text{ctnh}\frac{W_b}{L_p}$$

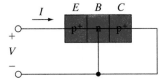

similarly,

(b) $$I_C = \frac{qAD_p}{L_p}\Delta p_E \,\text{csch}\frac{W_b}{L_p}$$

$$I_B = \frac{qAD_p}{L_p}\Delta p_E \tanh\frac{W_b}{2L_p}$$

where I_C and I_B are the components in the collector lead and base lead, respectively. Note that these results are analogous to those of Probs. 5.35 and 5.36 for the narrow base diode.

7.4.3 Approximations of the Terminal Currents

The general equations of the previous section can be simplified for the case of normal transistor biasing, and such simplification allows us to gain insight into the current flow. For example, if the collector is reverse biased, $\Delta p_C = -p_n$ from Eq. (7–9b). Furthermore, if the equilibrium hole concentration p_n is small (Fig. 7–8a), we can neglect the terms involving Δp_C. For $\gamma \simeq 1$, the terminal currents reduce to those of Example 7-2:

$$I_E \simeq qA\frac{D_p}{L_p}\Delta p_E \operatorname{ctnh}\frac{W_b}{L_p} \tag{7–20a}$$

$$I_C \simeq qA\frac{D_p}{L_p}\Delta p_E \operatorname{csch}\frac{W_b}{L_p} \tag{7–20b}$$

$$I_B \simeq qA\frac{D_p}{L_p}\Delta p_E \tanh\frac{W_b}{2L_p} \tag{7–20c}$$

Series expansions of the hyperbolic functions are given in Table 7–1. For small values of W_b/L_p, we can neglect terms above the first order of the argument. It is clear from this table and Eq. (7–20) that I_C is only slightly smaller than I_E, as expected. The first-order approximation of tanh y is simply y, so that the base current is

$$I_B \simeq qA\frac{D_p}{L_p}\Delta p_E\frac{W_b}{2L_p} = \frac{qAW_b\Delta p_E}{2\tau_p} \tag{7–21}$$

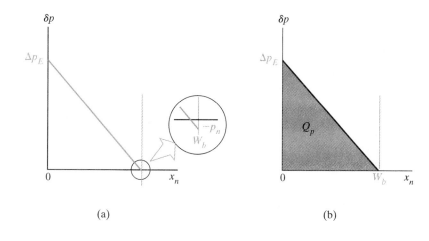

Figure 7–8
Approximate excess hole distributions in the base: (a) forward-biased emitter, reverse-biased collector; (b) triangular distribution for $V_{CB} = 0$ or for negligible p_n.

(a) (b)

Table 7–1   Expansions of several pertinent hyperbolic functions.

$$\operatorname{sech} y = 1 - \frac{y^2}{2} + \frac{5y^4}{24} - \cdots$$

$$\operatorname{ctnh} y = \frac{1}{y} + \frac{y}{3} - \frac{y^3}{45} + \cdots$$

$$\operatorname{csch} y = \frac{1}{y} - \frac{y}{6} + \frac{7y^3}{360} - \cdots$$

$$\tanh y = y - \frac{y^3}{3} + \cdots$$

The same approximate expression for the base current is found from the difference in the first-order approximations to I_E and I_C:

$$I_B = I_E - I_C$$

$$\simeq qA\frac{D_p}{L_p}\Delta p_E\left[\left(\frac{1}{W_b/L_p} + \frac{W_b/L_p}{3}\right) - \left(\frac{1}{W_b/L_p} - \frac{W_b/L_p}{6}\right)\right] \quad (7\text{–}22)$$

$$\simeq \frac{qAD_pW_b\Delta p_E}{2L_p^2} = \frac{qAW_b\Delta p_E}{2\tau_p}$$

This expression for I_B accounts for recombination in the base region. We must include injection into the emitter in many BJT devices, as discussed in Section 7.4.4.

If recombination in the base dominates the base current, I_B can be obtained from the charge control model, assuming an essentially straight-line hole distribution in the base (Fig. 7–8b). Since the hole distribution diagram appears as a triangle in this approximation, we have

$$Q_p \simeq \tfrac{1}{2}qA\,\Delta p_E W_b \quad (7\text{–}23)$$

If we consider that this stored charge must be replaced every τ_p seconds and relate the recombination rate to the rate at which electrons are supplied by the base current, I_B becomes

$$I_B \simeq \frac{Q_p}{\tau_p} = \frac{qAW_b\Delta p_E}{2\tau_p} \quad (7\text{–}24)$$

which is the same as that found in Eqs. (7–21) and (7–22).

Since we have neglected the collector saturation current and have assumed $\gamma = 1$ in these approximations, the difference between I_E and I_C is accounted for by the requirements of recombination in the base. In Eq. (7–24) we have a clear demonstration that the base current is reduced for small W_b and large τ_p. We can increase τ_p by using light doping in the base region, which of course also improves the emitter injection efficiency.

The straight-line approximation of the excess hole distribution (Fig. 7–8) is fairly accurate in calculating the base current. On the other hand, it does not give a valid picture of I_E and I_C. If the distribution were perfectly straight, the slope would be the same at each end of the base region. This would imply zero base current, which is not the case. There must be some "droop" to the distribution, as in the more accurate curve of Fig. 7–7. This slight deviation from linearity gives a steeper slope at $x_n = 0$ than at $x_n = W_b$, and the value of I_E is larger than I_C by the amount I_B. The reason we can use the straight-line approximation in the charge control calculation of base current is that the area under the hole distribution curve is essentially the same in the two cases.

7.4.4 Current Transfer Ratio

The value of I_E calculated thus far in this section is more properly designated I_{Ep}, since we have assumed that $\gamma = 1$ (the emitter current due entirely to hole injection). Actually, there is always some electron injection across the forward-biased emitter junction in a real transistor, and this effect is important in calculating the current transfer ratio. It is easy to show that the emitter injection efficiency of a p-n-p transistor can be written in terms of the emitter and base properties:

$$\gamma = \left[1 + \frac{L_p^n n_n \mu_n^p}{L_n^p p_p \mu_p^n} \tanh \frac{W_b}{L_p^n}\right]^{-1} \approx \left[1 + \frac{W_b n_n \mu_n^p}{L_n^p p_p \mu_p^n}\right]^{-1} \tag{7-25}$$

In this equation we use superscripts to indicate which side of the emitter–base junction is referred to. For example, L_p^n is the hole diffusion length in the n-type base region and μ_n^p is the electron mobility in the p-type emitter region. In an n-p-n the superscripts and subscripts would be changed along with the majority carrier symbols. Using Eq. (7–20a) for I_{Ep}, and Eq. (7–20b) for I_C, the base transport factor B is

$$B = \frac{I_C}{I_{Ep}} = \frac{\operatorname{csch} W_b/L_p}{\operatorname{ctnh} W_b/L_p} = \operatorname{sech} \frac{W_b}{L_p} \tag{7-26}$$

and the current transfer ratio α is the product of B and γ as in Eq. (7–3).

Assume that a p-n-p transistor is doped such that the emitter doping is ten times that in the base, the minority carrier mobility in the emitter is one-half that in the base, and the base width is one-tenth the minority carrier diffusion length. The carrier lifetimes are equal. Calculate α and β for this transistor.

EXAMPLE 7–3

From Eqs. (7–25) and (7–26), we have

SOLUTION

$$\alpha = B\gamma = \left[\cosh \frac{W_b}{L_p^n} + \frac{L_p^n n_n \mu_n^p}{L_n^p p_p \mu_p^n} \sinh \frac{W_b}{L_p^n}\right]^{-1}$$

Using the values given, and taking $L_p^n/L_n^p = \sqrt{\mu_p^n/\mu_n^p}$ for equal lifetimes,

$$\alpha = [\cosh 0.1 + \sqrt{2}(0.1)(0.5) \sinh 0.1]^{-1}$$
$$= [1.005 + 0.0707(0.1))]^{-1} = 0.988$$

We can find β from Eq. (7–6):

$$\beta = \frac{\alpha}{1 - \alpha} = 82$$

Thus an incremental change in I_B causes a significant change in I_C.

The expressions derived in Section 7.4 describe the terminal currents of the transistor, if the device geometry and other factors are consistent with the assumptions. Real transistors may deviate from these approximations, as we shall see in Section 7.7. The collector and emitter junctions may differ in area, saturation current, and other parameters, so that the proper description of the terminal currents may be more complicated than Eqs. (7–18) and (7–19) suggest. For example, if the roles of emitter and collector are reversed, these equations predict that the behavior of the transistor is symmetrical. Real transistors, on the other hand, are generally not symmetrical between emitter and collector. This is a particularly important consideration when the transistor is not biased in the usual way. We have discussed normal biasing (sometimes called the *normal active* mode), in which the emitter junction is forward biased and the collector is reverse biased. In some applications, particularly in switching, this normal biasing rule is violated. In these cases it is important to account for the differences in injection and collection properties of the two junctions. In this section we shall develop a generalized approach which accounts for transistor operation in terms of a coupled-diode model, valid for all combinations of emitter and collector bias. This model involves four measurable parameters that can be related to the geometry and material properties of the device. Using this model in conjunction with the charge control approach, we can describe the physical operation of a transistor in switching circuits and in other applications.

7.5.1 The Coupled-Diode Model

If the collector junction of a transistor is forward biased, we cannot neglect Δp_C; instead, we must use a more general hole distribution in the base region. Figure 7–9a illustrates a situation in which the emitter and collector junctions are both forward biased, so that Δp_E and Δp_C are positive numbers. We can handle this situation with Eqs. (7–18) and (7–19) for the symmetrical transistor. It is interesting to note that these equations can be considered as linear superpositions of the effects of injection by each junction. For example, the straight line hole distribution of Fig. 7–9a can be broken into the two components of Figs. 7–9b and 7–9c. One component (Fig. 7–9b) accounts for the holes injected by the emitter and collected by the collector. We can call the resulting currents (I_{EN} and I_{CN}) the *normal mode* components, since they are due to injection from emitter to collector. The component of the hole distribution illustrated by Fig. 7–9c results in currents I_{EI} and I_{CI}, which describe

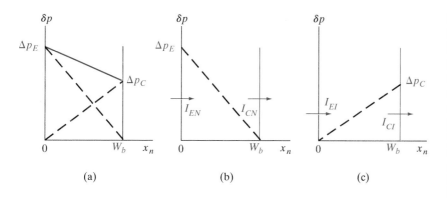

Figure 7–9
Evaluation of a hole distribution in terms of components due to normal and inverted modes: (a) approximate hole distribution in the base with emitter and collector junctions forward biased; (b) component due to injection and collection in the normal mode; (c) component due to the inverted mode.

injection in the *inverted mode* of injection from collector to emitter.[3] Of course, these inverted components will be negative, since they account for hole flow opposite to our original definitions of I_E and I_C.

For the symmetrical transistor, these various components are described by Eqs. (7–18). Defining $a \equiv (qAD_p/L_p)$ ctnh (W_b/L_p) and $b \equiv (qAD_p/L_p)$ csch (W_b/L_p), we have

$$I_{EN} = a\Delta p_E \quad \text{and} \quad I_{CN} = b\Delta p_E \quad \text{with } \Delta p_C = 0 \qquad (7\text{–}27\text{a})$$

$$I_{EI} = -b\Delta p_C \quad \text{and} \quad I_{CI} = -a\Delta p_C \quad \text{with } \Delta p_E = 0 \qquad (7\text{–}27\text{b})$$

The four components are combined by linear superposition in Eq. (7–18):

$$I_E = I_{EN} + I_{EI} = a\Delta p_E - b\Delta p_C$$
$$= \mathsf{A}(e^{qV_{EB}/kT} - 1) - \mathsf{B}(e^{qV_{CB}/kT} - 1) \qquad (7\text{–}28\text{a})$$

$$I_C = I_{CN} + I_{CI} = b\Delta p_E - a\Delta p_C$$
$$= \mathsf{B}(e^{qV_{EB}/kT} - 1) - \mathsf{A}(e^{qV_{CB}/kT} - 1) \qquad (7\text{–}28\text{b})$$

where $\mathsf{A} \equiv ap_n$ and $\mathsf{B} \equiv bp_n$.

We can see from these equations that a linear superposition of the normal and inverted components does give the result we derived previously for the symmetrical transistor. To be more general, however, we must relate the four components of current by factors which allow for asymmetry in the two junctions. For example, the emitter current in the normal mode can be written

$$I_{EN} = I_{ES}(e^{qV_{EB}/kT} - 1), \quad \Delta p_C = 0 \qquad (7\text{–}29)$$

where I_{ES} is the magnitude of the emitter saturation current in the normal mode. Since we specify $\Delta p_C = 0$ in this mode, we imply that $V_{CB} = 0$ in Eq. (7–8b). Thus we shall consider I_{ES} to be the magnitude of the emitter saturation current with

---

[3]Here the words *emitter* and *collector* refer to physical regions of the device rather than to the functions of injection and collection of holes.

the collector junction short circuited. Similarly, the collector current in the inverted mode is

$$I_{CI} = -I_{CS}(e^{qV_{CB}/kT} - 1), \quad \Delta p_E = 0 \qquad (7\text{–}30)$$

where I_{CS} is the magnitude of the collector saturation current with $V_{EB} = 0$. As before, the minus sign associated with I_{CI} simply means that in the inverted mode holes are injected opposite to the defined direction of I_C.

The corresponding collected currents for each mode of operation can be written by defining a new α for each case:

$$I_{CN} = \alpha_N I_{EN} = \alpha_N I_{ES}(e^{qV_{EB}/kT} - 1) \qquad (7\text{–}31a)$$

$$I_{EI} = \alpha_I I_{CI} = -\alpha_I I_{CS}(e^{qV_{CB}/kT} - 1) \qquad (7\text{–}31b)$$

where α_N and α_I are the ratios of collected current to injected current in each mode. We notice that in the inverted mode the injected current is I_{CI} and the collected current is I_{EI}.

The total current can again be obtained by superposition of the components:

$$\boxed{\begin{aligned} I_E &= I_{EN} + I_{EI} = I_{ES}(e^{qV_{EB}/kT} - 1) - \alpha_I I_{CS}(e^{qV_{CB}/kT} - 1) \\ I_C &= I_{CN} + I_{CI} = \alpha_N I_{ES}(e^{qV_{EB}/kT} - 1) - I_{CS}(e^{qV_{CB}/kT} - 1) \end{aligned}}$$

$$(7\text{–}32a)$$
$$(7\text{–}32b)$$

These relations were derived by J.J. Ebers and J.L. Moll and are referred to as the *Ebers–Moll equations.*[4] While the general form is the same as Eqs. (7–28) for the symmetrical transistor, these equations allow for variations in I_{ES}, I_{CS}, α_I, and α_N due to asymmetry between the junctions. Although we shall not prove it here, it is possible to show by reciprocity arguments that

$$\alpha_N I_{ES} = \alpha_I I_{CS} \qquad (7\text{–}33)$$

even for nonsymmetrical transistors.

An interesting feature of the Ebers–Moll equations is that I_E and I_C are described by terms resembling diode relations (I_{EN} and I_{CI}), plus terms which provide coupling between the properties of the emitter and collector (I_{EI} and I_{CN}). This *coupled-diode* property is illustrated by the equivalent circuit of Fig. 7–10. In this figure we take advantage of Eq. (7–8) to write the Ebers–Moll equations in the form

---

[4]J.J. Ebers and J.L. Moll, "Large-Signal Behavior of Junction Transistors," *Proceedings of the IRE 42,* pp. 1761–72 (December 1954). In the original paper and in many texts, the terminal currents are all defined as flowing *into* the transistor. This introduces minus signs into the expressions for I_c and I_B as we have developed them here.

$$I_E = I_{ES}\frac{\Delta p_E}{p_n} - \alpha_I I_{CS}\frac{\Delta p_C}{p_n} = \frac{I_{ES}}{p_n}(\Delta p_E - \alpha_N \Delta p_C) \qquad (7\text{--}34a)$$

$$I_C = \alpha_N I_{ES}\frac{\Delta p_E}{p_n} - I_{CS}\frac{\Delta p_C}{p_n} = \frac{I_{CS}}{p_n}(\alpha_I \Delta p_E - \Delta p_C) \qquad (7\text{--}34b)$$

It is often useful to relate the terminal currents to each other as well as to the saturation currents. We can eliminate the saturation current from the coupling term in each part of Eq. (7–32). For example, by multiplying Eq. (7–32a) by α_N and subtracting the resulting expression from Eq. (7–32b), we have

$$I_C = \alpha_N I_E - (1 - \alpha_N \alpha_I) I_{CS}(e^{qV_{CB}/kT} - 1) \qquad (7\text{--}35)$$

Similarly, the emitter current can be written in terms of the collector current:

$$I_E = \alpha_I I_C + (1 - \alpha_N \alpha_I) I_{ES}(e^{qV_{EB}/kT} - 1) \qquad (7\text{--}36)$$

The terms $(1 - \alpha_N \alpha_I)I_{CS}$ and $(1 - \alpha_N \alpha_I)I_{ES}$ can be abbreviated as I_{CO} and I_{EO}, respectively, where I_{CO} is the magnitude of the collector saturation current with the emitter junction *open* ($I_E = 0$), and I_{EO} is the magnitude of the emitter saturation current with the collector open. The Ebers–Moll equations then become

$$\boxed{\begin{aligned} I_E &= \alpha_I I_C + I_{EO}(e^{qV_{EB}/kT} - 1) \\ I_C &= \alpha_N I_E - I_{CO}(e^{qV_{CB}/kT} - 1) \end{aligned}} \qquad \begin{aligned} (7\text{--}37a) \\ (7\text{--}37b) \end{aligned}$$

and the equivalent circuit is shown in Fig. 7–11a. In this form the equations describe both the emitter and collector currents in terms of a simple diode characteristic plus a current generator proportional to the other current. For example, under normal biasing the equivalent circuit reduces to the form shown in Fig. 7–11b. The collector current is α_N times the emitter current plus

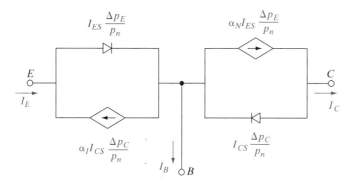

$$I_B = (1 - \alpha_N) I_{ES}\frac{\Delta p_E}{p_n} + (1 - \alpha_I) I_{CS}\frac{\Delta p_C}{p_n}$$

Figure 7–10
An equivalent circuit synthesizing the Ebers–Moll equations.

Figure 7–11
Equivalent circuits
of the transistor in
terms of the termi-
nal currents and
the open-circuit
saturation cur-
rents: (a) synthesis
of Eqs. (7–37); (b)
equivalent circuit
with normal bias-
ing; (c) collector
characteristics
with normal
biasing.

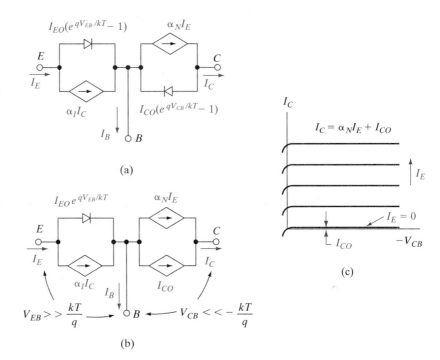

the collector saturation current, as expected. The resulting collector characteristics of the transistor appear as a series of reverse biased diode curves, displaced by increments proportional to the emitter current (Fig. 7–11c).

7.5.2 Charge Control Analysis

The charge control approach is useful in analyzing the transistor terminal
currents, particularly in a-c applications. Considerations of transit time effects and charge storage are revealed easily by this method. Following the
techniques of the previous section, we can separate an arbitrary excess hole
distribution in the base into the normal and inverted distributions of Fig. 7–9. The
charge stored in the normal distribution will be called Q_N and the charge
under the inverted distribution will be called Q_I. Then we can evaluate the
currents for the normal and inverted modes in terms of these stored charges.
For example, the collected current in the normal mode I_{CN} is simply the
charge Q_N divided by the mean time required for this charge to be collected. This time is the transit time for the normal mode τ_{tN}. On the other hand,
the emitter current must support not only the rate of charge collection by the
collector but also the recombination rate in the base Q_N/τ_{pN}. Here we use a
subscript N with the transit time and lifetime in the normal mode in contrast

to the inverted mode, to allow for possible asymmetries due to imbalance in the transistor structure. With these definitions, the normal components of current become

$$I_{CN} = \frac{Q_N}{\tau_{tN}} \quad , \quad I_{EN} = \frac{Q_N}{\tau_{tN}} + \frac{Q_N}{\tau_{pN}} \tag{7-38a}$$

Similarly, the inverted components are

$$I_{EI} = -\frac{Q_I}{\tau_{tI}} \quad , \quad I_{CI} = -\frac{Q_I}{\tau_{tI}} - \frac{Q_I}{\tau_{pI}} \tag{7-38b}$$

where the I subscripts on the stored charge and on the transit and recombination times designate the inverted mode. Combining these equations as in Eq. (7–32) we have the terminal currents for general biasing:

$$I_E = Q_N\left(\frac{1}{\tau_{tN}} + \frac{1}{\tau_{pN}}\right) - \frac{Q_I}{\tau_{tI}} \tag{7-39a}$$

$$I_C = \frac{Q_N}{\tau_{tN}} - Q_I\left(\frac{1}{\tau_{tI}} + \frac{1}{\tau_{pI}}\right) \tag{7-39b}$$

It is not difficult to show that these equations correspond to the Ebers–Moll relations [Eq. (7–34)], where

$$\alpha_N = \frac{\tau_{pN}}{\tau_{tN} + \tau_{pN}} \quad , \quad \alpha_1 = \frac{\tau_{pI}}{\tau_{tI} + \tau_{pI}}$$

$$I_{ES} = q_N\left(\frac{1}{\tau_{tN}} + \frac{1}{\tau_{pN}}\right) \quad , \quad I_{CS} = q_I\left(\frac{1}{\tau_{tI}} + \frac{1}{\tau_{pI}}\right) \tag{7-40}$$

$$Q_N = q_N\frac{\Delta p_E}{p_n} \quad , \quad Q_I = q_I\frac{\Delta p_C}{p_n}$$

The base current in the normal mode supports recombination, and the base-to-collector current amplification factor β_N takes the form predicted by Eq. (7–7):

$$I_{BN} = \frac{Q_N}{\tau_{pN}} \quad , \quad \beta_N = \frac{I_{CN}}{I_{BN}} = \frac{\tau_{pN}}{\tau_{tN}} \tag{7-41}$$

This expression for β_N is also obtained from $\alpha_N/(1 - \alpha_N)$. Similarly, I_{BI} is Q_I/τ_{pI}, and the total base current is

$$I_B = I_{BN} + I_{BI} = \frac{Q_N}{\tau_{pN}} + \frac{Q_I}{\tau_{pI}} \tag{7-42}$$

This expression for the base current is substantiated by $I_E - I_C$ from Eq. (7–39).

The effects of time dependence of stored charge can be included in these equations by the methods introduced in Section 5.5.1. We can include the proper dependencies by adding a rate of change of stored charge to each of the injection currents I_{EN} and I_{CI}:

$$i_E = Q_N\left(\frac{1}{\tau_{tN}} + \frac{1}{\tau_{pN}}\right) - \frac{Q_I}{\tau_{tI}} + \frac{dQ_N}{dt} \tag{7-43a}$$

$$i_C = \frac{Q_N}{\tau_{tN}} - Q_I\left(\frac{1}{\tau_{tI}} + \frac{1}{\tau_{pI}}\right) - \frac{dQ_I}{dt} \tag{7-43b}$$

$$i_B = \frac{Q_N}{\tau_{pN}} + \frac{Q_I}{\tau_{pI}} + \frac{dQ_N}{dt} + \frac{dQ_I}{dt} \tag{7-43c}$$

We shall return to these equations in Section 7.8, when we discuss the use of transistors at high frequencies.

**7.6
SWITCHING**

In a switching operation a transistor is usually controlled in two conduction states, which can be referred to loosely as the "on" state and the "off" state. Ideally, a switch should appear as a short circuit when turned on and an open circuit when turned off. Furthermore, it is desirable to switch the device from one state to the other with no lost time in between. Transistors do not fit this ideal description of a switch, but they can serve as a useful approximation in practical electronic circuits. The two states of a transistor in switching can be seen in the simple common-emitter example of Fig. 7–12. In this figure the collector current i_C is controlled by the base current i_B over most of the family of characteristic curves. The load line specifies the locus of allowable

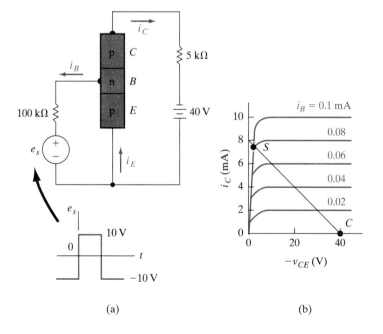

Figure 7–12
Simple switching circuit for a transistor in the common-emitter configuration: (a) biasing circuit; (b) collector characteristics and load line for the circuit, with cutoff and saturation indicated.

(a)   (b)

$(i_C, -v_{CE})$ points for the circuit, in analogy with Fig. 6-2. If i_B is such that the operating point lies somewhere between the two end points of the load line (Fig. 7–12b), the transistor operates in the normal active mode. That is, the emitter junction is forward biased and the collector is reverse biased, with a reasonable value of i_B flowing out of the base. On the other hand, if the base current is zero or negative, the point C is reached at the bottom end of the load line, and the collector current is negligible. This is the "off" state of the transistor, and the device is said to be operating in the *cutoff* regime. If the base current is positive and sufficiently large, the device is driven to the *saturation* regime, marked S. This is the "on" state of the transistor, in which a large value of i_C flows with only a very small voltage drop v_{CE}. As we shall see below, the beginning of the saturation regime corresponds to the loss of reverse bias across the collector junction. In a typical switching operation the base current swings from positive to negative, thereby driving the device from saturation to cutoff, and vice versa. In this section we shall explore the nature of conduction in the cutoff and saturation regimes; also we shall investigate the factors affecting the speed with which the transistor can be switched between the two states.

7.6.1 Cutoff

If the emitter junction is reverse biased in the cutoff regime (negative i_B), we can approximate the excess hole concentrations at the edges of the reverse-biased emitter and collector junctions as

$$\frac{\Delta p_E}{p_n} \simeq \frac{\Delta p_C}{p_n} \simeq -1 \qquad (7\text{–}44)$$

which implies $p(x_n) = 0$. With a straight-line approximation, the excess hole distribution in the base appears constant at $-p_n$, as shown in Fig. 7–13a. Actually, there will be some slope to the distribution at each edge to account for the reverse saturation current in the junctions, but Fig. 7–13a is approximately correct. The base current i_B can be approximated for a symmetrical transistor on a charge storage basis as $-qAp_n W_b/\tau_p$. In this calculation a negative excess hole concentration corresponds to *generation* in the same way that a positive distribution indicates recombination. This expression is also obtained by applying Eq. (7–44) to Eq. (7–19) with an approximation from Table 7–1. Physically, a small saturation current flows from n to p in each reverse-biased junction, and this current is supplied by the base current i_B (which is negative when flowing into the base of a p-n-p device according to our definitions). A more general evaluation of the currents can be obtained from the Ebers–Moll equations by applying Eq. (7–44) to Eq. (7–34):

$$i_E = -I_{ES} + \alpha_I I_{CS} = -(1 - \alpha_N)I_{ES} \qquad (7\text{–}45a)$$

$$i_C = -\alpha_N I_{ES} + I_{CS} = (1 - \alpha_I)I_{CS} \qquad (7\text{–}45b)$$

$$i_B = i_E - i_C = -(1 - \alpha_N)I_{ES} - (1 - \alpha_I)I_{CS} \qquad (7\text{–}45c)$$

Figure 7–13
The cutoff regime
of a p-n-p transis-
tor: (a) excess
hole distribution in
the base region
with emitter and
collector junctions
reverse biased;
(b) equivalent cir-
cuit correspond-
ing to Eq. (7–45).

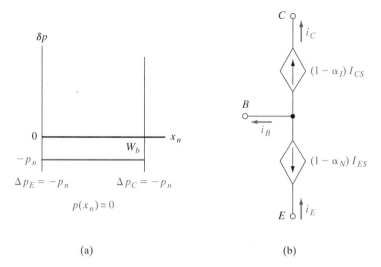

(a)   (b)

If the short-circuit saturation currents I_{ES} and I_{CS} are small and α_N and α_I are both near unity, these currents will be negligible and the cutoff regime will closely approximate the "off" condition of an ideal switch. The equivalent circuit corresponding to Eq. (7–45) is illustrated in Fig. 7–13b.

7.6.2   Saturation

The saturation regime begins when the reverse bias across the collector junction is reduced to zero, and it continues as the collector becomes forward biased. The excess hole distribution in this case is illustrated in Fig. 7–14. The device is saturated when $\Delta p_C = 0$, and forward bias of the collector junction (Fig. 7–14b) leads to a positive Δp_C, driving the device further into saturation. With the load line fixed by the battery and the 5-kΩ resistor in Fig. 7–12, saturation is reached by increasing the base current i_B. We can see how a large value of i_B leads to saturation by applying the reasoning of charge control to Fig. 7–14. Since a certain amount of stored charge is required to accommodate a given i_B (and vice versa), an increase in i_B calls for an increase in the area under the $\delta p(x_n)$ distribution.

In Fig. 7–14a the device has just reached saturation, and the collector junction is no longer reverse biased. The implication of this condition for the circuit of Fig. 7–12 is easy to state. Since the emitter junction is forward biased and the collector junction has zero bias, very little voltage drop appears across the device from collector to emitter. The magnitude of $-v_{CE}$ is only a fraction of a volt. Therefore, almost all of the battery voltage appears across the resistor, and the collector current is approximately 40 V/5 kΩ = 8 mA. As the device is driven deeper into saturation (Fig. 7–14b), the collector current stays essentially constant while the base current increases. In this saturation condition the transistor approximates the "on" state of an ideal switch.

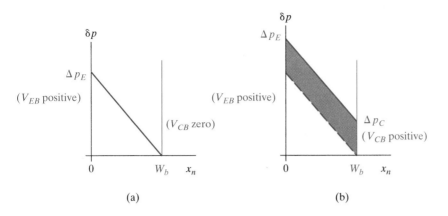

Figure 7–14
Excess hole distri-
bution in the base
of a saturated
transistor: (a) the
beginning of satu-
ration; (b) oversat-
uration.

Whereas the degree of "oversaturation" (indicated by the shaded area in Fig. 7–14b) does not affect the value of i_C significantly, it is important in determining the time required to switch the device from one state to the other. For example, from previous experience we expect the turn-off time (from saturation to cutoff) to be longer for larger values of stored charge in the base. We can calculate the various charging and delay times from Eq. (7–43). Detailed calculations are somewhat involved, but we can simplify the problem greatly with approximations of the type used in Chapter 5 for transient effects in p-n junctions.

7.6.3 The Switching Cycle

The various mechanisms of a switching cycle are illustrated in Fig. 7–15. If the device is originally in the cutoff condition, a step increase of base current to I_B causes the hole distribution to increase approximately as illustrated in Fig. 7–15b. As in the transient analysis of Chapter 5, we assume for simplicity of calculation that the distribution maintains a simple form in each time interval of the transient. At time t_s the device enters saturation, and the hole distribution reaches its final state at t_2. As the stored charge in the base Q_b increases, there is an increase in the collector current i_C. The collector current does not increase beyond its value at the beginning of saturation t_s, however. We can approximate this saturated collector current as $I_C \simeq E_{CC}/R_L$, where E_{CC} is the value of the collector circuit battery and R_L is the load resistor ($I_C \simeq 8$ mA for the example of Fig. 7–12). There is an essentially exponential increase in the collector current while Q_b rises to its value Q_s at t_s; this rise time serves as one of the limitations of the transistor in a switching application. Similarly, when the base current is switched negative (e.g., to the value $-I_B$), the stored charge must be withdrawn from the base before cutoff is reached. While Q_b is larger than Q_s, the collector current remains at the value I_C, fixed by the battery and resistor. Thus there is a storage delay time t_{sd} after the base current is switched and before i_C begins to fall toward zero. After the stored charge is reduced

Figure 7–15
Switching effects
in a common-
emitter transistor
circuit: (a) circuit
diagram; (b)
approximate
hole distributions
in the base
during switching
from cutoff to
saturation; (c)
base current,
stored charge,
and collector
current during
a turn-on and a
turn-off transient.

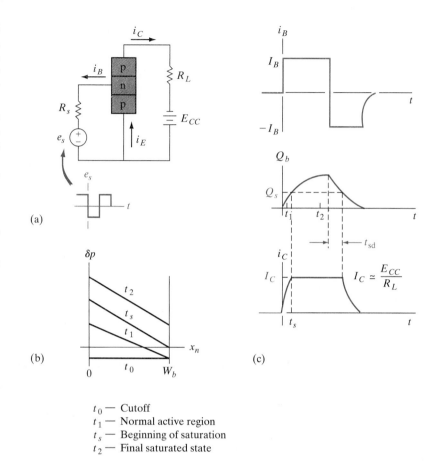

t_0 — Cutoff
t_1 — Normal active region
t_s — Beginning of saturation
t_2 — Final saturated state

below Q_s, i_C drops exponentially with the characteristic fall time. Once the stored charge is withdrawn, the base current cannot be maintained any longer at its large negative value and must decay to the small cutoff value described by Eq. (7–45c).

7.6.4 Specifications for Switching Transistors

We can determine t_s and t_{sd} by solving for the time-dependent base current, $i_B(t)$ given by an expression similar to Eq. (5–47). We must also not neglect the charging time of the emitter junction capacitance in going from cutoff to saturation. Since the emitter junction is reverse biased in cutoff, it is necessary for the emitter space charge layer to be charged to the forward bias condition before collector current can flow. Therefore, we should include a *delay time t_d* as in Fig. 7–16 to account for this effect. Typical values of t_d are given

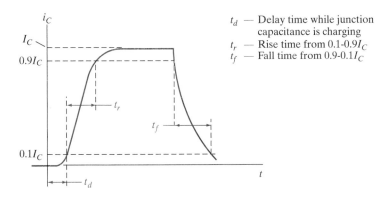

t_d — Delay time while junction capacitance is charging
t_r — Rise time from 0.1-$0.9I_C$
t_f — Fall time from 0.9-$0.1I_C$

Figure 7–16
Collector current during switching transients, including the delay time required for charging the junction capacitance; definitions of the rise time and fall time.

in the specification information of most switching transistors, along with a *rise time t_r*, defined as the time required for the collector current to rise from 10 to 90 per cent of its final value. A third specification is the *fall time t_f* required for i_C to fall through a similar fraction of its turn-off excursion.

The approach we have taken in analyzing the properties of transistors has involved a number of simplifying assumptions. Some of the assumptions must be modified in dealing with practical devices. In this section we investigate some common deviations from the basic theory and indicate situations in which each effect is important. Since the various effects discussed here involve modifications of the more straightforward theory, they are often labeled "secondary effects." This does not imply that they are unimportant; in fact, the effects described in this section can dominate the conduction in transistors under certain conditions of device geometry and circuit application.

In this section we shall consider the effects of nonuniform doping in the base region of the transistor. In particular, we shall find that graded doping can lead to a drift component of charge transport across the base, adding to the diffusion of carriers from emitter to collector. We shall discuss the effects of large reverse bias on the collector junction, in terms of widening the space charge region about the junction and avalanche multiplication. We shall see that transistor parameters are affected at high current levels by the degree of injection and by heating effects. We shall consider several structural effects that are important in practical devices, such as asymmetry in the areas of the emitter and collector junctions, series resistance between the base contact and the active part of the base region, and nonuniformity of injection at the emitter junction. All these effects are important in understanding the operation of transistors, and proper consideration of their interactions can contribute greatly to the usefulness of practical transistor circuits.

7.7 OTHER IMPORTANT EFFECTS

7.7.1 Drift in the Base Region

The assumption of uniform doping in the base breaks down for implanted junction transistors which usually involve an appreciable amount of impurity grading; for example, the implanted transistor of Fig. 7–5 has a doping profile similar to that sketched in Fig. 7–17. In this example there is a fairly sharp discontinuity in the doping profile, when the donor concentration in the base region becomes smaller than the constant p-type background doping in the collector. Similarly, the emitter is assumed to be a heavily doped (p^+) shallow region, providing a second rather sharp boundary for the base. Within the base region itself, however, the net doping concentration ($N_d - N_a \equiv N$) varies along a profile which decreases from the emitter edge to the collector edge. The most likely doping distribution in the base is a portion of a gaussian (see Section 5.1.4); however, we can clearly see the effect of an impurity gradient by assuming for simplicity that $N(x_n)$ varies exponentially within the base region (Fig. 7–17b).

One important result of a graded base region is that a built-in electric field exists from emitter to collector (for a p-n-p), thereby adding a drift component to the transport of holes across the base. We can demonstrate this effect very simply by considering the required balance of drift and diffusion in the base at equilibrium. If the net donor doping of the base is large enough to allow the usual approximation $n(x_n) \simeq N(x_n)$, the balance of electron drift and diffusion currents at equilibrium requires

$$I_n(x_n) = qA\mu_n N(x_n)\mathscr{E}(x_n) + qAD_n\frac{dN(x_n)}{dx_n} = 0 \qquad (7\text{–}46)$$

Therefore, the built-in electric field is

$$\mathscr{E}(x_n) = -\frac{D_n}{\mu_n}\frac{1}{N(x_n)}\frac{dN(x_n)}{dx_n} = -\frac{kT}{q}\frac{1}{N(x_n)}\frac{dN(x_n)}{dx_n} \qquad (7\text{–}47)$$

Figure 7–17
Graded doping in the base region of a p-n-p transistor: (a) typical doping profile on a semilog plot; (b) approximate exponential distribution of the net donor concentration in the base region on a linear plot.

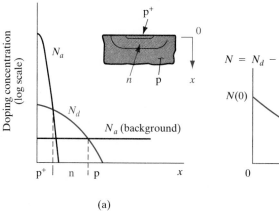

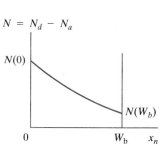

(a)

(b)

For a doping profile $N(x_n)$ that decreases in the positive x_n-direction, this field is positive, directed from emitter to collector.

For the example of an exponential doping profile, the electric field $\mathscr{E}(x_n)$ turns out to be constant with position in the base. We can represent an exponential distribution as

$$N(x_n) = N(0)e^{-ax_n/W_b} \qquad \text{where } a \equiv \ln \frac{N(0)}{N(W_b)} \qquad (7\text{--}48)$$

Taking the derivative of this distribution and substituting in Eq. (7–47), we obtain the constant field

$$\mathscr{E}(x_n) = \frac{kT}{q} \frac{a}{W_b} \qquad (7\text{--}49)$$

Since this field aids the transport of holes across the base region from emitter to collector, the transit time τ_t is reduced below that of a comparable uniform base transistor. Similarly, electron transport in an n-p-n is aided by the built-in field in the base. This shortening of the transit time can be very important in high-frequency devices (Section 7.8.2). Another approach for obtaining a built-in field is to vary the alloy composition x (and therefore E_g) in a base made of an alloy such as $Si_{1-x}Ge_x$ or $In_xGa_{1-x}As$. We will discuss this further in Section 7.9.

7.7.2 Base Narrowing

In the discussion of transistors thus far, we have assumed that the effective base width W_b is essentially independent of the bias voltages applied to the collector and emitter junctions. This assumption is not always valid; for example, the p$^+$-n-p$^+$ transistor of Fig. 7–18 is affected by the reverse bias applied to the collector. If the base region is lightly doped, the depletion region at the reverse-biased collector junction can extend significantly into the n-type base region. As the collector voltage is increased, the space charge layer takes up more of the metallurgical width of the base L_b, and as a result, the effective base width W_b is decreased. This effect is variously called *base narrowing, base-width modulation*, and the *Early effect* after J.M. Early, who first interpreted it. The effects of base narrowing are apparent in the collector characteristics for the common-emitter configuration (Fig. 7–18b). The decrease in W_b causes β to increase. As a result, the collector current I_C increases with collector voltage rather than staying constant as predicted from the simple treatment. The slope introduced by the Early effect is almost linear with I_C, and the common-emitter characteristics extrapolate to an intersection with the voltage axis at V_A, called the Early voltage.

For the p$^+$-n-p$^+$ device of Fig. 7–18 we can approximate the length l of the collector junction depletion region in the n material from Eq. (5–23b) with V_0 replaced by $V_0 - V_{CB}$ and V_{CB} taken to be large and negative:

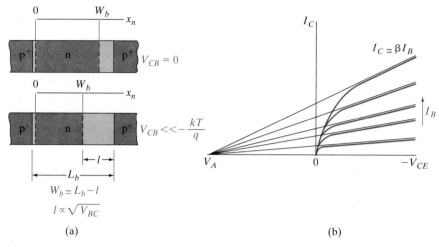

Figure 7–18
The effects of base narrowing on the characteristics of a p⁺-n-p⁺ transistor: (a) decrease in the effective base width as the reverse bias on the collector junction is increased; (b) common-emitter characteristics showing the increase in I_C with increased collector voltage. The black lines in (b) indicate the extrapolation of the curves to the Early voltage V_A.

$$l = \left(\frac{2\epsilon V_{BC}}{qN_d}\right)^{1/2} \tag{7–50}$$

If the reverse bias on the collector junction is increased far enough, it is possible to decrease W_b to the extent that the collector depletion region essentially fills the entire base. In this punch-through condition holes are swept directly from the emitter region to the collector, and transistor action is lost. Punch-through is a breakdown effect that is generally avoided in circuit design. In most cases, however, avalanche breakdown of the collector junction occurs before punch-through is reached. We shall discuss the effects of avalanche multiplication in the following section.

In devices with graded base doping, base narrowing is of less importance. For example, if the donor concentration in the base region of a p-n-p increases with position from the collector to the emitter, the intrusion of the collector space charge region into the base becomes less important with increased bias as more donors are available to accommodate the space charge.

7.7.3 Avalanche Breakdown

Before punch-through occurs in most transistors, avalanche multiplication at the collector junction becomes important (see Section 5.4.2). As Fig. 7–19 indicates, the collector current increases sharply at a well-defined breakdown voltage BV_{CBO} for the common-base configuration. For the common-emitter case, however, there is a strong influence of carrier multiplication over a fairly broad

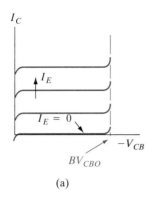

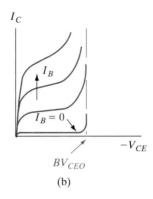

(a)   (b)

Figure 7–19
Avalanche break-
down in a transis-
tor: (a)
common-base
configuration; (b)
common-emitter
configuration.

range of collector voltage. Furthermore, the breakdown voltage in the common-emitter case BV_{CEO} is significantly smaller than BV_{CBO}. We can understand these effects by considering breakdown for the condition $I_E = 0$ in the common-base case and for $I_B = 0$ in the common-emitter case. These conditions are implied by the O in the subscripts of BV_{CEO} and BV_{CBO}. In each case the terminal current I_C is the current entering the collector depletion region multiplied by the factor M. Including multiplication due to impact ionization, Eq.(7–37b) becomes

$$I_C = (\alpha_N I_E + I_{CO})M = (\alpha_N I_E + I_{CO})\frac{1}{1 - (V_{BC}/BV_{CBO})^n} \quad (7\text{–}51)$$

where for M we have used the empirical expression given in Eq. (5-44).

For the limiting common-base case of $I_E = 0$ (the lowest curve in Fig. 7–23a), I_C is simply MI_{CO}, and the breakdown voltage is well defined, as in an isolated junction. The term BV_{CBO} signifies the collector junction breakdown voltage in common-base with the emitter open. In the common-emitter case the situation is somewhat more complicated. Setting $I_B = 0$, and therefore, $I_C = I_E$ in Eq. (7–51), we have

$$I_C = \frac{MI_{CO}}{1 - M\alpha_N} \quad (7\text{–}52)$$

We notice that in this case the collector current increases indefinitely when $M\alpha_N$ approaches unity. By contrast, M must approach infinity in the common-base case before BV_{CBO} is reached. Since α_N is close to unity in most transistors, M need be only slightly larger than unity for Eq. (7–52) to approach breakdown. Avalanche multiplication thus dominates the current in a common-emitter transistor well below the breakdown voltage of the isolated collector junction. The sustaining voltage for avalanching in the common-emitter case BV_{CEO} is therefore smaller than BV_{CBO}.

We can understand physically why multiplication is so important in the common-emitter case by considering the effect of M on the base current. When an ionizing collision occurs in the collector junction depletion region, a secondary hole and electron are created. The primary and secondary holes are swept into the collector in a p-n-p, but the electron is swept into the base by the junction field. Therefore, the supply of electrons to the base is increased, and from our charge control analysis we conclude that hole injection at the emitter must increase to maintain space charge neutrality. This is a regenerative process, in which an increased injection of holes from the emitter causes an increased multiplication current at the collector junction; this in turn increases the rate at which secondary electrons are swept into the base, calling for more hole injection. Because of this regenerative effect, it is easy to understand why the multiplication factor M need be only slightly greater than unity to start the avalanching process.

7.7.4   Injection Level; Thermal Effects

In discussions of transistor characteristics we have assumed that α and β are independent of carrier injection level. Actually, the parameters of a practical transistor may vary considerably with injection level, which is determined by the magnitude of I_E or I_C. For very low injection, the assumption of negligible recombination in the junction depletion regions is invalid (see Section 5.6.2). This is particularly important in the case of recombination in the emitter junction, where any recombination tends to degrade the emitter injection efficiency γ. Thus we expect that α and β should decrease for low values of I_C, causing the curves of the collector characteristics to be spaced more closely for low currents than for higher currents.

As I_C is increased beyond the low injection level range, α and β increase but fall off again at very high injection. The primary cause of this fall-off is the increase of majority carriers at high injection levels (see Section 5.6.1). For example, as the concentration of excess holes injected into the base becomes large, the matching excess electron concentration can become greater than the background n_n. This base conductivity modulation effect results in a decrease in γ as more electrons are injected across the emitter junction into the emitter region.

Large values of I_C may be accompanied by significant power dissipation in the transistor and therefore heating of the device. In particular, the product of I_C and the collector voltage V_{BC} is a measure of the power dissipated at the collector junction. This dissipation is due to the fact that carriers swept through the collector junction depletion region are given increased kinetic energy, which in turn is given up to the lattice in scattering collisions. It is very important that the transistor be operated in a range such that $I_C V_{BC}$ does not exceed the maximum power rating of the device. In devices designed for high power capability, the transistor is mounted on an efficient heat sink, so that thermal energy can be transferred away from the junction.

If the temperature of the device is allowed to increase due to power dissipation or thermal environment, the transistor parameters change. The most important parameters dependent on temperature are the carrier lifetimes and diffusion coefficients. In Si or Ge devices the lifetime τ_p increases with temperature for most cases, due to thermal reexcitation from recombination centers. This increase in τ_p tends to increase β for the transistor. On the other hand, the mobility decreases with increasing temperature in the lattice-scattering range, varying approximately as $T^{-3/2}$ (see Fig. 3-22). Thus from the Einstein relation, we expect D_p to decrease as the temperature increases, thereby causing a drop in β due to an increasing transit time τ_t. Of these competing processes, the effect of increasing lifetime with temperature usually dominates, and β becomes larger as the device is heated. It is clear from this effect that *thermal runaway* can occur if the circuit is not designed to prevent it. For example, a large power dissipation in the device can cause an increase in T; this results in a large β and therefore a large I_C for a given base current; the large I_C causes more collector dissipation and the cycle continues. This type of runaway of the collector current can result in overheating and destruction of the device.

7.7.5 Base Resistance and Emitter Crowding

A number of structural effects are important in determining the operation of a transistor. For example, the emitter and collector areas are considerably different in the implanted transistor of Fig. 7–20a. This and most other structural effects can be accounted for by differences in α_N, α_I, and other parameters in the Ebers–Moll model. Several effects caused by the structural arrangement of real transistors deserve special attention, however. One of the most important of these effects is the fact that base current must pass from the active part of the base region to the base contacts B. Thus, to be accurate, we should include a resistance r_b in equivalent models for the transistor to account for voltage drops which may occur between B and the active part of the base. Because of r_b, it is common to contact the base with the metallization pattern on both sides of the emitter, as in Fig. 7–20c.

If the transistor is designed so that the n-type regions leading from the base to the contacts are large in cross-sectional area, the base resistance r_b may be negligible. On the other hand, the distributed resistance r_b' along the thin base region is almost always important.[5] Since the width of the base between emitter and collector is very narrow, this distributed resistance is usually quite high. Therefore, as base current flows from points within the base region toward each end, a voltage drop occurs along r_b'. In this case the forward bias across the emitter-base junction is not uniform, but instead varies with position according to the voltage drop in the distributed base resistance. In particular, the forward bias of the emitter junction is largest at the corner of the emitter region near the base contact. We can see that this is the case by considering the simplified example of Fig. 7–20e. Neglecting variations in the base current along the path from point A to the contact B, the forward bias of the emitter junction above point A is approximately

---

[5]The distributed resistance r_b' is often called the *base spreading resistance*.

Figure 7–20
Effects of a base
resistance: (a)
cross section of
an implanted tran-
sistor; (b) and (c)
top view, showing
emitter and base
areas and metal-
lized contacts; (d)
illustration of base
resistance; (e) ex-
panded view of
distributed resis-
tance in the active
part of the base
region.

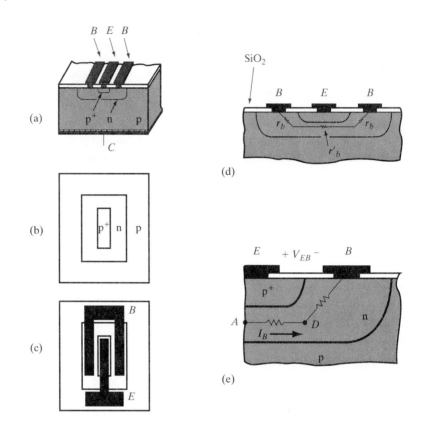

$$V_{EA} = V_{EB} - I_B(R_{AD} + R_{DB}) \qquad (7\text{–}53)$$

Actually, the base current is not uniform along the active part of the base region, and the distributed resistance of the base is more complicated than we have indicated. But this example does illustrate the point of nonuniform injection. Whereas the forward bias at A is approximately described by Eq. (7–53), the emitter bias voltage at point D is

$$V_{ED} = V_{EB} - I_B R_{DB} \qquad (7\text{–}54)$$

which can be significantly closer to the applied voltage V_{EB}.

Since the forward bias is largest at the edge of the emitter, it follows that the injection of holes is also greatest there. This effect is called *emitter crowding*, and it can strongly affect the behavior of the device. The most important result of emitter crowding is that high-injection effects described in the previous section can become dominant locally at the corners of the emitter before the overall emitter current is very large. In transistors designed to handle appreciable current, this is a problem which must be dealt with by proper structural design. The most effective approach to the problem of emitter crowding is to distribute the emitter current along a relatively large emitter edge, thereby reducing the current density at any one point. Clearly, what

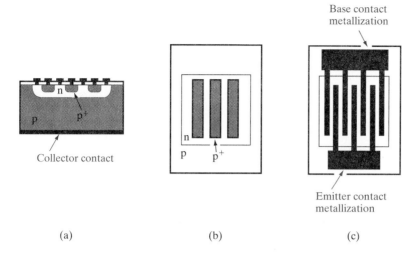

Emitter contact
metallization

(a)   (b)   (c)

Figure 7–21
An interdigitated
geometry to com-
pensate for the ef-
fects of emitter
crowding in a
power transistor:
(a) cross section;
(b) top view of im-
planted regions;
(c) top view with
metallized con-
tacts. The metal
interconnections
are isolated from
the device by an
oxide layer ex-
cept where they
contact the appro-
priate base and
emitter regions at
"windows" in the
oxide.

is needed is an emitter region with a large perimeter compared with its area.
A likely geometry to accomplish this is a long thin stripe for the emitter, with
base contacts on each side (Fig. 7–20b and c). With this geometry the total
emitter current I_E is spread out along a rather long edge on each side of the
stripe. An even better geometry is several emitter stripes, connected electrically by the metallization and separated by interspersing base contacts (Fig.
7–21). Many such thin emitter and base contact "fingers" can be interlaced
to provide for handling large current in a power transistor. This is often called
very descriptively an *interdigitated* geometry.

7.7.6   Gummel–Poon Model

The Ebers–Moll model runs into problems if a high degree of accuracy is required for very small BJTs, or where second order effects become important. The Gummel–Poon model, which is a charge control model, incorporates
much more physics. We present a simplified version of the model here.

As discussed in Section 7.7.1, for typical graded doping profiles in the
base, there is a built-in electric field that causes drift of minority carriers in
the same direction they are diffusing from emitter to collector. This forms the
starting point for the derivation of the Gummel–Poon model. As mentioned
in Section 7.7.1, this electric field aids the motion of the minority carriers in
the base, so the current can be written as

$$I_{Ep} = qA\mu_p p(x_n)\mathscr{E} - qAD_p\frac{dp(x_n)}{dx_n} \qquad (7\text{–}55)$$

We can replace the electric field in the base in Eq. (7–55) by the expression
in Eq. (7–47), assuming $n(x_n) = N_d(x_n)$

$$I_{Ep} = qA\mu_p p\left(\frac{-kT}{q}\frac{1}{n}\frac{dn}{dx_n}\right) - qAD_p\frac{dp}{dx_n}$$

$$= \frac{qAD_p}{n}\left(p\frac{dn}{dx_n} + n\frac{dp}{dx_n}\right) \tag{7–56}$$

where the Einstein relation is used. We recognize the expression in parenthesis as the derivative of the *pn* product.

$$I_{Ep} = \frac{-qAD_p}{n}\frac{d(pn)}{dx_n} \tag{7–57a}$$

$$\frac{-I_{Ep}n}{qAD_p} = \frac{d(pn)}{dx_n} \tag{7–57b}$$

We integrate both sides of Eq. (7–57b) from the emitter-base junction (0) to the base-collector junction (W_b), keeping in mind that the current I_{Ep}, flowing from the emitter to the collector is more or less constant in the narrow base (so that it can be pulled out of the integral).

$$-I_{Ep}\int_0^{W_b}\frac{ndx_n}{qAD_p} = \int_0^{W_b}\frac{d(pn)}{dx_n}dx_n = p(W_b)n(W_b) - p(0)n(0) \tag{7–58}$$

Now, as described in Sections 5.2.2 and 5.3.2, the *pn* product changes from the equilibrium value

$$pn = n_i^2 \tag{7–59a}$$

to the non-equilibrium expression

$$pn = n_i^2 e^{\frac{F_n - F_p}{kT}} = n_i^2 e^{\frac{qV}{kT}} \tag{7–59b}$$

where the separation of the Fermi levels is determined by the applied bias across the junction. Applying this to Eq. 7–58, we get

$$p(W_b)n(W_b) = n_i^2 e^{\frac{qV_{CB}}{kT}} \tag{7–60a}$$

$$p(0)n(0) = n_i^2 e^{\frac{qV_{EB}}{kT}} \tag{7–60b}$$

$$I_{Ep} = \frac{-qAD_p n_i^2\left(e^{\frac{qV_{CB}}{kT}} - e^{\frac{qV_{EB}}{kT}}\right)}{\int_0^{W_b} ndx_n} \tag{7–61}$$

We have assumed a constant hole diffusivity in the base, D_p. The integral in the denominator corresponds to the integrated majority carrier charge in the base, and is known as the *base Gummel number*, Q_B. In the normal active mode of operation, where the collector-base junction is reverse biased (V_{CB} is negative), and the emitter-base junction is forward biased, the emitter hole current flowing to the collector (which is the dominant current) becomes

$$I_{Ep} = \frac{qAD_p n_i^2 e^{\frac{qV_{EB}}{kT}}}{Q_B} \tag{7-62a}$$

We can similarly write the base electron current flowing back into the emitter as

$$I_{En} = \frac{qAD_n n_i^2 e^{\frac{qV_{EB}}{kT}}}{Q_E} \tag{7-62b}$$

where Q_E is the integrated majority carrier charge in the emitter, known as the *emitter Gummel number*. The crux of the Gummel–Poon model is that the currents are expressed in terms of the net integrated charges in the base and emitter regions, and can easily handle non-uniform doping. Also, since we have obtained expressions for I_{Ep} and I_{En} in terms of Q_B and Q_E, we can write down BJT parameters such as the emitter injection efficiency γ (see Eq. 7–2), in terms of the Gummel numbers.

We can also modify the Gummel–Poon model to handle several second order effects such as the Early effect and high level injection in the base, simply by writing the expression for the base Gummel number, Q_B, more precisely as follows:

$$Q_B = \int_{0(V_{EB})}^{W_b(V_{CB})} n(x_n)dx_n \tag{7-63}$$

where we explicitly account for the fact that the integration limits, the base-emitter junction ($x_n = 0$) and the base-collector junction (W_b) are bias dependent. This is, of course, the Early effect (Section 7.7.2).

Furthermore, we see that under high level injection (Sections 5.6.1 and 7.7.4), the integrated majority carrier charge becomes greater than the integrated base dopant charge:

$$\int_0^{W_b} n(x_n)dx_n > \int_0^{W_b} N_D(x_n)dx_n \tag{7-64}$$

Clearly, from Eq. 7–61, this will cause the current from the emitter-to-collector, I_{Ep}, to increase less rapidly with emitter-base voltage at high biases. Based on what we learned in Section 5.6.1 about high level injection in a diode, the emitter-to-collector current for high level injection in the base increases as

$$I_C \propto I_{Ep} \propto e^{\frac{qV_{EB}}{2kT}} \tag{7-65a}$$

On the other hand, because the emitter doping is typically higher than the base doping, one does not see high level injection effects in the emitter, and the base current injected into the emitter scales as

$$I_B \propto I_{En} \propto e^{\frac{qV_{EB}}{kT}} \tag{7-65b}$$

Hence, for high V_{EB},

$$\beta = \frac{I_C}{I_B} \propto \frac{e^{\frac{qV_{EB}}{2kT}}}{e^{\frac{qV_{EB}}{kT}}} \propto e^{\frac{-qV_{EB}}{2kT}} \propto I_C^{-1}$$

This result shows that the common emitter gain decreases at high injection levels due to excess majority carriers in the base.

The Gummel–Poon model also accounts for generation–recombination effects in the base-emitter depletion region at low current levels. As discussed in Section 5.6.2, such effects are accounted for by the diode ideality factor, **n**. Hence the base current injected into the emitter can be written as

$$I_B \propto I_{En} \propto e^{\frac{qV_{EB}}{nkT}} \tag{7-67a}$$

On the other hand, the generally large emitter current injected into the base is not likely to be affected by generation–recombination. Therefore,

$$I_{Ep} \propto e^{\frac{qV_{EB}}{kT}} \tag{7-67b}$$

Thus, for low V_{EB} or low I_C, the current gain

$$\beta = \frac{I_C}{I_B} \propto \frac{e^{\frac{qV_{EB}}{kT}}}{e^{\frac{qV_{EB}}{nkT}}} \propto e^{\frac{qV_{EB}}{kT}\left(1-\frac{1}{n}\right)} \propto I_C^{\left(1-\frac{1}{n}\right)} \tag{7-68}$$

The transistor collector current I_C and the base current I_B are plotted on a semi-log scale as a function of V_{EB} in Fig. 7–22a. This is referred to as a *Gummel* plot. The current gain β is shown as a function of I_C in Fig. 7–22b. We see the dependence of β on I_C in the different bias regions, as described by the Gummel–Poon model. At low injection levels β is degraded by poor emitter injection efficiency [Eq. (7–68)], and at high currents β decreases due to excess majority charge in the base, which degrades γ.

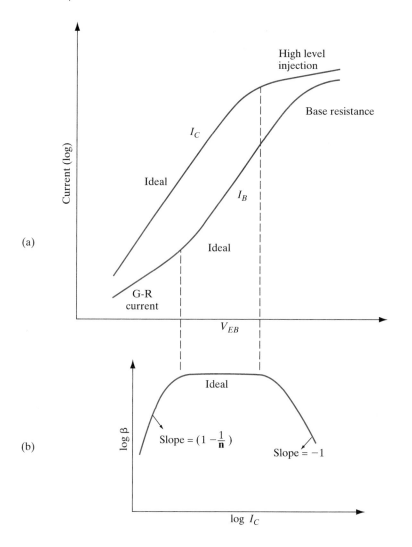

Figure 7–22
Current-voltage characteristics of BJTs: (a) Gummel plot of log of collector and base currents as a function of emitter-base forward bias; (b) d-c common-emitter current gain ($=I_C/I_B$) as a function of I_C. For the intermediate range, we have ideal behavior where both I_C and I_B increase exponentially with forward bias with ideality factor **n** = 1, leading to a current-independent gain. For very low forward biases, the generation-recombination (G–R) current increases I_B and the gain drops. For high forward biases, high level injection effects cause the I_C to increase more slowly (**n** = 2) than I_B, and the gain drops as I_C^{-1}.

7.7.7 Kirk Effect

The current gain drops at high collector currents due to yet another mechanism known as the *Kirk effect*. This involves an effective widening of the neutral base due to modification of the depletion space charge distribution at the reverse-biased base-collector junction. This is caused by the buildup of mobile carriers due to increased current flow from the emitter to the collector. This is illustrated in Fig. 7–23 for a p-n-p BJT. Notice that the polarity of these mobile charges adds to the fixed donor charges on the base side of the base-collector depletion region, but subtracts from the fixed acceptor charges

Figure 7–23
Kirk effect: (a)
cross-section of
p-n-p BJT; (b)
space charge dis-
tribution in the
base-collector re-
verse biased junc-
tion for very low
currents; (c)
space-charge
distribution at the
base-collector
junction for higher
current levels. We
see that the inject-
ed mobile holes
(shown in color)
add to the space
charge of the im-
mobile donors on
the base side of
the depletion re-
gion, but subtract
from the space
charge of the im-
mobile acceptors
on the collector
side. This leads to
a widening of the
neutral base width
from W_b to W_b'.

(a)
(b)
(c)

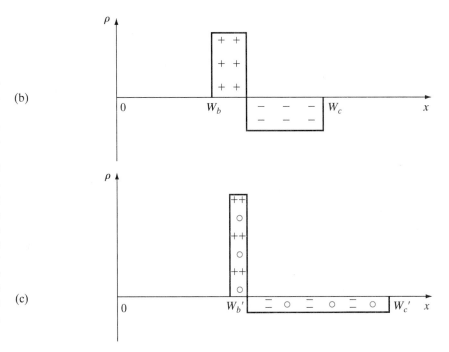

on the collector side of the junction (Fig. 7–23c). Therefore, fewer uncompensated donors (and thus a smaller depletion width) are needed to maintain the reverse voltage V_{CB} across this junction. As a result, the neutral base
width increases from W_b in Fig. 7–23b to W_b' in Fig. 7–23c. Also, the depletion
region extends more into the collector side. This is tantamount to moving
the base-collector junction deeper into the collector. This leads to an effective widening of the neutral base region (the Kirk effect) and to a drop of the
current gain and an increase of the base transit time.

The electric field profile in the collector depletion region in the presence of uncompensated dopant charges and mobile carriers (due to the current flow) is given by Poisson's equation.

$$\frac{d\mathscr{E}}{dx} = \frac{1}{\epsilon}\left[q(N_d^+ - N_a^-) + \frac{I_c}{Av_d}\right] \qquad (7\text{--}69)$$

where the mobile carrier charge concentration is given by the last term, and v_d is the drift velocity of the carriers.

The voltage across the reverse-biased collector-base junction, V_{CB}, is related to the electric field profile by:

$$V_{CB} = -\int_{W_b}^{W_c} \mathscr{E}dx \qquad (7\text{--}70)$$

Assuming that V_{CB} is fixed, and I_C increases, the last term in Eq. 7–69 becomes more important with respect to the ionized dopant charges. In Poisson's equation (Eq. 7–69), the extra holes injected into the depletion region have the same effect as if the doping level on the base side were increased and that on the collector side decreased. Since the integral of this field with respect to distance is fixed at V_{CB} (Eq. 7–70), this implies that the depletion region on the base side collapses.

Although we have chosen to illustrate the Kirk effect for a p-n-p BJT, similar results are obtained for the n-p-n transistor. Obviously, the treatment is identical except for the polarity of the various charges. From a more detailed analysis, for n-p-n devices with an n^+ sub-collector, it can be shown that the base widening can extend at even higher current levels all the way through the lightly doped collector region to the heavily doped buried sub-collector.

In this section we discuss the properties of bipolar transistors under high-frequency operation. Some of the frequency limitations are junction capacitance, charging times required when excess carrier distributions are altered, and transit time of carriers across the base region. Our aim here is not to attempt a complete analysis of high-frequency operation, but rather to consider the physical basis of the most important effects. Therefore, we shall include the dominant capacitances and charging times and discuss the effects of the transit time on high-frequency devices.

**7.8
FREQUENCY
LIMITATIONS OF
TRANSISTORS**

7.8.1 Capacitance and Charging Times

The most obvious frequency limitation of transistors is the presence of junction capacitance at the emitter and collector junctions. We have considered this type of capacitance in Chapter 5, and we can include junction capacitors C_{je} and C_{jc} in circuit models for the transistor (Fig. 7–24a). If there is some equivalent resistance r_b between the base contact and the active part of the base region, we can also include it in the model, along with r_c to account for

Figure 7–24
Models for a-c op-
eration: (a) inclu-
sion of base and
collector resis-
tances and junc-
tion capacitances;
(b) hybrid-pi
model systhesiz-
ing Eqs. (7–75)
and (7–76).

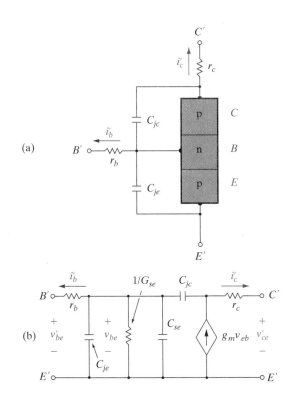

a series collector resistance.[6] Clearly, the combinations of r_b with C_{je} and r_c with C_{jc} can introduce important time constants into a-c circuit applications of the device.

From Section 5.5.4 we recall that capacitive effects can arise from the requirements of altering the carrier distributions during time-varying injection. In a-c circuits the transistor is usually biased to a certain steady state operating point characterized by the d-c quantities $V_{BE}, V_{CE}, I_C, I_B,$ and I_E; and then a-c signals are superimposed upon these steady state values. We shall call the a-c terms $v_{be}, v_{ce}, i_c, i_b,$ and i_e. Total (a-c + d-c) quantities will be lower-case with capitalized subscripts.

If a small a-c signal is applied to the emitter p-n junction along with a d-c level, we can show (Prob. 7-27) that

$$\Delta p_E(t) \simeq \Delta p_E(d\text{-}c)\left(1 + \frac{qv_{eb}}{kT}\right) \qquad (7\text{-}71)$$

[6]Since elements such as r_b and r_c in Fig. 7–24 are added to the basic transistor model we have previously analyzed, it is most convenient to refer here to the terminal voltages and currents as $v_{be}', i_c',$ and so on. In this way we can use previously derived expressions involving the internal quantities [i_b and v_{eb} in Eq. (7–75), for example]. In most circuits texts the primes are instead used for the internal quantities, just the opposite method from that used here.

We can relate this time-varying excess hole concentration to the stored charge in the base region, and then use Eq. (7–43) to determine the resulting currents. For simplicity we shall assume the device is biased in the normal active mode and use only $Q_N(t)$. Assuming an essentially triangular excess hole distribution in the base, Eq. (7–23) gives

$$Q_N(t) = \tfrac{1}{2}qAW_b\Delta p_E(t) = \tfrac{1}{2}qAW_b\Delta p_E(d\text{-}c)\left[1 + \frac{qv_{eb}}{kT}\right] \quad (7\text{-}72)$$

The terms outside the brackets constitute the d-c stored charge $I_B\tau_p$:

$$Q_n(t) = I_B\tau_p\left(1 + \frac{qv_{eb}}{kT}\right) \quad (7\text{-}73)$$

Now that we have a simple relation for the time-dependent stored charge, we can use Eq. (7–43c) to write the total base current as

$$i_B(t) = \frac{Q_N(t)}{\tau_p} + \frac{dQ_N(t)}{dt} \quad (7\text{-}74a)$$

As discussed in Section 5.5.4, we must be careful about boundary conditions in determining where the stored charges are extracted or "reclaimed" in a diode. For the emitter-base diode in a BJT, we have a "short" diode where Eq. (5–67b) is applicable. The result is that only 2/3 of the stored charge is reclaimed. Hence, we obtain

$$i_B(t) = I_B + \frac{q}{kT}I_Bv_{eb} + \frac{2}{3}\frac{q}{kT}I_B\tau_p\frac{dv_{eb}}{dt} \quad (7\text{-}74b)$$

The a-c component of the base current is

$$i_b = G_{se}v_{eb} + C_{se}\frac{dv_{eb}(t)}{dt} \quad (7\text{-}75)$$

where

$$G_{se} \equiv \frac{q}{kT}I_B \quad \text{and} \quad C_{se} \equiv \frac{2}{3}\frac{q}{kT}I_B\tau_p = \frac{2}{3}G_{se}\tau_p$$

Thus, as in the case of the simple diode, an a-c conductance and capacitance are associated with the emitter-base junction due to charge storage effects. From Eq. (7–43b) we have

$$i_C(t) = \frac{Q_N(t)}{\tau_t} = \beta I_B + \frac{q}{kT}\beta I_Bv_{eb}$$

$$i_c = g_m v_{eb} \quad \text{where} \quad g_m \equiv \frac{q}{kT}\beta I_B = \frac{3}{2}\frac{C_{se}}{\tau_t} \quad (7\text{-}76)$$

The quantity g_m is an a-c *transconductance*, which is evaluated at the steady-state value of collector current $I_C = \beta I_B$. We can synthesize Eqs. (7–75) and (7–76) in an equivalent a-c circuit as in Fig. 7–24b. In this equivalent circuit the voltage v_{be} used in the calculations appears "inside" the device, so that a new applied voltage v'_{be} must be used external to r_b to refer to the voltage applied between the contacts, and similarly for v'_{ce}. This equivalent model is discussed in detail in most electronic circuits texts; it is often called a *hybrid-pi* model.

From Fig. 7–24b it is clear that several charging times are important in the a-c operation of a transistor; the most important are the time required to charge the emitter and collector depletion regions and the delay time in altering the charge distribution in the base region. Other delay times included in a complete analysis of high-frequency transistors are the transit time through the collector depletion region and the charge storage time in the collector region. If all of these are included in a single delay time τ_d, we can estimate the upper frequency limit of the device. This is usually defined as the *cutoff frequency* for the transistor $f_T \equiv (2\pi\tau_d)^{-1}$. It is possible to show that f_T represents the frequency at which the a-c amplification for the device $[\beta(a\text{-}c) \equiv h_{fe} = \partial i'_c/\partial i'_b]$ drops to unity.

7.8.2 Transit Time Effects

In high-frequency transistors the ultimate limitation is often the transit time across the base. For example, in a p-n-p device the time τ_t required for holes to diffuse from emitter to collector can determine the maximum frequency of operation for the device. We can calculate τ_t for a transistor with normal biasing and $\gamma = 1$ from Eq. (7–20) and the relation $\beta \simeq \tau_p/\tau_t$:

$$\beta \simeq \frac{\operatorname{csch} W_b/L_p}{\tanh W_b/2L_p} = \frac{2L_p^2}{W_b^2} = \frac{2D_p\tau_p}{W_b^2} = \frac{\tau_p}{\tau_t}$$

$$\tau_t = \frac{W_b^2}{2D_p} \tag{7–77}$$

Another instructive way of calculating τ_t is to consider that the diffusing holes *seem* to have an average velocity $\langle v(x_n)\rangle$ (actually the individual hole motion is completely random, as discussed in Section 4.4.1). The hole current $i_p(x_n)$ is then given by

$$i_p(x_n) = qAp(x_n)\langle v(x_n)\rangle \tag{7–78}$$

The transit time is

$$\tau_t = \int_0^{W_b} \frac{dx_n}{\langle v(x_n)\rangle} = \int_0^{W_b} \frac{qAp(x_n)}{i_p(x_n)}dx_n \tag{7–79}$$

For a triangular distribution as in Fig. 7–8b, the diffusion current is almost constant at $i_p = qAD_p\Delta p_E/W_b$, and τ_t becomes

$$\tau_t = \frac{qA\,\Delta p_E W_b/2}{qAD_p\Delta p_E/W_b} = \frac{W_b^2}{2D_p} \tag{7-80}$$

as before. The average velocity concept should not be pushed too far in the case of diffusion, but it does serve to illustrate the point that a delay time exists between the injection and collection of holes.

We can estimate the transit time for a typical device by choosing a value of W_b, say 0.1 μm (10^{-5} cm). For Si, a typical number for D_p is about 10 cm²/s; then for this transistor $\tau_t = 0.5 \times 10^{-11}$ s. Approximating the upper frequency limit as $(2\pi\tau_t)^{-1}$, we can use the transistor to about 30 GHz. Actually, this estimate is too optimistic because of other delay times. The transit time can be reduced by making use of field-driven currents in the base. For the implanted transistor of Fig. 7–17, the holes drift in the built-in field from emitter to collector over most of the base region. By increasing the doping gradient in the base, we can reduce the transit time and thereby increase the maximum frequency of the transistor.

7.8.3 Webster Effect

While the transit time expression [Eq. (7–80)] is valid for low level injection, τ_t is reduced by up to a factor of 2 under high level injection. This occurs because the majority carrier concentration increases significantly above its equilibrium value in the base, to match the injected minority carrier concentration. Since the minority carrier concentration decreases from the base-emitter junction to the base-collector junction (see Fig. 7–8), so does the majority carrier concentration. This tends to create a diffusion of the majority carriers from emitter to base. Such majority carrier diffusion would upset the drift–diffusion balance required to maintain a quasi-equilibrium distribution in the base. Therefore, a built-in electric field develops in the base to create an opposing majority carrier drift current. The direction of this induced field then *aids* the minority carrier transport from the emitter to the collector, reducing the transit time τ_t in Eq. (7–80).

This is known as the Webster effect. It is interesting to note that this effect is similar to the drift field effects in the base region due to non-uniform base doping (Section 7.7.1). For the Webster effect, the induced field is not due to nonuniform doping but rather the nonuniform majority carrier concentration induced by carrier injection.

7.8.4 High-Frequency Transistors

The most obvious generality we can make about the fabrication of high-frequency transistors is that the physical size of the device must be kept small. The base width must be narrow to reduce the transit time, and the emitter and collector areas must be small to reduce junction capacitance. Unfortunately, the requirement of small size generally works against the requirements of power rating for the device. Since we usually require a trade-off between frequency and power, the dimensions and other design features of the transistor must be tailored to the specific circuit requirements. On the other hand, many of the fabrication techniques useful for power devices can be adapted to increase the frequency range. For example, the method of interdigitation (Fig. 7–21) provides a means of increasing the useful emitter edge length while keeping the overall emitter area to a minimum. Therefore, some form of interdigitation is generally used in transistors designed for high frequency and reasonable power requirements (Fig. 7–25).

Another set of parameters that must be considered in the design of a high-frequency device is the effective resistance associated with each region of the transistor. Since the emitter, base, and collector resistances affect the various RC charging times, it is important to keep them to a minimum. Therefore, the metallization patterns contacting the emitter and base regions must not present significant series resistance. Furthermore, the semiconductor regions themselves must be designed to reduce resistance. For example, the series base resistance r_b of an n-p-n device can be reduced greatly by performing a p^+ diffusion between the contact area on the surface and the active part of the base region. Further reduction of base resistance by heavy doping of the base requires the use of a heterojunction (Section 7.9) to maintain γ at an acceptable value.

Figure 7–25
A low-noise Si bipolar transistor with f_T = 8 GHz. This device has 9 interdigitated emitter stripes, each 1 μm X 20 μm. (Photograph courtesy of Motorola.)

In Si, n-p-n transistors are usually preferred, since the electron mobility and diffusion coefficient are higher than for holes. It is common to fabricate n-p-n transistors in n-type epitaxial material grown on an n⁺ substrate. The heavily doped substrate provides a low-resistance contact to the collector region, while maintaining low doping in the epitaxial collector material to ensure a high breakdown voltage of the collector junction (see Section 7.3). It is important, however, to keep the collector depletion region as small as possible to reduce the transit time of carriers drifting through the collector junction. This can be accomplished by making the lightly doped collector region narrow so that the depletion region under bias extends to the n⁺ substrate.

In addition to the various parameters of the device itself, the transistor must be packaged properly to avoid parasitic resistance, inductance, or capacitance at high frequencies. We shall not attempt to describe the many techniques for mounting and packaging transistors here, since methods vary greatly among manufacturers.

In Section 7.4.4 we saw that the emitter injection efficiency of a bipolar transistor is limited by the fact that carriers can flow from the base into the emitter region, over the emitter junction barrier, which is reduced by the forward bias. According to Eq. (7–25) it is necessary to use lightly doped material for the base region and heavily doped material for the emitter to maintain a high value of γ and, therefore, α and β. Unfortunately, the requirement of light base doping results in undesirably high base resistance. This resistance is particularly noticeable in transistors with very narrow base regions. Furthermore, degenerate doping can led to a slight shrinkage of E_g in the emitter as the donor states merge with the conduction band. This can decrease the emitter injection efficiency. Therefore, a more suitable BJT for high frequency would have a heavily doped base and a lightly doped emitter. This is just the opposite of the traditional BJT discussed thus far in this chapter. To accomplish such a radically different transistor design, we need some other mechanism instead of doping to control the relative amount of injection of electrons and holes across the emitter junction.

If transistors are made in materials that allow heterojunctions to be used, the emitter injection efficiency can be increased without strict requirements on doping. In Fig. (7–26) an n-p-n transistor made in a single material (*homojunction*) is contrasted with a *heterojunction bipolar transistor (HBT)*, in which the emitter is a wider band gap semiconductor. It is possible in such a structure for the barrier for electron injection (qV_n) to be smaller than the hole barrier (qV_p). Since carrier injection varies exponentially with the barrier height, even a small difference in these two barriers can make a very large difference in the transport of electrons and holes across the emitter

**7.9
HETEROJUNCTION
BIPOLAR
TRANSISTORS**

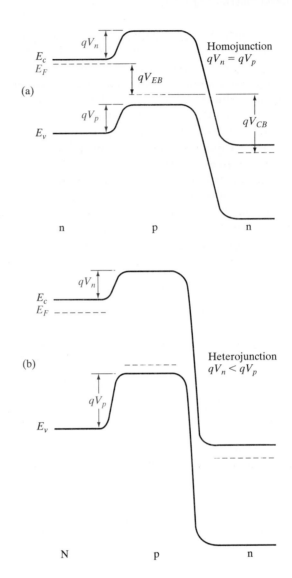

Figure 7–26
Contrast of carrier injection at the emitter of (a) a homojunction BJT and (b) a heterojunction bipolar transistor (HBT). In the forward-biased homojunction emitter, the electron barrier qV_n and the hole barrier qV_p are the same. In the HBT with a wide band gap emitter, the electron barrier is smaller than the hole barrier, resulting in preferential injection of electrons across the emitter junction.

junction. Neglecting differences in carrier mobilities and other effects, we can approximate the dependence of carrier injection across the emitter as

$$\frac{I_n}{I_p} \propto \frac{N_d^E}{N_a^B} e^{\Delta E_g / kT} \tag{7–81}$$

In this expression, the ratio of electron current I_n to hole current I_p crossing the emitter junction is proportional to the ratio of the doping in the emitter

N_d^E and the base N_a^B. In the homojunction BJT this doping ratio is all we have to work with in designing a useful emitter junction. However, in the HBT there is an additional factor in which the band gap difference ΔE_g between the wide band gap emitter and the narrow band gap base appears in an exponential factor. As a result, a relatively small value of ΔE_g in the exponential term can dominate Eq. (7–81). This allows us to choose the doping terms for lower base resistance and emitter junction capacitance. In particular, we can choose a heavily doped base to reduce the base resistance and a lightly doped emitter to reduce junction capacitance.

The heterojunction shown in Fig. 7–26 has a smooth barrier, without the spike and notch commonly observed for heterojunctions (see Fig. 5-46). It is possible to smooth out such discontinuities in the bands by grading the composition of the ternary or quaternary alloy between the two materials (Fig. 7–27). Clearly, grading out the conduction band spike improves the electron injection by reducing the barrier that electrons must overcome. There are some HBT designs, however, that make use of the spike as a "launching ramp" to inject hot electrons into the base.

Materials commonly used in HBTs obviously include the AlGaAs/GaAs system because of its wide range of lattice-matched composition. In addition, the InGaAsP system (including $In_{0.53}Ga_{0.47}As$) grown on InP has become popular in HBT design. InGaAs has much lower surface recombination than GaAs, and the Γ-L and Γ-X intervalley band separations are much larger than in GaAs (see Fig. 3-6). The lower rate of surface recombination reduces loss of injected carriers at the surface of the device. This is a particularly important effect in small-geometry devices, in which the base perimeter-to-area ratio, and thus the intersection of the base with the surface, is large. The larger intervalley band separation in InGaAs helps ensure that the electrons remain in the low-mass (high mobility) Γ valley during field-enhanced transport through the base region.

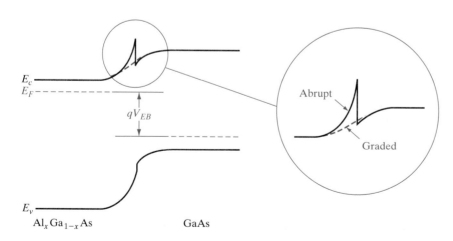

Figure 7–27
Removal of the conduction band spike by grading the alloy composition (x) in the heterojunction. In this example the junction is graded from the composition used in the AlGaAs emitter to x = 0 at the GaAs base. This grading typically takes place over a distance of 100 Å or less.

It is also possible to make HBTs using elemental semiconductor heterostructures such as $Si/Si_{1-x}Ge_x$. In this material system, the band gap difference ΔE_g between the Si emitter and the narrower band gap $Si_{1-x}Ge_x$ base occurs primarily in the valence band. As a result, a rather small addition of Ge to the base results in higher electron injection efficiency than is possible in homojunction silicon bipolar transistors.

If the alloy composition in the base of an n-p-n is varied such that E_g decreases slightly from the emitter side to the collector side of the base, a built-in electric field accelerates electrons through the base region. The resulting field-aided base transport is a major advantage of the HBT.

PROBLEMS

7.1 A bipolar junction transistor is fabricated by ion implantation as follows. A 50 keV boron ion implant (dose = 5×10^{14} cm^{-2}) is performed into an n-type silicon substrate with $N_d = 2 \times 10^{15}$ cm^{-3}, followed by a 30 keV phosphorus implant (dose = 1×10^{15} cm^{-2}). A rapid thermal anneal is then performed at 1100°C for 10 s in a N_2 environment. Assume the implants are 100 percent activated by this anneal. The diffusivity of boron at 1100°C can be assumed constant with $D = 10^{-12}$ cm^2/s, and for phosphorus $D = 5 \times 10^{-13}$ cm^2/s. Assume that dopants can outdiffuse from the substrate.

(a) Where are the peaks and what are the widths (at 0.607 of the peak) of the boron and phosphorus implant profiles prior to annealing?

(b) Where are the peaks and what are the widths of the implant profiles after annealing?

(c) What is the emitter junction depth and the collector junction depth (as measured from the silicon surface) after annealing?

7.2 Sketch the ideal collector characteristics $(i_C, -v_{CE})$ for the transistor of Fig. 7–4; let i_B vary from zero to 0.2 mA in increments of 0.02 mA, and let $-v_{CE}$ vary from 0 to 10 V. Draw a load line on the resulting characteristics for the circuit of Fig. 7–4, and find the steady state value of $-V_{CE}$ graphically for $I_B = 0.1$ mA.

7.3 Calculate and plot the excess hole distribution $\delta p(x_n)$ in the base of a p-n-p transistor from Eq. (7–14), assuming $W_b/L_p = 0.5$. The calculations are simplified if the vertical scale is measured in units of $\delta p/\Delta p_E$ and the horizontal scale in units of x_n/L_p. In good transistors, W_b/L_p is much smaller than 0.5; however, $\delta p(x_n)$ is quite linear even for this rather large base width.

7.4 Derive Eq. (7–19) from the charge control approach by integrating Eq. (7–11) across the base region and applying Eq. (7–13).

7.5 Extend Eq. (7–20a) to include the effects of nonunity emitter injection efficiency ($\gamma < 1$). Derive Eq. (7–25) for γ. Assume the emitter region is long compared with an electron diffusion length. Apply the charge control approach to the linear distribution in Fig. 7–8b to find the base transport factor B. Use Eq. (7–80) for the transit time. Is the result the same as Eq. (7–26) for small W_b/L_p?

7.6 A symmetrical p^+-n-p^+ Si bipolar transistor has the following properties:

	Emitter	Base
$A = 10^{-4}$ cm^{-2}	$N_a = 10^{17}$	$N_d = 10^{15}$ cm^{-3}
$W_b = 1$ μm	$\tau_n = 0.1$ μs	$\tau_p = 10$ μs
	$\mu_p = 200$	$\mu_n = 1300$ cm^2/V-s
	$\mu_n = 700$	$\mu_p = 450$

(a) Calculate the saturation current $I_{ES} = I_{CS}$.

(b) With $V_{EB} = 0.3$ V and $V_{CB} = -40$ V, calculate the base current I_B, assuming perfect emitter injection efficiency.

(c) Calculate the emitter injection efficiency γ and the amplification factor β, assuming the emitter region is long compared to L_n.

7.7 The symmetrical p$^+$-n-p$^+$ transistor of Fig. P7-7 is connected as a diode in the four configurations shown. Assume that $V \gg kT/q$. Sketch $\delta p(x_n)$ in the base region for each case. Which connection seems most appropriate for use as a diode? Why?

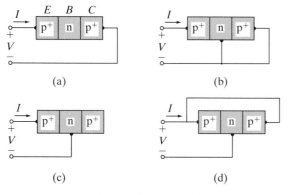

Figure P7–7

(a)   (b)   (c)   (d)

7.8 For the transistor connection in Fig. P7-7a, (a) show that $V_{EB} = (kT/q) \ln 2$; (b) find the expression for I when $V \gg kT/q$ and sketch I vs. V.

7.9 (a) Find the expression for the current I for the transistor connection of Fig. P7-7b; compare the result with the narrow base diode problem (Prob. 5.35).

(b) How does the current I divide between the base lead and the collector lead?

7.10 Suppose that V is negative in Fig. P7-7c.

(a) Find I from the Ebers–Moll equations.

(b) Find the expression for V_{CB}.

(c) Sketch $\delta p(x_n)$ in the base.

7.11 For the transistor connection of Fig. P7-7d, (a) find the expression for $\delta p(x_n)$ in the base region; (b) find the current I.

7.12 It is obvious from Eqs. (7–35) and (7–36) that I_{EO} and I_{CO} are the saturation currents of the emitter and collector junctions, respectively, with the opposite junction open circuited.

(a) Show that this is true from Eq. (7–32).

(b) Find expressions for the following excess concentrations: Δp_C with the emitter junction forward biased and the collector open; Δp_E with the collector junction forward biased and the emitter open.

(c) Sketch $\delta p(x_n)$ in the base for the two cases of part (b).

7.13 (a) Show that the definitions of Eq. (7–40) are correct; what does q_N represent?

(b) Show that Eqs. (7–39) correspond to Eqs. (7–34), using the definitions of Eqs. (7–40).

7.14 (a) How is it possible that the average time an injected hole spends in transit across the base τ_t is shorter than the hole lifetime in the base τ_p?

(b) Explain why the turn-on transient of a BJT is faster when the device is driven into oversaturation.

7.15 Use Example 5-6 to design an N-p-n heterojunction bipolar transistor with reasonable γ and base resistance.

7.16 The current amplification factor β of a BJT is very sensitive to the base width as well as to the ratio of the base doping to the emitter doping. Calculate and plot β for a p-n-p BJT with $L_p^n = L_n^p$, for:

(a) $n_n = p_p$, $W_b/L_p^n = 0.01$ to 1;

(a) $W_b = L_p^n$, $n_n/p_p = 0.01$ to 1.

Neglect mobility variations ($\mu_n^p = \mu_p^n$).

7.17 Calculate and plot the common-base characteristics at 300 K of a symmetrical Si p-n-p BJT with a base area of 10^2 cm^{-2}, a base width of 500 Å, base doping of 10^{14} cm^{-3}, emitter and collector doping of 10^{17} cm^{-3}, and carrier lifetimes $\tau_n = \tau_p = 10^{-6}$ s, for emitter currents of 0, 0.04, 0.08, and 0.12 A.

7.18 (a) How much charge (in coulombs) due to excess holes is stored in the base of the transistor shown in Fig. 7–4 at the d-c bias given?

(b) Why is the base transport factor B different in the normal and inverted modes for the transistor shown in Fig. 7–5?

7.19 A Si p-n-p transistor has the following properties at room temperature:

$\tau_n = \tau_p = 0.1$ μs

$D_n = D_p = 10$ cm^2/s

$N_E = 10^{19}$ cm^{-3} = emitter concentration

$N_B = 10^{16}$ cm^{-3} = base concentration

$N_C = 10^{16}$ cm^{-3} = collector concentration

W_E = emitter width = 3 μm

W = metallurgical base width = 1.5 μm = distance between base-emitter junction and base-collector junction

A = cross-sectional area = 10^{-5} cm^2

Calculate the neutral base width W_b for $V_{CB} = 0$ and $V_{EB} = 0.2$V, repeat for 0.6V.

7.20 For the BJT in Prob. 7.19, calculate the base transport factor and the emitter injection efficiency for $V_{EB} = 0.2$ and 0.6 V.

7.21 For the BJT in Prob. 7.19, calculate α, β, I_E, I_B and I_C for the two values of V_{EB}. What is the base Gummel number in each case?

7.22 A Si p-n-p BJT has the following parameters at room temperature.

Emitter	Base	Collector
$N_a = 5 \times 10^{18}$ cm^{-3}	$N_d = 10^{16}$	$N_a = 10^{15}$
$\tau_n = 1\mu s$	$\tau_p = 25$	$\tau_n = 2$
$\mu_n = 150$ cm^2/V-s	$\mu_n = 1500$	$\mu_n = 1500$
$\mu_p = 100$ cm^2/V-s	$\mu_p = 400$	$\mu_p = 450$

Base width, $W_b = 0.2$ μm

Area $= 10^{-4}$ cm^2

Calculate the β of the transistor from B and γ, and using the charge control model. Comment on the results.

7.23 For the BJT in Prob. 7.22, calculate the charge stored in the base when $V_{CB} = 0$ and $V_{EB} = 0.7$V. If the base transit time is the dominant delay component for this BJT, what is the f_T?

7.24 Three n-p-n transistors are identical except that transistor #2 has a base region twice as long as transistor #1, and transistor #3 has a base region doped twice as heavily as transistor #1. All other dopings and lengths are identical for the three transistors. Which transistors have the largest value of each parameter listed below?

Give clear mathematical reasons for each of your answers.

(a) Emitter injection efficiency

(b) Base transport factor

(c) Punch through voltage

(d) Collector junction capacitance with V_{CB} reverse biased at 10V.

(e) Common emitter current gain.

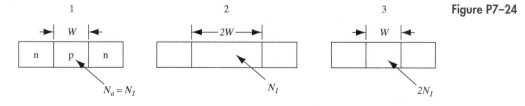

Figure P7-24

7.25 Assume the transit time for electrons across the base of an n-p-n transistor is 100 ps, and electrons cross the 1-μm depletion region of the collector junction at their scattering limited velocity. The emitter-base junction charging time is

30 ps and the collector capacitance and resistance are 0.1 pF and 10 Ω, respectively. Find the cutoff frequency f_T.

7.26 An n-p-n Si transistor has an emitter doping of 10^{18} donors/cm³ and a base doping of 10^{16} acceptors/cm³. At what forward bias of the emitter junction does high-level injection occur (injected electrons equal to the base doping)? Comment on the emitter injection efficiency for electrons.

7.27 Derive Eq. (7–71) for $\Delta p_E(t)$ assuming that the emitter has an applied voltage

$$v_{EB}(t) = V_{EB} + v_{eb}(t)$$

where $V_{EB} \gg kT/q$. For $v_{eb} \ll kT/q$, the approximation $e^x \simeq 1 + x$ can be employed.

READING LIST

Bardeen, J., and W.H. Brattain. "The Transistor, a Semiconductor Triode." *Phys. Rev.* 74 (1948), 230.

Inoue, K. "Recent Advances in InP-Based HEMT/HBT Device Technology." *Fourth International Conference on Indium Phosphide and Related Materials* (April 1992): 10–13.

Jaeger, R.C. *Modular Series on Solid State Devices: Vol V. Introduction to Microelectronic Fabrication.* Reading, MA: Addison-Wesley, 1988.

Levi, A. F. J., R. N. Nottenburg, and Y. K. Chen. "Ultrahigh-Speed Bipolar Transistors." *Physics Today* 43 (February 1990): 58–64.

Li, S. S. *Semiconductor Physical Electronics.* New York: Plenum Press, 1993.

Morgan, D. V., and R. H. Williams, eds. *Physics and Technology of Heterojunction Devices.* London: P. Peregrinus, 1991.

Muller, R. S., and T. I. Kamins. *Device Electronics for Integrated Circuits.* New York: Wiley, 1986.

Neamen, D. A. *Semiconductor Physics and Devices: Basic Principles.* Homewood, IL: Irwin, 1992.

Neudeck, G. W. *Modular Series on Solid State Devices: Vol. III. The Bipolar Junction Transistor.* Reading, MA: Addison-Wesley, 1983.

Pavlidis, D. "Current Status of Heterojunction Bipolar and High-Electron Mobility Transistor Technologies." *Microelectronic Engineering* 19 (September 1992): 305–12.

Shockey, W. "The Path to the Conception of the Junction Transistor." *IEEE Trans. Elec. Dev.* ED-23 (1976), 597.

Shur, M. *GaAs Devices and Circuits.* New York: Plenum Press, 1987.

Singh, J. *Semiconductor Devices.* New York: McGraw-Hill, 1994.

Sze, S. M. *High-Speed Semiconductor Devices.* New York: Wiley, 1990.

Sze, S. M. *Physics of Semiconductor Devices.* New York: Wiley, 1981.

Wang, S. *Fundamentals of Semiconductor Theory and Device Physics.* Englewood Cliffs, NJ: Prentice Hall, 1989.

Chapter 8

Optoelectronic Devices

So far we have primarily concentrated on electronic devices. There is also a wide variety of very interesting and useful device functions involving the interaction of photons with semiconductors. These devices provide the optical sources and detectors that allow broadband telecommunications and data transmission over optical fibers. This important area of device applications is called *optoelectronics*. In this chapter we will discuss devices that detect photons and those that emit photons. Devices that convert optical energy into electrical energy include photodiodes and solar cells. Emitters of photons include incoherent sources such as light-emitting diodes (LEDs) and coherent sources in the form of lasers.

In Section 4.3.4 we saw that bulk semiconductor samples can be used as photoconductors by providing a change in conductivity proportional to an optical generation rate. Often, junction devices can be used to improve the speed of response and sensitivity of detectors of optical or high-energy radiation. Two-terminal devices designed to respond to photon absorption are called *photodiodes*. Some photodiodes have extremely high sensitivity and response speed. Since modern electronics often involves optical as well as electrical signals, photodiodes serve important functions as electronic devices. In this section, we shall investigate the response of p-n junctions to optical generation of EHPs and discuss a few typical *photodiode detector* structures. We shall also consider the very important use of junctions as *solar cells*, which convert absorbed optical energy into useful electrical power.

8.1
PHOTODIODES

8.1.1 Current and Voltage in an Illuminated Junction

In Chapter 5 we identified the current due to drift of minority carriers across a junction as a generation current. In particular, carriers generated within the depletion region W are separated by the junction field, electrons being collected in the n region and holes in the p region. Also, minority carriers generated thermally within a diffusion length of each side of the junction diffuse to the depletion region and are swept to the other side by the electric field. If the junction is uniformly illuminated by photons with $hv > E_g$, an added generation rate g_{op} (EHP/cm³-s) participates in this current (Fig. 8–1). The number of holes created per second within a diffusion length of the transition region on the n side is $AL_p g_{op}$. Similarly $AL_n g_{op}$ electrons are generated per second within L_n of x_{p0} and $AW g_{op}$ carriers are generated within W. The resulting current due to collection of these optically generated carriers by the junction is

$$\boxed{I_{op} = qAg_{op}(L_p + L_n + W)} \qquad (8\text{–}1)$$

If we call the thermally generated current described in Eq. (5-37b) I_{th}, we can add the optical generation of Eq. (8–1) to find the total reverse current with illumination. Since this current is directed from n to p, the diode equation [Eq. (5-36)] becomes

$$I = I_{th}(e^{qV/kT} - 1) - I_{op}$$

$$I = qA\left(\frac{L_p}{\tau_p}p_n + \frac{L_n}{\tau_n}n_p\right)(e^{qV/kT} - 1) - qAg_{op}(L_p + L_n + W) \quad (8\text{–}2)$$

Thus the I–V curve is lowered by an amount proportional to the generation rate (Fig. 8–1c). This equation can be considered in two parts—the current described by the usual diode equation, and the current due to optical generation.

When the device is short circuited ($V = 0$), the terms from the diode equation cancel in Eq. (8–2), as expected. However, there is a short-circuit current from n to p equal to I_{op}. Thus the I–V characteristics of Fig. 8–1c cross the I-axis at negative values proportional to g_{op}. When there is an open circuit across the device, $I = 0$ and the voltage $V = V_{oc}$ is

Figure 8–1
Optical generation of carriers in a p-n junction: (a) absorption of light by the device; (b) current I_{op} resulting from EHP generation within a diffusion length of the junction on the n side; (c) I–V characteristics of an illuminated junction.

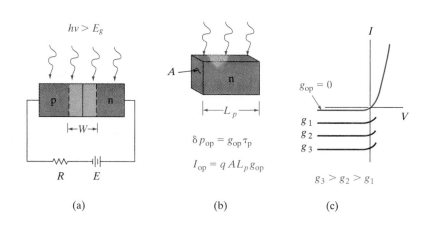

$$V_{oc} = \frac{kT}{q}\ln[I_{op}/I_{th} + 1]$$

$$= \frac{kT}{q}\ln\left[\frac{L_p + L_n + W}{(L_p/\tau_p)p_n + (L_n/\tau_n)n_p} \cdot g_{op} + 1\right] \qquad (8\text{-}3)$$

For the special case of a symmetrical junction, $p_n = n_p$ and $\tau_p = \tau_n$, we can rewrite Eq. (6-5) in terms of the thermal generation rate $p_n/\tau_n = g_{th}$ and the optical generation rate g_{op}. Neglecting generation within W:

$$V_{oc} \simeq \frac{kT}{q}\ln\frac{g_{op}}{g_{th}} \quad \text{for } g_{op} \gg g_{th} \qquad (8\text{-}4)$$

Actually, the term $g_{th} = p_n/\tau_n$ represents the *equilibrium* thermal generation–recombination rate. As the minority carrier concentration is increased by optical generation of EHPs, the lifetime τ_n becomes shorter, and p_n/τ_n becomes larger (p_n is fixed, for a given N_d and T). Therefore, V_{oc} cannot increase indefinitely with increased generation rate; in fact, the limit on V_{oc} is the equilibrium contact potential V_0 (Fig. 8–2). This result is to be expected, since the contact potential is the maximum forward bias that can appear across a junction. The appearance of a forward voltage across an illuminated junction is known as the *photovoltaic effect*.

Depending on the intended application, the photodiode of Fig. 8–1 can be operated in either the third or fourth quarters of its *I–V* characteristic. As Fig. 8–3 illustrates, power is delivered to the device from the external circuit when the current and junction voltage are both positive or both negative (first or third quadrants). In the fourth quadrant, however, the junction voltage is positive and the current is negative. In this case power is delivered from the junction to the external circuit (notice that in the fourth quadrant the current flows from the negative side of V to the positive side, as in a battery).

If power is to be extracted from the device, the fourth quadrant is used; on the other hand, in applications as a photodetector we usually reverse bias the junction and operate it in the third quadrant. We shall investigate these applications more closely in the discussion to follow.

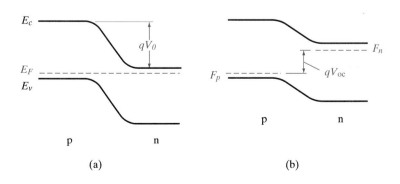

Figure 8–2
Effects of illumination on the open circuit voltage of a junction: (a) junction at equilibrium; (b) appearance of a voltage V_{oc} with illumination.

Figure 8–3
Operation of an
illuminated junc-
tion in the various
quadrants of its
I–V characteristic;
in (a) and (b)
power is deliv-
ered to the device
by the external
circuit; in (c) the
device delivers
power to the
load.

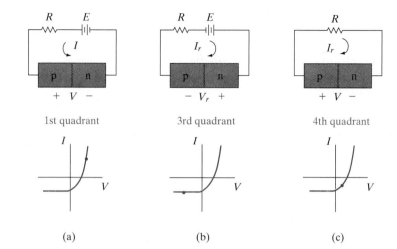

8.1.2   Solar Cells

Since power can be delivered to an external circuit by an illuminated junction, it is possible to convert solar energy into electrical energy. If we consider
the fourth quadrant of Fig. 8–3c, it appears doubtful that much power can be
delivered by an individual device. The voltage is restricted to values less than
the contact potential, which in turn is generally less than the band gap voltage E_g/q. For Si the voltage V_{oc} is less than about 1 V. The current generated
depends on the illuminated area, but typically I_{op} is in the 10–100 mA range
for a junction with an area of about 1 cm². However, if many such devices are
used, the resulting power can be significant. In fact, arrays of p-n junction
solar cells are currently used to supply electrical power for many space satel-
lites. Solar cells can supply power for the electronic equipment aboard a
satellite over a long period of time, which is a distinct advantage over bat-
teries. The array of junctions can be distributed over the surface of the satellite or can be contained in solar cell "paddles" attached to the main body of
the satellite (Fig. 8–4).

To utilize a maximum amount of available optical energy, it is necessary to
design a solar cell with a large area junction located near the surface of the device (Fig. 8–5). The planar junction is formed by diffusion or ion implantation,
and the surface is coated with appropriate materials to reduce reflection and to
decrease surface recombination. Many compromises must be made in solar cell
design. In the device shown in Fig. 8–5, for example, the junction depth d must
be less than L_p in the n material to allow holes generated near the surface to diffuse to the junction before they recombine; similarly, the thickness of the p region must be such that electrons generated in this region can diffuse to the
junction before recombination takes place. This requirement implies a proper
match between the electron diffusion length L_n, the thickness of the p region, and
the mean optical penetration depth $1/\alpha$ [see Eq. (4–2)]. It is desirable to have a
large contact potential V_0 to obtain a large photovoltage, and therefore heavy

Figure 8–4
Solar cell arrays
attatched to the
Mars fly-by satel-
lite *Mariner*. (Pro-
vided through the
courtesy of the
National Aero-
nautics and Space
Administration,
California Institute
of Technology,
Jet Propulsion
Laboratory.)

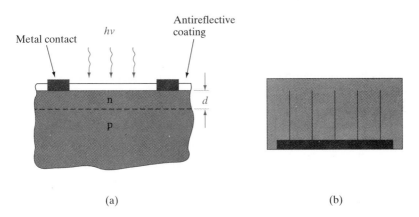

(a)   (b)

Figure 8–5
Configuration of a
solar cell: (a) en-
larged view of the
planar junction;
(b) top view,
showing metal
contact "fingers."

doping is indicated; on the other hand, long lifetimes are desirable and these are
reduced by doping too heavily. It is important that the series resistance of the device be very small so that power is not lost to heat due to ohmic losses in the device itself. A series resistance of only a few ohms can seriously reduce the output
power of a solar cell (Prob. 8.4). Since the area is large, the resistance of the
p-type body of the device can be made small. However, contacts to the thin n region require special design. If this region is contacted at the edge, current must
flow along the thin n region to the contact, resulting in a large series resistance.
To prevent this effect, the contact can be distributed over the n surface by providing small contact fingers as in Fig. 8–5b. These narrow contacts serve to reduce
the series resistance without interfering appreciably with the incoming light.

Figure 8–6 shows the fourth-quadrant portion of a solar cell characteristic, with I_r plotted upward for convenience of illustration. The open-circuit voltage V_{oc} and the short-circuit current I_{sc} are determined for a given light level by the cell properties. The maximum power delivered to a load by this solar cell occurs when the product VI_r is a maximum. Calling these values of voltage and current V_m and I_m, we can see that the maximum delivered power illustrated by the shaded rectangle in Fig. 8–6 is less than the $I_{sc}V_{oc}$ product. The ratio $I_mV_m/I_{sc}V_{oc}$ is called the *fill factor*, and is a figure of merit for solar cell design.

Applications of solar cells are not restricted to outer space. It is possible to obtain useful power from the sun in terrestrial applications using solar cells, even though the solar intensity is reduced by the atmosphere. About 1 kW/m² is available in a particularly sunny location, but not all of this solar power can be converted to electricity. Much of the photon flux is at energies less than the cell band gap, and is not absorbed. High-energy photons are strongly absorbed, and the resulting EHPs may recombine at the surface. A well-made Si cell can have about 10 percent efficiency for solar energy conversion, providing approximately 100 W/m² of electrical power under full illumination. This is a modest amount of power per unit solar cell area, considering the effort involved in fabricating a large area of Si cells. One approach to obtaining more power per cell is to focus considerable light onto the cell using mirrors. Although Si cells lose efficiency at the resulting high temperatures, GaAs and related compounds can be used at 100°C or higher. In such solar concentrator systems more effort and expense can be put into the solar cell fabrication, since fewer cells are required. For example, a GaAs–AlGaAs heterojunction cell provides good conversion efficiency and operates at the elevated temperatures common in solar concentrator systems.

8.1.3 Photodetectors

When the photodiode is operated in the third quadrant of its *I–V* characteristic (Fig. 8–3b), the current is essentially independent of voltage but is proportional to the optical generation rate. Such a device provides a useful means

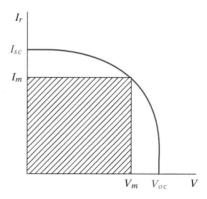

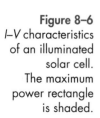

Figure 8–6
I–V characteristics of an illuminated solar cell. The maximum power rectangle is shaded.

of measuring illumination levels or of converting time-varying optical signals into electrical signals.

In most optical detection applications the detector's speed of response is critical. For example, if the photodiode is to respond to a series of light pulses 1 ns apart, the photogenerated minority carriers must diffuse to the junction and be swept across to the other side in a time much less than 1 ns. The carrier diffusion step in this process is time consuming and should be eliminated if possible. Therefore, it is desirable that the width of the depletion region W be large enough so that most of the photons are absorbed within W rather than in the neutral p and n regions. When an EHP is created in the depletion region, the electric field sweeps the electron to the n side and the hole to the p side. Since this carrier drift occurs in a very short time, the response of the photodiode can be quite fast. When the carriers are generated primarily within the depletion layer W, the detector is called a *depletion layer photodiode*. Obviously, it is desirable to dope at least one side of the junction lightly so that W can be made large. The appropriate width for W is chosen as a compromise between sensitivity and speed of response. If W is wide, most of the incident photons will be absorbed in the depletion region. Also, a wide W results in a small junction capacitance [see Eq. (5-62)], thereby reducing the RC time constant of the detector circuit. On the other hand, W must not be so wide that the time required for drift of photogenerated carriers out of the depletion region is excessive.

One convenient method of controlling the width of the depletion region is to build a *p-i-n photodetector* (Fig. 8–7). The "i" region need not be truly intrinsic, as long as the resistivity is high. It can be grown epitaxially on the n-type substrate, and the p region can be obtained by implantation. When this device is reverse biased, the applied voltage appears almost entirely across the i region. If the carrier lifetime within the i region is long compared with the drift time, most of the photogenerated carriers will be collected by the n and p regions.

If low-level optical signals are to be detected, it is often desirable to operate the photodiode in the avalanche region of its characteristic. In this mode each photogenerated carrier results in a significant change in the current because of avalanche multiplication. In the *avalanche photodiode* the

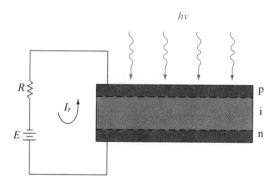

Figure 8–7
Schematic representation of a p-i-n photodiode.

junction must be uniform, and a guard ring is generally used to ensure against edge breakdown. With proper design a Si avalanche photodiode can have high sensitivity to low-level optical signals, and the response time is in the neighborhood of 1 ns. These devices are particularly useful in fiber optic communication systems (Section 8.2.2).

The type of photodiode described here is sensitive to photons with energies near the band gap energy (*intrinsic* detectors). If $h\nu$ is less than E_g, the photons will not be absorbed; on the other hand, if the photons are much more energetic than E_g, they will be absorbed very near the surface, where the recombination rate is high. Therefore, it is necessary to choose a photodiode material with a band gap corresponding to a particular region of the spectrum. Detectors sensitive to longer wavelengths can be designed such that photons can excite electrons into or out of impurity levels (*extrinsic* detectors). However, the sensitivity of such extrinsic detectors is much less than intrinsic detectors, where electron–hole pairs are generated by excitation across the band gap.

By using lattice-matched multilayers of compound semiconductors, the band gap of the absorbing region can be tailored to match the wavelength of light being detected. Wider band gap material can then be used as a window through which the light is transmitted to the absorbing region (Fig. 8–8). For example, we saw in Fig. 1–13 that InGaAs with an In mole fraction of 53 percent can be grown epitaxially on InP with excellent lattice matching. This composition of InGaAs has a band gap of about 0.75 eV, which is sensitive to a useful wavelength for fiber optic systems. (1.55 μm), as we shall see in Section 8.2.2. In making a photodiode using InGaAs as the active material, it is possible to bring the light through the wider band gap InP (1.35 eV), thus greatly reducing surface recombination effects. In the case of avalanche photodiodes requiring narrow band gap material, it is often advantageous to absorb the light in the narrow-gap semiconductor (e.g., InGaAs) and transport the resulting carriers to a junction made in wider band gap material (e.g., InP), where the avalanche multiplication takes place (Fig. 8–8b). Such a separation of the absorption and multiplication regions avoids the excessive leakage currents typical of reverse-biased junctions in narrow-gap materials.

8.1.4 Noise and Bandwidth of Photodetectors

In optical communication systems the sensitivity of the photodetector and its response time are of critical importance. Unfortunately, these two properties are generally difficult to optimize without making compromises between them. For example, in a photoconductor the gain depends on the ratio of carrier lifetime to transit time (see Prob. 8–6). On the other hand, the frequency response (and therefore the bandwidth) varies inversely with carrier lifetime. As a result, trade-offs must be made between these two desirable characteristics. It is common to express the *gain-bandwidth product* as a figure of merit for detectors. Designs which increase gain tend to decrease bandwidth and vice versa. Another important property of detectors is the

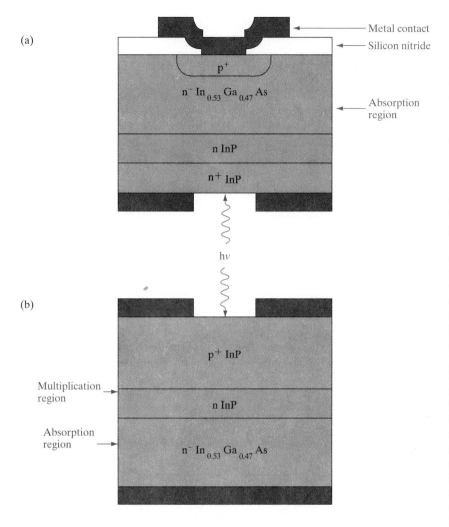

(a)

Metal contact
Silicon nitride

p^+

n^- In$_{0.53}$ Ga$_{0.47}$ As

Absorption region

n InP

n+ InP

hν

(b)

p^+ InP

Multiplication region

n InP

Absorption region

n^- In$_{0.53}$ Ga$_{0.47}$ As

Figure 8–8
Use of multilayer heterojunctions to enhance photodiode operation: (a) a p-i-n photodiode in which light near 1.55 μm is absorbed in a narrow band gap material (InGaAs, E_g = 0.75 eV) after passing through a wider-gap material (InP, E_g = 1.35 eV); (b) an avalanche photodiode in which light is absorbed in the InGaAs and holes are swept to an InP junction, where the avalanche multiplication takes place. This separation of the absorption and multiplication regions reduces the junction leakage current. In this figure, n$^-$ refers to lightly doped n-type material.

signal-to-noise ratio, which is the amount of usable information compared with the background noise in the detector.

In the case of photoconductors the major source of noise is random fluctuations in the dark current (called *Johnson noise*). The noise current increases with temperature and the conductance of the material in the dark. Therefore, the photoconductor noise at a given temperature can be reduced by increasing the dark resistance. Increased dark resistance also increases the gain of the photoconductor, thereby decreasing the bandwidth.

In a p-i-n diode there is no gain mechanism, since at most one electron–hole pair is collected by the junction for each photon absorbed. Thus the gain is essentially unity, and the gain-bandwidth product is determined by the bandwidth, or frequency response. In a p-i-n the response time is dependent on the width of the depletion region, and the main source of noise is random thermal generation of EHPs within this region (called *shot noise*). The noise in a p-i-n device is considerably lower than that in a photoconductor, which compensates for the lack of gain in the p-i-n.

Avalanche photodiodes have the advantage of providing gain through the avalanche multiplication effect. The disadvantage is increased noise relative to the p-i-n, due to random fluctuations in the avalanche process. This noise is reduced if the impact ionization in the high field region is due to only one type of carrier, since more fluctuations in the ionization process occur when both electrons and holes participate. In Si the ability of electrons to create EHPs in an impact ionization event is much higher than for holes. Therefore, Si avalanche photodiodes can be operated with high gain and relatively low noise. Unfortunately, Si avalanche photodiodes (APDs) cannot be used for most fiber optic transmission because Si is transparent at the wavelengths of low loss and low dispersion ($\lambda = 1.55$ and $1.3 \ \mu m$) for optical fibers. For these longer wavelengths, $In_{0.53}Ga_{0.47}As$ has become the material of choice. However, the ionization rates of electrons and holes in most compound semiconductors are comparable, which degrades their noise and frequency response relative to Si APDs. One creative way of overcoming this problem is shown in Fig. 8–9. The Si–InGaAs APD is fabricated by using wafer fusion to bond an InGaAs absorption layer to a Si avalanche photodiode. The advantage of this approach is that it utilizes the strengths of both materials systems. In operation, light is absorbed in the narrow bandgap InGaAs layer and the photogenerated electrons are injected into the Si avalanche region, which is better suited for the large fields applied. These

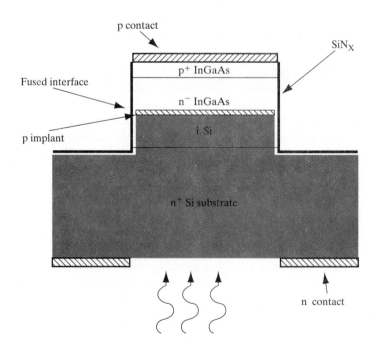

Figure 8–9
Schematic cross section of a silicon hetero-interface photo-detector (SHIP). The light passes through the wide-bandgap Si and is absorbed in the InGaAs.

APDs have achieved low dark current, a quantum efficiency of 60 percent at 1.3 μm, and a gain-bandwidth product of 300 GHz.

Another approach that has demonstrated excellent performance utilizes an APD structure in a resonant cavity. The resonant-cavity photodiode (Fig. 8–10) consists of a thin absorbing layer sandwiched between two *distributed Bragg reflector (DBR)* mirrors. The structure is similar to that of a vertical-cavity, surface-emitting laser (to be discussed in Section 8.4.4) except that the active region is an absorber instead of an emitter and the Q of the cavity is typically much lower than that of laser structures. The resonant-cavity structure can provide several performance advantages, one of which is that the tradeoff between responsivity and bandwidth that is inherent to conventional, single-pass p-i-n photodiode structures can be circumvented. For the typical normal-incidence photodetector, a wide bandwidth necessitates a thin absorption layer which, in turn, results in low quantum efficiency. The resonant-cavity structure, on the other hand, effectively decouples the responsivity from the transit-time component of the bandwidth because the optical signal makes multiple passes across the thin absorbing layer inside the microcavity. Consequently, high speed and high quantum efficiency can be achieved simultaneously. In the resonant-cavity APD shown in Fig. 8–10, light is absorbed in the thin InGaAs

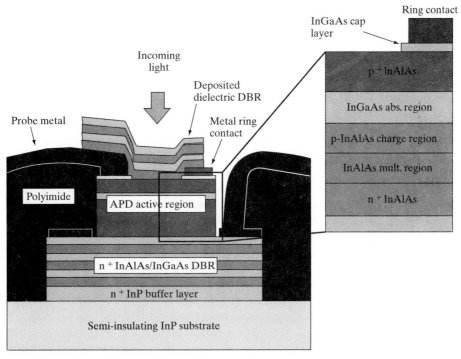

Figure 8–10
Cross section of InAlAs–InGaAs resonant-cavity avalanche photodiode, and a detail of the active region.

layer. The electrons then drift into the InAlAs multiplication region where the field is high enough to provide gain by impact ionization. For very thin multiplication layers there is a size effect that can lead to very low multiplication noise, comparable to that of Si APDs, and high gain-bandwidth products. At low gains where the bandwidth is limited by the transit time and the RC time constraint, bandwidths in excess of 30 GHz have been achieved. At higher gains the bandwidth is determined by the gain-bandwidth product which can be in excess of 300 GHz.

8.2
LIGHT-EMITTING
DIODES

When carriers are injected across a forward-biased junction, the current is usually accounted for by recombination in the transition region and in the neutral regions near the junction. In a semiconductor with an indirect band gap, such as Si or Ge, the recombination releases heat to the lattice. On the other hand, in a material characterized by direct recombination, considerable light may be given off from the junction under forward bias. This effect, called *injection electroluminescence* (Section 4.2.2), provides an important application of diodes as generators of light. The use of light-emitting diodes (LEDs) in digital displays is well known. There are also other important applications in communications and other areas. Another important device making use of radiative recombination in a forward-biased p-n junction is the *semiconductor laser*. As we shall see in Section 8.4, lasers emit coherent light in much narrower wavelength bands than LEDs, and with more collimation (directionality).

8.2.1   Light-Emitting Materials

The band gaps of various binary compound semiconductors are illustrated in Fig. 4–4 relative to the spectrum. There is a wide variation in band gaps and, therefore, in available photon energies, extending from the ultraviolet (GaN, 3.4 eV) into the infrared (InSb, 0.18 eV). In fact, by utilizing ternary and quaternary compounds the number of available energies can be increased significantly (see Figs. 1–13 and 3–6). A good example of the variation in photon energy obtainable from the compound semiconductors is the ternary alloy gallium arsenide–phosphide, which is illustrated in Fig. 8–11. When the percentage of As is reduced and P is increased in this material, the resulting band gap varies from the direct 1.43-eV gap of GaAs (infrared) to the indirect 2.26-eV gap of GaP (green). The band gap of $GaAs_{1-x}P_x$ varies almost linearly with x until the 0.45 composition is reached, and electron–hole recombination is direct over this range. The most common alloy composition used in LED displays is $x \simeq 0.4$. For this composition the band gap is direct, since the Γ minimum (at $\mathbf{k} = 0$) is the lowest part of the conduction band. This results in efficient radiative recombination, and the emitted photons (~1.9 eV) are in the red portion of the spectrum.

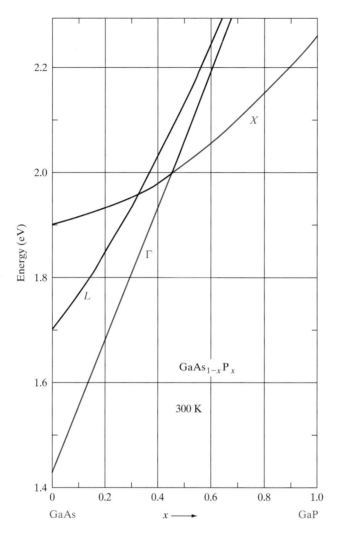

Figure 8–11
Conduction band
energies as a
function of alloy
composition for
$GaAs_{1-x}P_x$.

For $GaAs_{1-x}P_x$ with P concentrations above 45 percent, the band gap is due to the indirect X minimum. Radiative recombination in such indirect materials is generally unlikely, because electrons in the conduction band have different momentum from holes in the valence band (see Fig. 3–5). Interestingly, however, indirect $GaAs_{1-x}P_x$ (including GaP, $x = 1$) doped with nitrogen can be used in LEDs with light output in the yellow to green portions of the spectrum. This is possible because the nitrogen impurity binds an electron very tightly. This confinement in real space (Δx) means that the electron momentum is spread out in momentum space Δp by the Heisenberg uncertainty principle (see Eq. 2-18). As a result, the momentum conservation rules, which generally prevent radiative recombination in indirect materials, are circumvented. Thus nitrogen doping of $GaAs_{1-x}P_x$ is not only useful technologically, but also provides an interesting and practical illustration of the uncertainty principle.

In many applications light from a laser or an LED need not be visible to the eye. Infrared emitters such as GaAs, InP, and mixed alloys of these compounds are particularly well suited to optical communication systems. For example, a laser or light-emitting diode can be used in conjunction with a photodiode or other photosensitive device to transmit information optically between locations. By varying the current through the diode, the light output can be modulated such that analog or digital information appears in the optical signal directed at the detector. Alternatively, the information may be introduced between the source and detector. For example, a semiconductor laser-photodetector arrangement can be used in a compact disc system for reading digital information from the spinning disc. A light emitter and a photodiode form an *optoelectronic pair*, which provides complete electrical isolation between input and output, since the only link between the two devices is optical. In an *optoelectronic isolator*, both devices may be mounted on a ceramic substrate and packaged together to form a unit that passes information while maintaining isolation.

In view of the broad range of applications requiring semiconductor lasers and LEDs with visible and infrared wavelengths, the wide variety of available III–V materials is extremely useful. In addition to the AlGaAs and GaAsP systems shown in Figs. 3–6 and 8–11, the InAlGaP system is useful for yellow and green wavelengths, and GaN is a strong emitter in the blue. Even more wavelengths will be accessible as GaN and related materials become increasingly used in LEDs and lasers. For many years the II–VI semiconductors have been known as efficient light emitters in photoluminescence, but obtaining p-n junctions was extremely difficult. With traditional doping methods, crystal defects tend to compensate the doping impurities such that only n-type (ZnS, ZnSe, CdS, CdSe) or p-type (ZnTe) can be obtained. This frustrating problem prevented the formation of useful p-n junctions until 1990, when the use of nitrogen doping resulted in p-type ZnSe in MBE-grown material. Rapid progress has been made since then, including the use of multilayer heterostructures grown by MBE and OMVPE in the (Zn, Cd)(S, Se) system. Using a nitrogen plasma source, ZnTe can be doped p-type to acceptor concentrations above 10^{19} cm^{-3}. In spite of this research progress, however, II–VI LEDs and lasers lag behind III–V semiconductors in most applications. The availability of blue light from GaN is of particular importance in extending III–V light emission across the entire visible spectrum.

8.2.2 Fiber Optic Communications

The transmission of optical signals from source to detector can be greatly enhanced if an *optical fiber* is placed between the light source and the detector. An optical fiber is essentially a "light pipe" or waveguide for optical frequencies. The fiber is typically drawn from a boule of glass to a diameter of ~25 μm. The fine glass fiber is relatively flexible and can be used to guide optical signals over distances of kilometers without the necessity of perfect alignment between source and detector. This significantly increases the applications of optical communication in areas such as telephone and data transmission.

One type of optical fiber has an outer layer of very pure fused silica (SiO_2), with a core of germanium doped glass having a higher index of refraction (Fig. 8–12a).[1] Such a *step-index* fiber maintains the light beam primarily in the central core with little loss at the surface. The light is transmitted along the length of the fiber by internal reflection at the step in the refractive index.

Losses in the fiber at a given wavelength can be described by an attenuation coefficient α [similar to the absorption coefficient of Eq. (4–3)]. The intensity of the signal at a distance x along the fiber is then related to the starting intensity by the usual expression.

$$\mathbf{I}(x) = \mathbf{I}_0 e^{-\alpha x} \tag{8–5}$$

The attenuation is not the same for all wavelengths, however, and it is therefore important to choose a signal wavelength carefully. A plot of α vs. λ for a typical silica glass fiber is shown in Fig. 8–13. It is clear that dips in α near 1.3 and 1.55 μm provide "windows" in the attenuation, which can be exploited to reduce the degradation of signals. The overall decrease in attenuation with increasing wavelength is due to the reduced scattering from small random inhomogeneities which result in fluctuations of the refractive index on a scale comparable to the wavelength. This type of attenuation, called *Rayleigh scattering*, decreases with the fourth power of wavelength. This effect is observed at sunrise and sunset, when attenuation

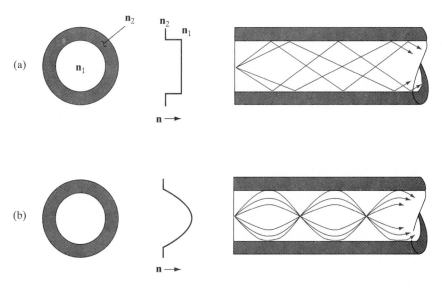

Figure 8–12
Two examples of multimode fibers: (a) *step-index,* having a core with slightly larger refractive index **n**; (b) *graded-index* having in this case a parabolic grading of **n** in the core. The figure illustrates the cross section (left) of the fiber, its index of refraction profile (center), and typical mode patterns (right).

[1]The *index of refraction* (or *refractive index*) **n** compares the velocity of light **v** in the material to its velocity c in a vacuum, **n** = c/**v**. Thus if $\mathbf{n}_1 > \mathbf{n}_2$ in Fig. 8–12a, the light velocity is greater in material 2 than in 1. The value of **n** varies somewhat with the wavelength of light.

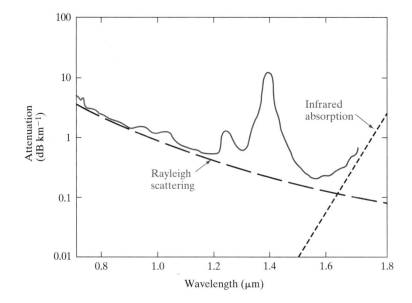

of short wavelength blue and green light results in red and orange sunlight. Obviously, Rayleigh scattering encourages operation at long wavelengths in fiber optic systems. However, a competing process of infrared absorption dominates for wavelengths longer than about 1.7 μm, due to vibrational excitation of the atoms making up the glass. Therefore, a useful minimum in absorption for silica fibers occurs at about 1.55 μm, where epitaxial layers in the (In, Ga) (As, P) system can be grown lattice-matched to InP substrates (see Fig. 1–13).

Another consideration in choice of operating wavelength is the *pulse dispersion*, or spreading of data pulses as they propagate down the fiber. This effect can be caused by the wavelength dependence of the refractive index, causing different optical frequencies to travel down the fiber with slightly different velocities. This effect, called *chromatic dispersion*, is much less pronounced at the 1.3 μm window in Fig. 8–13. Another cause of dispersion is the fact that different modes propagate with different path lengths (Fig. 8–12a). This type of dispersion can be reduced by grading the refractive index of the core (Fig. 8–12b) such that various modes are continually refocused, reducing the differences in path lengths.

In early optoelectronic systems for fiber optics, it was most convenient to use the well-established GaAs–AlGaAs system for making lasers and LEDs. These light sources are very efficient, and good detectors can be made using Si p-i-n or avalanche photodiodes. However, these sources operate in the wavelength range near 0.9 μm, where the attenuation is greater than for longer wavelengths. Modern systems, therefore, operate near the 1.3- and 1.55-μm minima in Fig. 8–13. At these wavelengths, sources can be made using InGaAs or InGaAsP grown on InP, and detectors can be made of the same materials (see Fig. 8–8), or using Ge.

8.2.3 Multilayer Heterojunctions for LEDs

The light source in a fiber optic system may be a laser or an LED. In the case of a laser, the light is of essentially a single frequency and allows a very large information bandwidth. Semiconductor lasers suitable for fiber optic communications will be discussed in Section 8.4. An LED designed for a fiber optic system is illustrated in Fig. 8–14. The LED is a multilayer structure of GaAs and AlGaAs. To take advantage of the 1.3- and 1.55-μm windows in Fig. 8–13, similar devices using InGaAs or InGaAsP can be used. The quaternary (four-element) alloy is particularly suitable, in that band gap (and therefore emission wavelength) can be adjusted along with choosing lattice constants for epitaxial growth on convenient substrates. In Fig. 8–14 the fiber is held in an etched well on the back side of the diode by an epoxy resin. This configuration, often called a "Burrus diode" after its developer, is particularly convenient for launching signals from an LED into a fiber, with good mechanical stability.

Although LEDs are less suited to transmission of digital information than are lasers, they are easily modulated by analog signals. The optical power emitted by a properly constructed LED varies linearly with the input current over a wide range. An LED is an *incoherent* light source, in that photons are emitted randomly from the junction in all directions and not in phase with each other. Therefore, transmission of LED-generated signals inherently involves many modes, as in Fig. 8–12. *Multimode* fibers are larger (~25 μm in diameter) than are *single-mode* fibers (~5 μm), which transmit a coherent laser beam.

By forming numerous optical fibers into a bundle, with an appropriate jacket for mechanical strength, an enormous amount of information can be transmitted over long distances.[2] Depending upon the losses in the fibers,

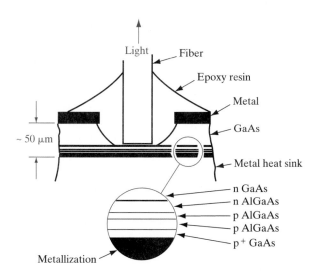

Figure 8–14
Cross section of a GaAs–AlGaAs LED for fiber-optic applications. [After C.A. Burrus and B.I. Miller, *Optics Communications*, vol. 4, p. 307 (1971).]

---

[2]Transmission rates of many G-bit/s have been achieved. As a convenient calibration of this rate, it is worth noting that the human eye is able to transmit about one G-bit/s to the brain.

repeater stations may be required periodically along the path. Thus many photodetectors and LED or laser sources are required in a fiber optic system. Semiconductor device development, including appropriate binary, ternary, and quaternary compounds for both emitters and detectors, is therefore crucial to the successful implementation of such optical communications systems.

8.3
LASERS

The word LASER is an acronym for *light amplification by stimulated emission of radiation,* which sums up the operation of an important optical and electronic device. The laser is a source of highly directional, monochromatic, coherent light, and as such it has revolutionized some longstanding optical problems and has created some new fields of basic and applied optics. The light from a laser, depending on the type, can be a continuous beam of low or medium power, or it can be a short burst of intense light delivering millions of watts. Light has always been a primary communications link between humans and the environment, but until the invention of the laser, the light sources available for transmitting information and performing experiments were generally neither monochromatic nor coherent, and were of relatively low intensity. Thus the laser is of great interest in optics; but it is equally important in optoelectronics, particularly in fiber optic communications. The last three letters in the word *laser* are intended to imply how the device operates: by the *stimulated emission* of *radiation.* In Chapter 2 we discussed the emission of radiation when excited electrons fall to lower energy states; but generally, these processes occur randomly and can therefore be classed as *spontaneous emission.* This means that the rate at which electrons fall from an upper level of energy E_2 to a lower level E_1 is at every instant proportional to the number of electrons remaining in E_2 (the *population* of E_2). Thus if an initial electron population in E_2 were allowed to decay, we would expect an exponential emptying of the electrons to the lower energy level, with a mean decay time describing how much time an average electron spends in the upper level. An electron in a higher or excited state need not wait for spontaneous emission to occur, however; if conditions are right, it can be *stimulated* to fall to the lower level and emit its photon in a time much shorter than its mean spontaneous decay time. The stimulus is provided by the presence of photons of the proper wavelength. Let us visualize an electron in state E_2 waiting to drop spontaneously to E_1 with the emission of a photon of energy $h\nu_{12} = E_2 - E_1$ (Fig. 8–15). Now we assume that this electron

Figure 8–15
Stimulated transition of an electron from an upper state to a lower state, with accompanying photon emission.

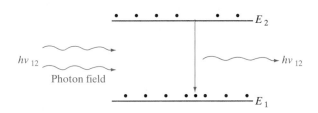

in the upper state is immersed in an intense field of photons, each having energy $hv_{12} = E_2 - E_1$, and in phase with the other photons. The electron is induced to drop in energy from E_2 to E_1, contributing a photon whose wave is *in phase* with the radiation field. If this process continues and other electrons are stimulated to emit photons in the same fashion, a large radiation field can build up. This radiation will be *monochromatic* since each photon will have an energy of precisely $hv_{12} = E_2 - E_1$ and will be *coherent*, because all the photons released will be in phase and reinforcing. This process of stimulated emission can be described quantum mechanically to relate the probability of emission to the intensity of the radiation field. Without quantum mechanics we can make a few observations here about the relative rates at which the absorption and emission processes occur. Let us assume the instantaneous populations of E_1 and E_2 to be n_1 and n_2, respectively. We know from earlier discussions of distributions and the Boltzmann factor that at *thermal equilibrium* the relative population will be

$$\frac{n_2}{n_1} = e^{-(E_2 - E_1)/kT} = e^{-hv_{12}/kT} \tag{8-6}$$

if the two levels contain an equal number of available states.

The negative exponent in this equation indicates that $n_2 \ll n_1$ at equilibrium; that is, most electrons are in the lower energy level as expected. If the atoms exist in a radiation field of photons with energy hv_{12}, such that the energy density of the field is $\rho(v_{12})$,[3] then stimulated emission can occur along with absorption and spontaneous emission. The rate of stimulated emission is proportional to the instantaneous number of electrons in the upper level n_2 and to the energy density of the stimulating field $\rho(v_{12})$. Thus we can write the stimulated emission rate as $B_{21}n_2\rho(v_{12})$, where B_{21} is a proportionality factor. The rate at which the electrons in E_1 absorb photons should also be proportional to $\rho(v_{12})$, and to the electron population in E_1. Therefore, the absorption rate is $B_{12}n_1\rho(v_{12})$, where B_{12} is a proportionality factor for absorption. Finally, the rate of spontaneous emission is proportional only to the population of the upper level. Introducing still another coefficient, we can write the rate of spontaneous emission as $A_{21}n_2$. For steady state the two emission rates must balance the rate of absorption to maintain constant populations n_1 and n_2 (Fig. 8–16).

$$
\begin{array}{lll}
B_{12}n_1\rho(v_{12}) & = A_{21}n_2 & + B_{21}n_2\rho(v_{12}) \\
\text{Absorption} & = \text{spontaneous} + & \text{stimulated} \\
& \text{emission} & \text{emission}
\end{array}
\tag{8-7}
$$

---

[3]The energy density $\rho(v_{12})$ indicates the total energy in the radiation field per unit volume and per unit frequency, due to photons with $hv_{12} = E_2 - E_1$.

Figure 8–16
Balance of ab-
sorption and emis-
sion in steady
state: (a) stimulat-
ed emission; (b)
absorption; (c)
spontaneous
emission.

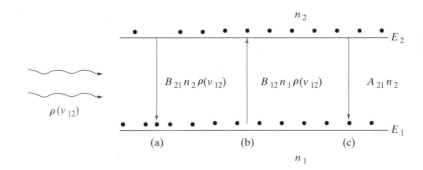

This relation was described by Einstein, and the coefficients B_{12}, A_{21}, B_{21} are called the *Einstein coefficients.* We notice from Eq. (8–7) that no energy density ρ is required to cause a transition from an upper to a lower state; spontaneous emission occurs without an energy density to drive it. The reverse is not true, however; exciting an electron to a higher state (absorption) requires the application of energy, as we would expect thermodynamically.

At equilibrium, the ratio of the stimulated to spontaneous emission rates is generally very small, and the contribution of stimulated emission is negligible. With a photon field present,

$$\frac{\text{Stimulated emission rate}}{\text{Spontaneous emission rate}} = \frac{B_{21}n_2\rho(\nu_{12})}{A_{21}n_2} = \frac{B_{21}}{A_{21}}\rho(\nu_{12}) \qquad (8\text{–}8)$$

As Eq. (8–8) indicates, the way to enhance the stimulated emission over spontaneous emission is to have a very large photon field energy density $\rho(\nu_{12})$. In the laser, this is encouraged by providing an *optical resonant cavity* in which the photon density can build up to a large value through multiple internal reflections at certain frequencies (ν).

Similarly, to obtain more stimulated emission than absorption we must have $n_2 > n_1$:

$$\frac{\text{Stimulated emission rate}}{\text{Absorption rate}} = \frac{B_{21}n_2\rho(\nu_{12})}{B_{12}n_1\rho(\nu_{12})} = \frac{B_{21}}{B_{12}}\frac{n_2}{n_1} \qquad (8\text{–}9)$$

Thus if stimulated emission is to dominate over absorption of photons from the radiation field, we must have a way of maintaining more electrons in the upper level than in the lower level. This condition is quite unnatural, since Eq. (8–6) indicates that n_2/n_1 is less than unity for any equilibrium case. Because of its unusual nature, the condition $n_2 > n_1$ is called *population inversion.* It is also referred to as a condition of *negative temperature.* This rather startling terminology emphasizes the nonequilibrium nature of population inversion, and refers to the fact that the ratio n_2/n_1 in Eq. (8–6) could be larger than unity only if the temperature were negative. Of course, this manner of speaking does not imply anything about temperature in the usual sense of that word. The fact is that Eq. (8–6) is a thermal equilibrium equation and cannot be applied to the situation of population inversion without invoking the concept of negative temperature.

In summary, Eqs. (8–8) and (8–9) indicate that if the photon density is to build up through a predominance of stimulated emission over both spontaneous emission and absorption, two requirements must be met. We must provide (1) an optical resonant cavity to encourage the photon field to build up and (2) a means of obtaining population inversion.

An optical resonant cavity can be obtained using reflecting mirrors to reflect the photons back and forth, allowing the photon energy density to build up. One or both of the end mirrors are constructed to be partially transmitting so that a fraction of the light will "leak out" of the resonant system. This transmitted light is the output of the laser. Of course, in designing such a laser one must choose the amount of transmission to be a small perturbation on the resonant system. The gain in photons per pass between the end plates must be larger than the transmission at the ends, scattering from impurities, absorption, and other losses. The arrangement of parallel plates providing multiple internal reflections is similar to that used in the Fabry–Perot interferometer;[4] thus the reflecting ends of the laser cavity are often referred to as Fabry–Perot faces. As Fig. 8–17 indicates, light of a particular frequency can be reflected back and forth within the resonant cavity in a reinforcing (coherent) manner if an integral number of half-wavelengths fit between the end mirrors. Thus the length of the cavity for stimulated emission must be

$$L = \frac{m\lambda}{2} \qquad\qquad (8\text{–}10)$$

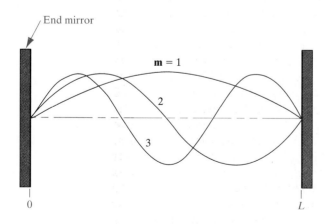

Figure 8–17
Resonant modes within a laser cavity.

[4]Interferometers are discussed in many sophomore physics texts.

where **m** is an integer. In this equation λ is the photon wavelength within the laser material. If we wish to use the wavelength λ_0 of the output light in the atmosphere (often taken as the vacuum value), the index of refraction **n** of the laser material must be considered

$$\lambda_0 = \lambda \mathbf{n} \qquad (8\text{--}11)$$

In practice, $L \gg \lambda$, and Eq. (8–10) is automatically satisfied over some portion of the mirror. An important exception occurs in the vertical cavity surface-emitting lasers discussed in Section 8.4.4, for which the cavity length is comparable to the wavelength.

There are ways of obtaining population inversion in the atomic levels of many solids, liquids, and gases, and in the energy bands of semiconductors. Thus the possibilities for laser systems with various materials are quite extensive. An early laser system used a ruby rod. In gas lasers, electrons are excited to metastable levels in molecules to achieve population inversion. These are interesting and useful laser systems, but in view of our emphasis on semiconductor devices in this book, we will move to the description of semiconductor lasers.

8.4 SEMICONDUCTOR LASERS

The laser became an important part of semiconductor device technology in 1962 when the first p-n junction lasers were built in GaAs (infrared)[5] and GaAsP (visible).[6] We have already discussed the incoherent light emission from p-n junctions (LEDs), generated by the spontaneous recombination of electrons and holes injected across the junction. In this section we shall concentrate on the requirements for population inversion due to these injected carriers and the nature of the coherent light from p-n junction lasers. These devices differ from solid, gas, and liquid lasers in several important respects. Junction lasers are remarkably small (typically on the order of 0.1 × 0.1 × 0.3 mm), they exhibit high efficiency, and the laser output is easily modulated by controlling the junction current. Semiconductor lasers operate at low power compared, for example, with ruby or CO_2 lasers; on the other hand, these junction lasers compete with He–Ne lasers in power output. Thus the function of the semiconductor laser is to provide a portable and easily controlled source of low-power coherent radiation. They are particularly suitable for fiber optic communication systems (Section 8.2.2).

8.4.1 Population Inversion at a Junction

If a p-n junction is formed between degenerate materials, the bands under forward bias appear as shown in Fig. 8–18. If the bias (and thus the current)

[5]R. N. Hall et al., *Physical Review Letters* 9, pp. 366–368 (November 1, 1962); M. I. Nathan et al., *Applied Physics Letters* 1, pp. 62–64 (November 1, 1962); T. M. Quist et al., *Applied Physics Letters* 1, pp. 91-92 (December 1, 1962).
[6]N. Holonyak, Jr., and S. F. Bevacqua, *Applied Physics Letters* 1, pp. 82–83 (December 1, 1962).

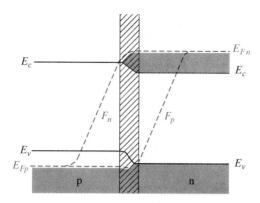

Figure 8–18
Band diagram of
a p-n junction
laser under for-
ward bias. The
cross-hatched re-
gion indicates the
inversion region
at the junction.

is large enough, electrons and holes are injected into and across the transition region in considerable concentrations. As a result, the region about the
junction is far from being depleted of carriers. This region contains a large
concentration of electrons within the conduction band and a large concentration of holes within the valence band. If these population densities are
high enough, a condition of population inversion results, and the region about
the junction over which it occurs is called an *inversion region.*[7]

Population inversion at a junction is best described by the use of the concept of *quasi-Fermi levels* (Section 4.3.3). Since the forward-biased condition of
Fig. 8–18 is a distinctly nonequilibrium state, the equilibrium equations defining
the Fermi level are not applicable. In particular, the concentration of electrons
in the inversion region (and for several diffusion lengths into the p material)
is larger than equilibrium statistics would imply; the same is also true for the
injected holes in the n material. We can use Eqs. (4–15) to describe the carrier concentrations in terms of the quasi-Fermi levels for electrons and holes in
steady state. Thus

$$n = N_c e^{-(E_c - F_n)/kT} = n_i e^{(F_n - E_i)/kT} \qquad (8\text{--}12a)$$

$$p = N_v e^{-(F_p - E_v)/kT} = n_i e^{(E_i - F_p)/kT} \qquad (8\text{--}12b)$$

Using Eqs. (8–12a) and (8–12b), we can draw F_n and F_p on any band diagram for which we know the electron and hole distributions. For example,
in Fig. 8–18, F_n in the neutral n region is essentially the same as the equilibrium Fermi level E_{Fn}. This is true to the extent that the electron concentration on the n side is equal to its equilibrium value. However, since large
number of electrons are injected across the junction, the electron concentration begins at a high value near the junction and decays exponentially to
its equilibrium value n_p deep in the p material. Therefore, F_n drops from E_{Fn}
as shown in Fig. 8–18. We notice that, deep in the neutral regions, the quasi-
Fermi levels are essentially equal. The separation of F_n and F_p at any point

---

[7]This is a different meaning of the term from that used in reference to MOS transistors.

Figure 8–19
Expanded view of
the inversion
region.

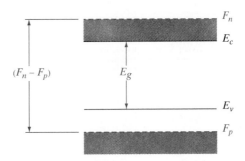

is a measure of the departure from equilibrium at that point. Obviously, this departure is considerable in the inversion region, since F_n and F_p are separated by an energy greater than the band gap (Fig. 8–19).

Unlike the case of the two-level system discussed in Section 8.3, the condition for population inversion in semiconductors must take into account the distribution of energies available for transitions between the bands. The basic definition of population inversion holds—for dominance of stimulated emission between two energy levels separated by energy $h\nu$, the electron population of the upper level must be greater than that of the lower level. The unusual aspect of a semiconductor is that bands of levels are available for such transitions. Population inversion obviously exists for transitions between the bottom of the conduction band E_c and the top of the valence band E_v in Fig. 8–19. In fact, transitions between levels in the conduction band up to F_n and levels in the valence band down to F_p take place under conditions of population inversion. For any given transition energy $h\nu$ in a semiconductor, population inversion exists when

$$(F_n - F_p) > h\nu \qquad (8\text{–}13a)$$

For band-to-band transitions, the minimum requirement for population inversion occurs for photons with $h\nu = E_c - E_v = E_g$

$$(F_n - F_p) > E_g \qquad (8\text{–}13b)$$

When F_n and F_p lie within their respective bands (as in Fig. 8–19), stimulated emission can dominate over a range of transitions, from $h\nu = (F_n - F_p)$ to $h\nu = E_g$. As we shall see below, the dominant transitions for laser action are determined largely by the resonant cavity and the strong recombination radiation occurring near $h\nu = E_g$.

In choosing a material for junction laser fabrication, it is necessary that electron-hole recombination occur directly, rather than through trapping processes such as are dominant in Si or Ge. Gallium arsenide is an example of such a "direct" semiconductor. Furthermore, we must be able to dope the material n-type or p-type to form a junction. If an appropriate resonant cavity can be constructed in the junction region, a laser results in which population inversion is accomplished by the bias current applied to the junction (Fig. 8–20).

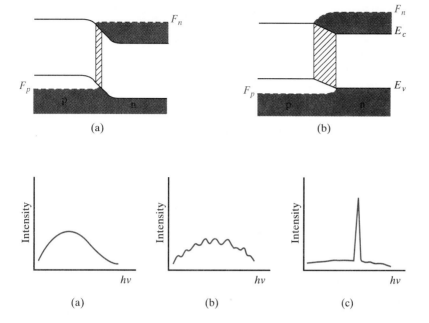

Figure 8–20
Variation of inversion region width with forward bias:
$V(a) < V(b)$.

Figure 8–21
Light intensity vs. photon energy $h\nu$ for a junction laser: (a) incoherent emission below threshold; (b) laser modes at threshold; (c) dominant laser mode above threshold. The intensity scales are greatly compressed from (a) to (b) to (c).

8.4.2 Emission Spectra for p-n Junction Lasers

Under forward bias, an inversion layer can be obtained along the plane of the junction, where a large population of electrons exists at the same location as a large hole population. A second look at Fig. 8–19 indicates that spontaneous emission of photons can occur due to direct recombination of electrons and holes, releasing energies ranging from approximately $F_n - F_p$ to E_g. That is, an electron can recombine over an energy from F_n to F_p, yielding a photon of energy $h\nu = F_n - F_p$, or an electron can recombine from the bottom of the conduction band to the top of the valence band, releasing a photon with $h\nu = E_c - E_v = E_g$. These two energies serve as the approximate outside limits of the laser spectra.

The photon wavelengths which participate in stimulated emission are determined by the length of the resonant cavity as in Eq. (8–10). Figure 8–21 illustrates a typical plot of emission intensity vs. photon energy for a semiconductor laser. At low current levels (Fig. 8–21a), a spontaneous emission spectrum containing energies in the range $E_g < h\nu < (F_n - F_p)$ is obtained. As the current is increased to the point that significant population inversion exists, stimulated emission occurs at frequencies corresponding to the cavity modes as shown in Fig. 8–21b. These modes correspond to successive numbers of integral half-wavelengths fitted within the cavity, as described by Eq. (8–10). Finally, at a still higher current level, a most preferred mode or set of modes will dominate the spectral output (Fig. 8–21c). This very intense mode represents the main laser output of the device; the output light will be composed of almost monochromatic radiation superimposed on a relatively weak radiation background, due primarily to spontaneous emission.

The separation of the modes in Fig. 8–21b is complicated by the fact that the index of refraction **n** for GaAs depends on wavelength λ. From Eq. (8–10) we have

$$\mathbf{m} = \frac{2L\mathbf{n}}{\lambda_0} \tag{8–14}$$

If **m** (the number of half-wavelengths in L) is large, we can use the derivative to find its rate of change with λ_0:

$$\frac{d\mathbf{m}}{d\lambda_0} = -\frac{2L\mathbf{n}}{\lambda_0^2} + \frac{2L}{\lambda_0}\frac{d\mathbf{n}}{d\lambda_0} \tag{8–15}$$

Now reverting to discrete changes in **m** and λ_0, we can write

$$-\Delta\lambda_0 = \frac{\lambda_0^2}{2L\mathbf{n}}\left(1 - \frac{\lambda_0}{\mathbf{n}}\frac{d\mathbf{n}}{d\lambda_0}\right)^{-1}\Delta\mathbf{m} \tag{8–16}$$

If we let $\Delta\mathbf{m} = -1$, we can calculate the change in wavelength $\Delta\lambda_0$ between adjacent modes (i.e., between modes **m** and **m** − 1).

8.4.3 The Basic Semiconductor Laser

To build a p-n junction laser, we need to form a junction in a highly doped, direct semiconductor (GaAs, for example), construct a resonant cavity in the proper geometrical relationship to the junction, and make contact to the junction in a mounting which allows for efficient heat transfer. The first lasers were built as shown in Fig. 8–22. Beginning with a degenerate n-type sample, a p region is formed on one side, for example by diffusing Zn into the n-type GaAs. Since Zn is in column II of the periodic table and is introduced substitutionally on Ga sites, it serves as an acceptor in GaAs; therefore, the heavily doped Zn diffused layer forms a p⁺ region (Fig. 8–22b). At

Figure 8–22
Fabrication of a simple junction laser: (a) degenerate n-type sample; (b) diffused p layer; (c) isolation of junctions by cutting or etching; (d) individual junction to be cut or cleaved into devices; (e) mounted laser structure.

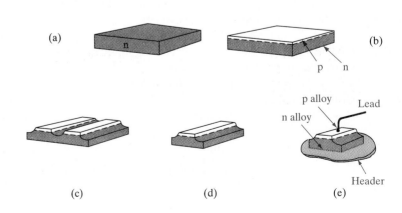

this point we have a large-area planar p-n junction. Next, grooves are cut or etched along the length of the sample as in Fig. 8–22c, leaving a series of long p regions isolated from each other. These p-n junctions can be cut or broken apart (Fig. 8–22d) and then cleaved into devices of the desired length.

At this point in the fabrication process, the very important requirements of a resonant cavity must be considered. It is necessary that the front and back faces (Fig. 8–22e) be flat and parallel. This can be accomplished by cleaving. If the sample has been oriented so that the long junctions of Fig. 8–22d are perpendicular to a crystal plane of the material, it is possible to cleave the sample along this plane into laser devices, letting the crystal structure itself provide the parallel faces. The device is then mounted on a suitable header, and contact is made to the p region. Various techniques are used to provide adequate heat sinking of the device for large forward current levels.

8.4.4 Heterojunction Lasers

The device described above was the first type used in the early development of semiconductor lasers. Since the device contains only one junction in a single type of material, it is referred to as a *homojunction* laser. To obtain more efficient lasers, and particularly to build lasers that operate at room temperature, it is necessary to use multiple layers in the laser structure. Such devices, called *heterojunction lasers,* can be made to operate continuously at room temperature to satisfy the requirements of optical communications. An example of a heterojunction laser is shown in Fig. 8–23. In this structure the injected carriers are confined to a narrow region so that population

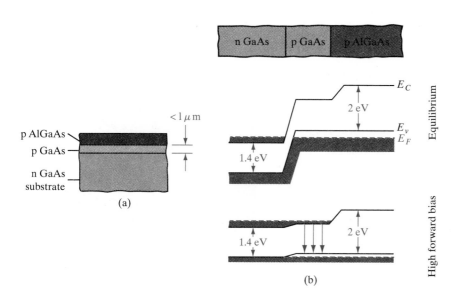

Figure 8–23
Use of a single heterojunction for carrier confinement in laser diodes: (a) AlGaAs heterojunction grown on the thin p-type GaAs layer; (b) band diagrams for the structure of (a), showing confinement of electrons to the thin p region under bias.

inversion can be built up at lower current levels. The result is a lowering of the *threshold current* at which laser action begins. Carrier confinement is obtained in this single-heterojunction laser by the layer of AlGaAs grown epitaxially on the GaAs.

In GaAs the laser action occurs primarily on the p side of the junction due to a higher efficiency for electron injection than for hole injection. In a normal p-n junction the injected electrons diffuse into the p material such that population inversion occurs for only part of the electron distribution near the junction. However, if the p material is narrow and terminated in a barrier, the injected electrons can be confined near the junction. In Fig. 8–23a, an epitaxial layer of p-type AlGaAs ($E_g \simeq 2$ eV) is grown on top of the thin p-type GaAs region. The wider band gap of AlGaAs effectively terminates the p-type GaAs layer, since injected electrons do not surmount the barrier at the GaAs–AlGaAs heterojunction (Fig. 8–23b). As a result of the confinement of injected electrons, laser action begins at a substantially lower current than for simple p-n junctions. In addition to the effects of carrier confinement, the change of refractive index at the heterojunction provides a waveguide effect for optical confinement of the photons.

A further improvement can be obtained by sandwiching the active GaAs layer between two AlGaAs layers (Fig. 8–24). This *double-heterojunction* structure further confines injected carriers to the active region, and the change in refractive index at the GaAs–AlGaAs boundaries helps to confine the generated light waves. In the double-heterojunction laser shown in Fig. 8–24b the injected current is restricted to a narrow stripe along the lasing direction, to reduce the total current required to drive the device. This type of laser was a major step forward in the development of lasers for fiber-optic communications.

Figure 8–24
A double-heterojunction laser structure: (a) multiple layers used to confine injected carriers and provide waveguiding for the light; (b) a stripe geometry designed to restrict the current injection to a narrow stripe along the lasing direction. One of many methods for obtaining the stripe geometry, this example is obtained by proton bombardment of the shaded regions in (b), which converts the GaAs and AlGaAs to semi-insulating form.

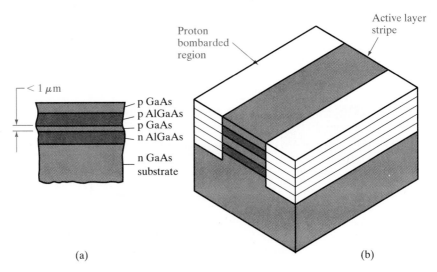

(a)   (b)

Separate Confinement and Graded Index Channels. One of the disadvantages of the double-heterostructure laser shown in Fig. 8–24 is the fact that the carrier confinement and the optical waveguiding both depend on the same heterojunctions. It is much better to optimize these two functions by using a narrow confinement region for keeping the carriers in a region of high recombination, and a somewhat wider optical waveguide region. In Fig. 8–25a we show a *separate confinement* laser in which the width of the optical waveguiding region (w) is optimized by using the refractive index step at a separate heterojunction from that used to confine the carriers. For example, in the GaAs–Al$_x$Ga$_{1-x}$As system the optical confinement (waveguiding) occurs at a boundary with much larger composition x (and therefore smaller refractive index) than is the case for the carrier confinement barrier. By grading the composition of the AlGaAs it is possible to obtain even better waveguiding. For example, in Fig. 8–25b a parabolic grading of the refractive index leads to a waveguide within the laser analogous to that shown in Fig. 8–12 for a fiber. This *graded index separate confinement heterostructure (GRINSCH)* laser also provides built in fields for better electron confinement.

Vertical Cavity Surface-Emitting Lasers (VCSELs). There are advantages to laser structures in which light is emitted normal to the surface, including ease of device testing on the wafer before packaging. An interesting approach is the VCSEL, in which the cavity mirrors are replaced by DBRs, which use many partial reflectors spaced to reflect light constructively. DBRs can be grown by MBE or OMVPE. In Fig. 8–26 the bottom DBR mirror of a VCSEL

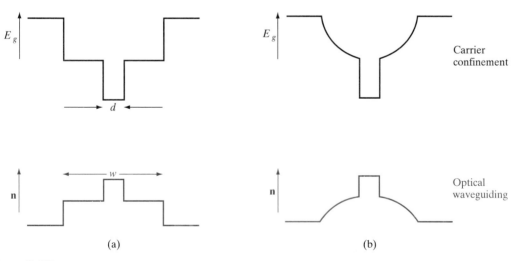

Figure 8–25
Separate confinement of carriers and waveguiding: (a) use of separate changes in AlGaAs alloy composition to confine carriers in the region (d) of smallest band gap, and to obtain waveguiding (w) at the larger step in refractive index; (b) grading the alloy composition, and therefore the refractive index, for better waveguiding and carrier confinement.

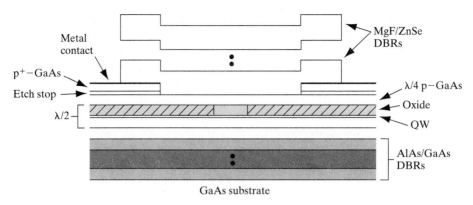

Figure 8–26
Schematic cross section of oxide-confined vertical cavity surface-emitting laser diode. [D.G.Deppe et al., *IEEE J. Selected Topics in Quantum Elec.*, 3(3) (June 1997): 893-904]

is composed of many alternating layers of AlAs and GaAs with thickness one-quarter of a wavelength in each material. The top mirror is composed of deposited dielectric layers (alternating ZnSe and MgF). Current is funneled into the active region from the top contact by using an oxide layer achieved by laterally oxidizing an AlGaAs layer to form an aluminum oxide. The active region of the laser employs InGaAs–GaAs quantum wells, and the GaAs cavity between the two DBRs is one wavelength long. The VCSEL can be made with much shorter cavity length than other structures, and as a result of Eq. (8–16) the laser modes are widely separated in wavelength. Thus single-mode laser operation is more easily achieved with the VCSEL. Lasing can be achieved at very low current ($< 50~\mu A$) with this device.

8.4.5 Materials for Semiconductor Lasers

We have discussed the properties of the junction laser largely in terms of GaAs and AlGaAs. However, as discussed in Section 8.2.2, the InGaAsP/InP system is particularly well suited for the type of lasers used in fiber optic communication systems. Lattice matching (Section 1.4.1) is important in creating heterostructures by epitaxial growth. The fact that the AlGaAs band gap can be varied by choice of composition on the column III sublattice allows the formation of barriers and confining layers such as those shown in Section 8.4.4. The quaternary alloy InGaAsP is particularly versatile in the fabrication of laser diodes, allowing considerable choice of wavelength and flexibility in lattice matching. By choice of composition, lasers can be made in the infrared range 1.3–1.55 μm required for fiber optics. Since four components can be varied in choosing an alloy composition, InGaAsP allows simultaneous choice of energy gap (and therefore emission wavelength) and lattice constant (for lattice matched growth on convenient substrates). In many applications, however, other wavelength ranges are required for laser output. For example, the use of lasers in pollution diagnostics requires wavelengths farther in the infrared than are available from InGaAsP and AlGaAs. In this application the ternary alloy PbSnTe provides laser output wavelengths from about 7 μm to more than 30 μm at low temperatures, depending on the material composition. For intermediate wavelengths, the InGaSb system can be used.

Materials chosen for the fabrication of semiconductor lasers must be efficient light emitters and also be amenable to the formation of p-n junctions and in most cases the formation of heterojunction barriers. These requirements eliminate some materials from practical use in laser diodes. For example, semiconductors with indirect band gaps are not sufficiently efficient light emitters for practical laser fabrication. The II–VI compounds, on the other hand, are generally very efficient at emitting light but junctions are difficult to form. By modern crystal growth techniques such as MBE and MOVPE it is possible to grow junctions in ZnS, ZnSe, ZnTe, and alloys of these materials, using N as the acceptor. Lasers can be made in these materials which emit in the green and blue-green regions of the spectrum.

In recent years much progress has been made in the growth of large bandgap semiconductors using GaN, and its alloys with InN and AlN. The InAlGaN system has direct bandgaps over the entire alloy composition range, and hence offers very efficient light emission. Bandgaps range from about 2eV for InN, to 3.4eV for GaN and 5 eV for AlN. This covers the wavelength range from about 620 nm to about 248 nm, which is from blue to UV. The resurgence of interest in this field was triggered by the work of Nakamura at Nichia Corporation in Japan who demonstrated very high efficiency blue light emitting diodes (LEDs) in GaN.

Two of the problems which had stymied progress in this field since pioneering work by Pankove in the 1970s was the absence of a suitable substrate having sufficient lattice match with GaN, and the inability to achieve p-type doping in this semiconductor. GaN bulk crystals cannot be grown easily because of the high vapor pressure of the nitrogen-bearing precursor (generally ammonia). This requires growth at high temperature and pressure. This precludes using bulk GaN wafers as substrates for epitaxial growth. However, epitaxial layers can be grown on other substrates with reasonable success, in spite of the lattice mismatch.

GaN exists in the cubic zincblende form (which is the preferred structure) as well as the hexagonal wurtzite form. It was demonstrated recently that cubic GaN could be grown heteroepitaxially on sapphire, even though it is not lattice matched to GaN. In fact, sapphire does not even have a cubic crystal structure—it is hexagonal. The lattice constant of GaN is about 4.5 Å, while that of sapphire is 4.8 Å, which is a huge lattice mismatch. Contrary to what would normally be expected, however, high quality epitaxial GaN films can be grown on sapphire by MOCVD using ammonia and tri-methyl gallium as the precursors. One possible reason for the high quality of the films, as evidence by blue LEDs and short wavelength lasers fabricated in these nitrides, is that these large-bandgap semiconductors have very high chemical bond strengths. This apparently precludes the easy propagation of dislocation defects from the heterointerface to the active part of the devices, where they would form traps and kill optical efficiency. Yet another lattice-mismatched substrate that has been successfully used for these nitride semiconductors is SiC.

The second breakthrough required in the nitrides was the ability to achieve high p-type doping so that p-n junctions could be formed. It has been demonstrated that Mg (which is a column II element) doping of MOCVD films, followed by high temperature annealing can be used to achieve high acceptor concentrations in these systems.

Why is there so much interest in short wavelength emitters such as blue LEDs and semiconductor lasers? As discussed in Section 8.2.1, high-efficiency red, green and yellow-green LEDs have existed for a long time in the GaAsP system, using concepts such as N isoelectronic doping. It has been a major goal of the optoelectronics community to achieve high efficiency blue emitters because, along with red and green, blue completes the list of three primary additive colors. In fact, blue LEDs made in GaN have been combined with the other color LEDs to form very intense white light sources with luminous efficiencies exceeding those of conventional light bulbs. Arrays of red, green and blue emitters can be used in outdoor displays and TV screens. Red, yellow and green LEDs are candidates for traffic lights because they have much higher reliability and lifetime than conventional light bulbs, and save energy.

Short wavelength emitters such as UV/blue semiconductor lasers are important for storage applications such as digital versatile discs (DVDs), which are higher density versions of compact discs (CDs). The storage density on these discs is inversely proportional to the square of the laser wavelength that is used to read the information. Thus reducing the laser wavelength by a factor of two leads to a four-fold increase of storage density. Such increased storage capacity opens up entirely new applications for DVDs that were not possible previously with conventional CDs, for example, the storage of full-length movies. A recent example of success in this rapidly progressing field is a 417 nm semiconductor laser made with InGaN multi–quantum-well heterostructures.

PROBLEMS

8.1 For the p-i-n photodiode of Fig. 8–7, (a) explain why this detector does not have gain; (b) explain how making the device more sensitive to low light levels degrades its speed; (c) if this device is to be used to detect light with $\lambda = 0.6$ μm, what material would you use and what substrate would you grow it on?

8.2 A Si solar cell 2 cm × 2 cm with $I_{th} = 32$ nA has an optical generation rate of 10^{18} EHP/cm^3-s within $L_p = L_n = 2$ μm of the junction. If the depletion width is 1 μm, calculate the short-circuit current and the open-circuit voltage for this cell.

8.3 A Si solar cell with dark saturation current I_{th} of 5 nA is illuminated such that the short-circuit current is 200 mA. Plot the I–V curve for the cell as in Fig. 8–6 (remember that I is negative but is plotted positive as I_r).

8.4 A major problem with solar cells is internal resistance, generally in the thin region at the surface, which must be only partially contacted, as in Fig. 8–5. Assume that the cell of Prob. 8.3 has a series resistance of 1 Ω, so that the cell voltage is reduced by the IR drop. Replot the I–V curve for this case and compare with the cell of Prob. 8.3.

8.5 Show schematically and discuss how several semiconductor materials might be used together to obtain a more efficient solar cell.

8.6 Assume that a photoconductor in the shape of a bar of length L and area A has a constant voltage V applied, and it is illuminated such that g_{op} EHP/cm³-s are generated uniformly throughout. If $\mu_n \gg \mu_p$, we can assume the optically induced change in current ΔI is dominated by the mobility μ_n and lifetime τ_n for electrons. Show that $\Delta I = qALg_{op}\tau_n/\tau_t$ for this photoconductor, where τ_t is the transit time of electrons drifting down the length of the bar.

8.7 What composition x of $Al_xGa_{1-x}As$ would produce red light emission at 680 nm? What composition of $GaAs_{1-x}P_x$? $In_xGa_{1-x}P$?

8.8 (a) Why must a solar cell be operated in the fourth quadrant of the junction I–V characteristic?

(b) What is the advantage of a quarternary alloy in fabricating LEDs for fiber optics?

(c) Why is a reverse-biased GaAs p-n junction not a good photodetector for light of $\lambda = 1$ μm?

8.9 For steady state optical excitation, we can write the hole diffusion equation as

$$D_p\frac{d^2\delta p}{dx^2} = \frac{\delta p}{\tau_p} - g_{op}$$

Assume that a long p⁺-n diode is uniformly illuminated by an optical signal, resuting in g_{op} EHP/cm³-s.

(a) Show that the excess hole distribution in the n region is

$$\delta p(x_n) = \left[p_n(e^{qVkT} - 1) - g_{op}\frac{L_p^2}{D_p}\right]e^{-x_n/L_p} + \frac{g_{op}L_p^2}{D_p}$$

(b) Calculate the hole diffusion current $I_p(x_n)$ and evaluate it at $x_n = 0$. Compare the result with Eq. (8–2) evaluated for a p⁺-n junction.

8.10 A Si solar cell has a short-circuit current of 100 mA and an open-circuit voltage of 0.8 V under full solar illumination. The fill factor is 0.7. What is the maximum power delivered to a load by this cell?

8.11 The maximum power delivered by a solar cell can be found by maximizing the I–V product.

(a) Show that maximizing the power leads to the expression

$$\left(1 + \frac{q}{kT}V_{mp}\right)e^{qV_{mp}/kT} = 1 + \frac{I_{sc}}{I_{th}}$$

where V_{mp} is the voltage for maximum power, I_{sc} is the magnitude of the short circuit current, and I_{th} is the thermally induced reverse saturation current.

(b) Write this equation in the form $\ln x = C - x$ for the case $I_{sc} \gg I_{th}$, and $V_{mp} = \gg kT/q$.

(c) Assume a Si solar cell with a dark saturation current I_{th} of 1.5 nA is illuminated such that the short-circuit current is $I_{sc} = 100$ mA. Use a graphical solution to obtain the voltage V_{mp} at maximum delivered power.

(d) What is the maximum power output of the cell at this illumination?

8.12 For a solar cell, Eq. (8–2) can be rewritten

$$V = \frac{kT}{q} \ln\left(1 + \frac{I_{sc} + I}{I_{th}} \right)$$

Given the cell parameters of Prob. 8.11, plot the I–V curve as in Fig. 8–6 and draw the maximum power rectangle. Remember that I is a negative number but is ploted positive as I_r in the figure. I_{th} and I_{sc} are positive magnitudes in the equation.

8.13 Solar cells are severely degraded by unwanted series resistance. For the cell described in Prob. 8–4, include a series resistance R, which reduces the cell voltage by the amount IR. Calculate and plot the fill factor for a series resistance R from 0 to 5 Ω, and comment on the effect of R on cell efficiency.

8.14 Based upon Fig. 1–13, what ternary alloy, composition, and binary substrate can be used for an LED at the 1.55-μm optical fiber window? What type of epitaxial layer/substrate combination would you use for an LED with emission at 1.3 μm?

8.15 The degenerate occupation of bands shown in Fig. 8–19 helps maintain the laser requirement that emission must overcome absorption. Explain how the degeneracy prevents band-to-band absorption at the emission wavelength.

8.16 Assume that the system described by Eq. (8–7) is in thermal equilibrium at an extremely high temperature such that the energy density $\rho(v_{12})$ is essentially infinite. Show that $B_{12} = B_{21}$.

8.17 The system described by Eq. (8–7) interacts with a blackbody radiation field whose energy density per unit frequency at v_{12} is

$$\rho(v_{12}) = \frac{8\pi h v_{12}^3}{c^3}[e^{hv_{12}/kT} - 1]^{-1}$$

from Planck's radiation law. Given the result of Prob. 8.16, find the value of the ratio A_{21}/B_{12}.

8.18 Assuming equal electron and hole concentrations and band-to-band transitions, calculate the minimum carrier concentration $n = p$ for population inversion in GaAs at 300 K. The intrinsic carrier concentration in GaAs is about 10^6 cm^{-3}.

Agrawal, G. P., and N. K. Dutta. *Long-Wavelength Semiconductor Lasers.* New York: Van Nostrand Reinhold, 1986.

Baughmann, M. G. D., J. C. Wright, A. B. Ellis, T. Kuech, and G. C. Lisensky. "Diode Lasers." *Journal of Chemical Education* 69 (February 1992): 89–95.

Bhattacharya, P. *Semiconductor Optoelectronic Devices.* Englewood Cliffs, NJ: Prentice Hall, 1994.

Buckley, D. N. "The Light Fantastic: Materials and Processing Technologies for Photonics." *The Electrochemical Society Interface* 1 (Winter 1992): 41+.

Campbell, J.C., A.G. Dentai, W.S. Holden and B.L. Kasper. "High Performance Avalanche Photodiode with Separate Absorption, Grading and Multiplication Regions." *Electronics Letters*,19 (1983): 818+.

Casey, Jr., H. C., and M. B. Panish. *Heterostructure Lasers: Part A. Fundamental Principles.* New York: Academic Press, 1978.

Cheo, P. K. *Fiber Optics and Optoelectronics,* 2nd ed. Englewood Cliffs, NJ: Prentice Hall, 1990.

Craford, M.G. "LEDs Challenge the Incandescents." *IEEE Circuits and Devices* 8(5) (1992): 24+.

Crow, J.D. "Optical Interconnects Speed Interprocessor Nets." *IEEE Circuits and Devices* 7 (March 1991): 20–5.

Dagenais, M., R. F. Leheny, H. Temkin, and P. Bhattacharya. "Applications and Challenges of OEICs" *Journal of Lightwave Technology* 8 (June 1990): 846–62.

Das, P. *Lasers and Optical Engineering.* New York: Springer-Verlag, 1991.

Denbaars, S.P. "Gallium Nitride Based Materials for Blue to Ultraviolet Optoelectronic Devices." *Proc. IEEE,* 85 (11) (November 1997): 1740–1749.

Desurvire, E. "The Golden Age of Optical Fiber Amplifiers." *Physics Today* 47 (January 1994): 20–7.

Dupuis, R. D. "AlGaAs-GaAs Lasers Grown by MOCVD—A Review." *Journal of Crystal Growth* 55 (October 1981): 213–22.

Han, J., L. He, R. L. Gunshor, and A. V. Nurmikko. "Blue/Green Lasers Focus on the Market." *IEEE Circuits and Devices* 10 (March 1994): 18–23.

Hecht, J. "Diode-Laser Performance Rises as Structures Shrink." *Laser Focus World* 28 (May 1992): 127–8+.

Hecht, J. "Laser Action in Fibers Promises a Revolution in Communications." *Laser Focus World* 29 (February 1993): 75–6+.

Hecht, J. "Semiconductor Lasers Shine Out." *Electronics and Wireless World* 97 (April 1992): 302–5.

Hummel, R. E. *Electronic Properties of Materials,* 2nd ed., Berlin: Springer-Verlag, 1993.

Ikegami, T. M. And M. Nakahara. "Optical Fiber Amplifiers." *Proceedings of the SPIE* 1362, pt. 1 (1991): 350–60.

Jahns, J., and S. H. Lee, eds. *Optical Computing Hardware.* Boston: Academic Press, 1993.

Jewell, J. L., and G. R. Olbright. "Surface-Emitting Lasers Emerge from the Laboratory." *Laser Focus World* 28 (May 1992): 217–23.

Jungbluth, E. D. "Crystal Growth Methods Shape Communications Lasers." *Laser Focus World* 29 (February 1993): 61–72.

Leheny, R. F. "Optoelectronic Integration: A Technology for Future Telecommunication Systems." *IEEE Circuits and Devices* 5 (May 1989): 38–41.

Neamen, D. A. *Semiconductor Physics and Devices: Basic Principles.* Homewood, IL: Irwin, 1992.

Palais, J. C. *Fiber Optic Communication*, 3rd ed. Englewood Cliffs, NJ: Prentice Hall, 1992.

Pankove, J. I. *Optical Processes in Semiconductors.* Englewood Cliffs, NJ: Prentice Hall, 1971.

Pollack, M. A. "Advances in Materials for Optoelectronic and Photonic Integrated Circuits." *Materials Science & Engineering B* B6 (July 1990): 233–45.

Saleh, B. E. A., and M. C. Teich. *Fundamentals of Photonics.* New York: Wiley, 1991.

Singh, J. *Semiconductor Devices.* New York: McGraw-Hill, 1994.

Verdeyen, J. T. *Laser Electronics,* 3rd ed. Englewood Cliffs, NJ: Prentice Hall, 1994.

Weisbuch, C., and B. Vinter. *Quantum Semiconductor Structures.* Boston: Academic Press, 1991.

Yamamoto, Y., and R. E. Slusher. "Optical Processes in Microcavities." *Physics Today* 46 (June 1993); 66–73.

Yariv, A. *Optical Electronics,* 3rd ed. New York: Holt, Rinehart, and Winston, 1985.

Zory, P. S., Jr. *Quantum Well Lasers.* Boston: Academic Press, 1993.

Chapter 9
Integrated Circuits

Just as the transistor revolutionized electronics by offering more flexibility, convenience, and reliability than the vacuum tube, the integrated circuit enables new applications for electronics that were not possible with discrete devices. Integration allows complex circuits consisting of many thousands of transistors, diodes, resistors, and capacitors to be included in a chip of semiconductor. This means that sophisticated circuitry can be miniaturized for use in space vehicles, in large-scale computers, and in other applications where a large collection of discrete components would be impractical. In addition to offering the advantages of miniaturization, the simultaneous fabrication of many ICs on a single Si wafer greatly reduces the cost and increases the reliability of each of the finished circuits. Certainly discrete components have played an important role in the development of electronic circuits; however, most circuits are now fabricated on the Si chip rather than with a collection of individual components. Therefore, the traditional distinctions between the roles of circuit and system designers do not apply to IC development.

In this chapter we shall discuss various types of ICs and the fabrication steps used in their production. We shall investigate techniques for building large numbers of transistors, capacitors, and resistors on a single chip of Si, as well as the interconnection, contacting, and packaging of these circuits in usable form. All the processing techniques discussed here are very basic and general. There would be no purpose in attempting a comprehensive review of all the subtleties of device fabrication in a book of this type. In fact, the only way to keep up with such an expanding field is to study the current literature. Many good reviews are suggested in the reading list at the end of this chapter; more important, current issues of those periodicals cited can be consulted for up-to-date information regarding IC technology. Having the background of this chapter, one should be able to read the current literature and thereby keep abreast of the present trends in this very important field of electronics.

In this section we provide an overview of the nature of integrated circuits and the motivation for using them. It is important to realize the reasons, both technical and economic, for the dramatic rise of ICs to their present role in

**9.1
BACKGROUND**

electronics. We shall discuss several main types of ICs and point out some of the applications of each. More specific fabrication techniques will be presented in later sections.

9.1.1 Advantages of Integration

It might appear that building complicated circuits, involving many interconnected components on a single Si substrate, would be risky both technically and economically. In fact, however, modern techniques allow this to be done reliably and relatively inexpensively; in most cases an entire circuit on a Si chip can be produced much more inexpensively and with greater reliability than a similar circuit built up from individual components. The basic reason is that many identical circuits can be built simultaneously on a single Si wafer (Fig. 9–1); this process is called *batch fabrication*. Although the processing steps for the wafer are complex and expensive, the large number of resulting integrated circuits makes the ultimate cost of each fairly low. Furthermore, the processing steps are essentially the same for a circuit containing millions of transistors as for a simpler circuit. This drives the IC industry to build increasingly complex circuits and systems on each chip, and use larger Si wafers (e.g., 8-inch diameter). As a result, the number of components in each circuit increases without a proportional increase in the ultimate cost of the system. The implications of this principle are tremendous for circuit designers; it greatly increases the flexibility of design criteria. Unlike circuits with individual transistors and other components wired together or placed on a circuit board, ICs allow many "extra" components to be included with-

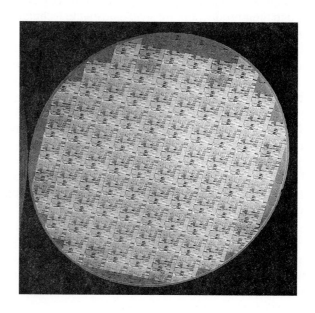

Figure 9–1
A 200 mm diameter (about 8 inch) wafer of integrated circuits. The circuits are tested on the wafer, and then sawed apart into individual chips for mounting into packages. (Photograph courtesy of IBM Corp.)

out greatly raising the cost of the final product. Reliability is also improved since all devices and interconnections are made on a single rigid substrate, greatly minimizing failures due to the soldered interconnections of discrete component circuits.

The advantages of ICs in terms of miniaturization are obvious. Since many circuit functions can be packed into a small space, complex electronic equipment can be employed in many applications where weight and space are critical, such as in aircraft or space vehicles. In large-scale computers it is now possible not only to reduce the size of the overall unit but also to facilitate maintenance by allowing for the replacement of entire circuits quickly and easily. Applications of ICs are pervasive in such consumer products as watches, calculators, automobiles, telephones, television, and appliances. Miniaturization and the cost reduction provided by ICs mean that we all have increasingly more sophisticated electronics at our disposal.

Some of the most important advantages of miniaturization pertain to response time and the speed of signal transfer between circuits. For example, in high-frequency circuits it is necessary to keep the separation of various components small to reduce time delay of signals. Similarly, in very high speed computers it is important that the various logic and information storage circuits be placed close together. Since electrical signals are ultimately limited by the speed of light (about 1 ft/ns), physical separation of the circuits can be an important limitation. As we shall see in Section 9.5, *large-scale integration* of many circuits on a Si chip has led to major reductions in computer size, thereby tremendously increasing speed and function density. In addition to decreasing the signal transfer time, integration can reduce parasitic capacitance and inductance between circuits. Reduction of these parasitics can provide significant improvement in the operating speed of the system.

We have discussed several advantages of reducing the size of each unit in the batch fabrication process, such as miniaturization, high-frequency and switching speed improvements, and cost reduction due to the large number of circuits fabricated on a single wafer. Another important advantage has to do with the percentage of usable devices (often called the *yield*) which results from batch fabrication. Faulty devices usually occur because of some defect in the Si wafer or in the fabrication steps. Defects in the Si can occur because of lattice imperfections and strains introduced in the crystal growth, cutting, and handling of the wafers. Usually such defects are extremely small, but their presence can ruin devices built on or around them. Reducing the size of each device greatly increases the chance for a given device to be free of such defects. The same is true for fabrication defects, such as the presence of a dust particle on a photolithographic mask. For example, a lattice defect or dust particle $\frac{1}{2}\mu m$ in diameter can easily ruin a circuit which includes the damaged area. If a fairly large circuit is built around the defect it will be faulty; however, if the device size is reduced so that four circuits occupy the same area on the wafer, chances are good that only the one containing the defect will be faulty and the other three will be good. Therefore, the percentage yield of usable circuits increases over a certain range of decreasing chip area. There is an optimum area for each circuit, above which defects are needlessly included and below which the elements are spaced too closely for reliable fabrication.

9.1.2 Types of Integrated Circuits

There are several ways of categorizing ICs as to their use and method of fabrication. The most common categories are *linear* or *digital* according to application, and *monolithic* or *hybrid* according to fabrication.

A linear IC is one that performs amplification or other essentially linear operations on signals. Examples of linear circuits are simple amplifiers, operational amplifiers, and analog communications circuits. Digital circuits involve logic and memory, for applications in computers, calculators, microprocessors, and the like. By far the greatest volume of ICs has been in the digital field, since large numbers of such circuits are required. Since digital circuits generally require only "on-off" operation of transistors, the design requirements for integrated digital circuits are often less stringent than for linear circuits. Although transistors can be fabricated as easily in integrated form as in discrete form, passive elements (resistors and capacitors) are usually more difficult to produce to close tolerances in ICs.

9.1.3 Monolithic and Hybrid Circuits

Integrated circuits that are included entirely on a single chip of semiconductor (usually Si) are called *monolithic* circuits (Fig. 9–1). The word monolithic literally means "one stone" and implies that the entire circuit is contained in a single piece of semiconductor. Any additions to the semiconductor sample, such as insulating layers and metallization patterns, are intimately bonded to the surface of the chip. A *hybrid* circuit may contain one or more monolithic circuits or individual transistors bonded to an insulating substrate with resistors, capacitors, or other circuit elements, with appropriate interconnections (Fig. 9–2). Monolithic circuits have the advantage that all components are contained in a single rigid structure which can be batch fabricated; that is, hundreds of identical circuits can be built simultaneously on a Si wafer. On the other hand, hybrid circuits offer excellent isolation between components and allow the use of more precise resistors and capacitors. Furthermore, hybrid circuits are often less expensive to build in small numbers.

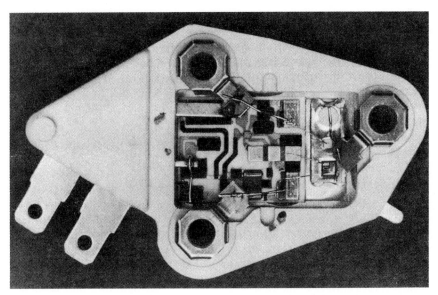

Figure 9–2
A hybrid circuit employing thick-film printing. This automobile voltage regulator circuit contains thick-film resistors and thermistors, a ceramic chip capacitor, and monolithic Si chips mounted on an insulating substrate. (Photograph courtesy of Delco Electronics Corp.)

When resistors and capacitors are made external to the monolithic Si chip, basically two types of technology are used; the passive elements are fabricated and interconnected by *thick-film* or *thin-film* processes. Although the dividing line between thin and thick films is not precise, they are fairly well separated in application to ICs: "thin" films are typically 0.1 to 0.5 μm, and "thick" films are about 25 μm.

The processing steps for the two hybrid techniques are quite different. In thick-film circuits the resistors and interconnection patterns are "printed" on a ceramic substrate (Fig. 9–2) by silk-screen or similar process. Conductive and resistive pastes consisting of metal powders in organic binders are printed on the substrate and cured in an oven. One advantage of this process is that resistors can be made below the rated values and then trimmed by abrasion, or by selective evaporation using a pulsed laser. These corrections can be made quickly with automated procedures while the resistance values are under test. Small ceramic chip capacitors can be bonded into place in the interconnection pattern, along with monolithic circuits or individual transistors.

Thin-film technology allows for greater precision and miniaturization, and is generally preferred when space is an important limitation. Thin-film interconnection patterns and resistors can be vacuum deposited on a glass or glazed ceramic substrate. The resistive films are usually made of tantalum or other resistive metal, and the conductors are often aluminum or gold. In general, the resistive materials must be deposited by sputtering. Pattern definition for the resistors and conductor paths can be achieved by depositing the films through metal shields which contain appropriate apertures. Better definition is obtained by metallizing the entire substrate, or large parts of it, and using photolithographic methods to remove the metal except in the desired pattern. Capacitors can be fabricated by thin-film techniques by depositing an insulating layer between two metal films or by oxidizing the surface of one film and then depositing a second film on top.

9.2 EVOLUTION OF INTEGRATED CIRCUITS

The IC was invented in February 1959 by Jack Kilby of Texas Instruments. The planar version of the IC was developed independently by Robert Noyce at Fairchild in July 1959. Since then the evolution of this technology has been extremely fast-paced. One way to gauge the progress of the field is to look at the complexity of ICs as a function of time. Figure 9–3a shows the number of transistors used in MOS *microprocessor* IC chips as a function of time. It is amazing that on this semi-log plot, where we have plotted log of the component count as a function of time, we get a straight line, indicating that there has been an exponential growth in the complexity of chips over three decades. The component count has roughly doubled every 18 months, as was noted early by Gordon Moore of Intel corporation. This regular doubling has become known as *Moore's law*.

The history of ICs can be described in terms of different eras, depending on the component count. Small Scale Integration (SSI) refers to the integration of $1–10^2$ devices, Medium Scale Integration (MSI) between $10^2–10^3$ devices, Large Scale Integration (LSI) between $10^3–10^5$ devices, Very Large Scale Integration (VLSI) between $10^5–10^6$ devices, and now Ultra Large Scale Integration (ULSI), where component count is between $10^6–10^9$. Of course, these boundaries are somewhat fuzzy. The next generation has been dubbed as Giga-Scale Integration. Wags have suggested that after that we will have RLSI or Ridiculously Large Scale Integration.

The main factor that has enabled this increase of complexity is the ability to shrink or scale devices. Typical dimensions or feature sizes (generally the MOSFET channel lengths) of state-of-the-art *dynamic random access memories (DRAMs)* at different times are also shown as a semi-log plot in Fig. 9–3b. Once again, we see a straight line, reflecting an exponential decrease of the typical feature sizes with time over three decades. Clearly, one can pack a larger number of components with greater functionality on an IC if they are smaller. As discussed in Section 6.5.9, scaling also has other advantages in terms of faster ICs which consume less power.

While scaling represents an opportunity, it also presents tremendous technological challenges. The most notable among these challenges lie in lithography and etching, as discussed in Section 5.1. However, since scaling of horizontal dimensions also requires scaling of vertical geometries as discussed in Section 6.5.9, there are also tremendous challenges in terms of doping, gate dielectrics and metallization. Small features and large chips also require device fabrication in extremely clean environments. Particles which may not have caused yield problems in a 1 μm IC technology can have catastrophic effects for a 0.25 μm process. This requires purer chemicals, cleaner equipment, and more stringent clean rooms. In fact, the levels of cleanliness required bypassed the best surgical operating rooms early in the evolution shown in Fig. 9–3. The cleanliness of these facilities is designated by the *Class* of the clean room. For instance, a Class 1 clean room, which is state-of-the-art in 2000, has less than 1 particle of size 0.2 μm or larger per cubic foot. There are more of the smaller particles, and fewer of the larger ones. Obviously, the lower the Class of a clean room, the better it is. A Class 1 clean room is much cleaner than a Class 100 fabrication facility or "fab." As one might expect, such high levels of cleanliness come with a hefty price tag. A state-of-the-art fab in 2000 comes equipped with a price tag of about two billion dollars.

In spite of the costs, the economic payoff for ULSI is tremendous. Just for calibration, let us examine some economic statistics at the dawn of the third millennium. The total annual economic output of all the countries in the world, or the so-called Gross World Product (GWP), is about 35 trillion US dollars. The US Gross National Product is about 8 trillion dollars, or about a quarter of the GWP. The worldwide IC industry is about 150 billion dollars, and the entire worldwide electronics industry in which these ICs participate is about 1 trillion dollars. As a single industry, electronics is one of the biggest in terms of the dollar amount. It has surpassed, for example, automobiles (worldwide sales of about 50 million cars annually) and petrochemicals. About 100–200 million personal computers are sold annually worldwide.

Perhaps even more dramatic than these raw economic numbers is the growth rate of these markets. If one were to plot IC sales as a function of time, one again finds a more or less exponential increase of sales with time over three decades. Of great importance to the consumer, the cost per electronic function has dropped dramatically over the same period of time. For example, the cost per bit of semiconductor memory (DRAM) has dropped from about 1 cent/bit in 1970 to about 10^{-4} cents per bit today, a cost improvement of four orders of magnitude in 30 years. There are no parallels in any other industry for this consistent improvement in functionality with such lowered cost.

While ICs started with bipolar processes in the 1960s, they were gradually supplanted by MOS and then CMOS devices, for reasons discussed in Chapters 6 and 7. Currently, about 88 percent of the IC market is MOS-based, and about 8 percent BJT-based. Optoelectronic devices based on compound semiconductors are still a relatively small component of the semiconductor market (about 4 percent), but are expected to grow in the future. Of the MOS ICs, the bulk are digital ICs. Of the entire semiconductor industry, only about 14 percent are analog ICs. Semiconductor memories such as DRAMs, SRAMs and non-volatile flash memories make up approximately 25 percent of the market, microprocessors about 25 percent, and other application-specific ICs (ASICs) about 20 percent.

Figure 9–3
Moore's law for integrated circuits: (a) Exponential increase of transistor count as a function of time for different generations of microprocessors; (b) exponential decrease of minimum transistor gate lengths with time, for different generations of dynamic random access memories (16 kb to 256 Mb DRAMs). For reference, sizes of blood cells and bacteria are shown on the μm scale.

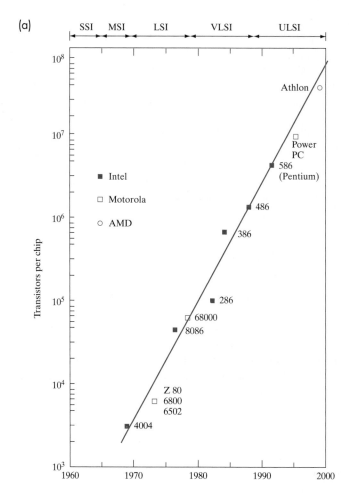

(a)

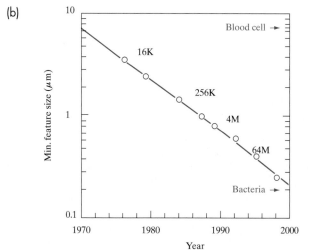

(b)

Now we shall consider the various elements that make up an integrated circuit, and some of the steps in their fabrication. The basic elements are fairly easy to name—transistors, resistors, capacitors, and some form of interconnection. There are some elements in integrated circuits, however, which do not have simple counterparts in discrete devices. We shall consider one of these, charge transfer devices, in Section 9.4. Discussion of fabrication technology is difficult in a book of this type, since device fabrication engineers seem to make changes faster than typesetters do! Since this important and fascinating field is changing so rapidly, the reader should obtain a basic understanding of device design and processing from this discussion and then search out new innovations in the current literature.

9.3.1 CMOS Process Integration

A particularly useful device for digital applications is a combination of n-channel and p-channel MOS transistors on adjacent regions of the chip. This *complementary MOS* (commonly called *CMOS*) combination is illustrated in the basic inverter circuit of Fig. 9–4a. In this circuit the drains of the two transistors are connected together and form the output, while the input terminal is the common connection to the transistor gates. The p-channel device has a negative threshold voltage, and the n-channel transistor has a positive threshold voltage. Therefore, a zero voltage input ($V_{in} = 0$) gives

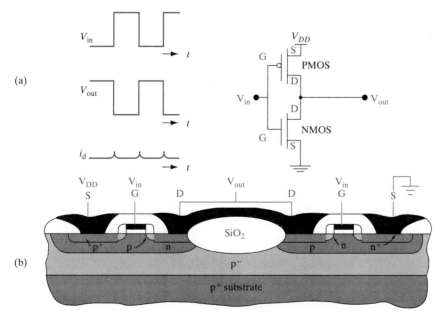

Figure 9–4
Complementary MOS structure: (a) CMOS inverter; (b) formation of p-channel and n-channel devices together.

zero gate voltage for the n-channel device, but the voltage between the gate and source of the p-channel device is $-V$. Thus the p-channel device is on, the n-channel device is off, and the full voltage V is measured at V_{out} (i.e., V appears across the nonconducting n-channel transistor). Alternatively, a positive value of V_{in} turns the n-channel transistor on, and the p-channel off. The output voltage measured across the "on" n-channel device is essentially zero. Thus, the circuit operates as an inverter—with a binary "1" at the input, the output is in the "0" state, whereas a "0" input produces a "1" output. The beauty of this circuit is that one of the devices is turned off for either condition. Since the devices are connected in series, no drain current flows, except for a small charging current during the switching process from one state to the other. Since the CMOS inverter uses ultra little power, it is particularly useful in applications such as electronic watch circuits which depend on very low power consumption. CMOS is also advantageous in ultra large scale integrated circuits (Section 9.5), since even small power dissipation in each transistor becomes a problem when millions of them are integrated on a chip.

The device technology for achieving CMOS circuits consists mainly in arranging for both n- and p-channel devices with similar threshold voltages on the same chip. To achieve this goal, a diffusion or implantation must be performed in certain areas to obtain n and p regions for the fabrication of each type of device. These regions are called *tubs, tanks,* or *wells* (Fig. 9–5). The critical parameter of the tub is its net doping concentration, which must be closely controlled by ion implantation. With the tub in place, source and drain implants are performed to make the n-channel and p-channel transistors. Matching of the two transistors is achieved by control of the surface doping in the tub and by threshold adjustment of both transistors by ion implantation.

Including bipolar transistors in the basic CMOS technology allows flexibility in circuit design, particularly for providing drive currents. The combination of bipolar and CMOS (called BiCMOS) provides circuits with increased speed.

Attention must be paid in CMOS designs to the fact that combining n-channel and p-channel devices in close proximity can lead to inadvertent (*parasitic*) bipolar structures. In fact, a p-n-p-n structure can be found in Fig. 9–4b, which can serve as an inefficient but troublesome *thyristor* (see Ch. 11). Under certain biasing conditions the p-n-p part of the structure can supply base current to the n-p-n structure, causing a large current to flow. This process, called *latchup*, can be a serious problem in CMOS circuits. Several methods have been used to eliminate the latchup problem, including using both n-type and p-type tubs, separated by trench isolation (Fig. 9–6). The use of two separate tubs (wells) also allows independent control of threshold voltages in both types of transistor.

We can illustrate most of the common fabrication steps for MOS integrated circuits by studying the flow of a twin well Self-Aligned siLICIDE (SALICIDE) CMOS process. This process is particularly important because most high-performance digital ICs, including microprocessors, memories and application specific ICs (ASICs), are made basically in this way. In order to make enhancement-mode n-channel devices, we need a p-type substrate, and vice versa. Since CMOS requires both, we must start either with an n-type or a p-type wafer and then make selected regions of the substrate have opposite doping by forming wells. For example, Fig. 9–5a shows a lightly doped p-epitaxial layer on a p$^+$-substrate. We can make n-channel devices in this layer.

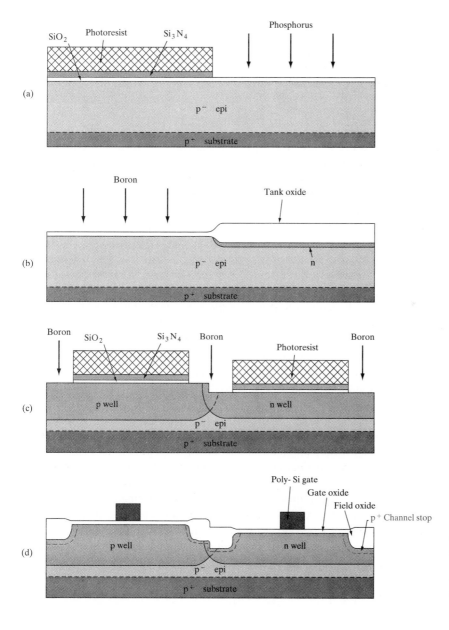

Figure 9–5
Self-aligned twin well process: (a) n-well formation using P donor implant and a photoresist mask; (b) p-well formation using B acceptor implant. A thick (~200 nm) "tank" oxide layer is grown wherever the silicon nitride–oxide stack is etched off, and the tank oxide is used to block the B implant in the n-wells in a self-aligned manner; (c) isolation pattern for field transistors showing B channel stop implant using photoresist mask; (d) local oxidation of silicon wherever nitride mask is removed, leading to thick LOCOS field oxide.

By implanting n-wells wherever needed, we can make p-channel devices also. This is an n-well CMOS process. Alternatively, if we start with an n-substrate and make p-wells in certain regions, we have a p-well CMOS process. For optimal device performance, however, it is usually desirable to separately implant both the n- and the p-well regions, which is called *twin-well* CMOS. The rationale for this can be appreciated if we keep in mind that for a state-of-the-art IC, typical doping levels are $\sim 10^{18}$ cm^{-3} and junction depths are ~ 1 μm in these wells. The doping levels have to be high enough to prevent punchthrough breakdown due to drain-induced barrier lowering (DIBL) in the MOSFETs, but low enough to keep the threshold voltages acceptably small. If we choose to use the p-substrate for n-channel devices as in Fig. 9–5a, the p-type epitaxial layer must be doped to 10^{18} cm^{-3}, and the implanted n-type layer must be achieved by counter-doping at a level of $\sim 2 \times 10^{18}$ cm^{-3}, resulting in a net n-type doping of $\sim 1 \times 10^{18}$ cm^{-3}, but a total doping in this region of $\sim 3 \times 10^{18}$ cm^{-3}. Such high levels of total doping are detrimental to carrier transport because they cause excessive ionized impurity scattering. Hence, for high performance ICs, the starting epitaxial doping level is generally very low ($\sim 10^{16}$ cm^{-3}). This layer is grown on a heavily doped substrate ($\sim 10^{19}$ cm^{-3}), to provide a highly conducting electrical ground plane. This helps with noise problems in ICs and helps minimize the problem of *latchup* by bypassing majority carriers (in this case holes) to the p$^+$ substrate.

To form the twin wells in a self-aligned fashion, we first grow thermally a "pad" oxide (~ 20 nm) on the Si substrate, followed by low pressure chemical vapor deposition (LPCVD) of silicon nitride (~ 20 nm). As shown in Fig. 9–5a this oxide–nitride stack is covered by photoresist, and a window is opened for the n-well. Reactive ion etching (RIE) is then used to etch the oxide–nitride stack. Using the photoresist as an implant mask, we then do an n-type implant using phosphorus. Phosphorus is preferred to As for this purpose because P is lighter and has a higher projected range; also, P diffuses faster. This fast diffusion is needed to drive the dopants fairly deep into the substrate to form the n-well. After the implant, the photoresist is removed, and the patterned wafer is subjected to wet oxidation to grow a "tank" oxide (~ 200 nm). It may be noted in Fig. 9–5b that the tank oxidation process consumes Si from the substrate, and the resulting oxide swells up. In fact, for every micron of thermally-grown oxide, the oxidation consumes 0.44 μm of Si, resulting in a 2.2× volume expansion. The oxide does not grow in the regions that are protected by silicon nitride because nitride has the property that it blocks the diffusion of oxygen and water molecules (and thereby prevents oxidation of the Si substrate). The pad oxide that is used under the nitride has two roles: it minimizes the thermal-expansion mismatch and concomitant stress between silicon nitride and the substrate; it also prevents chemical bonding of the silicon nitride to the silicon substrate.

Using the tank oxide as a *self-aligned* implant mask (i.e., without actually having to do a separate photolithographic step), one does a p-type well implant using boron (Fig. 9–5b). The tank oxide must thus be much thicker than the projected range of the B. The concept of self-alignment is very important, and is a recurring theme in IC processing. It is simpler and cheaper to use self-alignment than a separate lithographic step. It also allows tighter packing density of the twin wells, because it is not required to account for lithographic misalignment during layout. The P and the B are then diffused into the substrate to a well depth of typically a micron by a drive-in diffusion at very high tempatures (~1000 °C) for several hours. After this diffusion, the silicon nitride–oxide stack and the tank oxide are etched away. Since the tank oxidation consumes Si from the substrate, etching it off leads to a step in the Si substrate delineating the n-well and p-well regions. This step is important in terms of alignment of subsequent reticles, and is shown in an exaggerated fashion in Fig. 9–5c.

Next, we form the isolation regions or the field transistors which guarantee that there will be no electrical cross-talk between adjacent transistors, unless they are interconnected intentionally (Fig. 9–5c). This is achieved by ensuring that the threshold voltage of any parasitic transistor that may form in the isolation regions is much higher than the power supply voltage on the chip, so that the parasitic channel can never turn on under operating conditions. From the threshold voltage expression (Eq. 6-38), we notice that V_T can be raised by increasing substrate doping and increasing gate oxide thickness. However, a problem with that approach is the subthreshold slope S (Eq. 6-66), which degrades with increasing substrate doping and gate oxide thickness. One needs to optimize both V_T and S such that the off-state leakage current in the field between transistors is sufficiently low at zero gate bias.

A stack of silicon dioxide–silicon nitride is photolithographically patterned as in Fig. 9–5c and subjected to RIE. A boron *"channel stop"* implant between the twin wells increases the acceptor doping and thus increases the threshold voltage in the p-well between the n-channel transistors (the *field* threshold). However, B will compensate the donor doping on the n-well side, and thus reduce the threshold in the n-well between p-channel devices. The B channel stop dose must thus be optimized to have acceptably high field thresholds in both types of wells.

After the channel stop implant, the photoresist is removed and the wafer with the patterned nitride–oxide stack shown in Fig. 9–5c (without photoresist) is subjected to wet oxidation to selectively grow a field oxide ~300 nm thick. The nitride layer blocks oxidation of the Si substrate in the regions where we plan to make the transistors. This procedure, where Si is oxidized to form SiO_2 in regions not protected by nitride, is called LOCal Oxidation of Silicon (LOCOS) (Sec. 6.4.1). In this case, LOCOS provides electrical isolation between the two transistors, as shown in Fig. 9–5d.

The volume expansion of 2.2× upon oxidation is an important issue because the selective oxidation occurs in narrow, confined regions. The compressive stress, if excessive, can cause dislocation defects in the substrate. Another issue is the lateral oxidation near the nitride mask edges, which causes the nitride mask to lift up near the edges, forming what is known as a *bird's beak* and causing a *lateral moat encroachment* of ~0.2 μm into each active region, thereby wasting precious Si real estate. There have been various modified LOCOS and other isolation schemes proposed to minimize this lateral encroachment. A notable example is *Shallow Trench Isolation* (STI) which involves using RIE to etch a shallow (~1 μm) trench or groove in the Si substrate after the isolation pattern, filling it up completely by deposition of a dielectric layer of SiO_2 and polysilicon by Low Pressure Chemical Vapor Deposition (LPCVD), and then using Chemical Mechanical Polishing (CMP) to planarize the structure (Fig. 9–6). This consumes less Si real estate compared to LOCOS, but gives superior isolation because the sharp corners at the bottom of the trench give rise to potential barriers that block leakage currents (the *corner* effect).

The pad oxide between the nitride and the Si surface minimizes the stress due to the nitride, and prevents bonding of the nitride to the Si, as mentioned above. Any residual nitride on the Si would retard subsequent gate

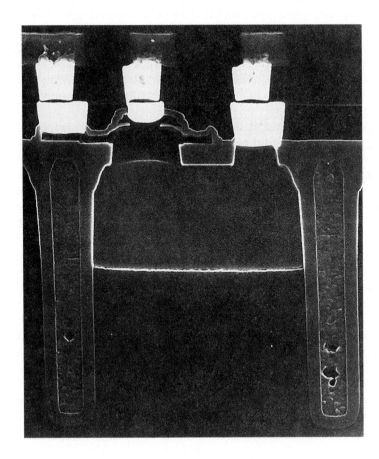

Figure 9–6
Trench isolation: A trench or groove is etched in the substrate using RIE, and re-filled with oxide and polysilicon, providing superior electrical isolation compared to LOCOS, using less Si real estate.

oxide formation, leading to weak spots in the gate region of the MOSFETs. This problem is known as the *white ribbon* effect or the *Kooi* effect, after the Dutch scientist who first identified it. The pad oxide mitigates this problem, but does not solve it completely. Therefore, very often a "sacrificial" or "dummy" oxide is grown to consume a layer of Si containing any residual nitride, and this oxide is wet etched prior to the growth of the actual gate oxide.

Next, an ultra-thin (~5–10 nm) gate oxide is grown on the substrate. Since the electrical quality of this oxide and its interface with the Si substrate is of paramount importance to the operation of the MOSFETs, dry oxidation is used for this step. It is common to incorporate some nitrogen at the Si–SiO$_2$ interface, forming oxy-nitrides which improve the interface quality in terms of hot electron effects. After the oxidation, it is immediately covered with LPCVD polysilicon in order to minimize contamination of the gate oxide. The polysilicon gate layer is doped very heavily (typically n$^+$ using a phosphorus dopant source, POCl$_3$ in a diffusion furnace) all the way to the polysilicon–oxide interface in order to make it behave electrically like a metal electrode. Alternatively, the LPCVD polysilicon film may also be *in situ* doped during the deposition itself by flowing in an appropriate dopant gas such as phosphine or diborane. Heavy doping of the gate material is very important, because otherwise a depletion layer can be formed in the polysilicon gate (the *poly depletion* effect). This could result in a depletion capacitance in series with the gate oxide capacitance, thereby reducing the overall gate capacitance and, therefore, the drive current (see Eq. 6-53). The high doping (~10^{20} cm^{-3}) in the polysilicon gate is also important for reducing the resistance of the gate and its *RC* time constant. The uniformly high doping in the polysilicon layer is facilitated by the presence of the grain boundary defects in the film, because diffusivity of dopants along grain boundaries is many orders of magnitude higher than in single crystal Si.

The doped polysilicon layer is then patterned to form the gates, and etched anisotropically by RIE to achieve vertical sidewalls. That is extremely important because this etched polysilicon gate is used as a self-aligned implant mask for the source/drain implants. As mentioned above, self-aligned processes are always desirable in terms of process simplicity and packing density. It is particularly useful in this case because we thereby guarantee that there will be *some* overlap of the gate with the source/drain but minimal overlap. The overlap is determined by the lateral scattering of the ions and by the lateral diffusion of the dopants during subsequent thermal processing (such as source/drain implant anneals). If there were no overlap, the channel would have to be turned on in this region by the gate fringing fields. The resulting potential barrier in the channel would degrade the device current. On the other hand, if there is too much overlap, it leads to an overlap capacitance between the source or drain and the gate. This is particularly bothersome near the drain end because it leads to the Miller overlap capacitance which causes undesired capacitive feedback between the output drain terminal and the input gate terminal (see Section 6.5.8).

Fabrication steps for the n-channel MOSFETs in the p-well are shown in Fig. 9–7. After the polysilicon gate is etched, we first do a self-aligned n-type source–drain implant, during which the tank masking level is used to protect the PMOS devices with a layer of photoresist. The NMOS source and drain implants are done in two stages. The first implant is a lightly doped drain (LDD) implant (Fig. 9–7a). This is typically a dose of $\sim 10^{13}$–10^{14} cm^{-2}, corresponding to a concentration of 10^{18}–10^{19} cm^{-3}, and an ultra-shallow junction depth of 50–100 nm. When a MOSFET is operated in the saturation region, the drain-channel junction is reverse biased, resulting in a very high electric field in the pinch-off region. As we saw in Section 5.4 for reverse-biased p-n junctions, reducing the doping level increases the depletion width and makes the peak electric field at the junction smaller. As discussed in Section 6.5.9, electrons traveling from the source to the drain in the channel can gain kinetic energy and thereby become hot electrons, which create damage. The low doping in the LDD helps reduce hot carrier effects at the drain end. The shallow junction depths in the LDD are also important for reducing short channel effects such as DIBL and charge sharing (Sections 6.5.10 and 6.5.11). The penalty that we pay with the use of an LDD region is that the source-to-drain series resistance goes up, which degrades the drive current.

As the technology is evolving towards lower power supply voltages, hot carrier effects are becoming less important. This, along with the need to reduce series resistance, has driven the trend towards increasing doping in the LDD to levels above 10^{19} cm^{-3}. In fact, the use of the term LDD then becomes a bit of a misnomer, and is often replaced by the term *source/drain extension* or *tip*.

After the LDD regions are formed alongside the polysilicon gate, we implant deeper (~ 200 nm) and more heavily doped (10^{20} cm^{-3}) source and drain junctions farther away from the gate edges (Fig. 9–7d). This more conductive region allows ohmic contacts to the source and drain to be formed more easily than they could be directly to the LDD regions, and reduces the source/drain series resistance. This implant is done using a self-aligned scheme by the formation of *sidewall oxide spacers*. After removing the photoresist covering the PMOS devices, we deposit *conformal* LPCVD oxide (~ 100–200 nm thick) using an organic precursor called tetra-ethyl-ortho-silicate (TEOS) over the entire wafer at fairly high temperatures (~ 700 °C) (Fig. 9–7b). The term conformal means that the deposited film has the same thickness everywhere, and follows the topography on the wafer. This oxide layer is then subjected to RIE, which is anisotropic (i.e., it etches predominantly in the vertical direction) (Section 5.1.7). If the RIE step is timed to just etch off the deposited oxide on the flat surfaces, it leaves oxide sidewall spacers on the edges of the polysilicon gate, as shown in Fig. 9–7c. This sidewall spacer is used as a self-aligned mask to protect the LDD regions very near the gate during the heavier, deeper n$^+$ source and drain implants (Fig. 9–7d).

Next, the NMOS devices are masked by photoresist, and a p$^+$ source and drain implant is done for the PMOSFETs (Fig. 9–8a). It may be noted that an

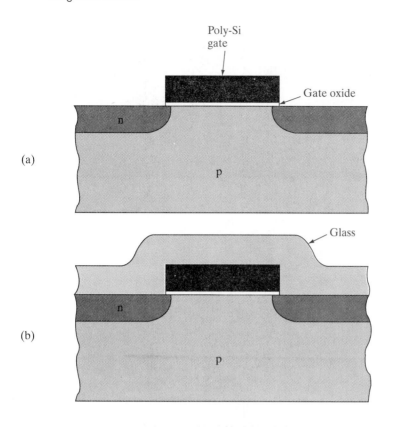

(a)

(b)

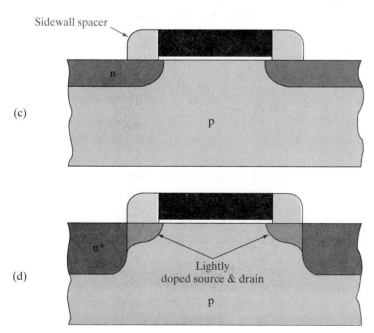

(c)

(d)

Figure 9–7
Fabrication of the lightly doped drain structure, using sidewall spacers. The poly-silicon gate covers the thin gate oxide and masks the first low-dose implant (a). A thick layer is deposited by CVD (b) and is anisotropically etched away to leave only the sidewall spacers (c). These spacers serve as a mask for the second, high-dose implant. After a drive-in diffusion, the LDD structure results (d).

Figure 9–8
Buried channel
PMOS: (a) self-
aligned p⁺
source/drain
implant with no
LDD using pho-
toresist to protect
NMOSFETs. A
p-type V_T adjust
implant is shown
in color in the
channel; (b) dop-
ing profile as a
function of depth
in the middle of
the channel show-
ing the p-type V_T
adjust implant
near the surface;
(c) electron poten-
tial energy as a
function of depth
in the middle of
the channel,
showing holes col-
lecting in the
"buried" channel.
For higher gate
bias the PMOS
operation
changes to sur-
face channel, as
indicated by the
dashed line.

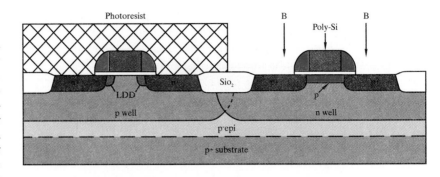

(a)

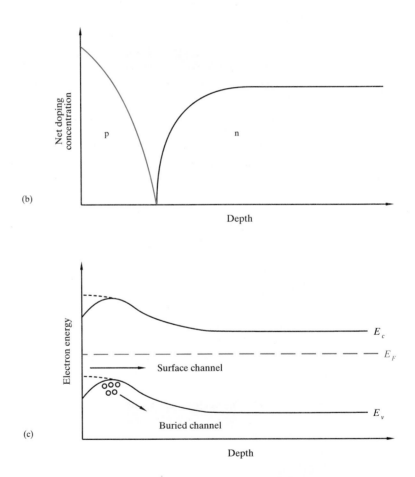

(b)

(c)

LDD was not used for the PMOS. This is due to the fact that hot hole effects
are less problematic than hot electron degradation, partly due to the lower
hole mobility and partly due to the higher Si–SiO₂ barrier in the valence band

(5 eV) than in the conduction band (3.1 eV). After the source–drain implants are done, the dopants are activated and the ion implant damage is healed by a furnace anneal, or more frequently by using a Rapid Thermal Anneal (RTA). In this anneal we use the minimum acceptable temperature and time combination (the thermal budget) because it is critically important to keep the dopant profiles as compact as possible in ultra-small MOSFETs.

We can now appreciate why most CMOS logic devices are made on p-type substrates, rather than n-type. The n-channel MOSFETs generate a lot more substrate current due to hot carrier effects than PMOSFETs. The holes, thus generated, can more easily flow to ground in a p-type substrate, than in an n-substrate. Also, it is easier to dope substrates p-type with B during Czochralski crystal growth than n-type with Sb. Antimony is the preferred donor, rather than As or P for bulk doping of the Si melt, because Sb evaporates less than the other species.

The use of n^+ polysilicon gates for both NMOS and PMOS devices raises some interesting device issues. Since the Fermi level in the n^+ gate is very close to the Si conduction band, its workfunction is well suited to achieving a low V_T for NMOS ($\Phi_{ms} \sim -1V$), but not for PMOS ($\Phi_{ms} \sim 0V$). From the V_T expression (Eq. 6-38), we notice that the second and third terms approach zero as thin-oxide technology evolves because C_i is getting larger. For high drive current we want V_T to be in the neighborhood of ~0.3 to ~0.7V for NMOS (–0.3 to –0.7 V for PMOS). We find from Eq. (6–38) that the p-well doping can be optimized to achieve the correct V_T for the NMOS transistor, while at the same time being high enough to prevent punchthrough breakdown between source and drain. For the PMOS transistor, on the other hand, an n-well doping of the order of 10^{18} cm^{-3} prevents punchthrough, but the Fermi potential, ϕ_F, is so large and negative that the V_T is too negative. That forces us to do a separate acceptor implant to adjust V_T for the PMOS devices (Fig. 9–8b). The acceptor dose is low enough that the p-layer is fully depleted at zero gate bias, leading to enhancement mode, rather than depletion mode transistors. In CMOS we try to make the negative V_T of the PMOS device about the same value as the positive V_T of the NMOS.

Close examination of the band diagram in the channel of the PMOSFET along the vertical direction (perpendicular to the gate oxide) shows that the energy minimum for the holes in the inversion layer is slightly below (~100 nm) the oxide–silicon interface, leading to what is known as *buried channel operation* for PMOS (Fig. 9–8c). On the other hand, for NMOS, the electron energy minimum in the inversion layer occurs right at the oxide–silicon interface, leading to surface channel operation.[1] There are good and bad aspects of this

---

[1]We can qualitatively understand why the acceptor implant in the channel leads to such buried channel behavior. Assume for a moment that the acceptor dose was high enough that the p-layer was not depleted at zero bias. In such a depletion mode device, a positive gate bias to turn off the device would first deplete the surface region of holes, still leading to hole conduction deeper in the substrate away from the oxide–Si interface.

buried channel behavior for PMOS. Since the holes in the inversion layer of the PMOSFET travel slightly away from the oxide–silicon interface, they do not suffer as much channel mobility degradation as the electrons in the NMOSFET due to surface roughness scattering. That is good, because hole mobilities in Si are generally lower than electron mobilities, which forces us to make PMOS devices wider than the NMOSFETs to get similar drive currents. However, buried channel devices have a greater propensity towards DIBL and punchthrough breakdown. Hence, as the size of MOSFETs is reduced, the DIBL problem becomes worse, and there is a desire to have surface channel operation for both NMOS and PMOS. Thus, there is interest in so-called *dual gate* CMOS, where n^+ polysilicon gates are used for NMOSFETs and p^+ gates are used for PMOSFETs. Such dual workfunction gates can be achieved by depositing the polysilicon undoped and then using the source and drain implants themselves to also dope the gates appropriately. This approach exploits the high polysilicon grain boundary diffusivities to degenerately dope the gates, while at the same time having ultra-shallow source and drain junctions to minimize DIBL.

As a historical footnote, it may be added that MOSFETs initially used Al gates which could not withstand high temperature processing. Hence, the source and drain regions had to be formed first, either by diffusion or by implant, and then the Al gate was deposited and patterned. Such non–self-aligned processes suffered from the Miller capacitance mentioned previously. What made self-aligned source and drain regions viable was the use of polysilicon, which has a sufficiently high melting point to withstand subsequent processing. Recent research on MOSFET technology is going back full circle to metal gates, but this time using refractory metals such as tungsten (W). These metals have better conductivity than heavily doped polysilicon, and a workfunction that is better suited for CMOS. The Fermi level of W is near the middle of the Si bandgap, which makes the flatband voltage and the threshold voltage more symmetric and better matched between NMOS and PMOS, and avoids the buried channel effect.

The next step is to form a metal-silicon alloy or silicide in the source/drain and gate regions of the MOSFETs in order to reduce the series resistance (and thereby the *RC* time constants), and increase the drive current. This involves depositing a thin layer of a refractory metal such as Ti over the entire wafer by sputtering, and reacting the Ti with Si wherever they come in *direct* contact, by doing a two-step heat treatment in a N ambient (Fig. 9–9a). A 600 °C anneal results in the formation of Ti_2Si (the C49 phase according to metallurgists, which has fairly high resistivity), followed by an 800 °C anneal which converts the Ti_2Si to the C54 phase, $TiSi_2$, which has an extremely low resistivity of ~17 $\mu\Omega$-cm, much lower than that of the most heavily doped Si. On the other hand, the Ti on top of the sidewall oxide spacers does not form a silicide, and stays as unreacted Ti or forms TiN because the process is done in a N ambient. The Ti and TiN can be etched off selectively using a wet hydrogen peroxide-based etch which does not attack the titanium disilicide, thereby electrically isolating the gates from the source/drains. It may be noted that this process results in a SALICIDE

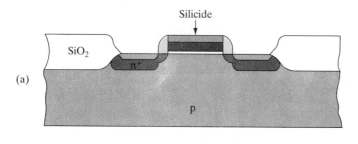

(a)

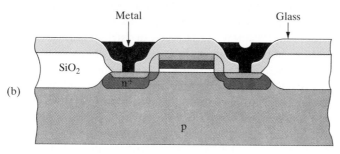

(b)

Figure 9–9
The formation of silicided source–drain and gate regions for low-resistance contacts: (a) a layer of refractory metal is deposited and reacted in regions of exposed Si to form a conducting silicide layer; (b) the unreacted metal is removed and a CVD glass is deposited and patterned for contact metallization.

without a separate masking level only on the source/drains and the polysilicon gate, where the silicide is often termed a *polycide*. This results in very high performance MOSFETs.

Finally, the MOSFETs have to be properly interconnected according to the circuit layout, using the metallization level. This involves LPCVD of an oxide dielectric layer doped with B and P, which is known as *boro-phospho-silicate glass* (BPSG) on the entire wafer, patterning it using the contact level reticle and using RIE to open up the contact holes to the substrate (Fig. 9–9b). The B and P allow the oxide layer to soften and reflow more readily upon annealing, thereby helping planarize or smooth out the topography on the wafer. This shaping of the millions of very small contact holes is critical on a ULSI chip, because otherwise metal deposited on the surface into the contact holes may not reach completely into the holes, leading to a catastrophic open circuit. In fact, sometimes a CVD tungsten layer is selectively deposited in the contact holes to form a contact *plug* before proceeding to the next step. Then a suitable metal layer such as Al (alloyed with ~1 percent Si, and ~4 percent Cu) is sputter deposited over the wafer, patterned using the metal interconnect level, and subjected to metal RIE. The Si is added to the Al to solve the junction *spiking* problem, where pure Al can incorporate the solid solubility limit of Si from the shallow source/drain regions. This would allow the Al to "spike" or short through the p-n junction. The Cu is added to enlarge the grain size in the Al interconnect films, which are polycrystalline, making it harder for the electrons moving during current flow to nudge the Al atoms along, thereby opening voids (open circuits) in the interconnect. This is an example of an *electromigration* phenomenon.

In a modern ULSI chip, the complexity of the device layout generally demands that multiple levels of metallization be used for interconnecting the devices (Fig. 9–10). Hence, after depositing the first metal, an inter-metal dielectric isolation layer such as SiO_2 is deposited by low temperature CVD. Low temperatures are very important in this *back-end* part of the processing because by now all the active devices are in place and one cannot allow the dopants to diffuse significantly. Also, the Al metallization cannot withstand temperatures higher than ~500 °C. The dielectric isolation layer must be suitably planarized prior to the deposition of the next layer of metal, and this is generally done by CMP. Planarization is important because if metal is deposited on a surface with rough topography and subjected to RIE, there can be residual metal sidewall filaments or "stringers" at the steps for the same reason one gets sidewall oxide spacers on either side of the MOSFET gate in Fig. 9–7. These metal stringers can cause short circuits between adjacent metal lines. Planarization is also important in maintaining good depth of focus during photolithography. After planarization of the isolation layer, one uses photolithography to open up a new set of contact holes called *vias*, followed by deposition, patterning and RIE of the next layer of metal, and so on for multi-level metallization. As mentioned previously, W metal plugs are sometimes selectively deposited to fill up the via holes prior to the metal deposition, and reduce the likelihood of an open circuit.

Finally, a protective overcoat is deposited on the IC to prevent contamination and failure of the devices due to the ambient (Fig. 9–10). This generally involves plasma CVD of silicon nitride, which has the nice attribute that it blocks the diffusion of water vapor and Na through it. Sodium, as mentioned in Section 6.4.3, causes a mobile ion problem in the gate dielectric of MOS devices. Sometimes, the protective overcoat is a BPSG layer. After the

Figure 9–10
Multi-level interconnect: Cross section of IC showing 5 levels of A1 interconections with suitably planarized inter-metal dielectrics. The transistors are at the very bottom, and are electrically isolated by shallow trench isolation (STI). (Photograph courtesy of Motorola.)

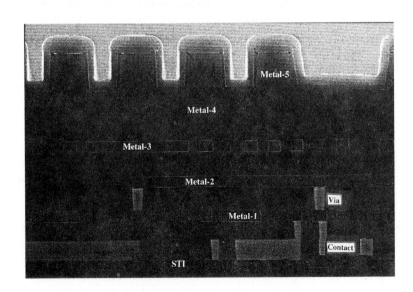

overcoat is deposited, openings are etched for the metal bond pads. After the chips are tested in an automated tester, the *known good dies* are packaged and wire bonded, as discussed in Section 9.6.

9.3.2 Silicon-on-Insulator (SOI)

An interesting and useful extension of the Si MOS process can be achieved by growing very thin films of single crystal Si on insulating substrates (Fig. 9–11). Two such substrates which have the appropriate thermal expansion match to Si are sapphire and spinel ($MgO–Al_2O_3$). Epitaxial Si films can be grown on these substrates by chemical vapor deposition (e.g., the pyrolysis of silane), with typical film thickness of about 1 μm. The film can be etched by standard photolithographic techniques into islands for each transistor. Implantation of p^+ and n^+ areas into these islands for source and drain regions result in the MOS devices. Since the film is so thin, the source and drain regions can be made to extend entirely through the film to the insulating substrate. As a result, the junction capacitance is reduced to the very small capacitance associated with the sidewalls between the source/drain and the channel region. In addition, since interconnections between devices pass over the insulating substrate, the usual interconnection–Si substrate capacitance is eliminated (along with the possibility of parasitic induced channels in the field between devices). These capacitance reductions improve considerably the high-frequency operation of circuits using such devices.

Other insulators can be used for SOI devices, including SiO_2. Since oxide can easily be grown on Si substrates, it serves as an attractive insulator for subsequent growth of thin-film Si. Since polycrystalline Si can be deposited directly over SiO_2, devices can be made in thin poly-Si films. However, to avoid grain boundaries and other defects typical of polycrystalline material, a variety of techniques have been developed to grow single crystal Si on oxide. For example, the oxide layer can be formed beneath the surface of a Si wafer by high-dose oxygen implantation. The thin Si layer remaining on the surface above the implanted oxide is usually about 0.1 μm thick, and can

Figure 9–11
Silicon on insulator. Both n-channel and p-channel enhancement transistors are made in islands of Si film on the insulating substrate. These devices can be interconnected for CMOS applications.

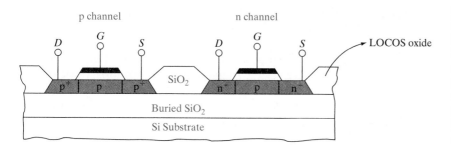

be used as the thin film for CMOS or other device fabrication. This process is called *separation by implantation of oxygen (SIMOX)*. In some cases a thicker Si film is grown epitaxially on the SIMOX wafer, using the thin Si crystalline layer over the oxide as a seed for the epitaxy.

Another approach for making SOI is to place two oxidized Si wafers face-to-face, and thermally bond the oxide layers by high temperature annealing in a furnace. One of the wafers is then chemically etched back almost completely, leaving about a micron of single crystal Si material on top of the SiO$_2$. This approach is known as *Bond-and-Etch-back* SOI (BE-SOI). Since it is challenging to etch off about 600 μm of Si and stop controllably so as to leave about a micron of Si, an initial p$^+$ implanted layer is often used to act as an etch stop near the end of the chemical etch. The chemical recipe that is used for etching Si has a much lower etch rate in a p$^+$ layer than in lightly doped Si.

Another recent approach is a modification of BE-SOI using very high dose H implantation into one of the oxidized Si wafers, such that the peak of the H profile is about a micron below the Si surface. During the high temperature annealing step, the H atoms coalesce to form tiny H bubbles which cause a thin layer of Si to cleave off from the implanted wafer, leaving the Si layer bonded to the oxidized Si surface. This way, one does not have to chemically etch off one Si wafer for every BE-SOI wafer. If properly done, the cleaved-off Si wafer has a smooth surface and can be reused.

Examining the two devices of Fig. 9–11 more closely, one of them is unlike the transistors we have considered thus far. The thin Si film is lightly doped p-type, and therefore the device labeled "p-channel" appears *junctionless*. Such a device is able to operate in the enhancement mode (normally off) because of the equilibrium effects of the work function difference and the interface charge. With the usual Φ_{ms} and Q_i a depletion region is formed in the central p material of each device with zero gate voltage. In fact, for a Si film of about 0.1 μm or less, this depletion region can extend all the way through the Si to the insulator. Such a device is called *fully depleted* and no drain-to-source current flows. In the n-channel device a positive gate voltage greater than V_T induces an inverted region at the surface, as usual for an n-channel enhancement device. For the p-channel case a small negative voltage V_G removes the depletion and causes hole accumulation beneath the gate. The result is the formation of a conducting channel by a small negative gate voltage, as is the case for a conventional p-channel enhancement device. Although the fully depleted type of p-channel device operates by a somewhat different mechanism, its current–voltage characteristics are similar to the conventional device. Since both p-channel and n-channel transistors can be made on the same insulating surface, the silicon-on-insulator technique is quite compatible with CMOS circuit fabrication.

The SOI approach does not suffer from the latchup problem of bulk CMOS because there is no p-n-p-n thyristor from the power supply to ground. For circuits requiring high speeds, low standby power (due to the elimination of junction leakage to the substrate), and radiation tolerance (due to the elimination of the Si substrate), the extra expense of preparing sapphire substrates or growing crystalline Si films on oxide is compensated by increased performance.

9.3.3 Integration of Other Circuit Elements

One of the most revolutionary developments of integrated circuit technology is the fact that integrated transistors are cheaper to make than are more mundane elements such as resistors and capacitors. There are, however, numerous applications calling for diodes, resistors, capacitors and inductors in integrated form. In this section we discuss briefly how these circuit elements can be implemented on the chip. We will also discuss a very important circuit element—the interconnection pattern which ties all of the integrated devices together in a working system.

Diodes. It is simple to build p-n junction diodes in a monolithic circuit. It is also common practice to use transistors to perform diode functions. Since many transistors are included in a monolithic circuit, no special diffusion step is required to fabricate the diode element. There are a number of ways in which a transistor can be connected as a diode. Perhaps the most common method is to use the emitter junction as the diode, with the collector and base shorted. This configuration is essentially the narrow base diode structure, which has high switching speed with little charge storage (Prob. 5.35). Since all the transistors can be made simultaneously, the proper connections can be included in the metallization pattern to convert some of the transistors into diodes.

Resistors. Diffused or implanted resistors can be obtained in monolithic circuits by using the shallow junctions used in forming the transistor regions (Fig. 9–12a). For example, during the base implant, a resistor can be implanted which is made up of a thin p-type layer within one of the n-type islands. Alternatively a p region can be made during the base implant, and an n-type

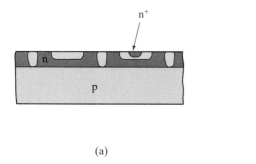

(a)

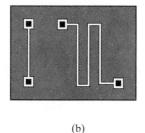

(b)

Figure 9–12
Monolithic resistors: (a) cross section showing use of base and emitter diffusions for resistors; (b) top view of two resistor patterns.

resistor channel can be included within the resulting p region during the emitter implant step. In either case, the resistance channel can be isolated from the rest of the circuit by proper biasing of the surrounding material. For example, if the resistor is a p-type channel obtained during the base implant, the surrounding n material can be connected to the most positive potential in the circuit to provide reverse-bias junction isolation. The resistance of the channel depends on its length, width, depth of the implant, and resistivity of the implanted material. Since the depth and resistivity are determined by the requirements of the base or emitter implant, the variable parameters are the length and width. Two typical resistor geometries are shown in Fig. 9–12b. In each case the resistor is long compared with its width, and a provision is made on each end for making contact to the metallization pattern.

Design of diffused resistors begins with a quantity called the *sheet resistance* of the diffused layer. If the average resistivity of a diffused region is ρ, the resistance of a given length L is $R = \rho L/w t$, where w is the width and t is the thickness of the layer. Now if we consider one square of the material, such that $L = w$, we have the sheet resistance $R_s \equiv \rho/t$ in units of ohms per square. We notice that R_s measured for a given layer is numerically the same for any size square. This quantity is simple to measure for a thin diffused layer by a four-point probe technique.[2] Therefore, for a given diffusion, the sheet resistance is generally known with good accuracy. The resistance then can be calculated from the known value of R_s and the ratio L/w (the *aspect ratio*) for the resistor. We can make the width w as small as possible within the requirements of heat dissipation and photolithographic limitations and then calculate the required length from w and R_s. Design criteria for diffused resistors include geometrical factors, such as the presence of high current density at the inside corner of a sharp turn. In some cases it is necessary to round corners slightly in a folded or zigzag resistor (Fig. 9–12b) to reduce this problem.

To reduce the amount of space used for resistors or to obtain larger resistance values, it is often necessary to obtain surface layers having larger sheet resistance than is available during the standard base or emitter implants. We can use a different implant, such as the V_T adjust implant to form shallow regions having very high sheet resistance ($\sim 10^5\ \Omega$/square). This procedure can provide a considerable saving of space on the chip. In integrated FET circuits it is common to replace load resistors with depletion-mode transistors, as mentioned in Section 6.5.5.

Capacitors. One of the most important elements of an integrated circuit is the capacitor. This is particularly true in the case of memory circuits, where charge is stored in a capacitor for each bit of information. Figure 9–13 illustrates a one-transistor DRAM cell, in which the n-channel MOS transistor provides

---

[2]This is a very useful method, in which current is introduced into a wafer at one probe, collected at another probe, and the voltage is measured by two probes in between. Special formulas are required to calculate resistivity or sheet resistance from these measurements.

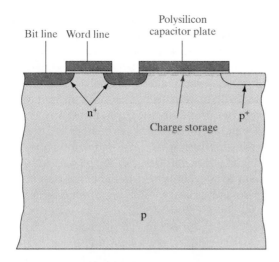

Bit line   Word line

Polysilicon
capacitor plate

n^+

Charge storage

p^+

p

Figure 9–13
Integrated capacitor for DRAM cells. A one-transistor memory cell in which the transistor stores and accesses charge in an adjacent planar MOS capacitor.

access to the adjacent MOS capacitor. The top plate of the capacitor is poly-silicon, and the bottom plate is an inversion charge contacted by an n^+ region of the transistor. The terms *bit line* and *word line* refer to the row and column organization of the memory (Section 9.5.2). One can also make use of the capacitance associated with p-n junctions, as discussed in Section 5.5.5.

Inductors. Inductors have not been incorporated into ICs in the past, because it is much harder to integrate inductors than the other circuit elements. Also, there has not been a great need for integrating inductors. Recently, that has changed because of the growing need for rf analog ICs for portable communication electronics. Inductors are very important for such applications, and can be made with reasonable Q factors using spiral wound thin metal films on an IC. Such spiral patterns can be defined by photolithography and etching techniques compatible with IC processing, or they can be incorporated in a hybrid IC.

Contacts and Interconnections. During the metallization step, the various regions of each circuit element are contacted and proper interconnection of the circuit elements is made. Aluminum is commonly used for the top metallization, since it adheres well to Si and to SiO_2 if the temperature is raised briefly to about 550°C after deposition. Gold is used on GaAs devices, but the adherence properties of Au to Si and SiO_2 are poor. Gold also creates deep traps in Si.

As mentioned throughout this chapter, silicide contacts and doped poly-silicon conductors are commonly used in integrated circuits. By opening windows through the oxide layers to these conductors, Al metallization can be used to contact them and connect them to other parts of the circuit. In cases where Al is used to contact the Si surface, it is usually necessary to use Al containing about 1 percent Si to prevent the metal from incorporating Si from the layer being contacted, thereby causing "spikes" in the surface. Thin diffusion barriers are also used between the Al and Si layers, to prevent migration between the two. The refractory silicides mentioned in Section 9.3.1 serve this purpose.

Increased complexity and packing density in integrated circuits inevitably leads to a need for multilayer metallization. Multiple levels of Cu metallization can be incorporated with interspersing dielectrics. In general, the metals may all be Al, Cu or they may be different conductors such as polysilicon or refractory metals (depending on the heat each is subjected to in subsequent processing). Also, the dielectrics may be deposited oxides, boro-phospho-silicate glass for reflow planarization, nitrides, etc. The planarization of the surface is extremely important to prevent breaks in the metallization, which can occur in traversing a step on the surface. Various approaches using reflow glass, poly-

(a)

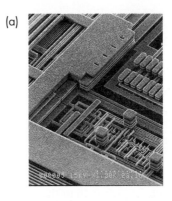

$Si_3 N_4$ overcoat

(b)

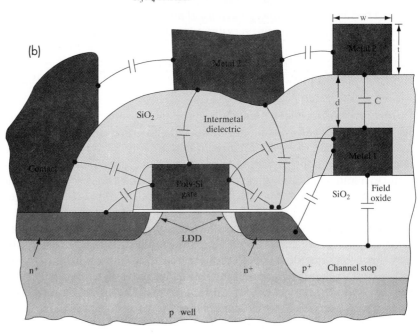

Figure 9–14
Multi-level interconnects: (a) Micrograph of multi-level interconnects in an IC. The inter-level dielectrics have been etched off to reveal the copper interconnect lines; (b) equivalent circuit illustrating the various parasitic capacitive elements associated with a multi-level interconnect. On the top right hand corner of the figure, we focus on the parallel plate capacitor model referred to in Eqs. (9–1) and (9–2). (Photograph courtesy of IBM.)

imide, and other materials to achieve planarization have been used, along with chemical mechanical polishing.

The most important challenge in designing interconnects is the RC time constant, which affects the speed and active power dissipation of the chip. A very simplistic model of two layers of interconnects with an inter-metal dielectric (Fig. 9–14b) shows that it can be regarded as a parallel plate capacitor. Regarding the interconnect as a rectangular resistor, its resistance is given by

$$R = \frac{\rho L}{tw} = R_s \frac{L}{w} \tag{9–1a}$$

where R_s is the sheet resistance, and the other symbols are defined in Fig. 9–14b. The capacitance is given by

$$C = \frac{\epsilon Lw}{d} \tag{9–1b}$$

The RC time constant is then

$$\left(\frac{\rho L}{wt}\right)\left(\frac{\epsilon Lw}{d}\right) = \frac{\rho \epsilon L^2}{td} = \frac{R_s \epsilon L^2}{d} \tag{9–2}$$

Interestingly, for this simple one-dimensional model, the width of the interconnect w cancels out. Therefore, it does not make sense to use wider conductors for high speed operation. It is also impractical to do so in terms of packing density. Of course, in reality we must account for the fringing electric fields, and therefore account for width dependence. From Eq. 9–2, it is clear we need as thick a metal layer (within practical limits of deposition times and etching times) and as low a resistivity as possible. Low resistivities are also important in minimizing ohmic voltage drops in metal bus lines that carry power from one end of a chip to the other. Aluminum is very good in this regard, and thus was a mainstay for Si technology for many years. Aluminum also has other nice attributes such as good ohmic contacts to both n and p-type Si, and good adhesion to oxides.

Copper has even lower resistivity (1.7 μΩ-cm) than Al (3 μΩ-cm) and is about two orders of magnitude less susceptible to electromigration. Hence, it is an excellent alternative to Al for very high speed ICs (Fig. 9–14a). The process breakthroughs that have made Cu viable for metallization include new electrodeposition and electroplating techniques because CVD or sputter deposition is not very practical for Cu. It is also very difficult to use reactive ion etching (RIE) for Cu because the etch byproducts for Cu are not very volatile. Hence, instead of RIE, Cu patterning is based on the so-called *Damascene* process where grooves are first etched in a dielectric layer, Cu is deposited on it, and the metal layer is chemically-mechanically polished down, leaving inlaid metal lines in the oxide grooves. In this method, the metal does not have to be etched directly using RIE, which is always a difficult process. The name "damascene" is derived from a metallurgical technique in ancient Turkey where metal artwork was inlaid into swords and other artifacts using this type of process. Copper can create traps deep in the bandgap of Si; hence, a suitable barrier layer such as Ti is needed between the Cu layer and the Si substrate.

Other parameters in Eq. (9–2) that can minimize the *RC* time constant are clearly the use of a thick inter-metal dielectric layer (once again within the limits of practicality in terms of deposition times), and as low a dielectric constant material as possible. Silicon dioxide has a relative dielectric constant of 3.9. There is active research in low dielectric constant materials (sometimes referred to as low-*K* materials). These include organic materials such as polyimides, or xerogels/aerogels which have air pockets or porosity purposely built in to minimize the dielectric constant.

In designing the layout of elements for a monolithic circuit, topological problems must be solved to provide efficient interconnection without *crossovers*—points at which one conductor crosses another conductor. If crossovers must be made on the Si surface, they can be accomplished easily at a resistor. Since the implanted or diffused resistor is covered by SiO_2, a conductor can be deposited crossing the insulated resistor. In cases requiring crossovers where no resistor is available, a low-value implanted resistor can be inserted in one of the conductor paths. For example, a short n^+ region can be implanted during the source/drain step and contacted at each end by one of the conductors. The other conductor can then cross over the oxide layer above the n^+ region. Usually, this can be accomplished without appreciable increase in resistance, since the n^+ region is heavily doped and its length can be made small.

During the metallization step, appropriate points in the circuit are connected to relatively large *pads* to provide for external contacts. These metal pads are visible in photographs of monolithic circuits as rectangular areas spaced around the periphery of the device. In the mounting and packaging process, these pads are contacted by small Au or Al wires or by special techniques such as those discussed in Section 9.6.

9.4 CHARGE TRANSFER DEVICES

One of the most interesting and broadly useful integrated devices is the *charge-coupled device (CCD)*. The CCD is part of a broader class of structures known generally as *charge transfer devices*. These are dynamic devices that move charge along a predetermined path under the control of clock pulses. These devices find applications in memories, various logic functions, signal processing, and imaging. In this section we lay the groundwork for understanding these devices, but their present forms and variety of applications must be found in the current literature.

9.4.1 Dynamic Effects in MOS Capacitors

The basis of the CCD is the dynamic storage and withdrawal of charge in a series of MOS capacitors. Thus we must begin by extending the MOS discussion of Chapter 6 to include the basics of dynamic effects. Figure 9–15

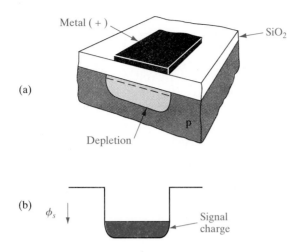

Metal (+)

SiO₂

(a)

Depletion

p

(b) ϕ_s ↓

Signal
charge

Figure 9–15
An MOS capaci-
tor with a positive
gate pulse: (a) de-
pletion region and
surface charge;
(b) potential well
at the interface,
partially filled
with electrons cor-
responding to the
surface charge
shown in (a).

shows an MOS capacitor on a p-type substrate with a large positive gate pulse
applied. A depletion region exists under the gate, and the surface potential increases considerably under the gate electrode. In effect, the surface potential
forms a *potential well*, which can be exploited for the storage of charge.

    If the positive gate bias has been applied for a sufficiently long time,
electrons accumulate at the surface and the steady state inversion condition
is established. The source of these carriers is the thermal generation of electrons at or near the surface. In effect, the inversion charge tells us the capacity of the well for storage charge. The time required to fill the well
thermally is called the *thermal relaxation time*, and it depends on the quality of the semiconductor material and its interface with the insulator. For good
materials the thermal relaxation time can be much longer than the charge
storage times involved in CCD operation.

    If instead of a steady state bias we apply a large positive pulse to the
MOS gate electrode, a deep potential well is first created. Before inversion
has occurred by thermal generation, the depletion width is greater than it
would be at equilibrium ($W > W_m$). This transient condition is sometimes
called *deep depletion*. If we can inject electrons into this potential well electrically or optically, they will be stored there.[3] The storage is temporary, however, because we must move the electrons out to another storage location
before thermal generation becomes appreciable.

    What is needed is a simple method for allowing charge to flow from one
potential well to an adjacent one quickly and without losing much charge in
the process. If this is accomplished, we can inject, move, and collect packets
of charge dynamically to do a variety of electronic functions.

---

[3]The potential well should not be confused with the depletion region, which extends into the bulk of the
semiconductor. The "depth" of the well is measured in electrostatic potential, not distance. Electrons stored
in the potential well are in fact located very near the semiconductor surface.

9.4.2 The Basic CCD

The original CCD structure proposed in 1969 by Boyle and Smith of Bell Laboratories consisted of a series of metal electrodes forming an array of MOS capacitors as shown in Fig. 9–16. Voltage pulses are supplied in three lines (L_1, L_2, L_3), each connected to every third electrode in the row (G_1, G_2, G_3). These voltages are clocked to provide potential wells, which vary with time as in Fig. 9–16. At time t_1 a potential well exists under each G_1 electrode, and we assume this well contains a packet of electrons from a previous operation. At time t_2 a potential is applied also to the adjacent electrode G_2, and the charge equalizes across the common G_1–G_2 well. It is easy to visualize this process by thinking of the mobile charge in analogy with a fluid, which flows to equalize its level in the expanding container. This fluid model continues at t_3 when V_1 is reduced, thus decreasing the potential well under G_1. Now the charge flows into the G_2 well, and this process is completed at t_4 when V_1 is zero. By this process the packet of charge has been moved from under G_1 to G_2. As the procedure is continued, the charge is next passed to the G_3 position, and continues down the line as time proceeds. In this way charge can be

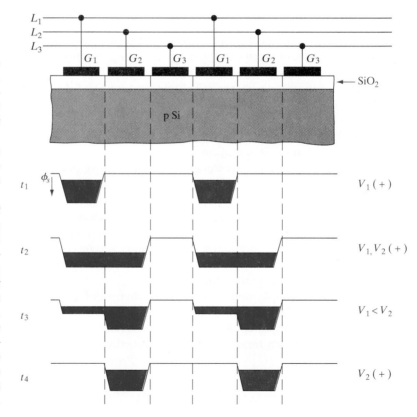

Figure 9–16
The basic CCD, composed of a linear array of MOS capacitors. At time t_1, the G_1 electrodes are positive, and the charge packet is stored in the G_1 potential well. At t_2 both G_1 and G_2 are positive, and the charge is distributed between the two wells. At t_3 the potential on G_1 is reduced, and the charge flows to the second well. At t_4 the transfer of charge to the G_2 well is completed.

injected using an input diode, transported down the line, and detected at the other end.

9.4.3 Improvements on the Basic Structure

Several problems arise in the implementation of the CCD structure of Fig. 9–16. For example, the separation between electrodes must be very small to allow coupling between the wells. An improvement can be made by using an overlapping gate structure such as that shown in Fig. 9–17. This can be done, for example, with poly-Si electrodes separated by SiO_2 or with alternating poly-Si and metal electrodes.

One of the problems inherent to the charge transfer process is that some charge is inevitably lost during the many transfers along the CCD. If the charges are stored at the Si–SiO_2 interface, surface states trap a certain amount of charge. Thus if the "0" logic condition is an empty well, the leading edge of a train of pulses is degraded by the loss of charge required in filling the traps which were empty in the "0" condition. One way of improving this situation is to provide enough bias in the "0" state to accommodate the interface and bulk traps. This procedure is colorfully referred to as using a *fat zero*. Even with the use of fat zeros, the signal is degraded after a number of transfers, by inherent inefficiencies in the transfer process.

Transfer efficiency can be improved by moving the charge transfer layer below the semiconductor–insulator interface. This can be accomplished by using ion implantation or epitaxial growth to create a layer of opposite type than the substrate. This shifts the maximum potential under each electrode into the semiconductor bulk, thus avoiding the semiconductor–insulator interface. This type of device is referred to as a *buried channel* CCD.

The three-phase CCD shown in Fig. 9–16 is only one example of a variety of CCD structures. Figure 9–18 illustrates one method for achieving a two-phase system, in which voltages are sequentially applied to alternating gate electrodes from two lines. A two-level poly-Si gate structure is used, in which the gate electrodes overlap, and a donor implant near the Si surface creates a built-in well under half of each electrode. When both gates are turned off (b), potential wells exist only under the implanted regions, and charge can be stored in any of these wells. With electrode G_2 pulsed positively, the charge

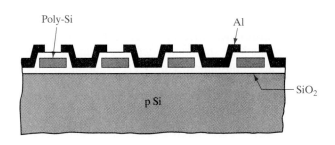

Figure 9–17
An overlapping gate CCD structure. One set of electrodes is polycrystalline Si, and the overlapping gates are Al in this case. SiO_2 separates the adjacent electrodes.

Figure 9–18
A two-phase CCD
with an extra po-
tential well built in
under the right
half of each elec-
trode by donor
implantation.

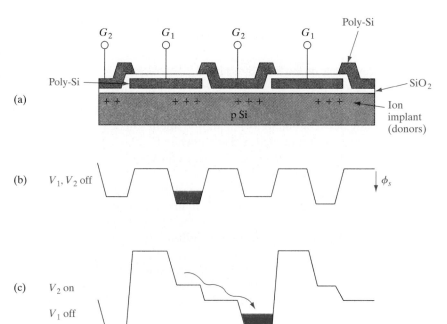

packet shown in (b) is transferred to the deepest well under G_2, which is its implanted region as (c) indicates. Then with both gates off, the wells appear as in (b) again, except that the charge is now under the G_2 electrode. The next step in the transfer process is obviously to pulse G_1 positively, so that the charge moves to the implanted region under the G_1 electrode to the right.

Other improvements to the basic structure are important in various applications. These include channel stops or other methods for achieving lateral confinement for the stored charge. Regeneration points must be included in the array to refresh the signal after it has been degraded.

9.4.4 Applications of CCDs

CCDs are used in a number of ways, including signal processing functions such as delay, filtering, and multiplexing several signals. Another interesting application of CCDs is in imaging for astronomy or in solid state TV cameras, in which an array of photosensors is used to form charge packets proportional to light intensity, and these packets are shifted to a detector point for readout. There are numerous ways of accomplishing this in CCDs, including the linear array line scanner, in which the second dimension is obtained by moving the scanner relative to the image. Alternatively, an area image sensor can be made which scans the image electronically in both dimensions. The latter device can be used as an alternative to the electron beam-addressed television imaging tube (Fig. 9–19).

Figure 9–19
A charge-coupled
device image sen-
sor shown as
large white rec-
tangular areas,
with peripheral
signal processing
circuitry. (Photo-
graph courtesy of
Texas Instruments.)

In the early development of integrated circuits it was felt that the inevitable defects that occur in processing would prevent the fabrication of devices containing more than a few dozen logic gates. One approach to integration on a larger scale tried in the late 1960s involved fabricating many identical logic gates on a wafer, testing them, and interconnecting the good ones (a process called *discretionary wiring*). While this approach was being developed, however, radical improvements were made in device processing which increased the yield of good chips on a wafer dramatically. By the early 1970s it was possible to build circuits with many hundreds of components per chip, with reasonable yield. These improvements made discretionary wiring obsolete almost as soon as it was developed. By reducing the number of processing defects, improving the packing density of components, and increasing the wafer size, it is now possible to place millions of device elements on a single chip of silicon and to obtain many perfect chips per wafer.

    A major factor in the development of integrated circuits has been the continual reduction in size of the individual elements (transistors, capacitors) within each circuit. Through improved design and better lithography, there has been a dramatic shrinking of the minimum feature size (e.g., a transistor gate) used in these devices. The results of shrinking the elements in a 16-Mb-DRAM are shown in Fig. 9–20. By reducing the minimum feature size in successive steps from 0.43 to 0.3 μm, the die area was reduced from about 135 mm^2 in the first-generation design to less than 42 mm^2 in the fifth-generation device. Obviously, more of the smaller chips can be made by batch fabrication on the wafer, and the effort in shrinking the design is rewarded in a more profitable device.

9.5
ULTRA LARGE-
SCALE
INTEGRATION
(ULSI)

Figure 9–20
Size reduction of
a 16 Mb DRAM
die as the mini-
mum feature size
is reduced from
0.43 μm for the
first generation
design (die on the
left) to 0.3 μm for
the die on the
right. (Photograph
courtesy of Mi-
cron Technology,
Inc.)

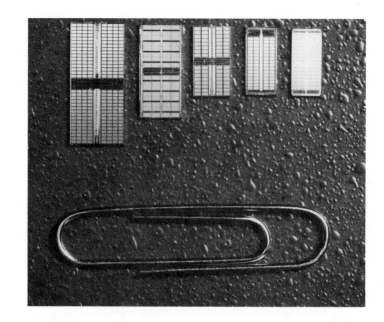

Figure 9–21
Three ways of ob-
taining 128
million bits of
dynamic random
access memory:
(a) one 128-Mb
chip; (b) two
64-Mb chips;
(c) eight 16-Mb
chips. The
128-Mb die
measures almost
1 by 1 inch.
(Photograph cour-
tesy of Micron
Technology, Inc.)

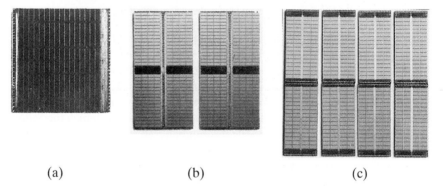

(a)   (b)   (c)

Successive designs using reduced feature sizes have made dramatically increased circuit complexity possible. DRAM design has set the pace over
the past two decades, in which successive 1-Mb, 4-Mb, and 16-Mb memories
led to similar powers of two increase to the 128-Mb range. Figure 9–21 illustrates the size comparison of a 128-Mb memory with an equivalent amount
of memory in the form of two 64-Mb and eight 16-Mb chips. These are examples of ultra large-scale integration (ULSI).

Although the achievement of many powers of two in memory is impressive and important, other ULSI chips are important for the integration
of many different system functions. A *microprocessor* includes functions for
a computer central processing unit (CPU), along with memory, control, tim-
ing, and interface circuits required to perform very complex computing func-

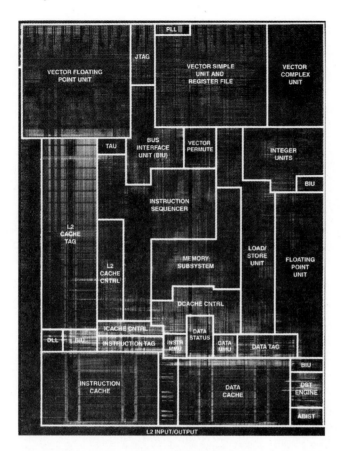

Figure 9–22
An example of ULSI, the Motorola Power PC microprocessor includes 10 million transistors employing CMOS technology with 0.35 μm minimum feature size. (Photograph courtesy of Motorola, Inc.)

tions. The complexity of such devices is shown in Fig. 9–22, which illustrates a microprocessor chip with various areas outlined by function.

Before leaving this section it might be useful to provide some calibration regarding the dimensions we have been discussing. Figure 9–23 compares the size of 64Mb DRAM circuit interconnect elements with a human hair, on the same scale. We can see that the densely packed 0.18 μm lines on this ULSI memory chip are dwarfed by the scanning electron micrograph of a human hair which has the diameter of about 50 microns. This makes it dramatically clear why ULSI chips must be fabricated in ultra-clean environments.

Although the focus of this book is devices and not circuits, it is important to look at some typical applications of MOS capacitors and FETs in semiconductor logic and memory ULSI, which constitute about 90 percent of all ICs. This should give the reader a better feel for why we have studied the physics of MOS devices in Chapter 6. This is clearly not a comprehensive discussion, because the design and analysis of circuits is a large subject covered in other books and courses. We will first look at some digital logic applications, followed by some typical memory devices.

Figure 9–23
Size comparison
of ULSI circuit ele-
ments with human
hair: (a) densely-
packed 0.18 μm
interconnect lines
in the array of a
64Mb DRAM
chip; (b) scanning
electron micro-
graph of a
human hair. Both
microgrqphs are
shown to the
same scale. (Pho-
tograph coutesy
of Micron
Technology, Inc.)

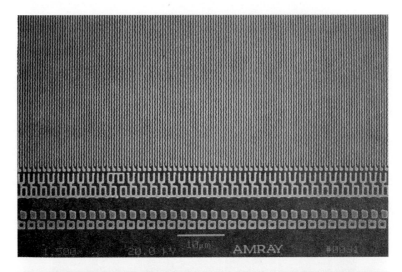

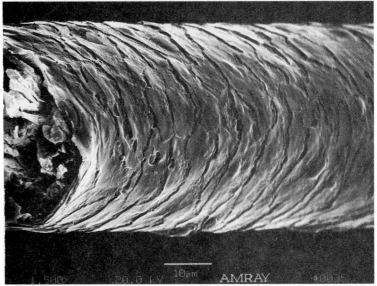

9.5.1 Logic Devices

A very simple and basic circuit element is the inverter, which serves to flip the logic state. When its input voltage is high (corresponding to logic "1"), its output voltage is low (logic "0"), and vice versa. Let us start the analysis with a resistor–loaded n-channel MOSFET inverter to illustrate the principles in the simplest possible manner (Fig. 9–24a). Then, we will extend the treatment to the slightly more complicated CMOS inverters which are much more useful and more common today.

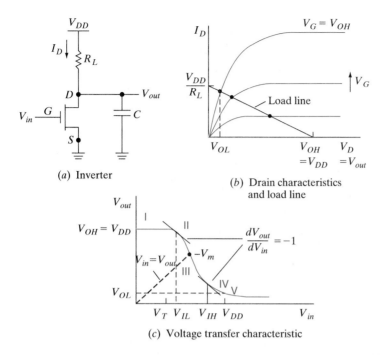

(a) Inverter

(b) Drain characteristics and load line

(c) Voltage transfer characteristic

Figure 9–24
Resistor load inverter voltage transfer characteristics (VTC): (a) NMOSFET with load resistor, R_L, and load parasitic capacitance, C; (b) determination of VTC by superimposing load line (linear I–V ohmic characteristics of resistor) on NMOSFET output characteristics; (c) VTC showing output voltage as a function of input voltage. The five key points on the VTC are logic high (V_{OH}), logic low (V_{OL}), unity gain points (V_{IL} and V_{IH}), and logic threshold where input equals output (V_m).

A key concept for inverters is the *voltage transfer characteristic* (VTC), which is a plot of the output voltage as a function of the input bias (Fig. 9–24c). The VTC gives us information, for example, about how much noise the digital circuit can handle, and the speed of switching of the logic gates. There are five key operating points (marked I through V) on the VTC. They include V_{OH}, corresponding to the logic high or "1", V_{OL}, corresponding to the logic low or "0", and V_m corresponding to the intersection of a line with unity slope (where $V_{out} = V_{in}$) with the VTC. V_m, known as the logic threshold (not to be confused with the V_T of the MOSFETs), is important when two inverters are cross-coupled in a flip-flop circuit because the output of one is fed to the input of the other, and vice versa. Two other key points are the unity gain points, V_{IL} and V_{IH}. The significance of these points is that if the input voltage is between them, the change of the input is amplified and we get a larger change of the output voltage. Outside of this operating range, the change of the input voltage is attenuated. Clearly, any noise voltage which puts the input voltage between V_{IL} and V_{IH} would be amplified, and lead to a potential problem with the circuit operation.

Let us see how to go about determining the VTC. From the circuit in Figure 9–24a, we see that in the output loop from the power supply to ground, the current through the resistor load is the same as the drain current of the MOSFET. The power supply voltage is equal to the voltage drop across the resistor plus the drain-to-source voltage. To determine the VTC, we superimpose the load line of the load element (in this case a straight line for an ohmic resistor) on the output characteristics of the MOSFET (Figure 9–24b). This is similar to our load line discussion in Section 6.1.1. The load line goes through V_{DD} on the voltage axis because when the current in the output loop is zero, there is no voltage drop across the resistor and all the voltage appears across the MOSFET. On the current axis, the load line goes through V_{DD}/R_L because when the voltage across the MOSFET is zero, the voltage across the resistor must be V_{DD}. As we change the input bias, V_{in}, we change the gate bias on the MOSFET, and thus in Figure 9–24b, we go from one constant V_G curve to the next. At each input bias (and a corresponding constant V_G curve) the intersection of the load line with that curve tells us what the drain bias V_D is, which is the same as the output voltage. This is because at the point of intersection, we satisfy the condition that for the d-c case where the capacitor does not play any role, the current through the resistor is the same as the MOSFET current. (Later on, we shall see that in the a-c case when the logic gates are switched, we need to worry about the displacement current through the capacitor when it is charged or discharged.) It can be clearly seen from Fig.9–24c that as the input voltage (or V_G) changes from low to high, the output voltage decreases from a high of V_{DD} to a low of V_{OL}. We can solve for any point on this VTC curve analytically simply by recognizing whether the MOSFET is in the linear region or in saturation, using the corresponding drain current expression [Eq. (6–49) or (6–53)] and setting it equal to the resistor current. As an illustration, suppose we want to determine the logic "0" level, V_{OL}. This occurs when the input V_G is high and the output V_D is low, putting the transistor in the linear region. Using equation (6–49), we can write

$$I_D = k\left[V_G - V_T - \frac{V_D}{2}\right]V_D = k\left[V_{DD} - V_T - \frac{V_{0L}}{2}\right]V_{0L} \qquad (9\text{–}3a)$$

Since in the d-c case the current through the MOSFET is the same as that through the resistor,

$$I_D = I_L = \frac{V_{DD} - V_{0L}}{R_L} \qquad (9\text{–}3b)$$

We can solve for V_{OL} if we know R_L and the MOSFET parameters. Alternatively, we can design the value of R_L to achieve a certain V_{OL}. What might dictate the choice of R_L? We shall see later in this section that for many applications we use two of these inverters in a cross-coupled manner to form a bistable flip-flop. The output of one flip-flop is fed back to the input of the other, and vice versa. Clearly, the V_{OL} must be designed to be significantly less than the V_T of the MOSFET. Otherwise, neither MOSFET will be fully turned off, and the flip-flop will not function properly. Similarly, all the other points on the VTC can be determined analytically by using the appropriate MOSFET drain current expression, and setting it equal to the current through the resistor.

We can make some general observations from this analysis. We want the transition region of the VTC (between V_{IL} and V_{IH}) to be as steep (i.e., high gain) as possible, and the transition should be around $V_{DD}/2$. High gain guarantees a high-speed transition from one logic state to the other. It is necessary to increase the load resistance to increase this gain in the transition region.

The transition around $V_{DD}/2$ guarantees high *noise immunity* or *margin* for both logic "1" and logic "0" levels. To appreciate the importance of noise immunity, we must recognize that in combinatorial or sequential digital circuits, the output of one inverter or logic gate is often fed into the input of the next stage. Noise immunity is a measure of how much noise voltage the circuit can tolerate at the input, and still have the digital outputs be at the correct logic level in the subsequent stages. For example in Fig. 9–24c, if the input is nominally at zero, the output should be high (logic "1"). If this is fed into another inverter stage, its output should be low, and so on. If a noise spike causes the input of the first stage to go above V_m, the output voltage decreases sufficiently to potentially create errors in the digital levels in subsequent stages. Having a symmetric transition of the VTC around $V_{DD}/2$ ensures that the noise margin is high for both logic levels.

One problem with the resistor load inverter is that the V_{OL} is low, but not zero. This, coupled with the fact that the load element is a passive resistor that cannot be turned off, causes high standby power dissipation in this circuit. These problems are addressed by the CMOS structure described next.

We can determine the VTC for the CMOS case exactly as for the resistor load, although the math is somewhat more messy (Fig. 9–25). As mentioned previously, for an input voltage V_{in}, the V_G of the NMOSFET is V_{in}, but that of the PMOSFET is $V_{in}-V_{DD}$. Similarly, if the output voltage is V_{out}, the V_D of the NMOSFET is V_{out}, but that of the PMOS is $V_{out}-V_{DD}$. The load element now is not a simple resistor with a linear current–voltage relationship, but instead is the PMOSFET device whose "load line" is a set of I_D–V_D output characteristics (Fig. 9–25b). The V_{out} can be determined as a function of the V_{in} by recognizing whether the NMOSFET and the PMOSFET are in the linear or saturation region of their characteristics, and using the appropriate current expressions. At each point, we would set the NMOSFET I_D equal to the PMOSFET I_D.

As in the case of the resistive load, there are five key points on the VTC (Fig. 9–25c). They are logic "1" equal to V_{DD}, logic "0" equal to 0, logic threshold V_m where $V_{in} = V_{out}$, and the two unity gain points, V_{IH} and V_{IL}. In region I in Fig. 9–25c, the NMOSFET is OFF, and $V_{out} = V_{DD}$. Similarly, in region V, the PMOSFET is OFF, and $V_{out} = 0$. We can illustrate the calculation in region II, where the NMOSFET is in saturation and the PMOSFET is in the linear region. In this case, we must use Eq. (6–53) for the saturation drain current of the NMOSFET.

$$I_{DN} = \frac{k_N}{2}(V_{in} - V_{TN})^2 \tag{9–4a}$$

On the other hand, we must use Eq. (6–49) for the PMOSFET in the linear region.

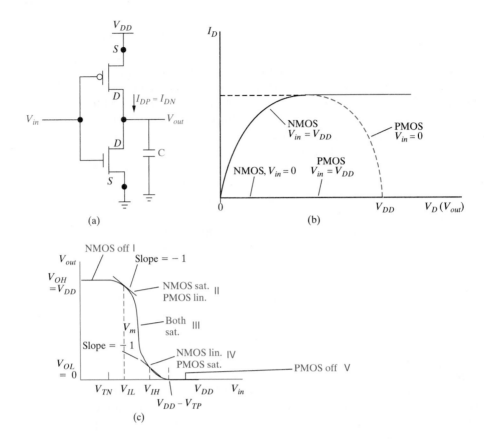

(a)

(b)

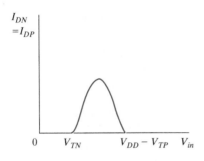

(c)

(d)

Figure 9–25

CMOS inverter voltage transfer characteristics: (a) NMOSFET with PMOSFET load and load parasitic capacitance, C; (b) determination of VTC by superimposing load line (output characterisitics of PMOSFET shown as dotted line) on NMOSFET output characteristics; (c) VTC showing output voltage as a function of input voltage. The five key points on the VTC are logic high (V_{OH}), logic low (V_{OL}), unity gain points (V_{IL} and V_{IH}), and logic threshold where input equals output (V_m); (d) switching current from V_{DD} to ground when the input voltage is in a range where both the NMOSFET and the PMOSFET are on.

$$I_{DP} = k_P \left[(V_{DD} - V_{in}) + V_{TP} - \frac{(V_{DD} - V_{out})}{2} \right] (V_{DD} - V_{out}) \qquad (9\text{–}4b)$$

Here V_{TN} and V_{TP} are the n- and p-channel threshold voltages. In the d-c case, since the output load capacitor does not play a role, the drain current through the PMOSFET device must be equal in magnitude to that through the NMOSFET. (However, for the a-c case, we need to consider the displacement current through the capacitor.)

$$I_{DN} = I_{DP} \qquad (9\text{–}5a)$$

Using Eq. (6–53) for the NMOSFET in saturation, and Eq. (6–49) for the PMOSFET in the linear region.

$$\frac{k_N}{2k_P}(V_{in} - V_{TN})^2 = \left[V_{DD} - V_{in} + V_{TP} - \frac{V_{DD} - V_{out}}{2} \right](V_{DD} - V_{out})$$

$$= \left[\frac{V_{DD}}{2} - V_{in} + V_{TP} + \frac{V_{out}}{2} \right](V_{DD} - V_{out}) \qquad (9\text{–}5b)$$

From Eq. (9–5b), we can get an analytical relation between the input and output voltages valid in Region II. We can get similar relationships in the other regions of the VTC.

Region IV is very similar to region II in Fig. 9–25c, except that now the NMOS is in the linear regime, while the PMOSFET is in saturation. In region III, both the NMOSFET and the PMOSFET are in saturation. Since the output impedance of a MOSFET is very high, this is tantamount to a semi-infinite load resistor, thereby resulting in a very steep transition region. That is why a CMOS inverter switches faster than the resistor load case. The CMOS inverter is also preferable because in either logic state (regions I or V), either the NMOSFET or the PMOSFET is OFF, and the standby power dissipation is very low. In fact, the current in either logic state corresponds to the (very low) source/drain diode leakage.

We want the transition region (region III) to be at $V_{DD}/2$ from the point of view of symmetry and noise immunity. Once again, by setting the NMOSFET I_D equal to that of the PMOSFET, it can be shown that the transition occurs at

$$V_{in} = (V_{DD} + \chi V_{TN} + V_{TP})/(1 + \chi) \qquad (9\text{–}6a)$$

where

$$\chi = \left(\frac{k_N}{k_P} \right)^{\frac{1}{2}} = \frac{\left[\overline{\mu}_n C_i \left(\frac{Z}{L} \right)_N \right]^{\frac{1}{2}}}{\left[\overline{\mu}_P C_i \left(\frac{Z}{L} \right)_P \right]^{\frac{1}{2}}} \qquad (9\text{–}6b)$$

We can design V_{in} to be at $V_{DD}/2$ by choosing $V_{TN} = -V_{TP}$ and $\chi = 1$. Since the effective electron mobility in the channel of a Si MOSFET is roughly twice that of the hole mobility, we must design CMOS circuits to have a $(Z/L)_P = 2(Z/L)_N$ to achieve the condition $\chi = 1$.

We can combine such CMOS inverters to form other logic gates for combinatorial circuits such as NOR gates and NAND gates (Fig. 9–26). The truth tables for these gates are shown in Fig. 9–27. By applying combinations of logic "high" or logic "low" to inputs A and B, we get the output states corresponding to the truth tables. The synthesis of logic circuits corresponding to these truth tables can be done using Boolean algebra and De Morgan's laws. The upshot of these laws is that any logic circuit can be made using inverters in conjunction with either NAND gates or NOR gates. Which would be preferable from a device physics point of view? We see from Fig. 9–26, that in the NOR gate the PMOSFET devices T_3 and T_4 are in series, while for the

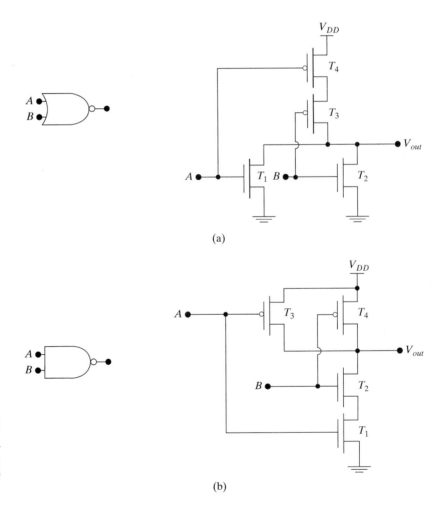

(a)

(b)

Figure 9–26
Logic gates and CMOS implementation of (a) NOR gate (b) NAND gate.

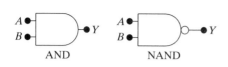

Input		Output Y	
A	B	AND	NAND
0	0	0	1
0	1	0	1
1	0	0	1
1	1	1	0

(a)

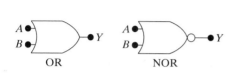

Input		Output Y	
A	B	OR	NOR
0	0	0	1
0	1	1	0
1	0	1	0
1	1	1	0

(b)

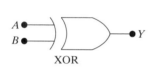

Input		Output Y
A	B	XOR
0	0	0
0	1	1
1	0	1
1	1	0

(c)

Figure 9–27
(a) AND/NAND logic symbols and truth table. (b) OR/NOR logic symbols and truth table. (c) XOR logic symbol and truth table.

NAND it is the NMOSFETs (T_1 and T_2). Since the electron channel mobilities are twice hole mobilities, we would obviously prefer NMOSFETs. Therefore, the preferred choice is NAND, along with inverters.

We can also estimate the power dissipation in the inverter circuit. We already know that the standby power dissipation is very small, being governed by the OFF state leakage current of either the NMOSFET or the PMOSFET, depending on the logic state. This leakage current depends on the source and drain diode leakage currents, or if the V_T is low, on the subthreshold leakage of the MOSFET that is turned OFF (see Section 6.5.7).

While the inverter is switching, there is also a transient current from the power supply to ground when both the transistors are ON (see Fig. 9–25d). This is known as the *switching current* or the *commutator current*. The magnitude of this current will clearly depend on the values of V_{TN} and V_{TP}. The higher the magnitudes of the thresholds, the less is the input voltage swing for which both the PMOSFET and the NMOSFET will be ON while the input voltage is being changed. The commutator current is then less during switching, which is desirable from a reduced power dissipation point of view. However, this reduction of power dissipation by increasing threshold voltages is obtained at the expense of reduced drive current and, therefore, overall speed of the circuit.

The speed penalty due to reduction of drive current is because in a digital circuit, while switching between logic states, the MOSFET drive currents must charge and discharge the parasitic capacitors that are inevitably associated with the output node (Fig. 9–25a). There is also some power dissipation involved in charging and discharging load capacitors attached to the output of the inverter. This load capacitance depends mostly on the input gate oxide capacitance of the MOSFETs of the next inverter stage (or logic gate) that this inverter (or logic gate) may be driving, along with some small parasitic capacitances. The input load capacitance of a single inverter is given by gate oxide capacitance per unit area C_i times the device areas.

$$C_{\text{inv}} = C_i \{(ZL)_N + (ZL)_P\} \tag{9–7}$$

The total load capacitance is then multiplied by a factor that depends on the *fan-out* of the circuit, which is the number of gates that are being driven in parallel by the inverter (or logic gate). It is necessary to add up the load capacitances for all the inverters or logic gates that are being driven by this inverter stage. The energy expended in charging up the equivalent load capacitor, C, is the integral of the product of the time-dependent voltage times the time-dependent displacement current through the capacitor during the charging cycle.

$$E_C = \int i_p(t)[V_{DD} - v(t)]dt$$
$$= V_{DD}\int i_p(t)dt - \int i_p(t)v(t)dt \tag{9–8a}$$

The energy stored in C is then obtained by considering the displacement current $(i_p(t) = C\,dv/dt)$ through the capacitor:

$$E_c = V_{DD}\int C\frac{dv}{dt}\,dt - \int Cv\frac{dv}{dt}\,dt = CV_{DD}\int_0^{V_{DD}} dv - C\int_0^{V_{DD}} v\,dv = CV_{DD}^2 - \frac{1}{2}CV_{DD}^2 \tag{9–8b}$$

Similarly, during one discharging cycle we get

$$E_d = \int i_n(t)v(t)dt = -\int_{V_{DD}}^0 Cv\,dv = \frac{1}{2}CV_{DD}^2 \tag{9–9}$$

If the inverter (or gate) is being charged and discharged at a frequency f, we get an active power disspation

$$P = CV_{DD}^2 f \tag{9–10}$$

In addition to power dissipation, we are also concerned with the speed of logic circuits. The speed of a gate, such as the one shown in Fig. 9–25, is determined by the *propagation delay* time t_P. We define the time required for the output to go from the logic high V_{OH} to $V_{OH}/2$ as t_{PHL}. The converse (to go from logic low $V_{OL}(=0)$ to $V_{OH}/2$) is defined as t_{PLH}. We can write down approximate estimates for these times by recognizing that for the output to go from high to low (or logic "1" to "0"), the NMOSFET has to discharge the output node towards ground. During this period, the NMOSFET will be in saturation. Assuming a constant saturation current as an approximation, we obtain from Eq. (6–53)

$$t_{PHL} = \frac{\frac{1}{2}CV_{DD}}{I_{DN}} = \frac{\frac{1}{2}CV_{DD}}{\frac{k_N}{2}(V_{DD} - V_{TN})^2} \qquad (9\text{--}11\text{a})$$

This is the decrease of charge on the capacitor divided by the discharging current. Conversely,

$$t_{PLH} = \frac{\frac{1}{2}CV_{DD}}{I_{DP}} = \frac{\frac{1}{2}CV_{DD}}{\frac{k_P}{2}(V_{DD} + V_{TP})^2} \qquad (9\text{--}11\text{b})$$

Knowing these times helps us considerably in designing circuits that meet the speed requirements of a design. Of course, for accurate numerical estimates of these propagation time delays or of the power dissipation we need to use computers. A very popular program to do so is the Simulation Program with Integrated Circuit Emphasis (SPICE). This discussion illustrates that the device physics plays an important role in the design and analysis of such circuits.

9.5.2 Semiconductor Memories

In addition to logic devices such as microprocessors, integrated circuits depend on semiconductor memories. We can illustrate many key MOS device physics issues by looking at three of the most important types of semiconductor memory cells: the *static random access memory (SRAM)*, the *dynamic random access memory (DRAM)*, and the non-volatile *flash memory cell*. SRAMs and DRAMs are volatile in the sense that the information is lost if the power supply is removed. For flash memories, however, information is stored indefinitely. For SRAMs, the information is static, meaning that as long as the power supply is on, the information is retained. On the other hand, the information stored in the cells of a DRAM must periodically be refreshed because stored charge representing one of the logic states leaks away rapidly. The refresh time must be short compared with the time needed for stored charge to degrade.

The overall organization of all these types of memories is rather similar, and is shown in Fig. 9–28. We will not describe the memory organization in great detail here, but will instead focus on the device physics. We need to know the type of cell that is used at the intersection of the rows or word-lines, and the columns or bitlines. These memories are all random access in the sense that the cells can be addressed for write or read operations in any order, depending on the row and column addresses provided to the address pins, unlike memories such as hard disks or floppy disks on a computer which can only be addressed sequentially. Generally, the same set of pins is used for both the row and the column addresses, in order to save pin count. This forces us to use what is known as address multiplexing. First, the row addresses are provided at the address pin, and decoded using row decoders. For N row addresses, we can have 2^N rows or wordlines. The row decoders then cause the

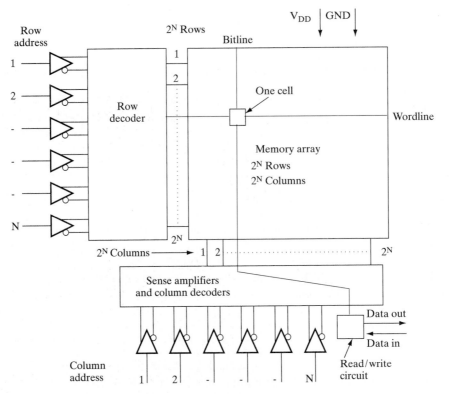

Figure 9–28
Organization of a random access memory (RAM): The memory array consists of memory cells arranged in an orthogonal array. There is one memory cell at the intersection of one row (wordline) and one column (bitline). To address a particular memory cell, the N row addresses are latched in from the N address pins, and decoded by the 2^N row decoders. All the memory cells on the selected row are read by the 2^N sense amplifiers. Of those, a cell (one bit) or group of cells (byte or word) is selected for transfer to the data output buffers depending on the column addresses that are decoded by the 2^N column decoders. Generally, to save pin count on the package, the N column addresses are provided in a multiplexed fashion to the same N address pins as the row addresses, *after* the row addresses have already been latched in.

selected wordline to go high, so that all the 2^N cells (corresponding to N column addresses) on this wordline are accessed for either read or write, through sense amplifiers at the end of the 2^N columns or bitlines. After the appropriate row has been decoded, the appropriate column addresses are provided to the same address pins, and the column decoders are used to select the bit or group of bits (known as *byte* or *word*) out of all the 2^N bits on the selected wordline. We can either write into or read from the selected bit (or group of bits) using the sense amplifiers, which are basically flip-flops used as differential amplifiers.

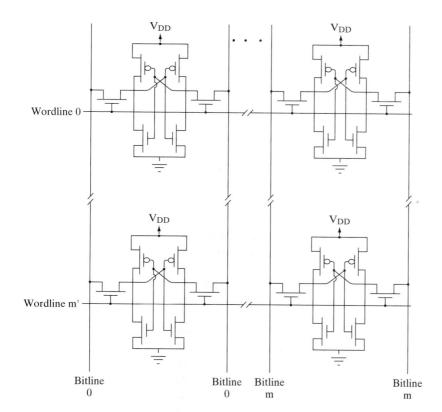

Figure 9–29
An array of 4 CMOS SRAM cells. The bitline and $\overline{\text{bitline}}$ (bitline-bar) are logical complements of each other.

Wordline 0

Wordline m'

Bitline 0   Bitline 0   Bitline m   Bitline m

SRAMs.   A group of four six-transistor CMOS SRAM cells is shown in Fig. 9–29. Each cell is found in this case at the intersection of a row or wordline, and a column or bitline (along with its logical complement known as bitline-bar). The cell is a flip-flop, consisting of two cross-coupled CMOS inverters. Clearly, it is bistable: if the output of one inverter is high (corresponding to the NMOSFET being OFF, and the PMOSFET ON), that high voltage is fed to the input of the other cross-coupled inverter, and the output of the other inverter will be low. This is one logic state (say "1") of the SRAM. Conversely, the other stable state of the flip-flop can be considered to be the other logic state (say "0"). Many of the device issues are identical to those described in Section 9.5.1 in connection with the VTC of inverters. We aim for a symmetric transition from V_{OH} to V_{OL} at $V_{DD}/2$ with a high gain in the transition region, to improve noise immunity and speed of convergence of the SRAM cell. The speed of convergence determines how fast the SRAM flip-flop latches into one stable logic state or the other. The cells are accessed through two access transistors whose gates are controlled by the wordline. That is why this is called a 6-transistor cell. Other SRAM cells use load resistors in the inverters, rather than PMOSFETs, leading to a 4-transistor, 2-resistor cell. As discussed in Section 9.5.1, the CMOS cell has superior performance, but at the expense of occupying more area.

Unless the row decoders cause a particular wordline to go high, the SRAM cells on that wordline are electrically isolated. By selecting a particular wordline, the access transistors on that row are turned ON and act as logic transmission gates between the output nodes of the SRAM cell and the bitline and its complement, the bitline-bar. During a read operation, the bitline and its complement are both precharged to the same voltage. Once the access transistors are turned ON, a small voltage differential develops between bitline and bitline-bar because the output nodes of the SRAM are at different voltages (0 and V_{DD}). The voltage differential that is established is due to a charge redistribution that occurs between the parasitic capacitance associated with the output nodes of the SRAM and the bitline capacitance. This voltage difference is amplified by the sense amplifiers. As mentioned previously, the sense amplifiers are differential amplifiers, very similar in configuration to the SRAM flip-flop cell itself. The bitline and bitline-bar (complement of the bitline) are fed to the two inputs of the sense amplifier, and the voltage differential is amplified until the voltage separation is V_{DD}.

DRAMs. The DRAM cell structure is shown in Fig. 9–30. The information is stored as charge on an MOS capacitor, which is connected to the bitline through a switch which is an MOS *pass transistor*, the gate of which is controlled by the wordline. There is one such cell at each intersection of the orthogonal array of wordlines and bitlines, exactly as for SRAMs. When the wordline voltage becomes higher than the V_T of the pass transistor (MOSFET between the bitline and the storage capacitor), the channel is turned ON, and connects the bitline to the MOS storage capacitor. The gate of this capacitor (or capacitor plate) is permanently connected to the power supply voltage V_{DD}, thereby creating a potential well under it which tends to be full of inversion electrons for a p-type substrate (Fig. 9–31a). We apply either 0V to the bitline (generally corresponding to logic "0"), or V_{DD} (corresponding

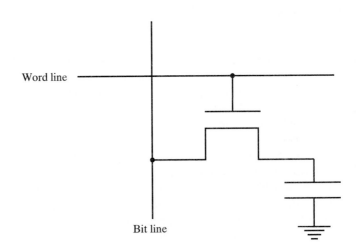

Figure 9–30
One transistor, one capacitor DRAM cell equivalent circuit: the storage MOS capacitor is connected to the bitline through the pass transistor (MOSFET switch) whose gate is controlled by the wordline.

Word line

Bit line

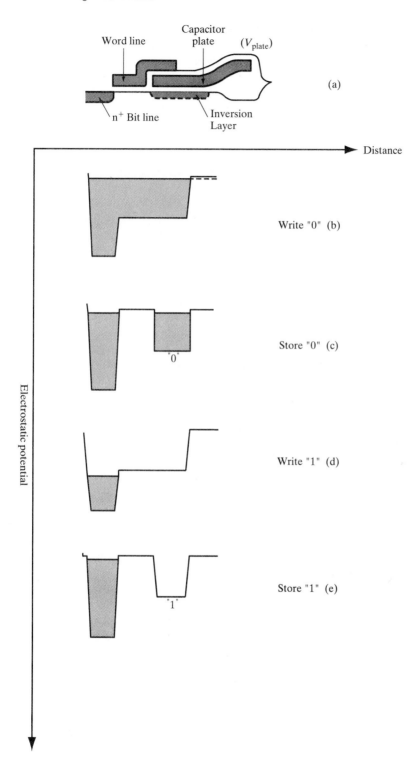

Figure 9–31
DRAM cell struc-
ture and cell oper-
ation: (a) cell
structure corre-
sponding to equiv-
alent circuit of
Fig. 9–30; (b)–(e)
potentials under
bitline, pass tran-
sistor channel and
storage capacitor
during write "0",
store "0", write
"1" and store "1"
operations. It
shows that the
logic state "0"
corresponds a
filled potential
well (stable state),
while the logic
state "1" corre-
sponds to an
empty potential
well (unstable
state) that is filled
up over time by
minority carriers
generated in the
substrate and
leakage through
the pass transistor.

to logic "1"), and the appropriate voltage appears as the substrate potential of the MOS capacitor. For a stored "0" in the cell, the potential appears as the substrate potential of the MOS capacitor. For a stored "0" in the cell, the potential well that is created under the MOS capacitor by the plate voltage is full of inversion charge (Fig. 9–31b,c). When the wordline voltage is turned low such that the MOS pass transistor is turned off, the inversion charge under the storage capacitor stays the same; this is the stable state of the capacitor. On the other hand, if a positive voltage (V_{DD}) is applied to the bitline, it draws out the inversion electrons through the pass transistor (Fig. 9–31d,e). When the pass transistor is cut off, we end up with an empty potential well under the MOS capacitor plate. Over a period of time, the potential well tends to be filled up by minority carrier electrons that are constantly created by thermal generation-recombination in the substrate and are collected under the charged MOS capacitor plate. Hence, the logic "1" degrades towards the logic "0". That is why a DRAM is considered to be "dynamic" unlike an SRAM. It is necessary to periodically restore the logic levels or "refresh" the stored information.

There are interesting device physics issues regarding the pass transistor. This is like the access transistor in the SRAM, or a logic transmission gate. We see that in this MOSFET, neither the source nor the drain is permanently grounded. In fact, which side acts as the source and which as the drain depends on the circuit operation. When we are writing a logic "1" into the cell, the bitline voltage is held high ($=V_{DD}$). As this voltage is written into the cell, it is as if the source of the pass transistor gets charged up to V_{DD}. Another way of looking at this is that with respect to the source, the substrate bias of the pass transistor is $-V_{DD}$. The body effect of the MOSFET (Section 6.5.6) causes its V_T to increase. This is very important because for the pass transistor to operate as a transmission gate it is necessary that it be in the linear regime throughout, and not get into saturation (with a concomitant voltage drop across the pinch-off region). Hence, the gate or the wordline voltage must be held at V_{DD} (which is the final voltage of the source/drains) *plus* the V_T of the MOSFET, taking body effect into account. It is also important to make sure that the leakage of the pass transistor is low enough to satisfy refresh requirements of the DRAM. Not only must the source/drain diodes be low leakage, but the V_T and the subthreshold slope must be optimized such that subthreshold leakage for the grounded wordline case is low enough.

The stored charge difference between the two logic states can be determined by looking at the capacitance–voltage (C–V) characteristics of the MOS capacitor (Fig. 9–32). For a stored "1", essentially there is a substrate bias applied to the MOS capacitor, which raises its V_T due to the body effect (Section 6.5.6). Hence, the C–V characteristics shift to the right for a stored "1". Since the MOS capacitance is not a fixed capacitance, but is voltage dependent, we saw earlier that it must be defined in a differential form (Eq. 6-34a). Alternatively, we can write down the stored charge under the capacitor as

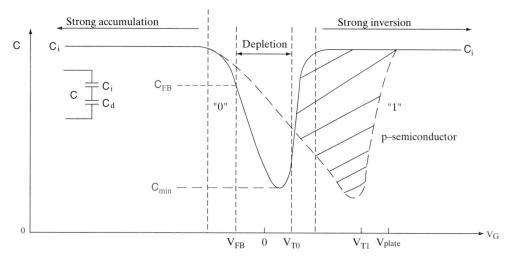

Figure 9–32
C–V characteristics of DRAM MOS capacitor in stored "0" and stored "1" states. The difference of area under the C–V curves shown by hatch-marked pattern reflects the charge differential between the two states.

$$Q = \int C(V)dV \qquad\qquad (9\text{–}12)$$

This is simply the area under the C–V curve. The charge differential that distinguishes the logic "1" and the logic "0" is the difference of areas under the capacitance–voltage curves in the two cases (Fig. 9–32).

When reading the cell, the pass transistor is turned on, and the MOS storage capacitor charge is dumped on the bitline capacitance C_B, precharged to V_B (typically $= V_{DD}$). The swing of the bitline voltage will clearly depend on the voltage V_C stored in the storage cell capacitance C_C. As in the case of the SRAM, the change of the bitline voltage depends on the capacitance ratio between the bitline and the cell. To do differential sensing in the case of DRAMs, we do not use two bitlines per cell as for SRAMs. Instead, we compare the bitline voltage for the selected cell with a reference bitline voltage to which is connected a dummy cell whose MOS capacitance, C_D, is roughly half that of the actual cell capacitance, C_C. Typical values of C_B, C_C and C_D in a DRAM are 800 fF, 50 fF and 20 fF, respectively. The voltage differential that is applied to the sense amplifier then becomes (Fig. 9–33)

$$\begin{aligned}
\Delta V &= \frac{C_C V_C + C_B V_B}{C_B + C_C} - \frac{C_C V_D + C_B V_B}{C_B + C_D} \\[2mm]
&= \frac{(V_B - V_D)C_B C_D - (V_B - V_C)C_B C_D - (V_C - V_D)C_C C_D}{(C_B + C_C)(C_B + C_D)} \qquad (9\text{–}13a)
\end{aligned}$$

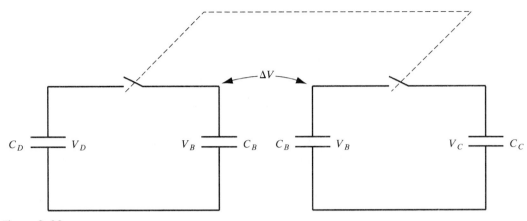

Figure 9–33
Equivalent circuit showing charge redistribution between cell capacitance (C_C) and bit line capacitance (C_B) on one side, versus dummy cell (C_D) and bit line capacitance (C_B) on other.

If V_D is set to zero, the expression simplifies to:

$$\Delta V = \frac{(V_B C_B + V_C C_C)C_D - (V_B - V_C)C_B C_C}{(C_B + C_C)(C_B + C_D)} \quad (9\text{–}13b)$$

Putting the cell voltage V_C equal to 0V or 5 V, and typical, acceptable bitline-to-cell capacitance ratios C_B/C_C (= 15–20) in Eq. (9–13b), we get different polarities of the differential voltage of the order of ± 100 mV for logic "1" and logic "0", respectively, which can be detected by sense amplifiers. From Eq. (9–13b), it can be seen that for much higher bitline-to-cell capacitance ratios, the swing of the bitline voltage will be negligible, regardless of the cell voltage. The minimum required cell capacitance C_C is about 50 fF, governed by so-called soft errors. DRAMs, like everything else on Earth, are constantly being bombarded by cosmic rays, and high energy alpha particles can create electron–hole pairs in semiconductors. A typical collected charge due to one of these events is about 100 fC. This spurious charge can be neglected if the cell capacitance is 50 fF and 5 V is applied to the cell, for which the stored charge is roughly 250 fC. The DRAM cell then becomes immune to typical alpha particle hits.

Maintaining a cell capacitance of 50 fF as the cell dimensions are reduced from one generation of DRAM to the next is a tremendous technological challenge. One way to look at this problem is shown in Fig. 9–34. The challenge is to store more charge per unit area on the planar surface (A_s) of the Si substrate. Approximating the MOS capacitance as a fixed, voltage-independent capacitor, one can write the stored charge Q as

$$Q = CV = (\epsilon A_C/d)V \quad (9\text{–}14)$$

Time	Past	Present	Future
Approaches	Scaled dielectric	Trench/ stacked capacitor	Alternate dielectric
Problems	Tunneling & wearout	Fabrication	Material properties

$$\frac{Q}{A_s} = \frac{CV}{A_s} = \frac{V}{d} \times \frac{A_c}{A_s} \times \epsilon$$

Figure 9–34

Various approaches (past, present and future) of achieving higher DRAM cell capacitance and charge storage density without increasing cell size. A_s = Area on wafer taken by capacitor; A_c = Area of capacitor. For a planar capacitor $A_c = A_s$; however, for non-planar structures $A_c > A_s$. $C = A_c \epsilon / d$ is the capacitance; and $Q = CV$ is the total stored charge in a fixed, voltage-independent capacitor.

where ϵ is the permittivity of the dielectric, d is its thickness, and A_C is the capacitor area. As shown in Fig. 9–34, the historical way of achieving the desired capacitance has been to scale the dielectric thickness, d. But that runs into the problems discussed in Section 6.4.7. Another approach, which is being taken currently, is to use fabrication schemes to increase the area devoted to the MOS storage capacitor, A_c, even as we reduce the planar surface area on the wafer, A_s, used for making this storage capacitance. Obviously, this can be done by moving away from a purely planar structure, and exploiting the third dimension. We can go down into the Si by digging "trenches" in the substrate with RIE and forming a trench storage capacitor on the sidewalls of the trench (Figure 9–35a). Alternatively, we can go up from the substrate by stacking multiple layers of capacitor electrodes to increase the "stacked" capacitor area (Fig. 9–35b). Other tricks that have been tried are to purposely create a rough polysilicon surface on the capacitor

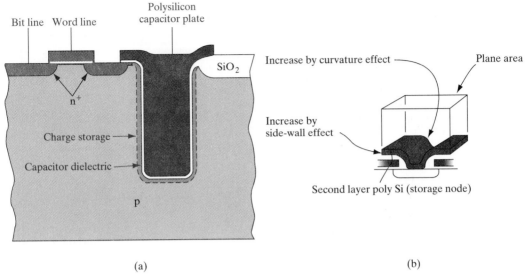

Figure 9–35
Increasing cell capacitance by exploiting the vertical dimension: (a) trench capacitors involve etching a trench in the substrate so that the larger area on the sidewalls can be used to increase capacitance; (b) stacked capacitors go "up" rather than "down" as in trenches, and increase capacitor area by using multiple polysilicon capacitor plates or "fins", as well as by exploiting the topgraphy of the cell surface.

plates to increase the surface area. In the future, alternative materials may be used. For example, the ferroelectrics have much higher dielectric constant than SiO_2 and offer larger capacitance without increasing area or reducing thickness. Promising materials include barium strontium titanate and zirconium oxide.

Flash Memories.   Another interesting MOS device is the flash memory, which is rapidly becoming the most important type of non-volatile memory. The memory cell structure is shown in Figure 9–36. It is very simple and compact, and looks just like a MOSFET, except that it has two gate electrodes, one on top of the other. The top electrode is the one that we have direct electrical access to, and is known as the control gate. Below that we have a so-called "floating" gate that is capacitively coupled to the control gate and the underlying silicon.

The capacitive coupling of the floating gate to the various terminals is illustrated in Fig. 9–36 in terms of the various coupling capacitance components. The floating gate and the control gate are separated by a stacked oxide–nitride–oxide dielectric in typical flash devices. The capacitance between these two gates is called C_{ONO} because of the oxide–nitride–oxide makeup of the dielectric stack. The total capacitance C_{TOT} is the sum of all the parallel components shown in Fig. 9–36.

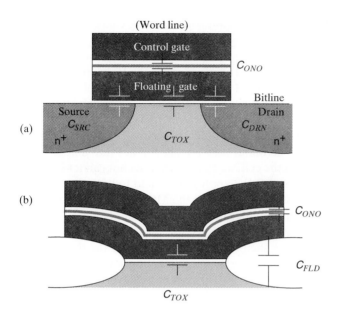

Figure 9–36
Flash memory cell
structure: (a) cell
structure shown
along the channel
length showing
the control gate
(wordline), float-
ing gate below it,
the source and the
drain (bitline); (b)
view of cell along
the width of the
MOSFET. The vari-
ous coupling
capacitors to the
floating gate are
shown.

$$C_{TOT} = C_{ONO} + C_{TOX} + C_{FLD} + C_{SRC} + C_{DRN} \qquad (9\text{--}15)$$

where C_{TOX} is the floating gate-to-channel capacitance through the tunnel oxide, C_{FLD} is the floating gate-to-substrate capacitance in the LOCOS field oxide region, and C_{SRC} and C_{DRN} are the gate-to-source/drain overlap capacitances.

Since it is isolated by the surrounding dielectrics, the charge on the floating gate Q_{FG} is not changed by (moderate) changes of the terminal biases.

$$Q_{FG} = 0 = C_{ONO}(V_{FG} - V_G) + C_{SRC}(V_{FG} - V_S) + C_{DRN}(V_{FG} - V_D) \quad (9\text{--}16)$$

We assume that the substrate bias is fixed, and hence ignore the contributions from C_{TOX} and C_{FLD}, which couple the floating gate to the substrate. The floating gate voltage can be indirectly determined by the various terminal voltages, in terms of the gate, drain and source coupling ratios as defined in Eq. (9–17).

$$V_{FG} = V_G \cdot GCR + V_S \cdot SCR + V_D \cdot DCR \qquad (9\text{--}17)$$

where

$$GCR = \frac{C_{ONO}}{C_{TOT}}$$

$$DCR = \frac{C_{DRN}}{C_{TOT}}$$

$$SCR = \frac{C_{SRC}}{C_{TOT}}$$

The basic cell operation involves putting charge on the floating gate or removing it, in order to program the MOSFET to have two different V_T's, corresponding to two logic levels. We can think of the stored charge on the floating gate to be like the fixed oxide charge in the V_T expression (equation 6-38). If many electrons are stored in the floating gate, the V_T of an NMOSFET is high; the cell is considered to have been "programmed" to exhibit the logic state "1". On the contrary, if electrons have been removed from the floating gate, the cell is considered to have been "erased" into a low V_T state or logic "0".

How do we go about transferring charges into and out of the floating gate? To program the cell, we can use channel hot carrier effects that we discussed in Section 6.5.9. We apply a high field to both the drain (bitline) and floating gate (wordline) such that the MOSFET is in saturation. It was discussed in Section 6.5.9 that the high longitudinal electric field in the pinch-off region accelerates electrons toward the drain and makes them energetic (hot). We maximize such hot carrier effects near the drain pinch-off region in a flash device by making the drain junction somewhat shallower than the source junction (Fig. 9–37a). This can be achieved by a separate higher energy source implant that is masked in the drain region. If the kinetic energy of electrons is high enough, a few can become hot enough to be scattered into the floating gate. They must surmount the 3.1 eV energy barrier that exists between the conduction band of Si and that of SiO_2, or hot electrons can tunnel through the oxide (Fig. 9–37b). Once they get into the floating gate, electrons become trapped in the 3.1 eV potential well between the floating polysilicon gate and the oxides on either side. This barrier is extremely high

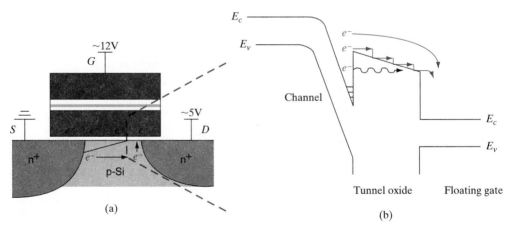

(a)

(b)

Figure 9–37
Hot carrier programming of the flash cell: (a) flash memory cell structure with typical biases required for writing into the cell. The channel of the MOSFET is pinched off in saturation; (b) band diagram along a vertical line in the middle of MOSFET channel showing hot electrons in the channel being injected across the gate oxide and getting trapped in the floating gate.

for a trapped (low kinetic energy) electron. Therefore the trapped electrons essentially stay in the floating gate forever, unless the cells is intentionally erased. That is why a flash memory is non-volatile.

To erase the cell, we use Fowler–Nordheim tunneling between the floating gate and the source in the overlap region (Fig. 9–38a). A high positive voltage (say ~12 V) is applied to the source with the control gate grounded. The polarity of the field is such that electrons tunnel from the floating gate into the source region, through the oxide barrier (Section 6.4.7). The band diagram (along a vertical line in this overlap region) during the operation is shown in Fig. 9–38b. Interestingly, in a flash device we make use of two effects that are considered to be "problems" in regular MOS devices: hot carrier effects and Fowler–Nordheim tunneling.

During the read operation, one applies a moderate voltage (~1 V) to the bitline (drain of the MOSFET), and a wordline (control gate) voltage V_{CG} that causes the capacitively coupled floating gate voltage to be between that of the high V_T and the low V_T state of the programmed flash memory cell (Fig. 9–39). There will be negligible drain current flow in the bit line (drain) for the high V_T case because the gate voltage is less than the threshold voltage. We will then interpret the selected cell as being in state "1". For the low V_T case, since the applied gate voltage is higher than the threshold voltage of the cell, there will be drain current flow in the bitline (drain), and this can be interpreted as state "0". The read operation can be understood by looking at the transfer characteristics of the MOSFET in the programmed and erased states (Fig. 9–39).

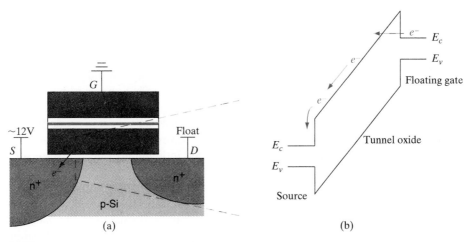

Figure 9–38
Fowler–Nordheim tunneling erasure: (a) flash memory cell structure with typical biases required for ersasing the cell; (b) band diagram as a function of depth in the gate/source overlap region of the MOSFET showing quantum mechanical tunneling of carriers from the floating gate into the oxide, and subsequent drift to the source.

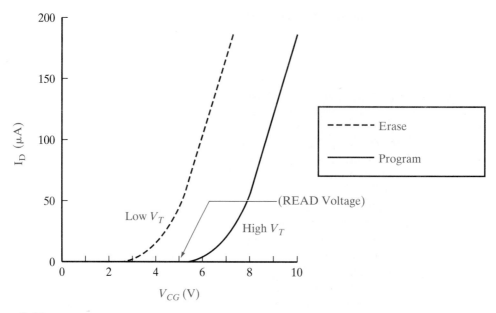

Figure 9–39
Drain (bitline) current versus control gate (wordline) voltage transfer characteristics of the MOSFET in a flash cell: if the cell is programmed to a high V_T (logic "1"), and a read voltage is applied to the wordline that is below this V_T, the MOSFET does not conduct, and there is negligible bitline current. On the other hand, if the cell had been erased to a low V_T state (logic "0") the MOSFET is turned ON, and there is significant bitline current.

9.6
TESTING,
BONDING, AND
PACKAGING

After the preceding discussions of rather dramatic fabrication steps in monolithic circuit technology, the processes of attaching leads and packaging the devices could seem rather mundane. Such an impression would be far from accurate, however, since the techniques discussed in this section are crucial to the overall fabrication process. In fact, the handling and packaging of individual circuits can be the most critical steps of all from the viewpoints of cost and reliability. The individual IC chip must be connected properly to outside leads and packaged in a way that is convenient for use in a larger circuit or system. Since the devices are handled individually once they are separated from the wafer, bonding and packaging are expensive processes. Considerable work has been done to reduce the steps required in bonding. We shall discuss the most straightforward technique first, which involves bonding individual leads from the contact pads on the circuit to terminals in the package. Then we shall consider two important methods for making all bonds simultaneously. Finally, we shall discuss a few typical packaging methods for ICs.

9.6.1 Testing

After the wafer of monolithic circuits has been processed and the final metallization pattern defined, it is placed in a holder under a microscope and is aligned for testing by a multiple-point probe (Fig. 9–40). The probe contacts the various pads on an individual circuit, and a series of tests are made of the electrical properties of the device. The various tests are programmed to be made

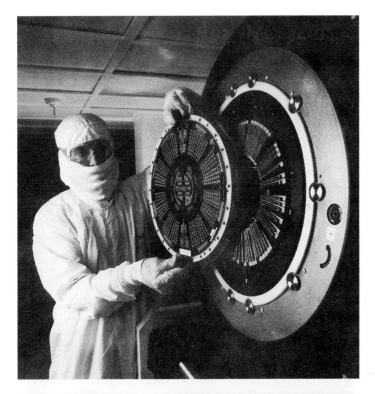

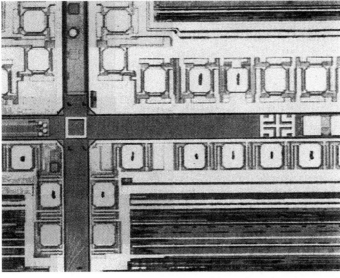

Figure 9–40 Automated probing of devices: (a) high speed testing of ICs is done using probe cards having a rigidly fixed array of probes that corresponds to the bond pad pattern on the IC to be tested. Many electrical signals are provided or measured by the automated tester at the various pins. After one chip is tested, the tester mechanically moves the wafer to the next die location; (b) array of Al bond pads near the chip periphery, some showing probe marks. The space between the arrays of bond pads is the "scribe line" along which the wafer will be sawed into individual chips after testing. (Photographs courtesy of Micron Technology.)

automatically in a very short time. These tests may take only milliseconds for a simple circuit, to several seconds for a complex ULSI chip. The information from these tests is fed into a computer, which compares the results with information stored in its memory, and a decision is made regarding the acceptability of the circuit. If there is some defect so that the circuit falls below specifications, the computer remembers that chip must be discarded. The probe automatically steps the prescribed distance to the next circuit on the wafer and repeats the process. After all of the circuits have been tested and the substandard ones noted, the wafer is removed from the testing machine, sawed between the circuits, and broken apart (Fig. 9–41). Then each die that passed the test is picked up and placed in the package. In the testing process, information from tests on each die can be stored to facilitate analysis of the rejected circuits or to evaluate the fabrication process for possible changes.

9.6.2 Wire Bonding

The earliest method used for making contacts from the monolithic chip to the package was the bonding of fine Au wires. Later techniques expanded wire bonding to include Al wires and several types of bonding processes. Here we shall outline only a few of the most important aspects of wire bonding.

If the chip is to be wire bonded, it is first mounted solidly on a metal lead frame or on a metallized region in the package. In this process a thin layer of Au (perhaps combined with Ge or other elements to improve the

Figure 9–41
Sawing of a wafer along scribe lines: After the wafer is tested, the "known good dies" are "inked" or identified. The wafer is then sawed into individual dies for subsequent packaging. (Photograph courtesy of Micron Technology.)

metallurgy of the bond) is placed between the bottom of the chip and the substrate; heat and a slight scrubbing motion are applied, forming an alloyed bond which holds the chip firmly to the substrate. This process is called *die bonding*. Generally, die bonding is done by a robotic arm that picks up each die, orients it, and places it for bonding. Once the chip is mounted, the interconnecting wires are attached from the various contact pads to posts on the package (Fig. 9–42).

In Au wire bonding, a spool of fine Au wire (about 0.007–0.002-inch diameter) is mounted in a *lead bonder* apparatus, and the wire is fed through a glass or tungsten carbide *capillary* (Fig. 9–43a). A hydrogen gas flame jet is swept past the wire to form a ball on the end. In *thermocompression bonding* the chip (or in some cases the capillary) is heated to about 360°C, and the capillary is brought down over the contact pad. When pressure is exerted by the capillary on the ball, a bond is formed between the Au ball and the Al pad (Fig. 9–43b). Then the capillary is raised and moved to a post on the package. The capillary is brought down again, and the combination of force and temperature bonds the wire to the post. After raising the capillary again, the hydrogen flame is swept past, forming a new ball (Fig. 9–43c); then the process is repeated for the other pads on the chip.

There are many variations in this basic method. For example, the substrate heating can be eliminated by *ultrasonic bonding*. In this method a tungsten carbide capillary is held by a tool connected to an ultrasonic transducer.

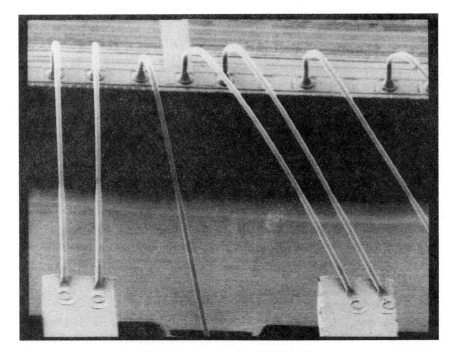

Figure 9–42
Attachment of leads from the Al pads on the periphery of the chip to posts on the package. (Photograph courtesy of Micron Technology.)

Figure 9–43
Wire bonding
techniques: (a)
capillary posi-
tioned over one of
the contact pads
for a ball (nail-
head) bond; (b)
pressure exerted
to bond the wire
to the pad; (c)
post bond and
flame-off; (d)
wedge bonding
tool; (e) pressure
and ultrasonic en-
ergy applied; (f)
post bond com-
pleted and wire
broken or cut for
next bond.

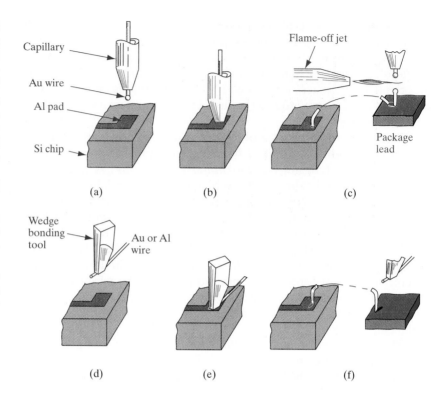

Figure 9–43
Wire bonding techniques: (a) capillary positioned over one of the contact pads for a ball (nailhead) bond; (b) pressure exerted to bond the wire to the pad; (c) post bond and flame-off; (d) wedge bonding tool; (e) pressure and ultrasonic energy applied; (f) post bond completed and wire broken or cut for next bond.

When it is in contact with a pad or a post, the wire is vibrated under pressure to form a bond. Other variations include techniques for automatically removing the "tail," which is left on the post in Fig. 9–43c. When the bond to the chip is made by exerting pressure on a ball at the end of the Au wire, it is called a *ball bond* or a *nail-head bond*, because of the shape of the deformed ball after the bond is made (Fig 9–44a).

Aluminum wire can be used in ultrasonic bonding; it has several advantages over Au, including the absence of possible metallurgical problems in bonds between Au and Al pads. When Al wire is used, the flame-off step is replaced by cutting or breaking the wire at appropriate points in the process. In forming a bond, the wire is bent under the edge of a wedge-shaped bonding tool (Fig. 9–43d). The tool then applies pressure and ultrasonic vibration, forming the bond (Fig. 9–43e and f). The resulting flat bond, formed by the bent wire wedged between the tool and the bonding surface, is called a *wedge bond*. A closeup view of ball and wedge bonds is given in Fig. 9–44.

9.6.3 Flip-chip Techniques

The time consumed in bonding wires individually to each pad on the chip can be overcome by several methods of simultaneous bonding. The *flip-chip* (a)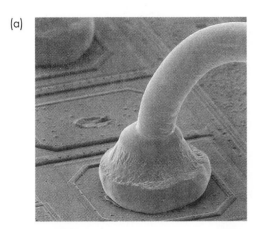

(b)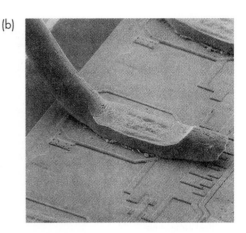

Figure 9–44
Scanning electron micrographs of a ball bond (a) and a wedge bond (b). (Photographs courtesy of Micron Technology, Inc.)

approach is typical of these methods. Relatively thick metal is deposited on the contact pads before the devices are separated from the wafer. After separation, the deposited metal is used to contact a matching metallized pattern on the package substrate.

In the flip-chip method, "bumps" of solder or special metal alloys are deposited on each contact pad. These metal bumps can be distributed over the die (Fig. 9–45). After separation from the wafer, each chip is turned upside down, and the bumps are properly aligned with the metallization pattern on the substrate. At this point, ultrasonic bonding or solder attaches each bump to its corresponding connector on the substrate. An obvious advantage of this method is that all connections are made simultaneously. Disadvantages include the fact that the bonds are made under the chip and therefore cannot be inspected visually. Furthermore, it is necessary to heat and/or exert pressure on the chip.

9.6.4 Packaging

The final step in IC fabrication is packaging the device in a suitable medium that can protect it from the environment of its intended application. In most cases this means the surface of the device must be isolated from moisture and contaminants and the bonds and other elements must be protected from corrosion and mechanical shock. The problems of surface protection are greatly minimized by modern passivation techniques, but it is still necessary to provide some protection in the packaging. In every case, the choice of package type must be made within the requirements of the application and cost considerations. There are many techniques for encapsulating devices, and the various methods are constantly refined and

Figure 9–45
Flip-chip bonding.
The Power PC
chip has metal-
lized "bumps" dis-
tributed over the
surface instead of
contact pads
around the periph-
ery. These bumps
are aligned with
the interconnec-
tion pattern on the
package and
bonded simultane-
ously. (Photo-
graph courtesy of
IBM Corp.)

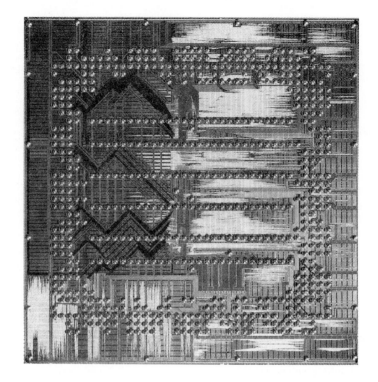

changed. Here we shall consider just a few general methods for the purpose of illustration.

In the early days of IC technology, all devices were packaged in metal headers. In this method the device is alloyed to the surface of the header, wire bonds are made to the header posts, and a metal lid is welded over the device and wiring. Although this method has several drawbacks, it does provide complete sealing of the unit from the outside environment. This is often called a *hermetically sealed* device. After the chip is mounted on the header and bonds are made to the posts, the header cap can be welded shut in a controlled environment (e.g., an inert gas), which maintains the device in a prescribed atmosphere.

Integrated circuits are now mounted in packages with many output leads (Fig. 9–46). In one version the chip is mounted on a stamped metal lead frame and wire bonding is done between the chip and the leads. The package is formed by applying a ceramic or plastic case and trimming away the unwanted parts of the lead frame.

Broadly speaking, packages can be through-hole-mount that involve inserting the package pins through holes on the printed circuit board (PCB) before soldering, or surface mount type where the leads do not pass through holes in the PCB. Instead, surface-mounted package leads are aligned to electrical contacts on the PCB, and are connected simultaneously by solder

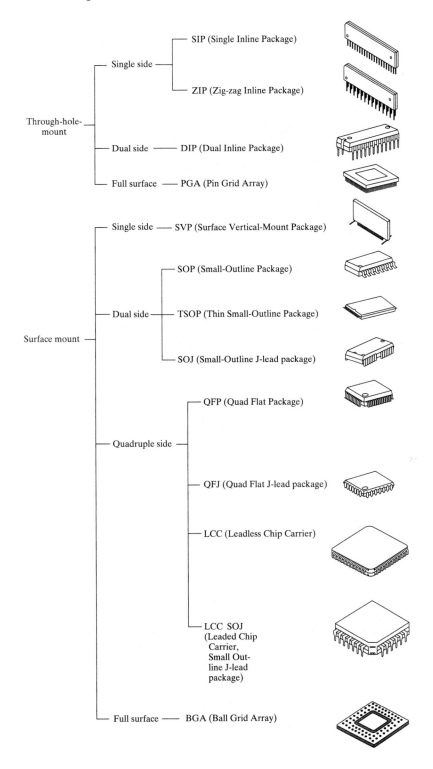

Figure 9–46
Various types of
packaging for
ICs: Packages can
be through-hole-
mount or surface
mount type, and
be made out of
plastic or ceramic.
The pins can be
on one side (SIP),
two sides (DIP) or
four side (quad)
of the package,
or distributed over
the surface of the
package (PGAs or
BGAs)

— SIP (Single Inline Package)

Single side

— ZIP (Zig-zag Inline Package)

Through-hole-
mount

— Dual side —— DIP (Dual Inline Package)

— Full surface —— PGA (Pin Grid Array)

— Single side —— SVP (Surface Vertical-Mount Package)

— SOP (Small-Outline Package)

— Dual side —— TSOP (Thin Small-Outline Package)

Surface mount

— SOJ (Small-Outline J-lead package)

— QFP (Quad Flat Package)

— Quadruple side

— QFJ (Quad Flat J-lead package)

— LCC (Leadless Chip Carrier)

— LCC SOJ
(Leaded Chip
Carrier,
Small Out-
line J-lead
package)

— Full surface —— BGA (Ball Grid Array)

reflow. Most packages can be made using ceramic or plastic (which is cheaper). The ICs are hermetically sealed for protection from the environment. The pins can be on one side (single inline or zig-zag pattern of leads), two sides (dual inline package or DIP) or four sides of the package (quad package) (Fig. 9–46). More advanced packages have leads distributed over a large portion of the surface of the package as in through-hole–mounted pin grid arrays (PGAs) (Fig. 9–47) or surface-mounted ball grid arrays (BGAs) (Fig. 9–48). By not restricting the leads to the edges of the package, the pin count can be increased dramatically, which is very attractive for advanced ULSI in which a large number of electrical leads must be accessed.

Since a sizable fraction of the cost of an IC is due to bonding and packaging, there have been a number of innovations for automating the process. These include the use of film reels that contain the metal contact pattern onto which the chips can be bonded. The film can then be fed into packaging equipment, where the position registration capabilities of a film reel can be used

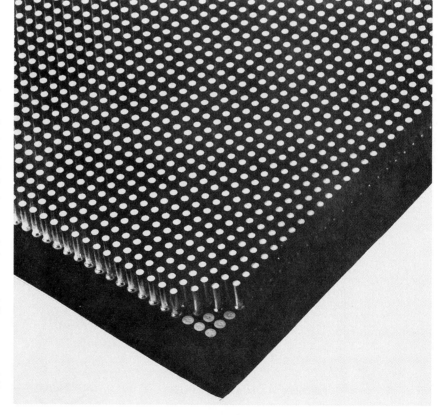

Figure 9–47 Ceramic column grid array (CCGA): This advanced ceramic package is a type of pin grid array made up of several hundred metal columns. Several ICs with metallized solder bumps on them as in Fig. 9–45 can be flip chip bonded on the back of this package, making this a multi-chip module (MCM). (Photograph courtesy of IBM.)

Figure 9–48
Ball grid array: In this package, the IC in the middle is wire bonded to electrical connections on the package. The package itself has an array of solder "balls" on the top, which can be properly aligned and surface-mount connected simultaneously to electrical sockets on a PCB using solder reflow. (Photograph courtesy of IBM.)

for automated handling. This process, called *tape-automated bonding (TAB)*, is particularly useful in mounting several chips on a large ceramic substrate having multilevel interconnection patterns (called a *multichip module*).

PROBLEMS

9.1 Assume that boron is diffused into a uniform n-type Si sample, resulting in a net doping profile $N_a(x)-N_d$. Set up an expression relating the sheet resistance of the diffused layer to the acceptor profile $N_a(x)$ and the junction depth x_j. Assume that $N_a(x)$ is much greater than the background doping N_d over most of the diffused layer.

9.2 A typical sheet resistance of a base diffusion layer is 200 Ω/square.

    (a) What should be the aspect ratio of a 10-k Ω resistor, using this diffusion?

    (b) Draw a pattern for this resistor (see Fig. 9–12b) which uses little area for a width $w = 5$ μm.

9.3 A 3-μm n-type epitaxial layer ($N_d = 10^{16}$ cm^{-3}) is grown on a p-type Si substrate. Areas of the n layer are to be junction isolated (see Fig. 9–12a) by a boron diffusion at 1200°C ($D = 2.5 \times 10^{-12}$ cm^2/s). The surface boron concentration is held constant at 10^{20} cm^{-3} (see Prob. 5.2).

    (a) What time is required for this isolation diffusion?

    (b) How far does an Sb-doped buried layer ($D = 2 \times 10^{-13}$ cm^2/s) diffuse into the epitaxial layer during this time, assuming the concentration at the substrate-epitaxial boundary is constant at 10^{20} cm^{-3}?

9.4 A 500 μm thick p-type Si wafer with a doping level of 1×10^{15} cm^{-3} has a certain region in which we do a constant source solid-solubility-limited P diffusion, resulting in a junction depth of 0.8 μm and a surface concentration of 6×10^{19} cm^{-3}. We do sheet resistance measurements on the two parts of the wafer. What is the measured sheet resistance of the p-type part? If we have a sheet resistance of 90 Ω/square in the n-type part, what is the *average* resistivity there? At what temperature was the P diffusion done, keeping in mind that typical diffusion temperatures are less than 1100°C? (Refer to Prob. 5.2.)

READING LIST   **Blouke, M. M.** "Charge-Coupled Devices Reach Maturity." *Laser Focus World* 27 (March 1991): A17– A19.

Campbell, S. A. *The Science and Engineering of Microelectronic Fabrication.* New York: Oxford, 1996.

Chang, C. Y., and S. M. Sze. *ULSI Technology.* New York: McGraw-Hill, 1996.

Ghandhi, S. K. *VLSI Fabrication Principles,* 2nd ed. New York: Wiley, 1994.

Hess, D. W., and K. F. Jensen, eds. *Microelectronics Processing: Chemical Engineering Aspects.* Washington, DC: American Chemical Society, 1989.

Hughes, W. A., A. A. Rezazadeh, and C. E. C. Wood. *GaAs Integrated Circuits.* Oxford: BSP Professional Books, 1988.

Jaeger, R. C. *Modular Series on Solid State Devices: Vol. V. Introduction to Microelectronic Fabrication.* Reading, MA: Addison-Wesley, 1988.

Levenson, M. D. "Wavefront Engineering for Photolithography." *Physics Today* 46 (July 1993): 28– 36.

Moreau, W. M. *Semiconductor Lithography: Principles, Practices, and Materials.* New York: Plenum Press, 1988.

Moslehi, M. M., R. A. Chapman, M. Wong, A. Paranjpe, H. N. Najm, J. Kuehne, R. F. Yeakley, and C. J. Davis. "Single-Wafer Integrated Semiconductor Device Processing." *IEEE Transactions on Electron Devices* 39 (January 1992):4– 32.

Oberai, A. S. "Lithography—Challenges of the Future." *Solid State Technology* 30 (September 1987): 123– 8.

Runyan, W. R., and K. E. Bean. *Semiconductor Integrated Circuit Processing Technology.* Reading, MA: Addison-Wesley, 1990.

Ruska, W. S. *Microelectronic Processing: An Introduction to the Manufacture of Integrated Circuits.* New York: McGraw-Hill, 1987.

Seraphim, D. P., R. C. Lasky, and C. Y. Li, eds. *Principles of Electronic Packaging.* New York: McGraw-Hill, 1989.

Tandon, U. S. "An Overview of X-ray Lithography for Use in Semiconductor Device Preparation." *Vacuum* 42 (1991): 1219– 28.

Uyemura, J. P. *Fundamentals of MOS Digital Integrated Circuits.* Reading, MA: Addison-Wesley, 1988.

Wolf, S., and R. N. Tauber. *Silicon Processing for the VLSI Era.* Sunset Beach, CA: Lattice Press, 1986.

Zucker, J. E. "Quantum Effects Enhance Integrated Optics Performance." *Laser Focus World* 29 (March 1993): 101–2+.

Chapter 10
Negative Conductance Microwave Devices

We have discussed a number of devices that are useful in microwave circuits, such as the varactor and specially designed high-frequency transistors, which can provide amplification and other functions at microwave frequencies up to 10^{11} Hz. However, transit time and other effects limit the application of transistors beyond the 10^{11} Hz range. Therefore, other devices are required to perform electronic functions such as switching and d-c to microwave power conversion at higher frequencies.

Several important devices for high-frequency applications use the instabilities that occur in semiconductors. An important type of instability involves *negative conductance.* Here we shall concentrate on three of the most commonly used negative conductance devices: *Esaki* or *tunnel* diodes, which depend on quantum-mechanical tunneling; transit time diodes, which depend on a combination of carrier injection and transit time effects; and *Gunn* diodes, which depend on the transfer of electrons from a high-mobility state to a low-mobility state. Each is a two-terminal device that can be operated in a negative conductance mode to provide amplification or oscillation at microwave frequencies in a proper circuit.

10.1 TUNNEL DIODES

The tunnel diode is a p-n junction device that operates in certain regions of its I–V characteristic by the quantum mechanical tunneling of electrons through the potential barrier of the junction (see Sections 2.4.4 and 5.4.1). The tunneling process for reverse current is essentially the Zener effect, although negligible reverse bias is needed to initiate the process in tunnel diodes. This device can be used in many applications, including high-speed switching and logic circuits. As we shall see in this section, the tunnel diode (often called the Esaki diode after L. Esaki, who in 1973 received the Nobel prize for his work on the effect) exhibits the important feature of *negative resistance* over a portion of its I–V characteristic.

10.1.1 Degenerate Semiconductors

Thus far we have discussed the properties of relatively pure semiconductors; any impurity doping represented a small fraction of the total atomic density of the material. Since the few impurity atoms were so widely spaced throughout the sample, we could be confident that no charge transport could take place within the donor or acceptor levels themselves. What happens, however, if we continue to dope a semiconductor with impurities of either type? As might be expected, a point is reached at which the impurities become so closely packed within the lattice that interactions between them cannot be ignored. For example, donors present in high concentrations (e.g., 10^{20} donors/cm^3) are so close together that we can no longer consider the donor level as being composed of discrete, noninteracting energy states. Instead, the donor states form a band, which may overlap the bottom of the conduction band. If the conduction band electron concentration n exceeds the effective density of states N_c, the Fermi level is no longer within the band gap but lies within the conduction band. When this occurs, the material is called *degenerate* n-type. The analogous case of degenerate p-type material occurs when the acceptor concentration is very high and the Fermi level lies in the valence band. We recall that the energy states below E_F are mostly filled and states above E_F are empty, except for a small distribution dictated by the Fermi statistics. Thus in a degenerate n-type sample the region between E_c and E_F is for the most part filled with electrons, and in degenerate p-type the region between E_v and E_F is almost completely filled with holes.

10.1.2 Tunnel Diode Operation

A p-n junction between two degenerate semiconductors is illustrated in terms of energy bands in Fig. 10–1a. This is the equilibrium condition, for which the Fermi level is constant throughout the junction. We notice that E_{Fp} lies below the valence band edge on the p side and E_{Fn} is above the conduction band edge on the n side. Thus the bands must overlap on the energy scale in order for E_F to be constant. This overlapping of bands is very important; it means that with a small forward or reverse bias, filled states and empty states appear opposite each other, separated by essentially the width of the depletion region. If the metallurgical junction is sharp, as in an alloyed junction, the depletion region will be very narrow for such high doping concentrations, and the electric field at the junction will be quite large. Thus the conditions for electron tunneling are met—filled and empty states separated by a narrow potential barrier of finite height.

As mentioned previously, the filled and empty states are distributed about E_F according to the Fermi distribution function; thus there are some filled states above E_{Fp} and some empty states below E_{Fn}. In Fig. 10–1 the bands are shown filled to the Fermi level for convenience of illustration, with the understanding that a distribution is implied.

Figure 10–1
Tunnel diode band diagrams and *I–V* characteristics for various biasing conditions: (a) equilibrium (zero bias) condition, no net tunneling; (b) small reverse bias, electron tunneling from p to n; (c) small forward bias, electron tunneling from n to p; (d) increased forward bias, electron tunneling from n to p decreases as bands pass by each other.

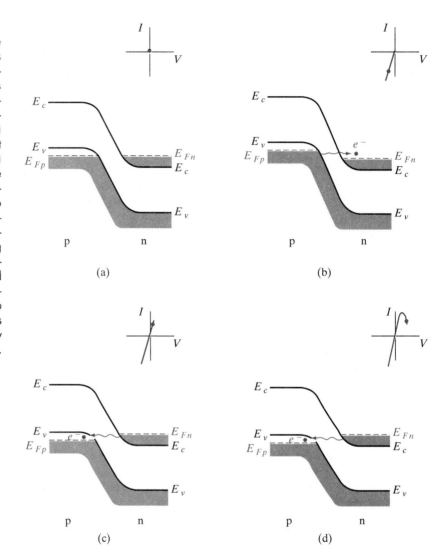

Since the bands overlap under equilibrium conditions, a small reverse bias (Fig. 10–1b) allows electron tunneling from the filled valence band states below E_{Fp} to the empty conduction band states above E_{Fn}. This condition is similar to the Zener effect except that no bias is required to create the condition of overlapping bands. As the reverse bias is increased, E_{Fn} continues to move down the energy scale with respect to E_{Fp}, placing more filled states on the p side opposite empty states on the n side. Thus the tunneling of electrons from p to n increases with increasing reverse bias. The resulting conventional current is opposite to the electron flow, that is, from n to p. At equilibrium (Fig. 10–1a) there is equal tunneling from n to p and from p to n, given a zero net current.

When a small forward bias is applied (Fig. 10–1c), E_{Fn} moves up in energy with respect to E_{Fp} by the amount qV. Thus electrons below E_{Fn} on the n side are placed opposite empty states above E_{Fp} on the p side. Electron tunneling occurs from n to p as shown, with the resulting conventional current from p to n. This forward tunneling current continues to increase with increased bias as more filled states are placed opposite empty states. However, as E_{Fn} continues to move up with respect to E_{Fp}, a point is reached at which the bands begin to pass by each other. When this occurs, the number of filled states opposite empty states decreases. The resulting decrease in tunneling current is illustrated in Fig. 10–1d. This region of the *I–V* characteristic is important in that the *decrease* of tunneling current with *increased* bias produces a region of negative slope; that is, the *dynamic resistance dV/dI* is negative. This negative resistance region is useful in a number of applications.

If the forward bias is increased beyond the negative resistance region, the current begins to increase again (Fig. 10–2). Once the bands have passed each other, the characteristic resembles that of a conventional diode. The forward current is now dominated by the diffusion current—electrons surmounting the potential barrier from n to p and holes surmounting their potential barrier from p to n. Of course, the diffusion current is present in the forward tunneling region, but it is negligible compared to the tunneling current.

The total tunnel diode characteristic (Fig. 10–3) has the general shape of an *N* (if a little imagination is applied); therefore, it is common to refer to this characteristic as exhibiting a *type N negative resistance*. It is also called a *voltage-controlled negative resistance*, meaning that the current decreases rapidly at some critical voltage (in this case the *peak voltage V_p*, taken at the point of maximum forward tunneling).

The values of *peak tunneling current I_p* and *valley current I_v* (Fig. 10–3) determine the magnitude of the negative resistance slope for a diode of given material. For this reason, their ratio I_p/I_v is often used as a figure of merit for

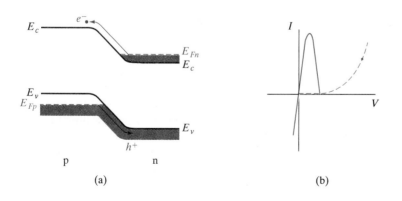

(a) (b)

Figure 10–2
Band diagram (a) and *I–V* characteristic (b) for the tunnel diode beyond the tunnel current region. In (b) the tunneling component of current is shown by the solid curve and the diffusion current component is dashed.

Figure 10–3
Total tunnel diode
characteristic.

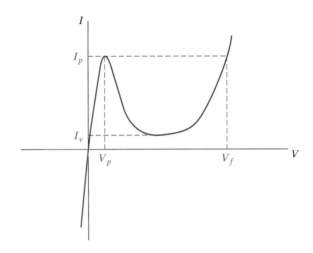

Figure 10–3
Total tunnel diode
characteristic.

the tunnel diode. Similarly, the ratio V_p/V_f is a measure of the voltage spread between the two positive resistance regions.

10.1.3 Circuit Applications

The negative resistance of the tunnel diode can be used in a number of ways to achieve switching, oscillation, amplification, and other circuit functions. This wide range of applications, coupled with the fact that the tunneling process does not present the time delays of drift and diffusion, makes the tunnel diode a natural choice for certain high-speed circuits. However, the tunnel diode has not achieved widespread application, because of its relatively low current operation and competition from other devices.

**10.2
THE IMPATT
DIODE**

In this section we describe a type of microwave negative conductance device that operates by a combination of carrier injection and transit time effects. Diodes with simple p-n junction structure, or with variations on that structure, are biased to achieve tunneling or avalanche breakdown, with an a-c voltage superimposed on the d-c bias. The carriers generated by the injection process are swept through a drift region to the terminals of the device. We shall see that the a-c component of the resulting current can be approximately 180° out of phase with the applied voltage under proper conditions of bias and device configuration, giving rise to negative conductance and oscillation in a resonant circuit. Transit time devices can convert d-c to microwave a-c signals with high efficiency and are very useful in the generation of microwave power for many applications.

The original suggestion for a microwave device employing transit time effects was made by W. T. Read and involved an n^+-p-i-p^+ structure such as that shown in Fig. 10–4. This device operates by injecting carriers into the drift region and is called an *impact avalanche transit time (IMPATT)* diode. Although IMPATT operation can be obtained in simpler structures, the Read diode is best suited for illustration of the basic principles. The device consists essentially of two regions: (1) the n^+-p region at which avalanche multiplication occurs and (2) the i (essentially intrinsic) region through which generated holes must drift in moving to the p^+ contact. Similar devices can be built in the p^+-n-i-n^+ configuration, in which electrons resulting from avalanche multiplication drift through the i region, taking advantage of the higher mobility of electrons compared to holes.

Although detailed calculations of IMPATT operation are complicated and generally require computer solutions, the basic physical mechanism is simple. Essentially, the device operates in a negative conductance mode when the a-c component of current is negative over a portion of the cycle during which the a-c voltage is positive, and vice versa. The negative conductance occurs because of two processes, causing the current to lag behind the voltage in time: (1) a delay due to the avalanche process and (2) a further delay due to the transit time of the carriers across the drift region. If the sum of these delay times is approximately one-half cycle of the operating frequency, negative conductance occurs and the device can be used for oscillation and amplification.

From another point of view, the a-c conductance is negative if the a-c component of carrier flow drifts opposite to the influence of the a-c electric field. For example, with a d-c reverse bias on the device of Fig. 10–4, holes drift from left to right (in the direction of the field) as expected. Now, if we superimpose an a-c voltage such that $\mathscr{E}$ decreases during the negative half-cycle, we would normally expect the drift of holes to decrease also. However,

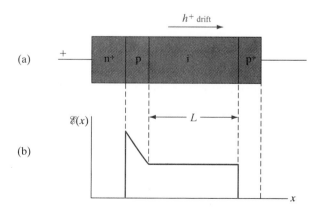

Figure 10–4
The Read diode: (a) basic device configuration; (b) electric field distribution in the device under reverse bias.

in IMPATT operation the drift of holes through the i region actually increases while the a-c field is decreasing. To see how this happens, let us consider the effects of avalanche and drift for various points in the cycle of applied voltage (Fig. 10–5).

To simplify the discussion, we shall assume that the p region is very narrow and that all the avalanche multiplication takes place in a thin region near the n^+-p junction. We shall approximate the field in the narrow p region by a uniform value. If the d-c bias is such that the critical field for avalanche $\mathscr{E}_a$ is just met in the n^+-p space charge region (Fig. 10–5a), avalanche multiplication begins at $t = 0$. Electrons generated in the avalanche move to the n^+ region, and holes enter the i drift region. We assume that device is mounted in a resonant microwave circuit so that an a-c signal can be maintained at a given frequency. As the applied a-c voltage goes positive, more and more holes are generated in the avalanche region. In fact, the pulse

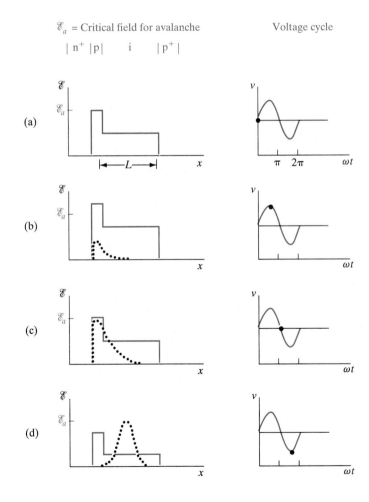

Figure 10–5
Time dependence of the growth and drift of holes during a cycle of applied voltage for the Read diode:
(a) $\omega t = 0$;
(b) $\omega t = \pi/2$;
(c) $\omega t = \pi$;
(d) $\omega t = 3\pi/2$.
The hole pulse is sketched as a dotted line on the field diagram.

of holes (dotted line) generated by the multiplication process continues to grow as long as the electric field is above $\mathscr{E}_a$ (Fig. 10–5b). It can be shown that the particle current due to avalanche increases exponentially with time while the field is above the critical value. The important result of this growth is that the hole pulse reaches its peak value not at $\pi/2$ when the voltage is maximum, but at π (Fig. 10–5c). Therefore, there is a phase delay of $\pi/2$ inherent in the avalanche process itself. A further delay is provided by the drift region. Once the avalanche multiplication stops ($\omega t > \pi$), the pulse of holes simply drifts toward the p^+ contact (Fig. 10–5d). But during this period the a-c terminal voltage is negative. Therefore, the dynamic conductance is negative, and energy is supplied to the a-c field.

If the length of the drift region is chosen properly, the pulse of holes is collected at the p^+ contact just as the voltage cycle is completed, and the cycle then repeats itself. The pulse will drift through the length L of the i region during the negative half-cycle if we choose the transit time to be one-half the oscillation period

$$\frac{L}{v_d} = \frac{1}{2}\frac{1}{f}, \quad f = \frac{v_d}{2L} \tag{10–1}$$

where f is the operating frequency and v_d is the drift velocity for holes.[1] Therefore, for a Read diode the optimum frequency is one-half the inverse transit time of holes across the drift region v_d/L. In choosing an appropriate resonant circuit for this device, the parameter L is critical. For example, taking $v_d = 10^7$ cm/s for Si, the optimum operating frequency for a device with an i region length of 5 μm is $f = 10^7/2(5 \times 10^{-4}) = 10^{10}$ Hz. Negative resistance is exhibited by an IMPATT diode for frequencies somewhat above and below this optimum frequency for exact 180° phase delay. A careful analysis of the small-signal impedance shows that the minimum frequency for negative conductance varies as the square root of the d-c bias current for frequencies in the neighborhood of that described by Eq. (10–1).

Although the Read diode of Fig. 10–4 displays most directly the operation of IMPATT devices, simpler structures can be used, and in some cases they may be more efficient. Negative conductance can be obtained in simple p-n junctions or in p-i-n devices. In the case of the p-i-n, most of the applied voltage occurs across the i region, which serves as a uniform avalanche region and also as a drift region. Therefore, the two processes of delay due to avalanche and drift, which were separate in the case of the Read diode, are distributed within the i region of the p-i-n. This means that both electrons and holes participate in the avalanche and drift processes.

---

[1] In general, v_d is a function of the local electric field. However, these devices are normally operated with fields in the i region sufficiently large that holes drift at their scattering limited velocity (Fig. 3–24). For this case the drift velocity does not vary appreciably with the a-c variations in the field.

10.3
THE GUNN DIODE

Microwave devices that operate by the *transferred electron* mechanism are often called *Gunn diodes* after J. B. Gunn, who first demonstrated one of the forms of oscillation. In the transferred electron mechanism, the conduction electrons of some semiconductors are shifted from a state of high mobility to a state of low mobility by the influence of a strong electric field. Negative conductance operation can be achieved in a diode[2] for which this mechanism applies, and the results are varied and useful in microwave circuits.

First, we shall describe the process of electron transfer and the resulting change of mobility. Then we shall consider some of the modes of operation for diodes using this mechanism.

10.3.1 The Transferred Electron Mechanism

In Section 3.4.4 we discussed the nonlinearity of mobility at high electric fields. In most semiconductors the carriers reach a scattering limited velocity, and the velocity vs. field plot saturates at high fields (Fig. 3–24). In some materials, however, the energy of electrons can be raised by an applied field to the point that they transfer from one region of the conduction band to another, higher-energy region. For some band structures, negative conductivity can result from this electron transfer. To visualize this process, let us recall the discussion of energy bands in Section 3.1. The band diagrams we usually draw vs. distance in the sample are good approximations when the conduction electrons exist near the minimum energy of the conduction band. However, in the more complete band diagram, electron energy is plotted vs. the propagation vector **k**, as in Fig. 3–5. It was shown in Example 3-1 that the **k** vector is proportional to the electron momentum in the vector direction; therefore, energy bands such as those in Fig. 3–5 are said to be plotted in *momentum space.*

A simplified band diagram for GaAs is shown in Fig. 10–6 for reference; some of the detail has been omitted in this diagram to isolate the essential features of electron transfer between bands. In n-type GaAs the valence band is filled, and the *central valley* (or *minimum*) of the conduction band at Γ ($\mathbf{k} = 0$) normally contains the conduction electrons. There is a set of *subsidiary minima* at L (sometimes called *satellite valleys*) at higher energy,[3] but these minima are many kT above the central valley and are normally unoccupied. Therefore, the direct band gap at Γ and the energy bands centered at $\mathbf{k} = 0$ are generally used to describe the conduction processes in GaAs. This was true of our discussion of GaAs lasers in Section 8.4, for example. The presence of the satellite valleys at L is crucial to the Gunn effect, however. If the material is subjected to an electric field above some critical value (about 3000 V/cm), the

---

[2]These devices are called diodes, since they are two-terminal devices. No p-n junction is involved, however. Gunn effect and related devices utilize bulk instabilities, which do not require junctions.

[3]We have shown only one satellite valley for convenience; there are other equivalent valleys for different directions in **k**-space. The effective mass ratio of 0.55 refers to the combined satellite valleys.

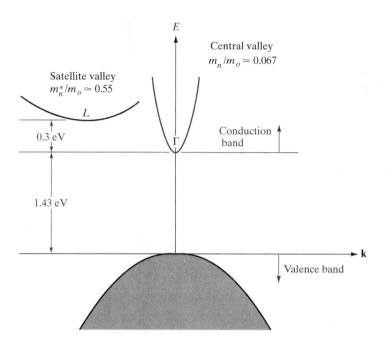

E

Central valley
$m_n/m_o \simeq 0.067$

Satellite valley
$m_n^*/m_o \simeq 0.55$

L

0.3 eV

Γ

Conduction band

1.43 eV

$\mathbf{k}$

Valence band

Figure 10–6
Simplified band diagram for GaAs, illustrating the lower (Γ) and upper (L) valleys in the conduction band.

electrons in the central Γ valley of Fig. 10–6 gain more energy than the 0.30 eV separating the valleys; therefore, there is considerable scattering of electrons into the higher-energy satellite valley at L.

Once the electrons have gained enough energy from the field to be transferred into the higher-energy valley, they remain there as long as the field is greater than the critical value. The explanation for this involves the fact that the combined effective density of states for the upper valleys is much greater than for the central valley (by a factor of about 24). Although we shall not prove it here, it seems reasonable that the probability of electron scattering between valleys should depend on the density of states available in each case, and that scattering from a valley with many states into a valley with few states would be unlikely. As a result, once the field increases above the critical value, most conduction electrons in GaAs reside in the satellite valleys and exhibit properties typical of that region of the conduction band. In particular, the effective mass for electrons in the higher L valleys is almost eight times as great as in the central valley, and the electron mobility is much lower. This is an important result for the negative conductivity mechanism: As the electric field is increased, the electron velocity increases until a critical field is reached; then the electrons *slow down* with further increase in field. The electron transfer process allows electrons to gain energy at the expense of velocity over a range of values of the electric field. Taking current density as $qv_d n$, it is clear that current also drops in this range of increasing field, giving rise to a negative differential conductivity $dJ/d\mathscr{E}$.

Figure 10–7
A possible char-
acteristic of elec-
tron drift velocity
vs. field for a
semiconductor ex-
hibiting the trans-
ferred electron
mechanism.

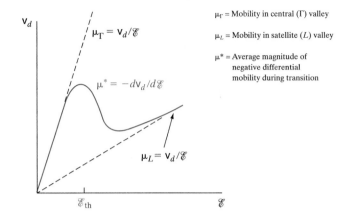

A possible dependence of electron velocity vs. electric field for a material capable of electron transfer is shown in Fig. 10–7. For low values of field, the electrons reside in the lower (Γ) valley of the conduction band, and the mobility ($\mu_\Gamma = v_d/\mathscr{E}$) is high and constant with field. For high values of field, electrons transfer to the satellite valleys, where their velocity is smaller and their mobility lower. Between these two states is a region of negative slope on the v_d vs. $\mathscr{E}$ plot, indicating a negative differential mobility $dv_d/d\mathscr{E} = -\mu^*$.

The actual dependence of electron drift velocity on electric field for GaAs and InP is shown in Fig. 10–8. The negative resistance due to electron transfer occurs at a higher field for InP, and the electrons achieve a higher peak velocity before transfer from Γ to L occurs.

The existence of a drop in mobility with increasing electric field and the resultant possibility of negative conductance were predicted by Ridley and Watkins and by Hilsum several years before Gunn demonstrated the effect in GaAs. The mechanism of electron transfer is therefore often called the Ridley–Watkins–Hilsum mechanism. This negative conductivity effect depends only on the bulk properties of the semiconductor and not on junction or surface effects. It is therefore called a *bulk negative differential conductivity (BNDC)* effect.

10.3.2   Formation and Drift of Space Charge Domains

If a sample of GaAs is biased such that the field falls in the negative conductivity region, space charge instabilities result, and the device cannot be maintained in a d-c stable condition. To understand the formation of these instabilities, let us consider first the dissipation of space charge in the usual semiconductor. It can be shown from treatment of the continuity equation that a localized space charge dies out exponentially with time in a homogeneous sample with positive resistance (Prob. 10.3). If the initial space charge is Q_0, the instantaneous charge is

$$Q(t) = Q_0 e^{-t/\tau_d} \tag{10–2}$$

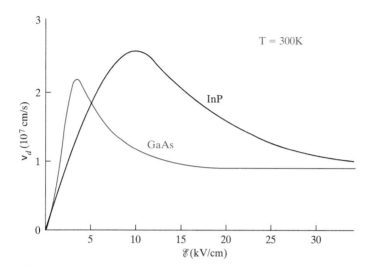

Figure 10–8
Electron drift ve-
locity vs. field for
GaAs and InP.

where $\tau_d = \epsilon/\sigma$ is called the *dielectric relaxation time*. Because of this process, random fluctuations in carrier concentration are quickly neutralized, and space charge neutrality is a good approximation for most semiconductors in the usual range of conductivities. For example, the dielectric relaxation time for a 1.0 Ω-cm Si or GaAs sample is approximately 10^{-12} s.

Equation (10–2) gives a rather remarkable result for cases in which the conductivity is negative. For these cases τ_d is negative also, and *space charge fluctuations build up* exponentially in time rather than dying out. This means that normal random fluctuations in the carrier distribution can grow into large space charge regions in the sample. Let us see how this occurs in a GaAs sample biased in the negative conductivity regime. The velocity–field diagram for n-type GaAs is illustrated in Fig. 10–9a. If we assume a small shift of electron concentration in some region of the device, a dipole layer can form as shown in Fig. 10–9b. Under normal conditions this dipole would die out quickly. However, under conditions of negative conductivity the charge within the dipole, and therefore the local electric field, builds up as shown in Fig. 10–9c. Of course, this buildup takes place in a stream of electrons drifting from the cathode to the anode, and the dipole (now called a *domain*) drifts along with the stream as it grows. Eventually the drifting domain will reach the anode, where it gives up its energy as a pulse of current in the external circuit.

During the initial growth of the domain, an increasing fraction of the applied voltage appears across it, at the expense of electric field in the rest of the bar. As a result, it is unlikely that more than one domain will be present in the bar at a time; after the formation of one domain, the electric field in the rest of the bar quickly drops below the threshold value for negative conductivity. If the bias is d-c, the field outside the moving domain will stabilize at a positive conductivity point such as *A* in Fig. 10–9a, and the field in the domain will stabilize at the high-field value *B*.

Figure 10–9
Buildup and drift
of a space charge
domain in GaAs:
(a) velocity–field
characteristic for
n-type GaAs; (b)
formation of a di-
pole; (c) growth
and drift of a di-
pole for condi-
tions of negative
conductivity.

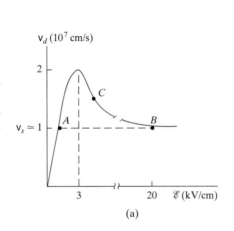

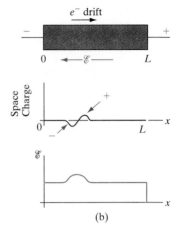

(a)

(b)

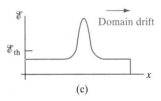

(c)

Let us follow the motion of a single domain as illustrated by Fig. 10–9.
A small dipole forms from a random noise fluctuation (or more likely at a
permanent *nucleation site* such as a crystal defect, a doping inhomogeneity,
or the cathode itself), and this dipole grows and drifts down the bar as a domain. During the early stages of domain development, we can assume a uniform electric field in the bar, except just at the small dipole layer. If the field
is in the negative mobility region, such as point C of Fig. 10–9a, the slightly
higher field within the dipole results in a lower value of electron drift velocity inside the dipole than outside. As a result, electrons on the right (down-
stream) of the domain drift away, while electrons pile up on the left
(upstream) side. This causes the accumulation and depletion layers of the dipole to grow, thereby further increasing the electric field in the domain. This
is obviously a runaway process, in which the electric field within the domain
grows while that outside the domain decreases. A stable condition is real-
ized when the domain field increases to point B in Fig. 10–9a, and the field
outside drops to point A. When this condition is met, the electrons drift at a
constant velocity v_s everywhere, and the domain moves down the bar without further growth.

In this discussion we have assumed that the domain has time to grow
to its stable condition before it drifts out of the bar. This is not always the case;
for example, in a short bar with a low concentration of electrons, a dipole

can drift the length of the bar before it develops into a domain. We can specify limits on the electron concentration n_0 and sample length L for successful domain formation by requiring the transit time (L/v_s) to be greater than the dielectric relaxation time (absolute value) in the negative mobility region. This requirement gives

$$\frac{L}{v_s} > \frac{\epsilon}{q\mu^* n_0}$$

$$Ln_0 > \frac{\epsilon v_s}{q\mu^*} \simeq 10^{12} \text{cm}^{-2} \tag{10-3}$$

for n-type GaAs, where the average negative differential mobility[4] is taken to be -100 cm^2/V-s. Therefore, for successful domain formation there is a critical product of electron concentration and sample length.

The type of domain motion we have described here was the first mode of operation observed by Gunn. In the observation of current vs. voltage for a GaAs sample, Gunn found a linear ohmic relation up to a critical bias, beyond which the current came in sharp pulses. The pulses were separated in time by an amount proportional to the sample length. This length dependence was due to the transit time L/v_s required for a domain nucleated at the cathode to drift the length of the bar. Gunn performed an interesting experiment in which he used a tiny capacitive probe to measure the electric field at various positions down the bar. By scanning the field distribution in the bar at various times in the cycle, he was able to plot out the growth and drift of the domains.

The formation of stable domains is not the only mode of operation for transferred electron devices. Nor is it the most desirable mode for most applications, since the resulting short pulses of current are inefficient sources of microwave power.

10.3.3 Fabrication

Devices utilizing the Gunn effect and its variations can be made in a number of materials which have appropriate band structures. Although GaAs and InP are the most common materials, transferred electron effects have been observed in CdTe, ZnSe, GaAsP, and other materials. The band structure of some materials can be altered to exhibit properties appropriate for electron transfer. For example, the energy bands of InAs can be distorted by the application of pressure to the crystal, such that a set of satellite valleys becomes available for electron transfer, although these upper valleys are too far above the lower valley at normal pressures. We have discussed the device behavior in terms of GaAs, since this material can be prepared with good purity and is most widely used in microwave applications.

---

[4]This is a rather crude approximation, since μ^* is not a constant but varies considerably with field; the negative dielectric relaxation time therefore changes with time as the domain grows.

Gunn diodes and related devices are simple structures in principle, since they are basically homogeneous samples with ohmic contacts on each end. In practice, however, considerable care must be taken in fabricating and mounting workable devices. In addition to the obvious requirements on doping, carrier mobility, and sample length, there are important problems with contacts, heat sinking, and parasitic reactances of the packaged device.

The samples must have high mobility, few lattice defects, and homogeneous doping in the range giving carrier concentrations $n_0 \simeq 10^{13} - 10^{16} \, \text{cm}^{-3}$. Devices can be made from GaAs or InP bulk samples cut from an ingot, but it is more common to use ingot material as a substrate for an epitaxial layer, which serves as the active region of the device. The material properties of epitaxial layers are often superior to bulk samples, and the precise control of layer thickness is helpful in these devices, which require exact sample lengths. In a typical configuration, an n-type epitaxial layer about 10 μm thick is grown on an n⁺ substrate wafer, which is perhaps 100 μm thick. The substrate serves as one of the contacts to the active region. A thin n⁺⁺ (very heavily doped) layer is grown on top of the n region, so that an n⁺-n-n⁺⁺ sandwiched structure results. External contacts can be made by evaporating a thin layer of Au–Sn or Au–Ge on each surface, followed by a brief alloying step in a hydrogen atmosphere. The wafer is divided into individual devices by cutting or cleaving (giving a cube structure) or by selective etching (giving a mesa structure). Each device is mounted with the n⁺⁺ side down on a copper stud or other heat sink, so that the active region can dissipate heat to the mount in one direction and to the substrate layer in the other direction. Then the substrate side can be contacted by a wire or pressure contact. In other configurations, planar fabrication techniques can be used to produce lateral devices in an n-type epitaxial layer grown on a high-resistivity substrate.

Removal of heat is a very serious problem in these devices. The power dissipation may be $10^7 \, \text{W/cm}^3$ or greater (Prob. 10.5), giving rise to considerable heating of the sample. As the temperature increases, the device characteristics vary because of changes in n_0 and mobility. As a result of such heating effects, these devices seldom reach their theoretical maximum efficiency. Pulsed operation allows better control of heat dissipation than does continuous operation, and efficiencies near the theoretical limits can sometimes be achieved in the pulsed mode. If the application does not require continuous operation, peak powers of hundreds of watts can be achieved in pulses of microwave oscillation.

PROBLEMS

10.1 Sketch the band diagram for an abrupt junction in which the doping on the p side is degenerate and the Fermi level on the n side is aligned with the bottom of the conduction band. Draw the forward and reverse bias band diagrams and sketch the *I–V* characteristic. This diode is often called a *backward diode*. Can you explain why?

10.2 What determines the peak tunneling voltage V_p of a tunnel diode? Explain.

If a large density of trapping centers is present in a tunnel diode (Fig. P10–2), tunneling can occur from the n-side conduction band to the trapping level (A–B). Then the electrons may drop to the valence band on the p side (B–C), thereby completing a two-step process of charge transport across the junction. In fact, if the density of trapping centers is large, it is possible to observe an increase in current as the states below E_{Fn} pass by the trapping level with increased bias. In Fig. P10–2, the trapping level E_t is located 0.3 eV above the valence band. Assume $E_g = 1$ eV, and $E_{Fn} - E_c$ on the n side equals $E_v - E_{Fp}$ on the p side, equals 0.1 eV.

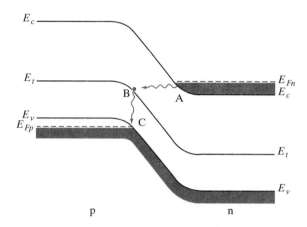

Figure P10–2

(a) Calculate the minimum forward bias at which tunneling through E_t occurs.

(b) Calculate the maximum forward bias for tunneling via E_t.

(c) Sketch the I–V curve for this tunnel diode. Assume the maximum tunneling current via E_t is about one-third of the peak band-to-band tunneling current.

10.3 (a) Use Poisson's equation, the continuity equation, and the definition of current density in terms of the gradient of electrostatic potential to relate the time variation of space charge density ρ to the conductivity σ and the permittivity ϵ of a material, neglecting recombination.

(b) Assuming a space charge density ρ_0 at $t = 0$, show that $\rho(t)$ decays exponentially with a time constant equal to the dielectric relaxation time τ_d.

(c) Given a sample of thickness L and area A, calculate the inherent RC time constant if the conductivity is σ and the permittivity is ϵ.

10.4 Assuming that n_Γ electrons/cm³ are in the lower (central) valley of the GaAs conduction band at time t and n_L are in the satellite (L) valleys, show that the criterion for negative differential conductivity ($dJ/d\mathscr{E} < 0$) is

$$\frac{\mathscr{E}(\mu_\Gamma - \mu_L)\dfrac{dn_\Gamma}{d\mathscr{E}} + \mathscr{E}\left(n_\Gamma\dfrac{d\mu_\Gamma}{d\mathscr{E}} + n_L\dfrac{d\mu_L}{d\mathscr{E}}\right)}{n_\Gamma\mu_\Gamma + n_L\mu_L} < -1$$

where μ_Γ and μ_L are the electron mobilities in the Γ and L valleys, respectively. *Note:* $n_0 = n_\Gamma + n_L$. Discuss the conditions for negative differential conductivity, assuming the mobilities are approximately proportional to $\mathcal{E}^{-1}$.

10.5 We wish to estimate the d-c power dissipated in a GaAs Gunn diode. Assume the diode is 5 μm long and operates in the stable domain mode.

(a) What is the minimum electron concentration n_0? What is the time between current pulses?

(b) Using data from Fig. 10–9a, calculate the power dissipated in the sample per unit volume when it is biased just below threshold, if n_0 is chosen from the calculation of part (a). In general, does operation at a higher frequency result in greater power dissipation?

10.6 (a) Calculate the ratio N_L/N_Γ of the effective density of states in the upper (L) valleys to the effective density of states in the lower (Γ) valley of the GaAs conduction band (Fig. 10–6).

(b) Assuming a Boltzmann distribution $n_L/n_\Gamma = (N_L/N_\Gamma) \exp(-\Delta E/kT)$, calculate the ratio of the concentration of conduction band electrons in the upper valley to the concentration in the central valley in equilibrium at 300 K.

(c) As a rough calculation, assume an electron at the bottom of the central valley has kinetic energy kT. After it is promoted to the satellite (L) valley, what is its approximate equivalent temperature?

READING LIST   **Bailey, M. J.** "Heterojunction IMPATT Diodes: Using New Material Technology in a Classic Device." *Microwave Journal* 36 (June 1993): 76+.

Bayraktaroglu, B. "Monolithic IMPATT Technology." *Microwave Journal* 32 (April 1989); 73–4+.

Bose, B. K. "Recent Advances in Power Electronics." *IEEE Transactions on Power Electronics* 7 (January 1992): 2–16.

Esaki, L. "Discovery of the Tunnel Diode." *IEEE Trans. Elec. Dev.*, ED-23 (1976): 644+.

Gunn, J. B. "Microwave Oscillations of Current in III-V Semiconductors." *Solid State Comm.*, 1 (1963): 88+.

Herman, M. A. *Molecular Beam Epitaxy: Fundamentals and Current Status.* Berlin: Springer-Verlag, 1989.

Hughes, W. A., A. A. Rezazadeh, and C. E. C. Wood. *GaAs Integrated Circuits.* Oxford: BSP Professional Books, 1988.

Kearney, M. J., N. R. Couch, and J. Stephens. "Heterojunction Impact Avalanche Transit-time Diodes Grown by Molecular Beam Epitaxy." *Semiconductor Science and Technology* 8 (April 1993) 560–7.

Lesurf, J. "The Rise and Fall of Negative Resistance." *New Scientist* 31 (31 March 1990): 56–60.

Neamen, D. A. *Semiconductor Physics and Devices: Basic Principles.* Homewood, IL: Irwin, 1992.

Read, W. T. "A Proposed High Frequency, Negative Resistance Diode." *Bell Syst. Tech. J.*, 37 (1958): 401+.

Ridley, B. K., and T. B. Watkins. "The Possibility of Negative Resistance Effects in Semiconductors." *Proc. Phys. Soc. Lond.* 78 (161): 293+.

Shockley, W. "Negative Resistance Arising From Transit Time in Semiconductor Diodes." *Bell Syst. Tech. J.*, 33 (1954): 799+.

Shur, M. *GaAs Devices and Circuits.* New York: Plenum Press, 1987.

Singh, J. *Semiconductor Devices.* New York: McGraw-Hill, 1994.

Sze, S. M. *High-Speed Semiconductor Devices.* New York: Wiley, 1990.

Sze, S. M. *Physics of Semiconductor Devices.* New York: Wiley, 1981.

Voelcker, J. "The Gunn Effect." *IEEE Spectrum* 26 (July 1989): 24.

Wang, S. *Fundamentals of Semiconductor Theory and Device Physics.* Englewood Cliffs, NJ: Prentice Hall, 1989.

Wood, J., and D. V. Morgan. "Gallium Arsenide and Related Compounds for Device Applications." *Acta Physica Polonica A* 79 (January 1991): 97–116.

Chapter 11

Power Devices

One of the most common applications of electronic devices is in switching, which requires the device to change from an "off" or *blocking* state to an "on" or *conducting* state. We have discussed the use of transistors in this application, in which base current drives the device from cutoff to saturation. Similarly, diodes and other devices can be used to serve as certain types of switches. There are a number of important switching applications that require a device remain in the blocking state under forward bias until switched to the conducting state by an external signal. Several devices which fulfill this requirement have been developed, and we shall discuss a family of switches in this chapter, the *semiconductor controlled rectifier (SCR)*[1] and related devices. These devices are typified by a high impedance ("off" condition) under forward bias until a switching signal is applied; after switching they exhibit low impedance ("on" condition). The signal required for switching can be varied externally; therefore, these devices can be used to block or pass currents at predetermined levels. In this chapter we shall discuss the physical operation of the SCR and a combination FET and SCR called an *insulated gate bipolar transistor.*

11.1
THE p-n-p-n
DIODE

The SCR is a four-layer (p-n-p-n) structure that effectively blocks current through two terminals until it is turned on by a small signal at a third terminal. There are many varieties of the basic p-n-p-n structure, and we shall not attempt to cover all of them; however, we can discuss the basic operation and physical mechanisms involved in these devices. We shall begin by investigating the current flow in a two-terminal p-n-p-n device and then extend the discussion to include triggering by a third terminal. We shall see that the p-n-p-n structure can be considered for many purposes as a combination of p-n-p and n-p-n transistors, and the analysis in Chapter 7 can be used as an aid in understanding its behavior.

Before discussing the control of an SCR using a third terminal, it is important to understand the basic transistor action at work in a p-n-p-n structure. Therefore, in this section we analyze the four-layer structure with only two terminals.

[1]Since Si is the material commonly used for this device, it is often called a *silicon controlled rectifier.*

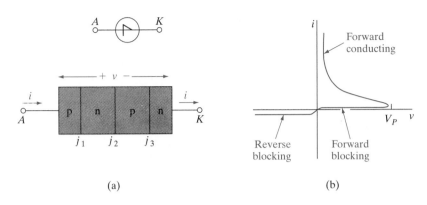

Figure 11–1
A two-terminal
p-n-p-n device: (a)
basic structure
and common cir-
cuit symbol; (b)
I–V characteristic.

11.1.1 Basic Structure

First we consider a four-layer diode structure with an *anode* terminal A at the outside p region with a *cathode* terminal K at the outside n region (Fig. 11–1a). We shall refer to the junction nearest the anode as j_1, the center junction as j_2, and the junction nearest the cathode as j_3. When the anode is biased positively with respect to the cathode (v positive), the device is forward biased. However, as the *I–V* characteristic of Fig. 11–b indicates, the forward-biased condition of this diode can be considered in two separate states, the high-impedance or *forward-blocking* state and the low-impedance or *forward-conducting* state. In the device illustrated here the forward *I–V* characteristic switches from the blocking to the conducting states at a critical peak forward voltage V_p.

We can anticipate the discussion of conduction mechanisms to follow by noting that an initial positive voltage v places j_1 and j_3 under forward bias and the center junction j_2 under reverse bias. As v is increased, most of the forward voltage in the blocking state must appear across the reverse-biased junction j_2. After switching to the conducting state, the voltage from A to K is very small (less than 1 V), and we conclude that in this condition all three junctions must be forward biased. The mechanism by which j_2 switches from reverse bias to forward bias is the subject of much of the discussion to follow.

In the *reverse-blocking* state (v negative), j_1 and j_3 are reverse biased and j_2 is forward biased. Since the supply of electrons and holes to j_2 is restricted by the reverse-biased junctions on either side, the device current is limited to a small saturation current arising from thermal generation of EHPs near j_1 and j_3. The current remains small in the reverse-blocking condition until avalanche breakdown occurs at a large reverse bias. In a properly designed device, with guards against surface breakdown, the reverse breakdown voltage can be several thousand volts.

We shall now consider the mechanism by which this device, often called a *Shockley diode,* switches from the forward-blocking state to the forward-conducting state.

11.1.2   The Two-Transistor Analogy

The four-layer configuration of Fig. 11–1a suggests that the p-n-p-n diode can be considered as two coupled transistors: j_1 and j_2 form the emitter and collector junctions, respectively, of a p-n-p transistor; similarly, j_2 and j_3 form the collector and emitter junctions of an n-p-n (note the emitter of the n-p-n is on the right, which is the reverse of what we usually draw). In this analogy, the collector region of the n-p-n is in common with the base of the p-n-p, and the base of the n-p-n serves as the collector region of the p-n-p. The center junction j_2 serves as the collector junction for both transistors.

This two-transistor analogy is illustrated in Fig. 11–2. The collector current i_{C1} of the p-n-p transistor drives the base of the n-p-n, and the base current i_{B1} of the p-n-p is dictated by the collector current i_{C2} of the n-p-n. If we associate an emitter-to-collector current transfer ratio α with each transistor, we can use the analysis in Chapter 7 to solve for the current i. Using Eq. (7–37b) with $\alpha_1 = \alpha_N$ for the p-n-p, $\alpha_2 = \alpha_N$ for the n-p-n, and with I_{CO1} and I_{CO2} for the respective collector saturation currents, we have

$$i_{C1} = \alpha_1 i + I_{C01} = i_{B2} \tag{11–1a}$$

$$i_{C2} = \alpha_2 i + I_{C02} = i_{B1} \tag{11–1b}$$

But the sum of i_{C1} and i_{C2} is the total current through the device:

$$i_{C1} + i_{C2} = i \tag{11–2}$$

Taking this sum in Eq. (11–1) we have

$$i(\alpha_1 + \alpha_2) + I_{C01} + I_{C02} = i$$

$$i = \frac{I_{C01} + I_{C02}}{1 - (\alpha_1 + \alpha_2)} \tag{11–3}$$

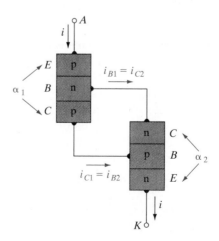

Figure 11–2
Two-transistor
analogy of the
p-n-p-n diode.

As Eq. (11–3) indicates, the current i through the devices is small (approximately the combined collector saturation currents of the two equivalent transistors) as long as the sum $\alpha_1 + \alpha_2$ is small compared with unity. As the sum of the alphas approaches unity, the current i increases rapidly. The current does not increase without limit as Eq. (11–3) implies, however, because the derivation is no longer valid as $\alpha_1 + \alpha_2$ approaches unity. Since j_2 becomes forward biased in the forward-conducting state, both transistors become saturated after switching. The two transistors remain in saturation while the device is in the forward-conducting state, being held in saturation by the device current.

11.1.3 Variation of α with Injection

Since the two-transistor analogy implies that switching involves an increase in the alphas to the point that $\alpha_1 + \alpha_2$ approaches unity, it may be helpful to review how alpha varies with injection for a transistor. The emitter-to-collector current transfer ratio α is given in Section 7.2 as the product of the emitter injection efficiency γ and the base transport factor B. An increase in α with injection can be caused by increases in either of these factors, or both. At very low currents (such as in the forward-blocking state of the p-n-p-n diodes), γ is usually dominated by recombination in the transition region of the emitter junction (Section 7.7.4). As the current is increased, injection across the junction begins to dominate over recombination within the transition region (Section 5.6.2) and γ increases. There are several mechanisms by which the base transport factor B increases with injection, including the saturation of recombination centers as the excess carrier concentration becomes large. Whichever mechanism dominates, the increase in $\alpha_1 + \alpha_2$ required for switching of the p-n-p-n diode is automatically accomplished. In general, no special design is required to maintain $\alpha_1 + \alpha_2$ smaller than unity during the forward-blocking state; this requirement is usually met at low currents by the dominance of recombination within the transition regions of j_1 and j_3.

11.1.4 Forward-Blocking State

When the device is biased in the forward-blocking state (Fig. 11–3a), the applied voltage v appears primarily across the reverse-biased junction j_2. Although j_1 and j_3 are forward biased, the current is small. The reason for this becomes clear if we consider the supply of electrons available to n_1 and holes to p_2. Focusing attention first upon j_1, let us assume a hole is injected from p_1 into n_1. If the hole recombines with an electron in n_1 (or in the j_1 transition region), that electron must be resupplied to the n_1 region to maintain space charge neutrality. The supply of electrons in this case is severely restricted, however, by the fact that n_1 is terminated in j_2, a reverse-biased junction. In a normal p-n diode the n region is terminated in an ohmic contact, so that the supply of electrons required to match recombination (and injection into p)

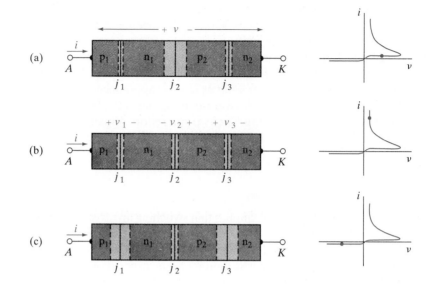

is unlimited. In this case, however, the electron supply is restricted essentially to those electrons generated thermally within a diffusion length of j_2. As a result, the current passing through the j_1 junction is approximately the same as the reverse saturation current of j_2. A similar argument holds for the current through j_3; holes required for injection into n_2 and to feed recombination in p_2 must originate in the saturation current of the center junction j_2. The applied voltage v divides appropriately among the three junctions to accommodate this small current throughout the device.

In this discussion we have tacitly assumed that the current crossing j_2 is strictly the thermally generated saturation current. This implies that electrons injected by the forward-biased junction j_3 do not diffuse across p_2 in any substantial numbers, to be swept across the reverse-biased junction into n_1 by transistor action. This is another way of saying that α_2 (for the "n-p-n transistor") is small. Similarly, the supply of holes to p_2 is primarily thermally generated, since few holes injected at j_1 reach j_2 without recombination (i.e., α_1 is small for the "p-n-p"). Now we can see physically why Eq. (11–3) implies a small current while $\alpha_1 + \alpha_2$ is small: Without the transport of charge provided by transistor action, the thermal generation of carriers is the only significant source of electrons to n_1 and holes to p_2.

11.1.5 Conducting State

The charge transport mechanism changes dramatically when transistor action begins. As $\alpha_1 + \alpha_2$ approaches unity by one of the mechanisms described above, many holes injected at j_1 survive to be swept across j_2 into p_2. This helps to feed the recombination in p_2 and to support the injection of holes into

n_2. Similarly, the transistor action of electrons injected at j_3 and collected at j_2 supplies electrons for n_1. Obviously, the current through the device can be much larger once this mechanism begins. The transfer of injected carriers across j_2 is regenerative, in that a greater supply of electrons to n_1 allows greater injection of holes at j_1 while maintaining space charge neutrality; this greater injection of holes further feeds p_2 by transistor action, and the process continues to repeat itself.

If $\alpha_1 + \alpha_2$ is large enough, so that many electrons are collected in n_1 and many holes are collected in p_2, the depletion region at j_2 begins to decrease. Finally the reverse bias disappears across j_2 and is replaced by a forward bias, in analogy with a transistor biased deep in saturation. When this occurs, the three small forward-bias voltages appear as shown in Fig. 11–3b. Two of these voltages essentially cancel in the overall v, so that the forward voltage drop of the device from anode to cathode in the conducting state is not much greater than that of a single p-n junction. For Si this forward drop is less than 1 V, until ohmic losses become important at high current levels.

We have discussed the current transport mechanisms in the forward-blocking and forward-conducting states, but we have not indicated how switching is initiated from one state to the other. Basically, the requirement is that the carrier injection at j_1 and j_2 must somehow be increased so that significant transport of injected carriers across j_2 occurs. Once this transport begins, the regenerative nature of the process takes over and switching is completed.

11.1.6 Triggering Mechanisms

There are several methods by which a p-n-p-n diode can be switched (or *triggered*) from the forward-blocking state to the forward-conducting state. For example, an increase in the device temperature can cause triggering, by sufficiently increasing the carrier generation rate and the carrier lifetimes. These effects cause a corresponding increase in device current and in the alphas discussed above. Similarly, optical excitation can be used to trigger a device by increasing the current through EHP generation.[2] The most common method of triggering a two-terminal p-n-p-n, however, is simply to raise the bias voltage to the peak value V_p. This type of *voltage triggering* results in a breakdown (or significant leakage) of the reverse-biased junction j_2; the accompanying increase in current provides the injection at j_1 and j_3 and transport required for switching to the conducting state. The breakdown mechanism commonly occurs by combination of *base-width narrowing* and *avalanche multiplication*.

When carrier multiplication occurs in j_2, many electrons are swept into n_1 and holes into p_2. This process provides the majority carriers to these regions needed for increased injection by the emitter junctions. Because of transistor action, the full breakdown voltage of j_2 need not be reached. As we

---

[2]Four-layer devices that can be triggered by a pulse of light are useful in many optoelectronic systems. This type of device is often called a *light-activated SCR*, or *LASCR*.

showed in Eq. (7-52), breakdown occurs in the collector junction of a transistor with $i_B = 0$ when $M\alpha = 1$. In the coupled transistor case of the p-n-p-n diode, breakdown occurs at j_2 when

$$M_p\alpha_1 + M_n\alpha_2 = 1 \tag{11-4}$$

where M_p is the hole multiplication factor and M_n is the multiplication factor for electrons.

As the bias v increases in the forward-blocking state, the depletion region about j_2 spreads to accommodate the increased reverse bias on the center junction. This spreading means that the neutral base regions on either side (n_1 and p_2) become thinner. Since α_1 and α_2 increase as these base widths decrease, triggering can occur by the effect of base-width narrowing. A true punch-through of the base regions is seldom required, since moderate narrowing of these regions can increase the alphas enough to cause switching. Furthermore, switching may be the result of a combination of avalanche multiplication and base-width narrowing, along with possible leakage current through j_2 at high voltage. From Eq. (11-4) it is clear that with avalanche multiplication present, the sum $\alpha_1 + \alpha_2$ need not approach unity to initiate breakdown of j_2. Once breakdown begins, the increase of carriers in n_1 and p_2 drives the device to the forward-conducting state by the regenerative process of coupled transistor action. As switching proceeds, the reverse bias is lost across j_2 and the junction breakdown mechanisms are no longer active. Therefore, base narrowing and avalanche multiplication serve only to start the switching process.

If a forward-bias voltage is applied rapidly to the device, switching can occur by a mechanism commonly called *dv/dt triggering*. Basically, this type of triggering occurs as the depletion region of j_2 adjusts to accommodate the increasing voltage. As the depletion width of j_2 increases, electrons are removed from the n_1 side and holes are removed from the p_2 side of the junction. For a slow increase in voltage, the resulting flow of electrons toward j_1 and holes toward j_3 does not constitute a significant current. If dv/dt is large, however, the rate of charge removal from each side of j_2 can cause the current to increase significantly. In terms of the junction capacitance (C_{j2}) of the reverse-biased junction, the transient current is given by

$$i(t) = \frac{dC_{j2}v_{j2}}{dt} = C_{j2}\frac{dv_{j2}}{dt} + v_{j2}\frac{dC_{j2}}{dt} \tag{11-5}$$

where v_{j2} is the instantaneous voltage across j_2. This type of current flow is often called *displacement current*. The rate of change of C_{j2} must be included in calculating current, since the capacitance varies with time as the depletion width changes.

The increase in current due to a rapid rise in voltage can cause switching well below the steady state triggering voltage V_P. Therefore, a dv/dt rating is usually specified along with V_P for p-n-p-n diodes. Obviously, dv/dt triggering can be a disadvantage in circuits subjected to unpredictable voltage transients.

The various triggering mechanisms discussed in this section apply to the two-terminal p-n-p-n diode. As we shall see in the following section, the semiconductor controlled rectifier is triggered by an external signal applied to a third terminal.

The semiconductor controlled rectifier (SCR) is useful in many applications, such as in power switching and in various control circuits. This device can handle currents from a few milliamperes to hundreds of amperes. Since it can be turned on externally, the SCR can be used to regulate the amount of power delivered to a load simply by passing current only during selected portions of the line cycle. A common example of this application is the light-dimmer switch used in many homes. At a given setting of this switch, an SCR is turned on and off repetitively, such that all or only part of each power cycle is delivered to the lights. As a result, the light intensity can be varied continuously from full intensity to dark. The same control principle can be applied to motors, heaters, and many other systems. We shall discuss this type of application in this section, after first establishing the fundamentals of device operation.

**11.2
THE
SEMICONDUCTOR
CONTROLLED
RECTIFIER**

11.2.1 Gate Control

The most important four-layer device in power circuit applications is the three-terminal SCR[3] (Fig. 11–4). This device is similar to the p-n-p-n diode,

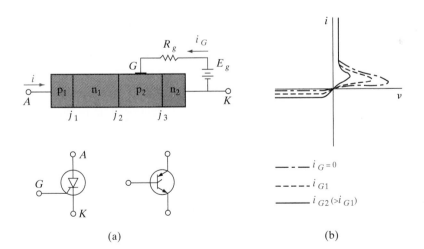

$i_{G}=0$

i_{G1}

$i_{G2}(>i_{G1})$

(a)   (b)

Figure 11–4
A semiconductor controlled rectifier: (a) four-layer geometry and common circuit symbols; (b) I–V characteristics.

[3]This device is often called a *thyristor* to indicate its function as a solid state analogue of the *gas thyratron*. The thyratron is a gas-filled tube that passes current when an arc discharge occurs at a critical firing voltage. Analogous to the gate current control of the SCR, this firing voltage can be varied by a voltage applied to a third electrode.

except that a third lead (*gate*) is attached to one of the base regions. When the SCR is biased in the forward-blocking state, a small current supplied to the gate can initiate switching to the conducting state. As a result, the anode switching voltage V_P decreases as the current i_G applied to the gate is increased (Fig. 11–4b). This type of turn-on control makes the SCR a useful and versatile device in switching and control circuits.

To visualize the *gate triggering* mechanism, let us assume the device is in the forward-blocking state, with a small saturation current flowing from anode to cathode. A positive gate current causes holes to flow from the gate into p_2, the base of the n-p-n transistor. This added supply of holes and the accompanying injection of electrons from n_2 into p_2 initiates transistor action in the n-p-n. After a transit time τ_{t2}, the electrons injected by j_3 arrive at the center junction and are swept into n_1, the base of the p-n-p. This causes an increase of hole injection of j_1, and these holes diffuse across the base n_1 in a transit time τ_{t1}. Thus, after a delay time of approximately $\tau_{t1} + \tau_{t2}$, transistor action is established across the entire p-n-p-n and the device is driven into the forward-conducting state. In most SCRs the delay time is less than a few microseconds, and the required gate current for turn-on is only a few milliamperes. Therefore, the SCR can be turned on by a very small amount of power in the gate circuit. On the other hand, the device current i can be many amperes, and the power controlled by the device may be very large.

It is not necessary to maintain the gate current once the SCR switches to the conducting state; in fact, the gate essentially loses control of the device after regenerative transistor action is initiated. For most devices a gate current pulse lasting a few microseconds is sufficient to ensure switching. Ratings of minimum gate pulse height and duration are generally provided for particular SCR devices.

11.2.2 Turning off the SCR

Turning off the SCR, changing it from the conducting state to the blocking state, can be accomplished by reducing the current i below a critical value (called the *holding current*) required to maintain the $\alpha_1 + \alpha_2 = 1$ condition. In some SCR devices, gate turn-off can be used to reduce the alpha sum below unity. For example, if the gate voltage is reversed in Fig. 11–4, holes are extracted from the p_2 base region. If the rate of hole extraction by the gate is sufficient to remove the n-p-n transistor from saturation, the device turns off. However, there are often problems involving the lateral flow of current in p_2 to the gate; nonuniform biasing of j_3 can result from the fact that the bias on this emitter junction varies with position when a lateral current flows. Therefore, SCR devices must be specifically designed for turn-off control; at best, this turn-off capability can be utilized only over a limited range for a given device.

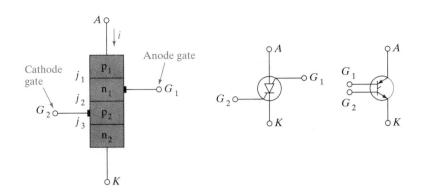

Figure 11–5
Semiconductor controlled switch: schematic configuration and common circuit symbols.

Some four-layer devices have two gate leads, one attached to n_1 and the other to p_2 (Fig. 11–5). This type of device is often called a *semiconductor controlled switch (SCS)*. The availability of the second gate electrode provides additional flexibility in circuit design. The SCS biased in the forward-blocking state can be switched to the conducting state by a positive current pulse applied to the *cathode gate* (at p_2) or by a negative pulse at the *anode gate* (n_1). If the device is designed for turn-off capability, separate circuits can be employed for turn-on at one gate and turn-off at the other. Other advantages of the SCS configuration include the possibility of minimizing unwanted dv/dt switching; for example, the gate not used for triggering can be capacitively coupled to the nearest current terminal (G_1 to A or G_2 to K) to allow a charging path for j_2 during a voltage transient without causing inadvertent switching.

11.2.3 Bilateral Devices

In many applications it is useful to employ devices which switch symmetrically with forward and reverse bias. This type of *bilateral* device is particularly useful in a-c circuits in which sinusoidal signals are switched on and off during positive and negative portions of the cycle. A typical bilateral p-n-p-n diode configuration is shown in Fig. 11–6a. This device differs from the p-n-p-n diode of Fig. 11–1 in that the n_2 region extends over only half the width of the cathode, and a new region n_3 is diffused into half of the anode region. In effect, this *bilateral diode switch* consists of two separate p-n-p-n diodes: the p_1-n_1-p_2-n_2 section and the p_2-n_1-p_1-n_3 section. We notice that the device shown in Fig. 11–6a is symmetrical. With the anode A biased positively with respect to the cathode K, junction j_1 is forward biased, while j_2 is reverse biased. Junction j_3 is shorted at one end (as is j_4) by the metal contact. When j_2 is biased to breakdown, however, a lateral current flows in p_2 biasing the left edge of j_3 into injection, and the device switches. This *shorted-emitter* design is commonly used in SCRs to enhance triggering control. During this operation, junction j_4 remains dormant. Because the device is symmetrical, j_4 serves

Figure 11–6
A bilateral diode
switch: (a)
schematic of the
device configura-
tion and a com-
mon circuit
symbol; (b) typical
I–V characteristic.

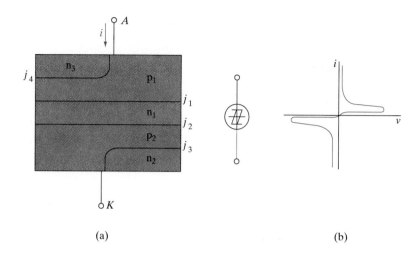

(a)   (b)

the shorted emitter function when the polarity is reversed (K positive with respect to A), and j_1 is the junction which is biased to break down in initiating the switching operation. If the bilateral diode is constructed properly, the forward and reverse characteristics are symmetrical as shown in Fig. 11–6b.

Bilateral triode switches (sometimes called *triacs*) can be constructed with SCR characteristics that can be triggered in either the forward- or reverse-bias mode. A good discussion of these devices is presented in the book by Gentry et al. in the reading list for this chapter.

11.2.4 Fabrication and Applications

Many variations of diffusion, implantation, and epitaxial growth are used in the fabrication of p-n-p-n devices. The type of fabrication process depends largely on the power rating and intended use of the device. For high-current devices the anode is attached to a heavy copper stud, and the cathode is contacted by a large cable. In high-current operation, heat is carried away from the junction by the massive metal substrate. The entire device is hermetically sealed in a housing, which provides protection from the atmosphere and from thermal and mechanical shock. Devices with this type of mounting can be rated at several hundred amperes in the conducting state. Of course, SCR devices intended for small-signal applications can be made in simpler and smaller packages.

Applications of SCRs and other four-layer devices are quite varied and extend into many fields of electronics, switching, and control. As a simple example, let us consider the problem of delivering variable power to a load from a constant line source (Fig. 11–7). The load may be the heater windings of a furnace, a light bulb, or another circuit. The amount of power delivered to the load during each half-cycle depends on the switching of the SCR. If pulses are delivered to the gate near the beginning of each half-cycle, essen-

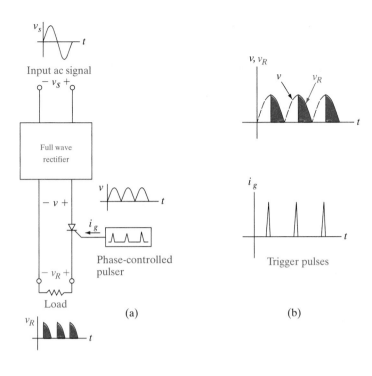

Figure 11–7
Example of the use of an SCR to control the power delivered to a load: (a) schematic diagram of the circuit; (b) waveforms of the delivered signal and the phase-variable trigger pulse.

tially the full power of the input is delivered to the load. On the other hand, if the trigger pulses are delayed, the SCR does not turn on until later in the half cycle. As a result, the amount of power delivered to the load can be varied from almost full power to no power.

Many examples of SCR and other four-layer device applications can be found in the reading list of this chapter and in the current literature.

We saw in Section 11.2 that the SCR has difficulty in efficiently turning off the device using the gate. We need to use additional circuitry to reduce the anode-to-cathode current below the holding current to change the SCR from the conducting state to the blocking state. This is, of course, clumsy and expensive.

Hence, the *insulated-gate bipolar transistor* (*IGBT*) was invented by Baliga in 1979 to address this issue. This variation on the SCR can easily be turned off from the conducting to the blocking state by the action of the gate. This device is also known by several other names such as conductivity-modulated FET (COMFET), insulated gate transistor (IGT), insulated gate rectifier (IGR), gain-enhanced MOSFET (GEMFET) and bipolar FET (BiFET).

The basic structure is shown for an n-channel device in Fig. 11–8. It basically combines an SCR with a MOSFET able to connect or disconnect the

11.3
INSULATED GATE
BIPOLAR
TRANSISTOR

Figure 11–8
Structure of an insulated gate bipolar transistor.

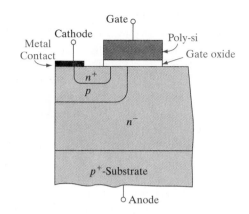

Figure 11–9
Output current–voltage characteristics of an insulated gate bipolar transistor (n-channel).

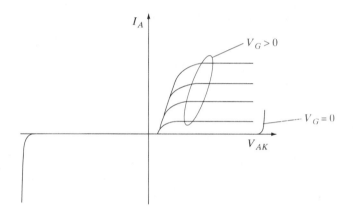

n^+ cathode to the n^- base region, depending on the gate bias of the MOSFET. The MOSFET channel length is determined by the p-region, which is formed by diffusion of the acceptors implanted in the same region as the n^+ cathode. In other words, the channel length is not determined by lithography of the gate, as in a conventional MOSFET, but rather by the diffusion of the acceptors. Such a MOSFET structure is known as a double-diffused MOSFET (DMOS). The DMOS device is essentially an NMOSFET.

The main part of the IGBT is the n^- region, which acts as the drain of the DMOS device. This is generally a thick (~50 μm) epitaxial region with a low doping (~10^{14} cm^{-3}) grown on a heavily p^+-doped substrate which forms the anode. This n^- region can therefore support a large blocking voltage in the OFF state. In the ON state, the conductivity of this lightly doped region is modulated (increased) by the electrons injected from the n^+ cathode and the holes injected from the p^+ anode; hence, the alternative name, conductivity modulated FET (COMFET). This increased conductivity allows the voltage drop across the device to be minimal in the ON state.

The current–voltage characteristics are shown in Fig. 11–9. If the DMOS gate voltage is zero (or below the threshold voltage), an n-type inversion re-

gion is not formed in the p-type channel region and the n⁺ cathode is not shorted to the n⁻ base. The structure then looks exactly like a conventional SCR which
allows minimum current flow in either polarity until breakdown is reached. For
positive anode-to-cathode bias V_{AK}, avalanche breakdown occurs at the n⁻-p
junction, while for negative V_{AK} avalanche occurs at the n⁻-p⁺ junction.

When a gate bias is applied to the DMOS gate, for positive V_{AK}, we see
that there is significant current flow (Fig. 11–9). The characteristics look like
that of a MOSFET, with one difference. Instead of the current starting to increase from the origin, there is an offset or cut-in voltage of ~0.7V, just like
for a diode. The reason for this can be understood by looking at the equivalent circuit in Fig. 11–10a. For small V_{AK} up to the offset voltage, the structure looks like a DMOS in series with a p-i-n diode made up of the p⁺
substrate (anode), the n⁻ blocking region (base) which is essentially like an
intrinsic region, and the n⁺ cathode. In this regime, there is negligible voltage
drop across the DMOS device, and the p-i-n device is in forward bias. The injected carriers from the anode and the cathode recombine in the n⁻ region.
As we saw in Chapter 5, for a diode dominated by recombination in the depletion region, the current–voltage characteristics show an exponential behavior, with a diode ideality factor of **n** = 2. Therefore, we get in this region

$$I_A \propto \exp(qV_{AK}/2\,kT) \qquad (11\text{–}6)$$

On the other hand when V_{AK} is larger than the offset voltage (~0.7 V),
the characteristics look like that of a MOSFET, multiplied by a p-n-p bipolar junction transistor gain term. The equivalent circuit in this region is shown
in Fig. 11–10b. In this regime, not all the injected carriers recombine in the
near-intrinsic n⁻ region. This current, which is essentially the DMOSFET
current, I_{MOS}, acts as the base current of the vertical p-n-p BJT formed between the p⁺ substrate (anode), the n⁻ base and the p⁻ channel of the DMOS
device. Hence, the current now is given by

$$I_A = (1 + \beta_{pnp})I_{MOS} \qquad (11\text{–}7)$$

The shape of the characteristics looks like that of the DMOS device. This is
the preferred mode of operation of the IGBT.

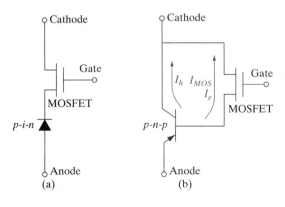

Figure 11–10
IGBT equivalent
circuit: (a) below
the offset voltage,
for low V_{AK}; (b)
above the offset
voltage, for high
V_{AK}.

Finally, if the current levels are too high, the IGBT latches into a low impedance state like that of a conventional SCR in the ON state. This is undesirable because it means that the gate of the DMOS device has now lost control.

The IGBT clearly incorporates some of the best features of MOSFETs and BJTs. Like a MOSFET, it has high input impedance and low input capacitance. On the other hand, in the ON state, it has low resistance and high current handling capability, like a BJT or SCR. Because of these factors, and because it can turn off more easily than a SCR, the IGBT is gradually becoming the power device of choice, in place of the more traditional SCR.

PROBLEMS

11.1 Explain why two separate transistors cannot be connected as in Fig. 11–2 to achieve the p-n-p-n switching action of Fig. 11–1.

11.2 In the p-n-p-n diode (Fig. 11–3a), the junction j_3 is forward biased during the forward-blocking state. Why, then, does the forward bias provided by the gate-to-cathode voltage in Fig. 11–4 cause switching?

11.3 (a) Sketch the energy band diagrams for the p-n-p-n diode in equilibrium; in the forward-blocking state; and in the forward-conducting state.

(b) Sketch the excess minority carrier distributions in regions n_1 and p_2 when the p-n-p-n diode is in the forward-conducting state.

11.4 Use schematic techniques such as those illustrated in Fig. 7–3 to describe the hole flow and electron flow in a p-n-p-n diode for the forward-blocking state and for the forward-conducting state. Explain the diagrams and be careful to define any new symbols (e.g., those representing EHP generation and recombination).

11.5 Using the coupled transistor model, rewrite Eqs. (11–1) to include avalanche multiplication in j_2, and show that Eq. (11–4) is valid for the p-n-p-n diode.

READING LIST

Baliga, B. J. *Power Semiconductor Devices.* Boston, MA: PWS, 1996.

Gentry, F. E., F. W. Gutzwiller, N., Holonyak, Jr., and E. E. Von Zastrow. *Semiconductor Controlled Rectifiers: Principles and Application of p-n-p-n Devices.* Englewood Cliffs, NJ: Prentice Hall, 1964.

Jaeklin, A. A., ed. *Power Semiconductor Devices and Circuits.* New York: Plenum Press, 1992.

Moll, J. L., M. Tanenbaum, J. M. Goldey and N. Holonyak. "P-N-P-N Transistor Switches." *Proc. IRE*, 44 (1956), 1174+.

Taylor, P. D. *Thyristor Design and Realization.* Chichester: Wiley, 1992.

Wang, S. *Fundamentals of Semiconductor Theory and Device Physics.* Englewood Cliffs, NJ: Prentice Hall, 1989.

Appendix I
Definitions of Commonly Used Symbols[1]

a	Chapter 1: unit cell dimension (Å); Chapter 6: metallurgical channel half-width for an FET (cm)
a, b, c	basis vectors
A	area (cm^2)
$\mathcal{B}$	magnetic flux density (Wb/cm^2)
B	base transport factor for a BJT
B, E, C	base, emitter, collector of a BJT
c	speed of light (cm/s)
C	capacitance/area in MOS (F/cm^2)
C_i, C_d, C_{it}	insulator, depletion, interface-state MOS capacitance/area (F/cm^2)
C_j	junction capacitance (F)
C_s	charge storage capacitance (F)
D, D_n, D_p	diffusion coefficient for dopants, electrons, holes (cm^2/s)
D, G, S	drain, gate, source of an FET
e	Napierian base
e^-	electron
$\mathcal{E}$	electric field strength (V/cm)
E	energy[2] (J, eV); battery voltage (V)
E_a, E_d	acceptor, donor energy level (J, eV)
E_c, E_v	conduction band, valence band edge (J, eV)
E_F	equilibrium Fermi level (J, eV)
E_g	band gap energy (J, eV)
E_i	intrinsic level (J, eV)
E_r, E_t	recombination, trapping energy level (J, eV)
$f(E)$	Fermi–Dirac distribution function
F_n, F_p	quasi-Fermi level for electrons, holes (J, eV)
g, g_{op}	EHP generation rate, optical generation rate (cm^{-3}-s^{-1})

[1]This list does not include some symbols that are used only in the section where they are defined. Units are given in common semiconductor usage, involving cm where appropriate; it is important to note, however, that calculations should be made in the MKS system in some formulas.

[2]In the Boltzmann factor exp $(-\Delta E/kT)$, ΔE can be expressed in J or eV if k is expressed in J/K or eV/K, respectively.

g_m	mutual transconductance (Ω^{-1}, S)
h	Planck's constant (J-s, eV-s); Chapter 6: FET channel half-width (cm)
$\hbar$	Planck's constant divided by 2π (J-s, eV-s)
$h\nu$	photon energy (J, eV)
h, k, l	Miller indices
h^+	hole
i, I	current[3] (A)
I (subscript)	inverted mode of a BJT
i_B, i_C, i_E	base, collector, emitter current in a BJT (A)
I_{CO}, I_{EO}	magnitude of the collector, emitter saturation current with the emitter, collector open (A)
I_{CS}, I_{ES}	magnitude of the collector, emitter saturation current with the emitter, collector shorted (A)
I_D	channel current in an FET, directed from drain to source (A)
I_0	reverse saturation current in a p-n junction (A)
j	$\sqrt{-1}$
J	current density (A/cm^2)
k	Boltzmann's constant (J/K, eV/K)
k_N, k_P	transconductance of NMOSFET, PMOSFET divided by V_D (A/V^2)
k	wave vector (cm^{-1})
k_d	distribution coefficient
K	scaling factor
K	$4\pi\epsilon_0$ (F/cm)
l, L	length (cm)
L_D	Debye length (cm)
$\bar{l}$	mean free path for carriers in random motion (cm)
m, m^*	mass, effective mass (kg)
m_n^*, m_p^*	effective mass for electrons, holes (kg)
m_l, m_t	longitudinal, transverse electron effective mass (kg)
m_{lh}, m_{hh}	light, heavy hole effective mass (kg)
m_0	rest mass of the electron (kg)
M	avalanche multiplication factor
m, n	integers; exponents
n	concentration of electrons in the conduction band (cm^{-3})
n	n-type semiconductor material
n$^+$	heavily doped n-type material
n_i	intrinsic concentration of electrons (cm^{-3})
n_n, n_p	equilibrium concentration of electrons in n-type, p-type material (cm^{-3})
n_0	equilibrium concentration of electrons (cm^{-3})
N (subscript)	normal mode of a BJT
N_a, N_d	concentration of acceptors, donors (cm^{-3})
N_a^-, N_d^+	concentration of ionized acceptors, donors (cm^{-3})
N_c, N_v	effective density of states at the edge of the conduction band, valence band (cm^{-3})
p	concentration of holes in the valence band (cm^{-3})
p	p-type semiconductor material

[3]See note at the end of this list.

p^+	heavily doped p-type material
p	momentum (kg-m/s)
p_i	intrinsic hole concentration (cm^{-3}) = n_i
p_n, p_p	equilibrium concentration of holes in n-type, p-type material (cm^{-3})
p_0	equilibrium hole concentration (cm^{-3})
q	magnitude of the electronic charge (C)
Q_+, Q_-	total positive, negative charge (C)
Q_d	depletion region charge/area (C/cm^2)
Q_f	oxide fixed charge/area (C/cm^2)
Q_i	effective MOS interface charge/area (C/cm^2)
Q_{it}	interface trap charge/area (C/cm^2)
Q_m	mobile ionic charge/area (C/cm^2)
Q_n, Q_p	charge stored in an electron, hole distribution (C)
Q_n	mobile charge/area in FET channel (C/cm^2)
Q_{ot}	oxide trapped charge/area (C/cm^2)
$R_p, \Delta R_p$	projected range, straggle (cm)
r, R	resistance (Ω)
R_H	Hall coefficient (cm^3/C)
S	subthreshold slope (mV/decade)
t	time (s)
t	sample thickness (cm)
$\bar{t}$	mean free time between scattering collisions (s)
t_{sd}	storage delay time (s)
T	temperature (K)
v, V	voltage[4] (V)
V	potential energy (J)
$\mathcal{V}$	electrostatic potential (V)
V_{CB}, V_{EB}	voltage from collector to base, emitter to base in a BJT (V)
V_D, V_G	voltage from drain to source, gate to source in an FET (V)
$\mathcal{V}_n, \mathcal{V}_p$	electrostatic potential in the neutral n, p material (V)
V_0	contact potential (V)
V_P	Chapter 6: pinch-off voltage for an FET; Chapter 11: forward breakover voltage for an SCR (V)
V_T, V_{FB}	MOS threshold voltage, flat-band voltage (V)
v, v$_d$	velocity, drift velocity (cm/s)
w	sample width (cm)
W	depletion region width (cm)
W_b	base width in a BJT, measured between the edges of the emitter and collector junction depletion regions (cm)
x	distance (cm), alloy composition
x_n, x_p	distance in the neutral n region, p region of a junction, measured from the edge of the transition region (cm)
x_{n0}, x_{p0}	penetration of the transition region into the n region, p region, measured from the metallurgical junction (cm)
Z	atomic number; dimension in z-direction (cm)
α	emitter-to-collector current transfer ratio in a BJT

[4]See note at the end of this list.

α	optical absorption coefficient (cm^{-1})
α_r	recombination coefficient (cm^3/s)
β	base-to-collector current amplification factor in a BJT
γ	emitter injection efficiency; in a p-n-p, the fraction of i_E due to the hole current i_{Ep}
δ, Δ	incremental change
$\delta n, \delta p$	excess electron, hole concentration (cm^{-3})
$\Delta n_p, \Delta p_n$	excess electron, hole concentration at the edge of the transition region on the p side, n side (cm^{-3})
$\Delta p_C, \Delta p_E$	excess hole concentration in the base of a BJT, evaluated at the edge of the transition region of the collector, emitter junction (cm^{-3})
$\epsilon, \epsilon_r, \epsilon_0$	permittivity, relative dielectric constant, permittivity of free space (F/cm); $\epsilon = \epsilon_r \epsilon_0$
λ	wavelength of light $(\mu m, \text{Å})$
μ	mobility $(cm^2/\text{V-s})$
v	frequency of light (s^{-1})
ρ	resistivity $(\Omega\text{-cm})$; charge density (C/cm^3)
σ	conductivity $(\Omega\text{-cm})^{-1}$
τ_d	dielectric relaxation time (s); in a BJT, delay time (s)
τ_n, τ_p	recombination lifetime for electrons, holes (s)
τ_t	transit time (s)
ϕ	flux density $(cm^2\text{-s})^{-1}$; potential (V), dose (cm^{-2})
ϕ_F	$(E_i - E_F)/q$ (V)
ϕ_s	surface potential (V)
Φ	work function potential (V)
Φ_B	metal–semiconductor barrier height (V)
Φ_{ms}	metal–semiconductor work function potential difference (V)
ψ, Ψ	time-independent, time-dependent wave function
ω	angular frequency (s^{-1})
$\langle \rangle$	average of the enclosed quantity

Note: For d-c voltage and current, capital symbols with capital subscripts are used; lowercase symbols with lowercase subscripts represent a-c quantities; lowercase symbols with capital subscripts represent total (a-c + d-c) quantities. For voltage symbols with double subscripts, V is positive when the potential at the point referred to by the first subscript is higher than that of the second point. For example, V_{GD} is the potential difference $V_G - V_D$.

Appendix II
Physical Constants and Conversion Factors[1]

Avogadro's number	$N_A = 6.02 \times 10^{23}$ molecules/mole
Boltzmann's constant	$k = 1.38 \times 10^{-23}$ J/K
	$= 8.62 \times 10^{-5}$ eV/K
Electronic charge (magnitude)	$q = 1.60 \times 10^{-19}$ C
Electronic rest mass	$m_0 = 9.11 \times 10^{-31}$ kg
Permittivity of free space	$\epsilon_0 = 8.85 \times 10^{-14}$ F/cm
	$= 8.85 \times 10^{-12}$ F/m
Planck's constant	$h = 6.63 \times 10^{-34}$ J-s
	$= 4.14 \times 10^{-15}$ eV-s
Room temperature value of kT	$kT = 0.0259$ eV
Speed of light	$c = 2.998 \times 10^{10}$ cm/s

Prefixes:

1 Å (angstrom) $= 10^{-8}$ cm	milli-,	m-	$= 10^{-3}$
1 μm (micron) $= 10^{-4}$ cm	micro-,	μ-	$= 10^{-6}$
1 nm $= 10$Å $= 10^{-7}$ cm	nano-,	n-	$= 10^{-9}$
2.54 cm = 1 in.	pico-,	p-	$= 10^{-12}$
1 eV $= 1.6 \times 10^{-19}$ J	kilo-,	k-	$= 10^{-3}$
	mega-,	M-	$= 10^{6}$
	giga-,	G-	$= 10^{9}$

A wavelength λ of 1 μm corresponds to a photon energy of 1.24 eV.

[1]Since cm is used as the unit of length for many semiconductor quantities, caution must be exercised to avoid unit errors in calculations. When using quantities involving length in formulas which contain quantities measured in MKS units, it is usually best to use all MKS quantities. Conversion to standard semiconductor usage involving cm can be accomplished as a last step. Similar caution is recommended in using J and eV as energy units.

Appendix III
Properties of Semiconductor Materials

		E_g (ev)	μ_n (cm^2/V-s)	μ_p (cm^2/V-s)	m^*_n/m_o (m_l, m_t)	m^*_p/m_o (m_{lh}, m_{hh})	a (Å)	ϵ_r	Density (g/cm^3)	Melting point (°C)
Si	(i/D)	1.11	1350	480	0.98, 0.19	0.16, 0.49	5.43	11.8	2.33	1415
Ge	(i/D)	0.67	3900	1900	1.64, 0.082	0.04, 0.28	5.65	16	5.32	936
SiC (α)	(i/W)	2.86	500	—	0.6	1.0	3.08	10.2	3.21	2830
AlP	(i/Z)	2.45	80	—	—	0.2, 0.63	5.46	9.8	2.40	2000
AlAs	(i/Z)	2.16	1200	420	2.0	0.15, 0.76	5.66	10.9	3.60	1740
AlSb	(i/Z)	1.6	200	300	0.12	0.98	6.14	11	4.26	1080
GaP	(i/Z)	2.26	300	150	1.12, 0.22	0.14, 0.79	5.45	11.1	4.13	1467
GaAs	(d/Z)	1.43	8500	400	0.067	0.074, 0.50	5.65	13.2	5.31	1238
GaN	(d/Z, W)	3.4	380	—	0.19	0.60	4.5	12.2	6.1	2530
GaSb	(d/Z)	0.7	5000	1000	0.042	0.06, 0.23	6.09	15.7	5.61	712
InP	(d/Z)	1.35	4000	100	0.077	0.089, 0.85	5.87	12.4	4.79	1070
InAs	(d/Z)	0.36	22600	200	0.023	0.025, 0.41	6.06	14.6	5.67	943
InSb	(d/Z)	0.18	10^5	1700	0.014	0.015, 0.40	6.48	17.7	5.78	525
ZnS	(d/Z, W)	3.6	180	10	0.28	—	5.409	8.9	4.09	1650*
ZnSe	(d/Z)	2.7	600	28	0.14	0.60	5.671	9.2	5.65	1100*
ZnTe	(d/Z)	2.25	530	100	0.18	0.65	6.101	10.4	5.51	1238*
CdS	(d/W, Z)	2.42	250	15	0.21	0.80	4.137	8.9	4.82	1475
CdSe	(d/W)	1.73	800	—	0.13	0.45	4.30	10.2	5.81	1258
CdTe	(d/Z)	1.58	1050	100	0.10	0.37	6.482	10.2	6.20	1098
PbS	(i/H)	0.37	575	200	0.22	0.29	5.936	17.0	7.6	1119
PbSe	(i/H)	0.27	1500	1500	—	—	6.147	23.6	8.73	1081
PbTe	(i/H)	0.29	6000	4000	0.17	0.20	6.452	30	8.16	925

All values at 300 K.   *Vaporizes

The first column lists the semiconductor, the second indicates band structure type and crystal structure. Definitions of symbols: *i* is indirect; *d* is direct; *D* is diamond; *Z* is zincblende; *W* is wurtzite; *H* is halite (NaCl). Values of mobility are for material of high purity.

Crystals in the wurtzite structure are not described completely by the single lattice constant given here, since the unit cell is not cubic. Several II–VI compounds can be grown in either the zincblende or wurtzite structures.

Many values quoted here are approximate or uncertain, particularly for the II–VI and IV–VI compounds. The gaps indicate that the values are unknown.

For electrons, the first set of band curvature effective masses is the longitudinal mass, the second set the transverse. For holes, the first set is for light holes, the second for heavy holes.

Appendix IV

Derivation of the Density of States in the Conduction Band

In this derivation we shall consider the conduction band electrons to be essentially free. Constraints of the particular lattice can be included in the effective mass of the electron at the end of the derivation. For a free electron, the three-dimensional Schrödinger wave equation becomes

$$-\frac{\hbar^2}{2m}\nabla^2\psi = E\psi \qquad (IV–1)$$

where ψ is the wave function of the electron and E is its energy. The form of the solution to Eq. (IV–1) is

$$\psi = (\text{const.})e^{j\mathbf{k}\cdot\mathbf{r}} \qquad (IV–2)$$

We must describe the electron in terms of a set of boundary conditions within the lattice. A common approach is to use periodic boundary conditions, in which we quantize the electron energies in a cube of material of side L. This can be accomplished by requiring that

$$\psi(x + L, y, z) = \psi(x, y, z) \qquad (IV–3)$$

and similarly for the y- and z-directions. Thus our wave function can be written as

$$\psi_n = A \exp\left[j\frac{2\pi}{L} (\mathbf{n}_x x + \mathbf{n}_y y + \mathbf{n}_z z) \right] \qquad (IV–4)$$

where the $2\pi\mathbf{n}/L$ factor in each direction guarantees the condition described by Eq. (IV–3), and A is a normalizing factor. Substituting ψ_n into the Schrödinger equation (IV–1), we obtain

$$-\frac{\hbar^2}{2m}A\nabla^2 \exp\left[j\frac{2\pi}{L}(\mathbf{n}_x x + \mathbf{n}_y y + \mathbf{n}_z z) \right] = EA \exp\left[j\frac{2\pi}{L} (\mathbf{n}_x x + \mathbf{n}_y y + \mathbf{n}_z z) \right] \quad (IV–5)$$

Let us determine the number of allowed states per unit volume as a function of energy [the density of states, $N(E)$] in various cases such as 1, 2, or 3- dimensions. We first count states in $\mathbf{k}$-space, then we can use the band-structure, $E(\mathbf{k})$, to convert to $N(E)$.

For the the 3-D case in Eq. (IV–5), the components of the **k**-vector are $\mathbf{k}_x = 2\pi\,\mathbf{n}_x/L$, $\mathbf{k}_y = 2\pi\,\mathbf{n}_y/L$, and $\mathbf{k}_z = 2\pi\,\mathbf{n}_z/L$. Since there is one **k**-state for every distinct choice of integer quantum numbers, $(\mathbf{n}_x, \mathbf{n}_y, \mathbf{n}_z)$, the volume per **k**-state is $(2\pi)^3/L^3 = (2\pi)^3/V$, where $V = L^3$ is the three-dimensional volume. Hence, the number of states for 3-D in a **k**-space of $\Delta\mathbf{k}$, taking into account the factor of 2 spin degeneracy, is

$$\left\{ \frac{L^3}{(2\pi)^3}\Delta\mathbf{k} \right\} \times (2)\,spin \qquad\qquad \text{(IV–6a)}$$

The number of states per unit volume for 3-D is

$$\frac{2}{(2\pi)^3}\,(\Delta\mathbf{k}) \qquad\qquad \text{(IV–6b)}$$

In general, for p-dimensions we can generalize this expression as

$$\text{Number of states per unit volume} = \frac{2}{(2\pi)^p}(\Delta\mathbf{k}) \qquad \text{(IV–7a)}$$

We can then transform from **k**-space to E-space using the $E(\mathbf{k})$ band-structure relationship by setting

$$N(E)\,\Delta E = \frac{2}{(2\pi)^p}(\Delta\mathbf{k}) \qquad\qquad \text{(IV–7b)}$$

As described in Sec. 3.2.2, the simplest bandstructure is parabolic:

$$E(\mathbf{k}) = \frac{\hbar^2 k^2}{2m^*} \qquad\qquad \text{(IV–8a)}$$

This is often a good approximation, particularly near the bottom of the conduction band or top of the valence band. Using this, we get the relation between **k** and E as follows:

$$k = \sqrt{\frac{2m^* E}{\hbar^2}} \qquad\qquad \text{(IV–8b)}$$

$$dk = \left\{ \sqrt{\frac{m^*}{2}}\,\frac{1}{\hbar} \right\}\frac{1}{\sqrt{E}}\,dE \qquad\qquad \text{(IV–8c)}$$

For $p = 3$ we have the 3-D case, which is typical of bulk semiconductors. The volume in **k**-space between two constant-k spherical surfaces at k and $k + dk$ is (Figure IV–1a):

$$\Delta\mathbf{k} = 4\pi k^2 dk \qquad\qquad \text{(IV–9a)}$$

neglecting terms with dk multiplied by itself.
The density-of-states then becomes:

$$N(E)dE = \frac{2}{(2\pi)^3}\,4\pi k^2 dk = \frac{\sqrt{2}}{\pi^2}\left(\frac{m^*}{\hbar^2}\right)^{3/2} E^{1/2}dE \qquad \text{(IV–9b)}$$

We see that if we plot $N(E)$ versus E, we get a parabolic density-of-states function in 3-D for a parabolic bandstructure relationship (Figure IV–2a).

For $p = 2$, we get a so-called 2-D electron gas (2-DEG) or hole gas. This can arise, for example, in a quantum well (Section 3.2.5) or in the inversion layer of a MOSFET.

In this case, the "volume" in **k**-space is the annular region between two circles, k and $k+dk$, as shown in Figure (IV–1b), where

$$\Delta \mathbf{k} = (2\pi k)dk \qquad (IV–10a)$$

again neglecting dk^2.

Using Eq. IV–7a, this leads to a density of states (per unit area)

$$N(E)dE = \frac{2}{(2\pi)^2}(2\pi k)dk = \frac{m^*}{\pi \hbar^2}dE \qquad (IV–10b)$$

We see that for 2-D, the density of states is a constant in energy, unlike the parabolic density of states for 3-D (Figure IV–2b). Actually, for the 2-DEG

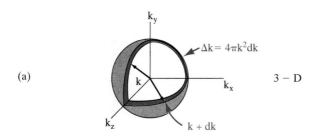

(a) 3 – D

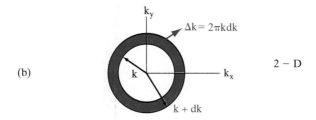

(b) 2 – D

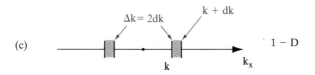

(c) 1 – D

Figure IV–1a-c
Volume in
k-space:
(a) 3-D systems;
(b) 2-D systems;
(c) 1-D systems.

Figure IV–2
Density of states:
(a) in 3-D or bulk;
(b) in 2-D electron
or hole gases;
(c) in 1-D
quantum "wires".

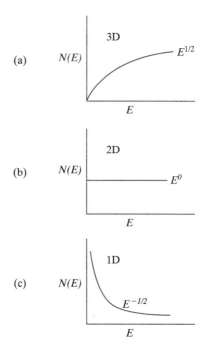

in a quantum well or inversion layer (see Chapter 6) we must add the various constant 2-D densities-of-states for the different "particle-in-a-box" levels that were discussed in Sections 2.4.3 and 3.2.5, leading to a so-called "staircase" density of states.

For $p = 1$, we get 1-D quantum "wires." These more esoteric structures can be grown, for example, by MBE or MOCVD. In this case, the "volume" in **k**-space in the region between k and $k+dk$ in 1-D is (Figure IV–1c):

$$\Delta \mathbf{k} = 2(dk) \tag{IV–11a}$$

Using Eq. IV–7a, this leads to a density of states

$$N(E)dE = \frac{2}{(2\pi)^1}(2dk) = \frac{\sqrt{2m^*}}{\pi\hbar\sqrt{E}}dE \tag{IV–11b}$$

By examining the density of states in 3, 2 and 1-D (Eqs. IV–9b, IV–10b and IV–11b, respectively) we notice a very interesting trend. Every time we go to a lower dimensionality system, the dependence of density of states on energy changes by $1/\sqrt{E}$. In fact, one finds that for 0-D quantum "dots" the density of states is indeed proportional to $1/E$. In the 1 and 0-D cases, we see that the density of states has singularities in energy, which has very important implications for semiconductor devices. Unfortunately, those discussions are beyond the scope of this book.

To include the probability of occupation of any energy level E, we use the Fermi–Dirac distribution function:

$$f(E) = \frac{1}{e^{(E-E_F)/kT} + 1} \qquad \text{(IV–12)}$$

The concentration of electrons in the range dE is given by the product of the density of allowed states in that range and the probability of occupation. Thus the density of occupied electron states N_e in dE is

$$N_e dE = N(E)f(E)dE \qquad \text{(IV–13)}$$

For the 3-D case we may calculate the concentration of electrons in the conduction band at a given temperature by integrating Eq. (IV–13) across the band:

$$n = \int_0^\infty N(E)f(E)dE = \frac{1}{2\pi^2}\left(\frac{2m}{\hbar^2}\right)^{3/2} e^{E_F/kT} \int_0^\infty E^{1/2} e^{-E/kT} dE \qquad \text{(IV–14)}$$

In this integration we have referred the energies in the conduction band to the band edge (E_c taken as $E = 0$). Furthermore, we have taken the function $f(E)$ to be

$$f(E) = e^{(E_F - E)/kT} \qquad \text{(IV–15)}$$

for energies such that $(E - E_F) \gg kT$.

The integral in Eq. (IV–14) is of the standard form:

$$\int_0^\infty x^{1/2} e^{-ax} dx = \frac{\sqrt{\pi}}{2a\sqrt{a}} \qquad \text{(IV–16)}$$

Thus Eq. (IV–14) gives

$$n = 2\left(\frac{2\pi mkT}{h^2}\right)^{3/2} e^{E_F/kT} \qquad \text{(IV–17)}$$

If we refer to the bottom of the conduction band as E_c instead of $E = 0$, the expression for the electron concentration is

$$n = 2\left(\frac{2\pi m_n^* kT}{h^2}\right)^{3/2} e^{(E_F - E_C)/kT} \qquad \text{(IV–18)}$$

which corresponds to Eq. (3–15). We have included constraints of the lattice through the effective mass of the electron in the crystal, m_n^*.

Appendix V
Derivation of Fermi–Dirac Statistics

In this section, we will give a simplified derivation of Fermi–Dirac statistics. We will not go through all the details, but will instead point out the physical assumptions involved. The distribution function is determined by calculating the number of distinct ways (W_k) we can put n_k indistinguishable electrons in g_k states at an energy level E_k, subject to the Pauli exclusion principle.

The assumptions are:

1. Each allowed state has a maximum of one electron (Pauli principle).
2. The probability of occupancy of each allowed (degenerate) quantum state is the same.
3. All electrons are indistinguishable.

The number of distinct ways we can put the electrons in a particular level is

$$W_k = \frac{(g_k)(g_k - 1)(g_k - \overline{n_k - 1})}{n_k!} = \frac{g_k!}{(g_k - n_k)!n_k!} \tag{V–1}$$

For N levels in a band, the number of distinct ways we can put in the various electrons gives us the so-called "multiplicity function,"

$$W_b = \prod_k W_k = \prod_k \frac{g_k!}{(g_k - n_k)!n_k!} \tag{V–2}$$

If we ask, "What is the most probable distribution of the n_k electrons in the various E_k levels (degeneracy of g_k in level E_k)?", the statistical mechanical answer is:

In thermal equilibrium, the distribution which is most disordered (i.e., has the maximum entropy, or which can occur in the largest number of ways) is the most probable.

We therefore have to maximize W_b with respect to n_k.

We assume here that the total number of electrons in the band is fixed.

$$\sum_k n_k = n = \text{constant} \Rightarrow \sum_k dn_k = 0 \tag{V–3}$$

We also assume that the total energy in the band is constant.

$$E_{tot} = \sum_k E_k n_k = \text{constant, implying } \sum_k E_k dn_k = 0 \qquad \text{(V–4)}$$

To maximize or minimize some function $f(x_i)$ of q variables $x_i(i = 1, \ldots, q)$ subject to the constraints that $g(x_i)$ and $h(x_i)$ are constant, we use the method of Lagrange undetermined multipliers.

We have

$$df = 0 \quad \text{(for extremal value of } f) \qquad \text{(V–5)}$$

$$dg = 0, dh = 0 \quad \text{(because } g \text{ and } h \text{ are constant)} \qquad \text{(V–6)}$$

Introducing two Lagrange undetermined multipliers α and β, we get

$$\sum_i \frac{\partial}{\partial x_i}[f(x_i) + \alpha g(x_i) + \beta h(x_i)]dx_i = 0$$

$$\frac{\partial}{\partial x_i}[f(x_i) + \alpha g(x_i) + \beta h(x_i)] = 0 \qquad \text{(V–7)}$$

for $i = 1, \ldots q$

$$g(x_i) = \text{const.} \quad h(x_i) = \text{const.} \qquad \text{(V–8)}$$

We thus get $(q + 2)$ equations in $(q + 2)$ unknowns of (x_i, α, β)

We apply this technique to our problem at hand. Instead of maximizing W_b, we maximize $\ln W_b$ instead because it makes the mathematics simpler. Since the log function increases monotonically with the argument, maximizing one is the same as maximizing the other.

$$\ln W_b = \sum_k [\ln(g_k)! - \ln(g_k - n_k)! - \ln(n_k)!] \qquad \text{(V–9)}$$

To simplify these terms, we use Stirling's approximation for factorials of large numbers. $\ln x! = x \ln x - x$ for large x.

$$\ln W_b = \sum_k [g_k \ln(g_k) - g_k - (g_k - n_k)\ln(g_k - n_k) + (g_k - n_k) - n_k \ln(n_k) + n_k]$$

$$= \sum_k [g_k \ln(g_k) - (g_k - n_k)\ln(g_k - n_k) - n_k \ln(n_k)] \qquad \text{(V–10)}$$

Now $dg_k = 0$ because these are system constraints. We then get

$$d(\ln W_b) = \sum_k \frac{\partial[\ln W_b]}{\partial n_k} dn_k = \sum_k \ln\left(\frac{g_k}{n_k} - 1\right) dn_k = 0 \qquad \text{(V–11)}$$

Also, from the two constraints we get

$$\sum_k dn_k = 0 \text{ and } \sum_k E_k dn_k = 0 \qquad \text{(V–12)}$$

Then,

$$\sum_k \left[\ln\left(\frac{g_k}{n_k} - 1\right) - \alpha - \beta E_k \right] dn_k = 0 \qquad \text{(V–13)}$$

$$\ln\left(\frac{g_k}{n_k} - 1\right) - \alpha - \beta E_k = 0 \qquad \text{(V–14)}$$

From this,

$$\frac{n_k}{g_k} = f(E_k) = \frac{1}{1 + e^{\alpha + \beta E_k}} \qquad \text{(V–15)}$$

From basic thermodynamics, it can be shown that

$$\alpha = -\frac{E_F}{kT}, \quad \beta = \frac{1}{kT} \qquad \text{(V–16)}$$

to get the Fermi–Dirac distribution function,

$$f(E_k) = \frac{1}{\exp\left[\dfrac{E_k - E_F}{kT}\right] + 1} \qquad \text{(V–17)}$$

For the limit of high energies,

$$E \gg E_F, \quad f(E) \simeq \exp\frac{E_F - E}{kT}. \qquad \text{(V–18)}$$

This is the classical Maxwell–Boltzmann limit of the Fermi–Dirac distribution function. Once we have the probabilities of electron occupancy, the probability of hole occupancy becomes

$$1 - f(E) = \frac{1}{\exp\dfrac{E_F - E}{kT} + 1} \qquad \text{(V–19)}$$

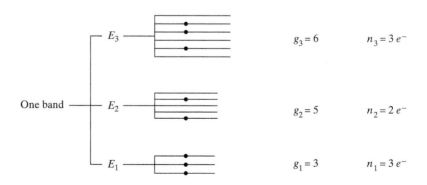

Figure V–1
Example showing three energy levels in a band, having different degeneracies, g, and electron occupancies, n, as shown.

Appendix VI

Dry and Wet Thermal Oxide Thickness Grown on Si (100) as a Function of Time and Temperature[1]

[1]From B. Deal. "The Oxidation of Silicon in Dry Oxygen, Wet Oxygen and Steam." *J. Electrochem. Soc.* 110 (1963): 527.

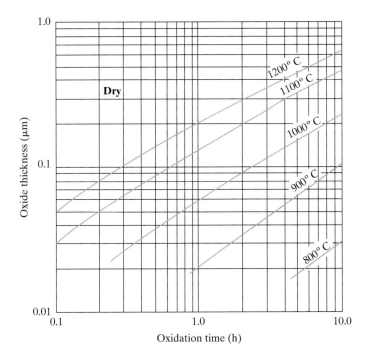

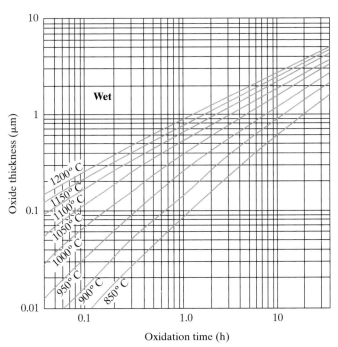

Appendix VII
Solid Solubilities of Impurities in Si[1]

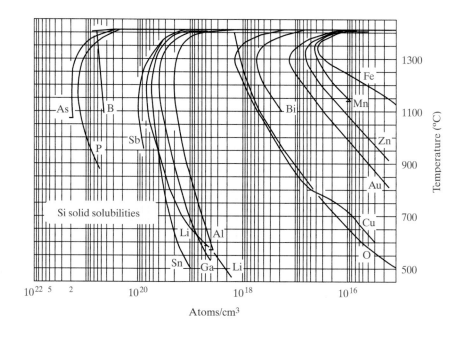

[1]From **F. A. Trumbore.** "Solid Solubilities of Impurity Elements in Si and Ge," *Bell System Technical Journal* 39, no. 1, pp. 205–233 (January 1960) copyright 1960, The American Telephone and Telegraph Co., reprinted by permission. Alterations have been made to include later data.

Appendix VIII
Diffusivities of Dopants in Si and SiO₂[1]

[1]Silicon diffusivity data from **C. S. Fuller and J. A. Ditzenberger.** "Diffusion of Donor and Acceptor Elements in Silicon." *J. Appl. Physics*, 27 (1956), 544.

SiO₂ diffusivity data from **M. Ghezzo and D. M. Brown.** "Diffusivity Summary of B, Ga, P, As and Sb in SiO₂," *J. Electrochem. Soc.* 120 (1973), 146.

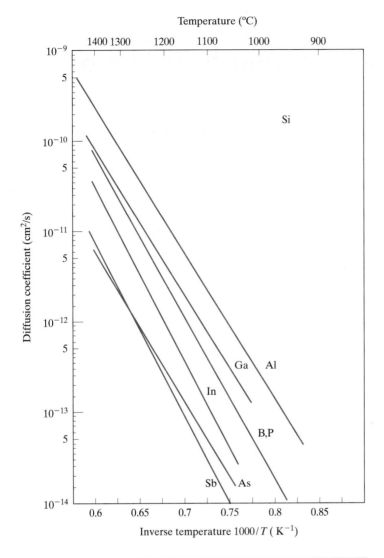

Diffusivity of various impurities in SiO$_2$		
Element	D_o(cm^2/sec)	E_A(eV)
Boron	3×10^{-4}	3.53
Phosphorus	0.19	4.03
Arsenic	250	4.90
Antimony	1.31×10^{16}	8.75

Appendix IX
Projected Range and Straggle as Function of Implant Energy in Si[1]

[1]From **J. F. Gibbons, W. S. Johnson and S. W. Mylroie.** *Projected Range Statistics: Semiconductors and Related Materials.* Stroudsburg: Dowden, Hutchison and Ross, 1975.
The projected ranges in SiO_2 are very close to those in Si.

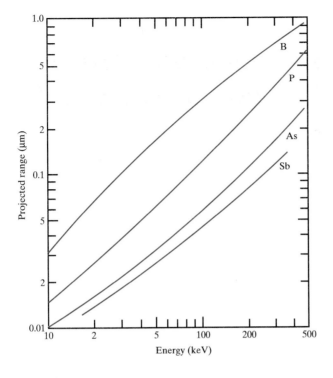

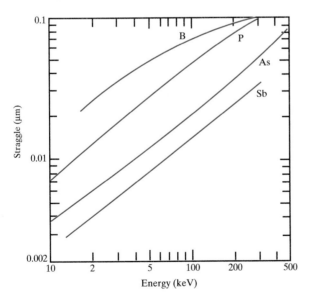

Index

Boldfaced numbers refer to illustrations.

A

Abrupt junctions, 210–11, **211**, 218–19, **219**, **220**
Absorption
  energy, 33
  excess carriers, 108–141
  lasers, 397–98, **398**
  light, 28, 397–98, **398**
  optical absorption, 108–11, **109**, **110**
Absorption coefficient, 110
A-c conductance, junctions, 210
Acceleration tube in ion implantation system, 149, **149**
Acceptors/acceptor level in doping, 77
  trapping, 118–20, 118
Affinity rule, 227–32
Aluminum (Al), 11, 12, 294
  contacts/interconnections, 225, 441–44, **442**
  heteroepitaxy, 18
  metallization, 156–57, **157**
  ultrasonic bonding, 478
AlGaAs, 2, 11
  alloy composition and variations of energy bands, 64, 66
  fiber optic communications, 394, **395**
  heteroepitaxy, 18
  heterojunction bipolar transistors (HBT), 373–74, **373**
  heterojunctions, 227–32
  lasers, 407, 409, 410
  light-emitting diodes (LED), 392
  modulation doping, 252–54
  molecular beam epitaxy (MBE), 23, **24**, **25**
  quantum well, potential well problems, 79–80, **79**
Alloy composition and variations of energy bands, **19**, 64, **65**, 66
Amorphous solids, 3, **4**
Amphoteric impurities in semiconductors, 78
Amplifiers/amplification
  bipolar junction transistor (BJT), 325–29
  field-effect transistors (FET), 242–44
Anderson affinity rule, 227–22
Angular momentum of electron in orbit, 33, 45–46
Anisotropic etching, 155
Annealing, 150
  ICs, 433
  rapid thermal anneal (RTA), 433
Anode gate, power devices, 513
Anode terminal, power devices, 505
Anti-punchthrough, MOSFET, 313
Anti-bonding states, 58
Antimony (Sb), 75, 225
Application-specific ICs (ASIC), 421, 425
Arrhenius dependence of diffusivity, 144
Arsenic (As), 75
Aspect ratios, 440
Asymmetrical effects, bipolar junction transistor (BJT), 351
Atoms and electrons (*See also* quantum mechanics), 28–54
Attenuation of light, fiber optic communications, 393–94, **394**
Au (*See* gold)

B

Back-end processing, 156, 436
Ball bond, 477, 478, **479**
Ball grid arrays (BGA), 482
Balmer series, 32, 35
Band gap (*See also* energy gaps), 2–3, 79
  absorption coefficient, 110
  junctions, 191–92, 227–32
  optical absorption, 108–11, **109**, **110**
Bardeen, 241–42, 329
Base current, bipolar junction transistor (BJT), 324–25, **325**
Base narrowing, bipolar junction transistor (BJT), 353–54, **354**
Base resistance, bipolar junction transistor (BJT), 357–59, **358**
Base spreading resistance, bipolar junction transistor (BJT), 357
Base transport factor, bipolar junction transistor (BJT), 326
Base-to-collector current amplification factor, bipolar junction transistor (BJT), 327
Base-width modulation, bipolar junction transistor (BJT), 353–54

Avalanche breakdown, 186, 188–90, **188**, **190**, 193, 354–56, **355**
Avalanche multiplication, 100, 188–90, **188**, **190**, 193
  bipolar junction transistor (BJT), 354–56, **355**
  power devices, 509–510
Avalanche photodiodes (APD), 385–86, 388, 394

543

Base-width narrowing, power devices, 509
Basis vectors in crystal lattice, 4
Batch fabrication methods for ICs, 416
Beryllium (Be), acceptors, 78
Beveling of edge, **192**
Biasing
  bipolar junction transistor (BJT), 340–46
  MOSFET, 300–1
Bias-temperature stress test, metal oxide transistor/MOSFET, 280
Bilateral devices/diodes, 513–14, **514,**
Binary-compound semiconductors, 2
Binding energy of donor materials, 77–78
Bipolar FET (BiFET), 515
Bipolar ICs, 421
Bipolar junction transistor (BJT), 241–42, 322–78, 349
  amplification, 325–29
  asymmetrical effects, 351
  avalanche breakdown, 354–56, **355**
  base current, 324–25, **325**
  base narrowing, 353–54, **354**
  base resistance, 357–59, **358**
  base spreading resistance, 357
  base transport factor, 326
  base-to-collector current amplification factor, 327
  base-width modulation, 353–54
  biasing, 340–46
  buried or sub-collector, 331
  capacitance and charging time, 365–68, **366**
  charge control analysis, 344–46
  collector junction, 324
  collector/base current relation, 326–27
  common base bias, 324
  common-emitter circuits, 328–29
  coupled-diode model, 340–44, **341**
  current transfer ratio, 326, 339–40
  cutoff frequency, 347–48, **348**, 368
  delay time, **351**
  diffusion equation, in base region, 333–34, **333**
  doping, 331–32, 351, 352–53, **352**, 371
  drift in base region, 352–353
  Early effect, 353–**54**
  Ebers-Moll equation/model, 342–44, **343**, **344**, 350, 359
  emitter crowding, 357–359
  emitter injection efficiency, 326
  emitter junction, 324
  equivalent circuits, 342–44, **343**, **344**
  fabrication processes, 329–32
  fall time, **351**
  frequency limitations, 365–71
  graded doping, 352–53, **352**
  Gummel numbers, base, emitter, 361
  Gummel plot, 362, **362**
  Gummel-Poon model, 359–62, **363**
  heterojunction bipolar transistors (HBT), 371–74
  high-frequency transistors, 369–71, **370**
  hybrid-pi model, **336**, 368
  injection levels, 356–57
  interdigitated geometry, 359, **359**
  inverted mode, 341–42
  Kirk effect, 363–65, **364**
  LOCOS, **330**, 331
  minority carrier distribution, 332–40
  nonuniformity effects, 351
  normal active mode bias, 340
  polysilicon emitter (polyemitter), 332
  recombination of base current, 326
  rise time, **351**
  saturation, 347, 348–49, **349**
  self alignment, **330**, 332
  series resistance effects, 351, 357–59, **358**
  spacers, 331
  switching, 346–51, **346**
  terminal current approximation, 337–39, **337**
  terminal currents, 334–39
  thermal effects, 356–57
  transconductance, 367–68
  transit time, 328, 368–69
  Webster effect, 369
Bipolar/CMOS (BiCMOS), 424
Bird's beak defect, 428
Bit line/word line, 441
Blocking state, power devices, 504, 505, 507–8, **508**
Body effect, MOSFET, 300–1, 466
Body-centered cubic (bcc) lattice, 5
Bohr model for atoms and electrons, 33–36, **35**, 43, 46, 47
Boltzmann factor, 80, 397
Bond-and-etch-back SOI (BE-SOI), ICs, 438
Bonding, ICs, 476–78, **478**
Bonding forces in solids, 55–58, **56**
Boro-phospho-silicate glass (BPSG), 435
Boron (B), 77, 145–46
  B_2O_3, 13
Bose-Einstein statistics, 80
Brattain, 241–42, 329
Breakdown diodes, 186, 193–94, **194**
Built-in fields and diffusion, 1 27–30
Bulk negative differential conductivity (BNDC), microwave devices, 496
Buried channel CCDs, 447
Buried channel operation, 433–34
Buried or sub-collector, bipolar junction transistor (BJT), 331
Burrus diodes, fiber optic communications, 395
Bytes or words in memory, 462

C

Capacitance/capacitors
  bipolar junction transistor (BJT) and charging time, 365–68, **366**
  ICs, 418, 429, 440–41, **441**, 444–45, **445**
  ideal MOS capacitor, 260–72, **261**
  MOSFET, time-dependent measurement, 280
  Miller overlap capacitance MOS, 444–45
  MOSFET, 304
  p-n junctions, 202–10
  Zerbst plotting of capacitance, 281, **282**
Capillary, in IC bonding, 477
Carbon Dioxide (CO_2) lasers, 400
Carrier injection, junctions, 174–83
Carrier lifetime, minority carrier lifetime, 116
Cathode gate, power devices, 513
Cathode terminal, power devices, 505
Cathodoluminescence, 111
Cadmium (Cd) acceptors, 78

Cadmium sulfide (CdS)
  band gap, 111, **111**
  light-emitting diodes
    (LED), 392
CdSe, 2
  light-emitting diodes
    (LED), 392
CdTe
  Gunn effect, 499
Central processing units (CPU),
  ICs, 450
Central valley or minimum,
  conduction band, 494
Centripetal force, 34
Ceramic column grid arrays, **483**
Channel length modulation,
  MOSFET, 313
Channel length reduction, 3
Channel stops
  ICs, 427
  metal oxide transistor
    (MOS), 260
  MOSFET, 260, 299–300
Charge carriers, 66–104
  avalanche multiplication, 100
  binding energy of donor
    materials, 77–78
  Boltzmann factor, 80
  Bose-Einstein statistics, 80
  compensation, 90–92, **91**
  concentrations of carriers, 80–92
  conduction bands, 68–70,
    79–80, **79**
  continuum of allowed states, 79
  density of states, effective,
    84–88, **85**
  doping effect on electron
    mobility, 97–98, **99**
  doping in semiconductors,
    75–79, **76**
  drift and resistance, 96–97
  drift, electric and magnetic
    fields, 92–102, **92**
  effective mass, 70–74, 77
  electron-hole pairs (EHP),
    67–70, **67**, 74–75, **74**,
    79–80, **79**
  equilibrium, at Fermi level,
    102–4
  equilibrium, electron-hole
    pairs and carrier
    concentrations, 83–88
  excess carriers, 108–41
  extrinsic semiconductor
    material, 75–79, **76**
  Fermi levels, 80–83, **81**, **82**, 102–4
  Fermi-Dirac statistics, 80–83,
    81, **82**

Hall effect, 100–102, **101**
  heavy hole bands, 73
  heterojunctions, 79–80, **79**
  high-field effects, 99–100
  hot carrier effects, 100
  impurity scattering, 97–98, **98**
  intrinsic semiconductor
    materials, 74–75, 87
  lattice scattering, 97–98, **98**
  light hole bands, 73
  Maxwell-Boltzmann
    statistics, 80
  mean free time, 93
  minority vs. majority carriers, 79
  mobility of electrons, 92–102, **92**
  n-dopants in semiconductors,
    75–79
  p-dopants in semiconductors,
    75–79
  quantum wells, 79–80, **79**
  recombination processes, 75
  resistance, 96–97
  scattering mechanisms, 93, 97–
    98, **98**, 100
  space charge neutrality, 90–92
  split-off band, 73
  steady state, 93, 94
  temperature dependence of
    carriers, 88–90, **89**, **90**
  temperature effects on
    electron mobility, 97–98, **98**
  true vs. effective mass, 73
  valence bands, 67–70, **69**, 79–
    80, **79**
  wave-particle motion of
    electrons, 70
Charge control approximation,
  junctions, 180
Charge coupled device (CCD),
  444–48, **446**, **447**
Charge sharing, MOSFET, 313–14
Charge storage capacitance,
  junctions, 202
Charge transfer devices, 444–48
Chemical beam epitaxy, 23
Chemical mechanical polishing
  (CMP), 154, 428
Chemical vapor deposition
  (CVD), 18, 150–51, 332, 435, 436
Chromatic dispersion, fiber optic
  communications, 394
Classes of clean rooms, in IC
  fabrication, 421
Cleavage planes of crystals, 11
Coherent light, 379, 396, 397
Collector, collector junction,
  bipolar junction transistor
    (BJT), 324

Collector/base current relation,
  bipolar junction transistor
    (BJT), 326–27
Common-base bias, bipolar
  junction transistor (BJT), 324
Common-emitter circuits,
  bipolar junction transistor
    (BJT), 328–29
Compact disc (CD), 404
Compensation, 90–92, **91**
Complementary error function, 145
Complementary MOS (CMOS),
  421, 423–39, **423**
  back-end processing, 436
  bipolar/CMOS (BiCMOS), 424
  bird's beak defect, 428
  boro-phospho-silicate glass
    (BPSG), 435
  buried channel operation,
    433–34
  capacitance, 429
  channel stops, 427
  chemical mechanical
    polishing (CMP), 428
  conformal LPCVD, 430
  corner effect, 428
  CVD, 435, 436
  defects, 428–29
  doping, 429, 435
  drain induced barrier lowering
    (DIBL), 426, 430
  dual-gate CMOS, 434
  electromigration
    phenomenon, 435
  gates, 429
  inverters, 451
  isolation regions, 427
  Kooi effect, 429
  latchup, 424, 426
  lateral moat encroachment
    defect, 428
  lightly doped drain (LDD),
    430, **431**
  LOCOS, 427, 428
  LPCVD, 426, 428, 429, 430, 435
  Miller overlap capacitance, 429
  MOSFETs, 430, **432**, 433–36
  NMOS devices, 430, 433–34
  overcoating, 436
  overlapping capacitance, 429
  parasitic bipolar structures, 424
  plugs, 435
  PMOS devices, 430, **432**, 432–34
  polycide, 435
  reactive ion etching (RIE), 426
  self-aligned silicide
    (SALICIDE), 424–25, **425**,
    434–35

self-alignment, 426–27
shallow trench isolation
    (STI), 428
sidewall oxide spacers, 430
source/drain extension or
    tip, 430
source/drain, 434–35, **435**
spiking problems, 435
tanks, 425–26, **425**
threshold voltages, 427
tubs or wells, 424
twin-well CMOS, 426
vias, 436
voltage transfer characteristics
    (VTC), 455, **456**
white ribbon effect, 429
Compound semiconductors, 1, 2
Concentrations of carriers, 80–92
Conductance
    MOSFET, 287
    negative, in microwave
        devices, 486
Conducting state, power devices,
    504, 505, 507–8, **508**, 508–9
Conduction bands, 60–61, 68–70,
    79–80, **79**
Conductivity, 28
    effective mass, 95
Conductivity modulation, 193, 217
Conductivity-modulated FET
    (COMFET), 516–17
Conformal LPCVD, 430
Constants, physical constants and
    conversion factors, 523
Contact potential, 159–63, 232
    junctions and, carrier injection
        vs., 212–14, **212**, **213**
Contacts, ICs, 441
Continuity equation, 130–32, **131**
Continuum of allowed states, 79
Conversion factors, physical
    constants, 523
Copper (Cu), contacts/
    interconnections, 441–44, **442**
Corner effect, 428
Coulomb potential, 43–44, 48, **50**
Coupled-diode model, bipolar
    junction transistor (BJT),
    340–44, **341**
Covalent bonding, 57–58, 77
Critical layer thickness, **20**
Crossovers, ICs, 444
Crystal properties, 1–27
    basis vectors in lattice, 4
    body-centered cubic (bcc)
        lattice, 5
    bulk crystal growth processes,
        12–17, **14**, **15**

chemical beam epitaxy, 23
chemical vapor deposition
    (CVD) epitaxy, 18
cleavage planes of crystals, 11
covalent bonding, 57–58
cubic lattice structures, 5–7, **6**
Czochralski single-crystal
    growth process, 13–14, **14**,
    16–17
diamond lattice structures,
    9–11, **10**, **12**
doping in semiconductors,
    75–79, **76**
effective mass, 70–74
epitaxial crystal growth
    process (epitaxy), 17–25
face-centered cubic (fcc)
    lattice, 5, **6**
gas-source molecular beam
    epitaxy (MBE), 23
heteroepitaxy, 18–21, **19**, **20**
lattice constant, 5, 10–11
lattice matching in epitaxial
    growth process, 18–21, **19**, **20**
lattice structures, 3–11
liquid-encapsulated
    Czochralski (LEC) single-
    crystal growth, 13
liquid-phase epitaxy (LPE), 18
metal-organic vapor-phase
    epitaxy (MOVPE), 23
metallic bonding, 57
Miller indices for crystal
    lattices, 8–9
misfit dislocations in
    heteroepitaxy, 20–21, **20**
molecular beam epitaxy
    (MBE), 18), 23, **24**, **25**
organometallic vapor-phase
    epitaxy (OMVPE), 23
periodic structures in crystal
    lattices, 3–4, **4**, **5**
planes and directions in crystal
    lattice, 7–9, **7**, **8**, **9**
polycrystalline solids, 3, **4**
primitive cell in crystal lattice, 4
pseudomorphic layers, 21
reaction chamber or reactor
    for epitaxy, 21–22, **22**
silicon dioxide as starting
    material for
    semiconductors, 12–13
simple cubic lattice, 5
single-crystal ingot growth
    processes, 13–14, **14**
strained-layer superlattice
    (SLS) structures, 21
sublattices, 9–10, **10**

unit cells in lattice, 4, **5**
vapor-phase epitaxy (VPE),
    21–23, **22**
wafer manufacture,
    semiconductors, 14–16, **15**
wurtzite lattice, 10
zincblende lattice, 9, 11
Cu (*See* Copper)
Cubic lattice structures, 5–7, **6**
Current flow, 28, 61, 169–74
Current transfer ratio, bipolar
    junction transistor (BJT), 326
Cutoff frequency, BJT, 347–48,
    348, 368
Czochralski single-crystal growth
    process, 13–14, **14**, 16–17

D

Damascene copper etching
    process, ICs, 443–44
Dark current, photodetectors, 387
Debye screening length, metal
    oxide transistor/MOSFET,
    265, 279
Decay, photoconductive decay, 120
Decoders, row/column, **462**
Deep depletion
    ICs, 445
    metal oxide transistor/
        MOSFET, 281
Degenerate semiconductors, 487
Delay time, bipolar junction
    transistor (BJT), 351
Density of state, 84–88, **85**,
    525–29, **528**
    effective mass, 86
Depletion approximation, 165
Depletion capacitance,
    junctions, **204**, 205
Depletion layer photodiodes, 385
Depletion mode
    MOSFET, 299
    transistors, 257, **258**
Depth-of-focus in
    photolithography, 154
Diamond lattice structures, 9–11,
    10, **12**
Dichlorosilene (SiH$_2$Cl$_3$), 13
Die, 151
Dielectric constant, MOSFET, 297
Dielectric relaxation time, 497
Diffraction-limited minimum
    geometry, 154
Diffused junctions, 157
Diffusion capacitance, junctions,
    206, **207**, 208–10
Diffusion currents, junctions,
    171–72

Diffusion equation, bipolar junction transistor (BJT), in base region, 333–34, **333**

Diffusion processes, 124–38, **125**
  Arrhenius dependence of diffusivity, 144
  built-in fields, 127–30
  carrier injection, junctions, 174–83
  coefficient, 144
  complementary error function, 145
  continuity equation, 130–32, **131**
  diffusion currents, junctions, 171–72
  dopant diffusivity, Si and SiO$_2$, 537–38
  drift, 127–30
  Einstein relation, 129
  electron diffusion coefficient, 126
  gradients in quasi-Fermi levels, 137–38
  Haynes-Shockley experiment in drift and diffusion, 134–37
  length of diffusion, 132–34
  p-n junctions, 144–46, **145**
  random motion and diffusion, 124–26, **125**
  recombination, 130–32, **131**
  silicon, 144–46, **145**
  SiO$_2$, 144–46, **145**

Diffusivity, 144

Digital ICs, 418

Digital versatile disc (DVD), 404

Diode equation, 177, **178**

Diodes
  bilateral devices/diodes, 513–14, **514**
  breakdown diodes, 186, 193–94, **194**
  Burrus diodes, 395
  diode equation, 177, **178**
  Esaki diodes, 486
  Gunn diodes, 486, 500
  ICs, 439
  ideal diodes, 190
  impact avalanche transit time (IMPATT) diodes, 491–93, **491**
  long diode, **207**
  narrow base diodes, 198, 202
  photodiodes, 173, 379–90
  p-n-p-n devices, 504–11, **505**
  Read diode, 491–93, **491**
  reference diodes, 194
  Schottky barrier diodes, 220–24
  Shockley diodes, 505

short diode, **207**
switching diodes, 201–2
tunnel diodes, 486–90, **488**
varactor diodes, 210–11
voltage regulators, 194
Zener diodes, 194

Direct semiconductors, 62–64, **63, 65**

Dispersion, fiber optic communications, 394

Displacement current, 206, 510

distributed Bragg reflector (DBR) mirrors, 389–90, **389**, **408**–9

Distribution coefficient K$_d$ in semiconductor doping, 16–17

Donor/donor level in doping, 76–79
  trapping, 118–20

Doping in semiconductors, 16–17, 75–79, **76**
  amphoteric impurities in semiconductors, 78
  bipolar junction transistor (BJT), 331–32, 351–53, **352**, 371
  diffusivity for Si and SiO$_2$, 537–38
  electron mobility vs. doping, 97–98, **99**
  heterojunction bipolar transistors (HBT), 371–74, **372**
  ICs, 425–26, 429, 435
  junctions, 192
  modulation doping, 252–54
  ohmic contacts, 225
  trapping, 118–20
  tunnel diodes, 487

Double-heterojunction lasers, 407–9, **408**

Double-diffused MOSFET (DMOS), 516–18

Drain
  ICs, 434–35, **435**
  junction FETs (JFET), 245
  metal oxide transistor (MOS), 256–57
  MOSFET, 256–57, 287, 288, **289**

Drain induced barrier lowering (DIBL), 289, 311–13, **312**, 426, 430

Drift
  bipolar junction transistor (BJT), 352–53
  built-in fields and diffusion, 127–30
  diffusion, 127–30

drift current, junctions, 172, 180–81
electric and magnetic fields, 92–102, **92**
equilibrium, 160
Haynes-Shockley experiment in drift and diffusion, 134–37, **134–36**
high-field effects, 99–100
microwave devices, 496–99, **497, 498**
resistance, 96–97

Drift current, junctions, 172, 180–81

Drift tube in ion implantation system, 149, **149**

Dual-gate CMOS, 434

Dual-inline package (DIP) ICs, 482

dv/dt triggering, power devices, 510–11

Dynamic RAM (DRAM), 420, 421, 449–51, **450**, 461, 464–70, **464, 465**

Dynamic resistance, tunnel diodes, 489

E

Early effect, bipolar junction transistor (BJT), 353–**54**

Ebers-Moll equations, bipolar junction transistor (BJT), 342–44, **343, 344**, 350, 359

Effective channel length in MOSFET, 304

Effective mass, 70–74, 77

Einstein coefficients, lasers, 398

Einstein relation, 129, 161

Electric fields
  diffusion, **128**
  drift, 93–**96**, 128
  Hall effect, 100–102, **101**
  high-field effects, 99–100
  junctions, 170

Electroluminescence, 111, 114, 390

Electromigration phenomenon, ICs, 435

Electron affinity, 222, 227–32

Electron-hole pairs (EHP), 67–70, **67**, 74–75, 79–80, **79**
  equilibrium, carrier concentrations, 83–88
  indirect recombination, 117–20, **118**
  photoluminescence, 111–14, **113, 114**
  quasi-Fermi levels, 120–23

recombination processes, 115–20

steady state carrier generation, 120–23

Electronic structure of atoms, 28

Electrostatic force, 34, 35

Electrostatic potential barrier, junctions, 170

Elemental semiconductors, 1, 2

Emission of energy/light, 28, 33

Emission spectra of lasers, 404–5, **405**

Emitter crowding, bipolar junction transistor (BJT), 357–59

Emitter injection efficiency, bipolar junction transistor (BJT), 326

Emitter, emitter junction, bipolar junction transistor (BJT), 324

Energy bands, 55–66
  alloy composition and variations of energy bands, 18, 64, **65**
  bandstructures for semiconductors, 72–73, **72**
  bonding forces in solids, 55–58, **56**
  conduction bands, 60–61
  covalent bonding, 57–58
  direct semiconductors, 62–64, **63**, **65**
  energy gaps, 60–61
  energy levels and bonding/anti-bonding states, 58
  forbidden bands, 61
  heavy hole bands, 73
  indirect semiconductors, 62–64, **63**, **65**
  insulators, 61–62
  ionic bonding, 55–57
  light emission, 66
  light hole bands, 73
  linear combinations of atomic orbitals (LCAO), 58–59, **59**
  metallic bonding, 57
  metals, 61–62
  mixed bonding, 58
  Pauli exclusion principle, 59, 81
  potential energy, 55
  separation in junctions, 170
  split-off band, 73
  valence bands, 60–61
  wave vectors, 62–64, **63**
  Zener breakdown, 186–87, **187**

Energy gaps, 60–61

Energy levels of electrons, 28, 33, 43, 48, **50**
  bonding/anti-bonding states, 58

Energy of electrons, 35, 39, 43, 69

Enhancement-mode transistors, 257, **258**

Epitaxial crystal growth process (epitaxy), 17–25

Equilibrium, 120
  contact potential, 159–63
  diffusion and drift, 160
  Einstein relations, 161
  electron-hole pairs and carrier concentrations, 83–88
  Fermi levels, 83–88, 102–4, 129, 161, 163–64
  junctions, 157–69, **159**, 227–32
  metal oxide transistor/MOSFET, 273, **273**
  space charge, 164–69, **164**

Esaki diodes, 486

Etching, 155–56
  Damascene copper etching process, 443–44

Excess carriers, 108–41
  carrier lifetime, minority carrier lifetime, 116
  diffusion processes, 124–38, **125**
  gradients in quasi-Fermi levels, 137–38
  Haynes-Shockley experiment in drift and diffusion, 134–37
  indirect recombination, 117–20, **118**
  luminescence, 111–14
  photoconductive decay, 120
  photoconductive devices, 123–24
  photoconductivity, 114–15
  quasi-Fermi levels, 120–23
  steady state carrier generation, 120–23
  steady state carrier injection, 132–34
  trapping, 118–20

Expectation values, 36

Extraction of carriers, 183

Extrinsic semiconductor material, 75–79, **76**

F

Face-centered cubic (fcc) lattice, 5, 6

Fall time, bipolar junction transistor (BJT), 351

Fan out, ICs, 460

Fast interface state density, metal oxide transistor/MOSFET, 279–80, **281**

Fat zero inCCDs, 447

Fermi levels, 80–83, **81**, **82**, 530–32
  bipolar junction transistor (BJT), 360
  degenerate semiconductors, 487
  equilibrium, 83–88, 102–4, 129, 161, 163–64
  gradients in quasi-Fermi levels, 137–38
  junctions, 161, 170, 225–26, **226**, 227–32
  Maxwell-Boltzmann limit, 532
  quasi-Fermi levels, 120–23, 179–80
  temperature dependence of carriers, 88–90, **89**, **90**

Fermi-Dirac statistics, 80–83, **81**, **82**, 530–32

Ferroelectrics, 297

Fiber optic communications, 392–94, **393**
  attenuation of light, 393–94, **394**
  avalanche photodiodes, 394
  chromatic dispersion, 394
  dispersion, 394
  graded-index optical fiber, 393, **393**
  index of refraction, 393
  lasers, 394
  losses, 393
  multilayer heterojunction LEDs, 395–96
  multimode fibers, 395
  p-i-n photodectors, 394
  pulse dispersion, 394
  Rayleigh scattering of light, 393–94, **394**
  single-mode fibers, 395
  step-index optical fiber, 393, **393**

Field ionization (*See* Zener effect)

Field-effect transistors (FET), 241
  amplification, 242–44
  bipolar FET (BiFET), 515
  bipolar junction transistor (BJT), 241–42
  conductivity-modulated FET (COMFET), 516–17
  depletion-mode transistors, 257, **258**
  double-diffused MOSFET (DMOS), 516–18
  enhancement-mode transistors, 257, **258**
  GaAs MESFET, 251–52, **252**
  gain-enhanced MOSFET (GEMFET), 515
  high electron mobility transistor (HEMT), 252–54
  input impedance in, 242

insulated gate FET (IGFET), 255
insulated gate bipolar transistor (IGBT), 515–18, **516**
insulated gate rectifier (IGR), 515
insulated gate transistor (IGT), 515
junction FETs (JFET), 241, 244–51, **244**
    load lines, 242–43, **242**, **243**
    metal oxide semiconductor FET (MOSFET), 241, 255–316
    metal-insulator semiconductor (MISFET), 241, 255
    metal-semiconductor FET (MESFET), 241, 251–55, **252**
    modulation doped FET (MODFET), 252–54
    on-off states, 242
    pseudomorphic HEMTs, 254
    separately doped FET (SEDFET), 254
    switching, 242–44
    two-dimensional electron gas FET (2–DEG FET or TEGFET), 254
    unipolar transistors, 241
Fill factor, solar cells, 384
Fixed oxide charge, **274**
Flash memory, 421, 461, 470–73, **471**, **472**, **474**
Flat band condition, MOSFET, **261**, 263, 273, **273**, 286
Flip-flops, 454, 463, 478–79, **480**
Floating gates, 470–72, **471**
Fluorescence, 2, 112
Flux density, 126
Forbidden bands, 61
Forward resistance, junctions, 191
Forward-blocking state, power devices, 505, 507–8, **508**
Forward-conducting state, power devices, 505, 507–8, **508**
Fowler-Nordheim tunneling
    flash memory, 473, **473**
    metal oxide transistor/ MOSFET, 283, **284**, 285
Fully depleted SOI devices, ICs, 438
Furnaces
    diffusion/oxidation, **14**
    horizontal/vertical, **14**

G

Gallium (Ga), 10, 11, 77
Gallium-Arsenide (GaAs), 2, 10, 13

alloy composition and variations of energy bands, 64, 66
annealing process, 150
band diagram, **495**
band gap, 110–11, **111**
conduction bandstructures, 72–73, **72**
donor/acceptors, 78
drift, space charge domains, 496–99, **497**, **498**
effective and true mass, 73
energy bands, 63
fiber optic communications, 394, **395**
Gunn effect, 499
heteroepitaxy, 18, 19–20
heterojunction bipolar transistors (HBT), 373–74, **373**
heterojunctions, 227–32
high-field effects, 99
intrinsic carrier concentration vs. temperature, **89**
lasers, 400, 402, 405–7, 409, 410
MESFET, 251–52, **252**
microwave devices, 500
mobility variation vs. doping impurity concentration, **98**
modulation doping, 252–54
molecular beam epitaxy (MBE), 23, **24**, **25**
photodetectors, 388
photoluminescence, 113–14, **114**
quantum well, 79–80, **79**
vapor-phase epitaxy (VPE), 21–23
GaAsP, 2, 11
    Gunn effect, 499
    heteroepitaxy, 19–20
    lasers, 400, 404
    light-emitting diodes (LED), 390–92
    vapor-phase epitaxy (VPE), 22
Gain-bandwidth product, photodetectors, 386–87
Gain-enhanced MOSFET (GEMFET), 515
GaN, 2
    lasers, 403
    light-emitting diodes (LED), 390, 392
GaP, 2
    band gap, 111, **111**
    heteroepitaxy, 20
    vapor-phase epitaxy (VPE), 22
Gas-source molecular beam epitaxy (MBE), 23

Gate-triggering, SCRs, 512
Gate-induced drain leakage (GIDL), 315–16, **316**
Gates
    ICs, 429
    junction FETs (JFET), 245, 247–49, **247**, **248**
    MOS, 433, 434
    SCR, 511–12
Gates, logic gates, 458–61, **458**
Gauss's law, 165
Generation, in junctions, 172–73, 215–16, **215**
Germanium (Ge), 2, 9
    band gap, 110–11, **111**
    covalent bonding, 57
    effective and true mass, 73
    extrinsic material, 75, 77
    fiber optic communications, 394
    heteroepitaxy, 20
    high-field effects, 99
    intrinsic carrier concentration vs. temperature, **89**
    mobility variation vs. doping impurity concentration, **98**
    recombination processes, 117
    rectification properties, 192
Gold contacts/interconnections, 441–44
Graded doping, bipolar junction transistor (BJT), 352–53, **352**
Graded index separate confinement heterostructure (GRINSCH) lasers, 408
Graded junctions, 157, 218–19, **219**, **220**
Graded-index optical fiber, 393, **393**
Gradients, 124
Gradual channel approximation, junction FETs (JFET), 249
Ground state, 46, 48, **49**
Guard ring, 192, **192**
Gummel-Poon model, bipolar junction transistor (BJT), 359–62, **362**, **363**
Gunn diodes, 486, 494–95, 499, 500

H

Hall effect, 100–2, **101**
Halo implants, MOSFET, 313
Haynes-Shockley experiment in drift and diffusion, 134–37, **134–36**
Heavy hole bands, 73
Heisenberg uncertainty principle, 37, 391

He-Ne lasers, 400
Heteroepitaxy, 18–21, **19**, **20**
Heterojunction bipolar
    transistors (HBT), 371–74, **372**
Heterojunction lasers, 406–9,
    407, **408**
Heterojunctions, 79–80, **79**,
    226–32, **229**, 370
HgCdTe, 2
High electron mobility transistor
    (HEMT), 252–54, **253**
High-frequency transistors,
    bipolar junction transistor
    (BJT), 369–71, **370**
High-field effects, 99–100
Holding current, SCRs, 512
Holes (*See* electrons-hole pairs
    (EHP))
Homojunctions, 226, 406
Hot carrier/hot electron, 100
    flash memory, 472–73, **472**
    MOSFET, 307–11, **309**
Hybrid ICs, 418–20, **419**
Hybrid-pi model, bipolar junction
    transistor (BJT), **366**, 368
Hydrogen atom model, 43–46, **44**
Hyperabrupt junctions, 210–11, **211**

I

Ideal diodes, 190
Ideality factor, 214–15
Impact avalanche transit time
    (IMPATT) diodes, 491–93, **491**
Impact ionization, 188–90, **188**,
    190, 193
Implanted junctions, 157
Impurity scattering, 97–98, **98**
InAlGaN lasers, 403
Incoherent light, 395
Index of refraction, fiber optic
    communications, 393
Indirect recombination,
    117–20, **118**
Indirect semiconductors, 62–64,
    63, 65
Inductor for ICs, 441
InGaAs
    fiber optic communications, 394
    heteroepitaxy, 18
    lasers, 409
    photodetectors, 388
InGaAsP
    fiber optic communications, 394
    heteroepitaxy, 19
    lasers, 410
    modulation doping, 253–54

InGaN lasers, 404
Injection electroluminescence,
    114, 390
Injection of carriers, 174–83
    contact potential vs. carrier
        injection, 212–14, **212**, **213**
InP
    fiber optic communications, 394
    Gunn effect, 499
    heteroepitaxy, 18, 19
    lasers, 410
    microwave devices, 500
InSb, 2
    band gap, 110–11, **111**
Insulated gate bipolar transistor
    (IGBT), 515–18, **516**
Insulated gate FET (IGFET), 255
Insulated gate transistor
    (IGT), 515
Insulators, energy bands, 61–62
Integrated circuits, 415–85, **416**
    amplifiers, 418
    analog communication
        circuits, 418
    application-specific ICs
        (ASIC), 421, 425
    back-end processing, 436
    batch fabrication methods, 416
    bipolar ICs, 421
    bipolar/CMOS (BiCMOS), 424
    bird's beak defect, 428
    bond-and-etch-back SOI
        (BE-SOI), 438
    boro-phospho-silicate glass
        (BPSG), 435
    buried channel CCDs, 447
    buried channel operation,
        433–34
    capacitors, 418, 440–41, **441**,
        444–45, **445**
    central processing units
        (CPU), 450
    channel stops, 427
    charge coupled device (CCD),
        444–48, **446**, **447**
    chemical mechanical
        polishing (CMP), 428
    complementary MOS (CMOS)
        ICs, 421, 423–39, **423**
    conformal LPCVD, 430
    contacts, 441
    corner effect, 428
    crossovers, 444
    CVD, 435, 436
    Damascene copper etching
        process, 443–44
    deep depletion, 445

    defects, 417, 428–29
    depletion, 438, 445
    developmental history and
        evolution of ICs, 420–22
    digital ICs, 418
    diodes, 439
    doping, 425–26, 429, 435
    drain induced barrier lowering
        (DIBL), 426, 430
    dual-gate CMOS, 434
    dynamic RAM (DRAM), 420,
        421, 449–51, **450**, 461,
        464–70, **464**, **465**
    electromigration
        phenomenon, 435
    expansion during fabrication,
        428–29
    fabrication, 425–37
    fan out, 460
    fat zero in CCDs, 447
    flash memory, 421, 461, 470–73,
        471, **472**, **474**
    flip-chips, 454, 463, 478–79, **480**
    fully depleted SOI devices, 438
    gates, 429
    hybrid ICs, 418–20, **419**
    inductors, 441
    integration benefits, 416–18,
        439–44
    interconnections, 441–44, **442**
    isolation regions, 427
    junctionless devices, 438
    Kooi effect, 429
    large scale integration (LSI),
        417, 420
    latchup, 424, 426
    lateral moat encroachment
        defect, 428
    lightly doped drain (LDD),
        430, **431**
    linear ICs, 418
    LOCOS, 427, 428
    logic circuits, 418, 451, 452–61
    logic gates, 458–61, **458**
    LPCVD, 426, 428, 429, 430, 435
    market for ICs, 421
    medium scale integration
        (MSI), 420
    memory circuits, 418, 420, 421,
        450–51, 461–74
    microprocessors, 450
    Miller overlap capacitance, 429
    miniaturization, 417
    monolithic ICs, 418–20, 423–44
    Moore's Law, 420, **422**
    MOS ICs, 421
    MOSFETs, 430, **432**, 433–36

NMOS devices, 430, 433–34
noise immunity or margin, 455
overcoating, 436
overlapping capacitance, 429
packaging, 479–82, **481**
pads, 444
parasitic bipolar structures, 424
plugs, 435
PMOS devices, 430, **432**, 432–34
polycide, 435
potential wells, 445
propagation delay, 460
random access memory
    (RAM), 461, **462**
rapid thermal anneal
    (RTA), 433
RC time constants, 443–44
reactive ion etching (RIE), 426
resistors, 418, 439–40, **439**
sacrificial or dummy oxide
    growth, 429
scaling and dimensions, 420–22
self-aligned silicide
    (SALICIDE), 424–25,
    434–35
self-alignment, 426–27
separation by implantation of
    oxygen (SIMOX), 437–38
shallow trench isolation
    (STI), 428
sheet resistance, 440, 443
Si/SiO$_2$ contacts/
    interconnections,
    441–44, **442**
sidewall oxide spacers, 430
silicon on insulator (SOI),
    437–39, **437**
simulation program with
    integrated circuit
    emphasis, 461
small scale integration
    (SSI), 420
source/drain extension or
    tip, 430
source/drain, 434–35, **435**
spiking problems, 435
static RAM (SRAM), 421, 461,
    463–64, **463**
tanks in CMOS ICs, 425–26, **425**
testing, 474–76
thermal relaxation time, 445
thick- vs. thin-film
    technology, 419–20
threshold voltages, 427
transistors, 418
tubs or wells in CMOS ICs, 424
twin-well CMOS, 426

ultra large scale integration
    (ULSI), 420–21, 449–73
very large scale integration
    (VLSI), 420
vias, 436
voltage transfer characteristics
    (VTC), 453–58, **454**
white ribbon effect, 429
wire bonding, 476–78, **478**
yield of ICs, 417
Interconnections, ICs, 441–44, **442**
Interdigitated geometry, bipolar
    junction transistor (BJT),
    359, **359**
Interface charge, metal oxide
    transistor/MOSFET,
    274–75, **274**
Interface states, 274
Intrinsic semiconductor
    materials, 74–75, 87
Intrinsic vs. Extrinsic
    photodetectors, 386
Inversion/inversion regions
    MOSFET, 291–93
    lasers, 401–2, **402**, 403
Inversion, strong inversion, metal
    oxide transistor/MOSFET,
    263–70, **263**, **268**, 273
Inverters, 451
Ion implantation, 147–50, **148**,
    149, 297–300, **298**
Ionic bonding, 55–57
Ionization
    field ionization (*See* Zener
        effect)
    impact ionization, 188–90,
        188, **190**, 193
    ionization region, 90
Isolation or field regions, 260, 427
Isotropic etching, 155

J

Johnson noise, photodetectors, 387
Junction capacitance, 202–5
Junction FETs (JFET), 241,
    244–51, **244**
    current-voltage
        characteristics, 249–51
    drain, 245
    gate, 245
    gate control, 247–49, **247**, **248**
    gradual channel
        approximation, 249
    mutual transconductance, 250
    pinch-off, 245–46, **246**

source, 245
Junctions, 142–40

K

Kinetic energy, 35, 39, 69
Kirk effect, 363–65, **364**
Kooi effect, 429

L

Large scale integration (LSI),
    417, 420
Lasers, 28, 379, 396–10
    absorption of light, 397–98, **398**
    AlGaAs lasers, 407, 409, 410
    CO$_2$ lasers, 400
    coherent light, 396, 397
    distributed Bragg reflector
        (DBR) mirrors, 408–9
    double-heterojunction lasers,
        407–9, **408**
    Einstein coefficients, 398
    emission spectra, 404–5, **405**
    energy density, 397
    fabrication of semiconductor
        laser, 405–6, **406**
    fiber optic communications, 394
    GaAs lasers, 400, 402, 405–7,
        406, 409, 410
    GaAsP lasers, 400, 404
    GaN lasers, 403
    graded index separate
        confinement
        heterostructure
        (GRINSCH) lasers, 408
    He-Ne lasers, 400
    heterojunction lasers, 406–9,
        407, **408**
    homojunction lasers, 406
    InAlGaN lasers, 403
    InGaAs lasers, 409
    InGaAsP lasers, 410
    InGaN lasers, 404
    InP lasers, 410
    inversion regions, 401–2,
        402, 403
    materials used in lasers, 410
    monochromatic light, 397
    negative temperature, 398–99
    optical resonant cavities,
        398–400, **399**
    p-n junction lasers, 400
    population inversion,
        398–404, **401**
    quasi-Fermi levels, 401
    ruby lasers, 400

semiconductor lasers, 390, 400–10

separate confinement lasers, 408, **409**

spontaneous vs. stimulated emission of light, 396–97, **396**

vertical cavity surface-emitting lasers (VCSELs), 408–9, **409**

Latchup, 424, 426

Lateral moat encroachment defect, 428

Lattice constant, 5, 10–11

Lattice matching in epitaxial growth process, 18–21, **19**, **20**

Lattice scattering, 97–98, **98**

Lattice structures, crystal (*See also* crystal properties), 3–11

Lead bond, 477

Light emission, 35–36, 66

Light hole bands, 73

Light-emitting diodes (LED), 66, 379, 390–96

fiber optic communications, 392–94, **393**

injection electroluminescence, 390

multilayer heterojunctions, 395–96

optoelectronic isolators, 392

semiconductor lasers, 390

Lightly doped drain (LDD), 308–9, 430, **431**

Linear combinations of atomic orbitals (LCAO), 51–52, 58–59, **59**

Linear ICs, 418

Linear regime, metal oxide transistor/MOSFET, 259

Liquid-encapsulated Czochralski (LEC) growth, 13–14, **14**

Liquid-phase epitaxy (LPE), 18

Load lines, field-effect transistors (FET), 242–43, **242**, **243**

Local oxidation of silicon (LOCOS) fabrication, 428

bipolar junction transistor (BJT), 330, 331

ICs, 427

metal oxide, 260

MOSFET, 304, 314–15

Logic circuits, 418

Logic gates, 458–61, **458**

Longitudinal effective mass, **72**

Low pressure chemical vapor deposition (LPCVD), 150–51, **150**, 294, 428

bipolar junction transistor (BJT), 331

ICs, 426, 429, 430, 435

Luminescence, 111–14

Lyman series, 32, 35

M

Magnetic fields

Hall effect, 100–2, **101**

Masks in photolithography, 151–52, 153

Mass separator in ion implantation system, 149, **149**

Mass, effective mass, 70–74

Mass, true vs. Effective mass, 73

Matrix mechanics, 36

Maxwell-Boltzmann limit, 80, 532

Mean free time, 93

Medium scale integration (MSI), 420

Memory circuits, 418, 420, 421, 450–51, 461–74

Metal-oxide semiconductor FET (MOSFET), 241–42, 286–16, **287**, 430, **432**, 433–35, 515–18

anti-punchthrough, 313

bias-temperature stress test, 280

body effect, 300–1, 466

capacitance, time-dependent, 280

capacitance-voltage relationship, 270–71, 277–80, **278**

capacitor, ideal MOS, 260–72, **261**

channel length modulation, 313

channel stop implants, 260, 299–300

charge sharing, 313–14

C_i control, 295–96

cross-sectional view, **295**

current-voltage characteristics, 283

Debye screening length, 265, 279

deep depletion, 281

depletion-mode, 257, **258**, 299

dielectric constant materials, 297

direct tunneling, 283, 285, **285**

double-diffused MOSFET (DMOS), 516–18

drain, 287, 288, **289**

drain-induced barrier lowering (DIBL), 289, 311–13, **312**

enhancement mode, 257, **258**

equivalent circuits, 304–5, **305**

equilibrium, 273, **273**

fabrication, 259

fast interface state density, 279–80

flat band condition, **261**, 263, 273, **273**, 286

Fowler-Nordheim tunneling, 283, **284**

gain-enhanced MOSFET (GEMFET), 515

gate electrode choice vs. threshold, 294–95

gate-induced drain leakage (GIDL), 315–16, **316**

halo implants, 313

hot carrier degradation, 309–11, **309**

hot electron effects, 307–11

ICs, 436

insulated gate bipolar transistor (IGBT), 515–18, **516**

interface charge, 274–75, **274**

inversion, strong inversion, 263–70, 291–93

inverters, 451

ion implantation for threshold adjustment, 297–300, **298**

isolation or field regions, 260

lightly doped drain (LDD), 308–9

linear regime, 259

local oxidation of silicon (LOCOS), 260, 304, 314–15

LPCVD, 294

Miller overlap capacitance, 304

mobile ion determination, 279–80

mobility degradation parameters, 292–93

mobility models, 290–93

narrow width effect, 314–15, **315**

noise immunity or margin, 455

output characteristics, 286–88

pinch-off, 293, 313

pinning of bandbending, 265

pocket implants, 313

real surface effects, 272–75

reverse short channel effect (RSCE), 314

roll off, 314

saturation, 259, 288, 289, **291**, 293

scaling, 307, **307**

self-aligned fabrication process, 259–60

self-aligned gates, 304

short channel effect (SCE), 313–15, **314**

short channel, 289, 293, **294**, 307, **308**, 313–15, **314**

SiO$_2$ layers, 260
Si-SiO$_2$ interface, 272
source, 287
source/drain series
    resistance, 304–6, **306**
space-charge density,
    264–66, **265**
substrate bias effects, 300–1
substrate current, 310–11, **310**
subthreshold conduction,
    301–4, **303**
threshold voltage, 257, 269–70,
    275–77, **276**, 286, 287, 293–300
time-dependent dielectric
    breakdown, 283
transconductance, 287
transfer characteristics,
    288–89, **290**
transverse field, 291–93, **292**
tunneling, 283, 285, **285**
universal mobility degradation
    curves, 291
voltage transfer characteristics
    (VTC), 453–58, 454
work function, 272–73, **273**
work function, modified, 260,
    262–64
Zerbst plotting of
    capacitance, 281, **282**
Metal-oxide transistor (MOS)
    (See metal oxide semiconductor
    FET (MOSFET)
Metal-semiconductor FET
    (MESFET), 241, 251–55, **252**
GaAs MESFET, 251–52, **252**
high electron mobility
    transistor (HEMT), 252
modulation doped FET
    (MODFET), 252
modulation doping, 252–54
pseudomorphic HEMTs, 254
separately doped FET
    (SEDFET), 254
short channel effects,
    254–55, **255**
two-dimensional electron gas
    FET (2–DEG or
    TDEGFET), 254
Metal-insulator semiconductor
    (MISFET), 241
Metal-organic vapor-phase
    epitaxy (MOVPE), 23, 410
Metal-semiconductor junctions,
    220–27
Metallic bonding, 57
Metallization, 156–57, **157**
Metals, energy bands, 61–62
Microprocessor ICs, 450

Microwave devices, 486–503
    bulk negative differential
        conductivity (BNDC), 496
    central valley or minimum,
        conduction band, 494
    degenerate semiconductors, 487
    Esaki diodes, 486
    GaAs devices, 500
    Gunn diodes, 486, 494, 500
    Gunn effect, 494–95
    impact avalanche transit time
        (IMPATT) diodes,
        491–93, **491**
    InP devices, 500
    momentum space, 494
    negative conductance, 486
    Read diode, 491–93, **491**
    satellite valleys, 494
    space charge domains,
        496–99, **497**, **498**
    subsidiary minima, 494
    transferred electron
        mechanism, 494
    transit time devices, 490–93
    tunnel diodes, 486–90, **488**
Miller indices for crystal
    lattices, 8–9
Miller overlap capacitance,
    304, 429
Minority carrier lifetime, 116
Minority vs. Majority carriers, 79
Misfit dislocations in
    heteroepitaxy, 20–21, **20**
Mobile ions, **274**
Mobility of electrons, 92–102, **92**
    high electron mobility transistor
        (HEMT), 252–54, **253**
    junctions, 232
    metal oxide transistor/
        MOSFET, 279–80, **281**
    mobility degradation
        parameters, 292–93
    MOSFET, 290–93
    universal mobility degradation
        curves, 291
Modified Ohm's law, 137–38, 137
Modulation doped FET
    (MODFET), 252
Modulation doping, 252–54
Molecular beam epitaxy (MBE),
    18, 23, **24**, **25**, 79, 392, 410
Momentum, relation to wave
    vector, 64, 73
Monochromatic light, lasers, 397
Monolithic ICs, 418–20, 423–44
Moore, Gordon, 420
Moore's Law, ICs, 420, **422**
Multichip modules, ICs, 482

Multimode optical fibers, 395
Mutual transconductance,
    junction FETs (JFET), 250

N

Na (See Sodium)
Nail-head bond, 478
NAND gates, 458–61, **458**
NaOH solution used in
    photolithography, 16, 153–54
Narrow base diodes, 198, 202
Narrow width effect, MOSFET,
    314–15, **315**
n-dopants in semiconductors,
    75–79
Negative conductance,
    microwave devices, 486
Negative temperature, lasers,
    398–99
NMOS devices, 430, 433–34
Noise and bandwidth,
    photodetectors, 386–90
Noise immunity or margin,
    ICs, 455
Nonrectifying contacts in
    schottky diodes, 222, 224
Nonuniformity effects, bipolar
    junction transistor (BJT), 351
NOR gates, 458–61, **458**, **459**
Normalization, 37–38, 41, 44–45
Numerical aperture, 154

O

Offset voltage, 190
Ohmic contacts, 224–26, **225**
Ohmic loss, 217, **218**
Ohm's law
    drift of electrons, 94
    high-field effects, 99–100
    modified Ohm's law, 137–38
On-off states, field-effect
    transistors (FET), 242
Optical absorption, 108–11,
    109, **110**
Optical fiber (See fiber optic
    communications)
Optical resonant cavities, lasers,
    398–400, **399**
Optoelectronic devices, 379–414
OR gates, 459, 459
Orbit of electrons, 33–35, **34**, **35**,
    43, 46, 47, 51–52
Organometallic vapor-phase
    epitaxy (omvpe), 23, 79, 392
Overcoating, ICs, 436
Overlapping capacitance, ICs, 429

Oxidation rates of silicon, 533–34
Oxidation, thermal oxidation process, 142–44, **143**

P

p-i-n photodectors, fiber optic communications, 394
p-n junctions, 142–57
Packaging for ICs, 479–82, **481**
Pad oxide, 428
Pads, ICs, 444
Parasitic bipolar structures, ICs, 424
Paschen series, 32, 35
Pass transistors, 464
Pauli exclusion principle, 46–48, 59, 81, 530
PbTe, 2
p-dopants in semiconductors, 75–79
Peak tunneling current, tunnel diodes, 489
Periodic structures in crystal lattices, 3–4, **4**, **5**
Periodic table, 46–52, **47**
Phosphorescence, 112
Phosphors, 112
Phosphorus, 145–46
Photoconductive decay, 120
Photoconductivity, 114–15, 123–24
Photodetectors, 379, 384–90, **385**
  avalanche photodiodes (APD), 388
  avalanche photodiodes, 385–86
  dark current, 387
  distributed Bragg reflector (DBR) mirrors, 389–90, **389**
  gain-bandwidth product, 386–87
  intrinsic vs. extrinsic detectors, 386
  Johnson noise, 387
  noise and bandwidth, 386–90
  p-i-n photodetectors, 385–90
  resonant-cavity APD, 389
  shot noise, 388
  signal-to-noise ratio (SNR), 387
  silicon heterointerface photodetector (SHIP), 388, **388**
Photodiodes, *see* photodetectors
Photoelectric effect, 30–31, **31**
Photolithography, 151–55, **151**, **152**
Photoluminescence, 111–14, **113**, **114**
Photons, 31

Photoresists in photolithography, 153
Photovoltaic effect, 381, **381**
Physical models, 28–30, 43
Piecewise-linear equivalents, 190–91, **191**
p-i-n photodetectors, 385–90
Pinch-off
  junction FETs (JFET), 245–46, **246**
  MOSFET, 293, 313
Pinning of bandbending, metal oxide transistor/MOSFET, 265
Planck's constant, 30, 33, 37
Planes and directions in crystal lattice, 7–9, **7**, **8**, **9**
Plasma-enhanced chemical vapor deposition (PECVD), 150–51
Plugs, ICs, 435
PMOS devices, 430, 432, 432–34
p-n junction diodes, 322–23, **323**, **324**
p-n junction lasers, 400
p-n junctions
  anisotropic etching, 155
  annealing, 150
  Arrhenius dependence of diffusivity, 144
  back-end processing, 156
  chemical mechanical polishing (CMP), 154
  chemical vapor deposition (CVD), 150–51
  complementary error function, 145
  diffraction-limited minimum geometry, 154
  diffusion processes, 144–46, **145**
  etching, 155–56
  ion implantation, 147–50, **148**, **149**
  isotropic etching, 155
  low pressure chemical vapor deposition (LPCVD), 150–51, **150**
  metallization, 156–57, **157**
  photolithography, 151–55, **151**, **152**
  plasma-enhanced chemical vapor deposition (PECVD), 150–51
  rapid thermal processing (RTP), 146–47, **146**
  reactive ion etching, 155–56, **155**
  sputtering in metallization, 156–57, **157**

thermal oxidation process, 142–44, 143
p-n-p-n devices, 504–11, **505**
Pocket implants, MOSFET, 313
Point contact transistors as early bjts, 329
Poisson's equation, 165
Polycide, ICs, 435
Polycrystalline solids, 3, **4**
Poly depletion, 429
Polysilicon emitter (polyemitter), bipolar junction transistor (BJT), 332
Population inversion, 398–4, **401**
Potential energy, 35, 39, 43–44, 55, 69
Potential well, 40–42, **41**, 48, 79–80, **79**, 232, **232**, 445
Power devices, 504–18
  alpha variation with injection, 507
  anode gate, 513
  anode terminal, 505
  avalanche multiplication, 509–10
  base-width narrowing, 509
  bilateral devices/diodes, 513–14, **514**
  bilateral triode switches (triacs), 514
  blocking state, 504, 505, 507–8, **508**
  cathode gate, 513
  cathode terminal, 505
  conducting state, 504, 505, 507–9, **508**
  displacement current, 510
  dv/dt triggering, 510–11
  fabrication and applications for SCRs, 514–15, **515**
  forward-blocking state, 505, 507–8, **508**
  forward-conducting state, 505, 507–8, **508**
  insulated gate bipolar transistor (IGBT), 515–18, **516**
  p-n-p-n devices, 504–11, **505**
  reverse-blocking state, 505
  semiconductor controlled rectifier (SCR), 504, 511–15, **511**
  semiconductor controlled switch (SCS), 513
  Shockley diodes, 505
  shorted-emitter designs, 513–14
  switching mechanisms, 504–11
  triggering mechanisms, 509–11

two-transistor analogy, p-n-p-n diodes, 506–7, **506**
voltage triggering, 509
Primitive cell in crystal lattice, 4
Printed circuit boards (pcb), 480–81
Probability density function, 37–38
Probability, 36–38
Propagation delay, 460
Pseudomorphic layers, 21, 259
Pulse dispersion, fiber optic communications, 394
Punch-through, 192

Q

Quad package, **481**
Quantization of energy, 30, 41–42
Quantum mechanics, 28, 29–30, 36–43
Quantum numbers, 41, 45, 46, 47–48, **47**
Quantum state, 41, 47
Quantum wells, 79–80, **79**
Quasi-Fermi levels, 120–23
gradients in quasi-Fermi levels, 137–38
junctions, 179–80
Quasi-steady state, junctions, 197–98, 197
Quaternary-compound semiconductors, 2

R

Random access memory (RAM), 461, **462**
Random motion and diffusion, 124–26, **125**
Range and straggle as function of silicon implant energy, 539–40
Rapid thermal anneal (RTA), ICs, 433
Rapid thermal processing (RTP), 146–47, **146**
Rayleigh scattering of light, 393–94, **394**
RC time constants, ICs, 443–44
Reaction chamber or reactor for epitaxy, 21–22, **22**
Reactive ion etching (RIE), 155–56, **155**, 426
Read diode, 491–93, **491**
Reclaimable charge, 209
Recombination lifetime, 116
Recombination processes, 75, 115–20, **118**

bipolar junction transistor (BJT), 326
continuity equation, 130–32, **131**
diffusion, 130–32, **131**
Haynes-Shockley experiment in drift/diffusion, 134–37, **134–36**
length of diffusion, 132–34
photoconductive decay, 120
transition region, junctions, 214–15
Rectifiers, 190–93
insulated gate rectifier (IGR), 515
junctions used as rectifiers, 169
rectifying contacts in Schottky diodes, 222–24
semiconductor controlled rectifier (SCR), 511–15, **511**
Reference diodes, 194
Refresh of memory, 466
Resistance/resistors, 96–97
aspect ratios, 440
ICs, 418, 439–40, **439**
photoconductive devices, 123–24
sheet resistance, 440, 443
source/drain series resistance, MOSFET, 304–6, **306**
Resonant cavities, lasers, 398–400, **399**
Resonant-cavity APD, **389**
Reticle in photolithography, **151**
Reverse bias, 183–94, **184**
bipolar junction transistor (BJT), 351
rectification, 190–93, **192**
Reverse breakdown, 185–86, **186**
Reverse recovery transient, 198–201, **199**, **200**
Reverse saturation current, 174
Reverse short channel effect (RSCE), MOSFET, 314
Reverse-blocking state, power devices, 505
Richardson constant of thermionic emission, 226
Rise time, bipolar junction transistor (BJT), 351
Ritz combination principle, 32–33
Roll off, MOSFET, 314
Ruby lasers, 400
Rydberg constant, 32, 36

S

Sacrificial or dummy oxide growth, ICs, 429

Satellite valleys, **65**, 494
Saturation
bipolar junction transistor (BJT), 347, 348–49, **349**
metal oxide transistor/ MOSFET, 259
MOSFET, 288, 289, **291**, 293
Sb (*See* Antimony)
Scaling, 307, **307**, 420–22
Scattering mechanisms, 93, 97–98, **98**, 100
Schottky barriers, 220–22, **221**, **222**, 226–27
Schottky diodes, 220–24
nonrectifying contacts in Schottky diodes, 222, 224
rectifying contacts in Schottky diodes, 222–24
Schottky effect, 220–22
Schrodinger wave equation, 38–40, 43, 51–52, 55, 58, 525
Selectivity in etching, 155
Self-aligned fabrication process, metal oxide transistor/MOSFET, 259–60
Self-aligned gates, MOSFET, 304
Self-aligned silicide (SALICIDE), 424–25, **425**, 434–35
Self-alignment, 332, 426–27
Semiconductor controlled rectifier (SCR), 504, 511–15, **511**
Semiconductor controlled switch (SCS), 513
Semiconductor lasers, 390, 400–10
Separate confinement lasers, 408, **409**
Separately doped FET (SEDFET), 254
Separation by implantation of oxygen (SIMOX), 437–38
Series resistance effects, BJT, 351, 357–59, **358**
Shallow trench isolation (STI), 428
Sheet resistance, 440, 443
Shells of electrons, 46, 47, 51–52
Shockley diodes, 505
Shockley, 241–42
Short channel/short channel effect (SCE)
MESFETs, 254–55, **255**
MOSFET, 289, 293, **294**, 307, **308**, 313–15, **314**
Shorted-emitter designs, power devices, 513–14
Shot noise, photodetectors, 388
Silicon (Si), 2, 9, 12, 13
band gap, 110–11, **111**

bipolar junction transistor (BJT), 329
chemical vapor deposition (CVD), 150–51
conduction bandstructures, 72–73, **72**
covalent bonding, 57
diffusion processes, 144–46, **145**
donor/acceptors, 78
dopant diffusivity, 537–38
effective and true mass, 73
electron-hole pairs (EHP), 67
energy bands, 58, 63
etching, 155–56
extrinsic material, 75, 77
high-field effects, 99
high-frequency transistors, 371
IC contacts/interconnections, 441–44, **442**
impurity energy levels, 119–20, **119**
impurity solubility in, 535–36
integrated circuits, 415–85, **416**
interface charge, metal oxide transistor/MOSFET, 274–75, **274**
intrinsic carrier concentration vs. temperature, **89, 90**
ion implantation, 147–50, **148, 149**
metallization, 156–57, **157**
mobility variation vs. doping impurity concentration, **98**
ohmic contacts and doping, 225
oxidation rates, dry and wet thermal, 533–34
photodetectors, 388
photolithography, 153–55, **151, 152**
range and straggle as function of implant energy, 539–40
recombination processes, 117
silicon on insulator (SOI), 437–39, **437**
thermal oxidation process, 142–44, **143**
vapor-phase epitaxy (VPE), 21–23
Silicon dioxide (SiO$_2$), 4, 12, 16, 144–46, **145**
chemical vapor deposition (CVD), 150–51
etching, 155–56
interface charge, metal oxide transistor/MOSFET, 274–75, **274**

metal oxide transistor/MOSFET, 255, 260, 272
thermal oxidation process, 142–44, **143**
SiC, l4
vapor-phase epitaxy (VPE), 22
Sidewall oxide spacers, ICs, 430
SiGe, heteroepitaxy, 20
Signal-to-noise ratio (SNR), photodetectors, 387
SiH$_4$, vapor-phase epitaxy (VPE), 22
Silicon heterointerface photodetector (SHIP), 388, **388**
Silicon on insulator (SOI), 437–39, **437**
Simple cubic lattice, 5
Simulation program with integrated circuit emphasis (SPICE), 461
Single-crystal ingot growth processes, 13–14, **14**
Single-inline package (SIP) ICs, 482
Single-mode optical fibers, 395
Sinker, **330**
Small scale integration (SSI), 420
Sodium (Na), 55
interface charge, metal oxide transistor/MOSFET, 274–75, **274**
NaCl and ionic bonding, 55–57
Solar cells, 379, 382–84, **383, 384**
Source
ICs, 434–35, **435**
junction FETs (JFET), 245
metal oxide transistor/MOSFET, 256–57
MOSFET, 287
Source/drain extension or tip, ICs, 430
Series resistor, **305**
Space charge
density of space charge, metal oxide transistor/MOSFET, 264–66
depletion approximation, 165
equilibrium, 164–69, **164**
Gauss's laws, 165
junctions, 164–69, **164**
microwave devices, 496–99, **497, 498**
neutrality, 90–92
Poisson's equation, 165
Spacers, bipolar junction transistor (BJT), 331

Spectrum of atoms, atomic spectra, 31–33, **32**
Spectrum of emission, lasers, 404–5, **405**
Spiking problems, ICs, 435
Spin of electrons, 45–47
Split-off band, 73
Spontaneous vs. Stimulated emission of light, 396–97, **396**
Sputtering in metallization, 156–57, **157**
Stacked capacitor, **469**
Static RAM (SRAM), 421, 461, 463–64, **463**
Steady state, 93, 94, 120
Steady state carrier generation/injection, 120–23, 132–34
Step junctions, 157, 158
Step-index optical fiber, 393, **393**
Steppers in photolithography, 153
Storage delay time, junctions, 200–1, **200**
Stored charge, time variation of stored charge, 195–98, **196**
Straggle, imlantation, 147–48
Strained-layer superlattice (SLS) structures, 21
Sub-collector, **330**
Sublattices, 9–10, **10**
Substrate bias effects, MOSFET, 300–1
Substrate current, MOSFET, 310–11, **310**
Subthreshold conduction, MOSFET, 301–4, **303**
Surface states, junctions, 225
Surface-mounted ICs, 480
Switching
bilateral triode switches (triacs), 514
bipolar junction transistor (BJT), 346–51, **346, 350**
field-effect transistors (FET), 242–44
power devices, 504–11
semiconductor controlled switch (SCS), 513
Switching current, 459
Switching diodes, 201–2
Symbols, definitions, 519–22

T

Ta$_2$O$_5$, 297
Tanks in CMOS ICs, 425–26, **425**

Tape-automated bonding
  (TAB), ICs, 482
Target chamber in ion
  implantation system, 149, **149**
Temperature dependence of
  carriers, 88–90, **89**, **90**
Temperature effects on electron
  mobility, 97–98, **98**
Terminal currents, bipolar junction
  transistor (BJT), 334–39
Ternary-compound
  semiconductors, 2, 11
Thermal budget, 144
Thermal oxidation process,
  142–44, **143**
Thermal relaxation time,
  ICs, 445
Thermal runaway, bipolar
  junction transistor (BJT), 357
Thermionic emission, 226
Thermocompression bond, 477
Thick- vs. Thin-film technology,
  419–20
Threshold voltage
  ICs, 427
  metal oxide transistor/
    MOSFET, 257, 269–70,
    275–77, **276**
  roll-off, 286, 287, 293–300, **315**
Through-hole mounted ICs, 480
Thyristors, 424, 511
Time-dependent dielectric
  breakdown (TDDB), 283
Transconductance
  bipolar junction transistor
    (BJT), 367–68
  MOSFET, 287
Transconductance, mutual
  transconductance, junction
  FETs (JFET), 250
Transfer characteristics,
  MOSFET, 288–89, **290**
Transferred electron mechanism,
  microwave devices, 494
Transient and a-c conditions,
  junctions, 194–211
Transistors
  bipolar FET (BiFET), 515
  bipolar junction transistor
    (BJT), 241–42, 322–78
  conductivity-modulated FET
    (COMFET), 516–17
  depletion-mode transistors,
    257, **258**
  double-diffused MOSFET
    (DMOS), 516–18

enhancement-mode
  transistors, 257, **258**
field-effect transistors
  (FET), 241
frequency limitations, 365–71
GaAs MESFET, 251–52, **252**
gain-enhanced MOSFET
  (GEMFET), 515
heterojunction bipolar
  transistors (HBT),
  371–74, **372**
ICs, 418
insulated gate bipolar transistor
  (IGBT), 515–18, **516**
insulated gate FET
  (IGFET), 255
insulated gate transistor
  (IGT), 515
junction FETs (JFET), 241,
  244–51, **244**
metal-oxide semiconductor
  FET (MOSFET), 241–42,
  286–16, **287**
metal-oxide transistor (MOS),
  255–85, **256**
metal semiconductor FET
  (MESFET), 241, 251–55, **252**
metal-insulator semiconductor
  (MISFET), 241
metal-insulator-semiconductor
  (MIS) transistor, 255
modulation doped FET
  (MODFET), 252
point contact transistors as
  early BJTs, 329
pseudomorphic HEMTs, 254
separately doped FET
  (SEDFET), 254
two-dimensional electron gas
  FET (2–DEG FET or
  TEGFET), 254
unipolar transistors, 241
Transit time, 328, 368–69, 490–93
Transition region, junctions, 160,
  170, 214–16, **215**
Transitions between energy
  levels (orbits) of electrons, 28,
  33, 34–35, **34**, **35**
Transverse effective mass, **72**
Trench capacitor, **470**
  isolation, **428**
Trapping, 118–20
Triacs, 514
Triggering mechanisms, power
  devices, 509–11
Tubs or wells in CMOS ICs, 424

Tunnel diodes, 486–90, **488**
Tunneling
  flash memory, 473, **473**
  Fowler-Nordheim tunneling,
    283, **284**, 285
  metal oxide transistor/
    MOSFET, 283, 285, **285**
  quantum mechanics, 42–43, **42**
Twin-well CMOS, 426
Two-dimensional electron gap, 232
Two-dimensional electron gas
  FET (2–DEG FET or
  TEGFET), 254
Type N negative resistance,
  tunnel diodes, 489

U

Ultra large scale integration
  (ULSI), 420–21, 449–73
Ultrasonic bonding, 477
Uncertainty principle, 36–38
Unipolar transistors, 241
Unit cells in crystal lattice, 4, **5**
Universal mobility degradation
  curves, 291

V

Vacuum level, **229**
Valence bands, 60–61, 67–70, **69**,
  79–80, **79**
Valley current, tunnel diodes, 489
Vapor-phase epitaxy (VPE),
  21–23, **22**
Varactor diodes, 210–11
Vertical cavity surface-emitting
  lasers (VCSELS), 408–9, **409**
Very large scale integration
  (VLSI), 420
Vias, in ICs, 436
Voltage regulators, 194
Voltage transfer characteristics
  (VTC), ICs, 453–58, **454**
Voltage triggering in power
  devices, 509
Voltage-variable capacitance,
  junctions, 205
Voltage-controlled negative
  resistance, tunnel diodes, 489

W

Wafer manufacture,
  semiconductors, 14–16, **15**
Wave mechanics, 36

Wave nature of light, 31
Wave vectors, energy bands,
  62–64, **63**
Wavelengths of spectral
  emissions, 32
Wave-particle motion, 70
Weak inverse, **266**
Webster effect, bipolar junction
  transistor (BJT), 369
Wedge bond, 478, **479**
Well problem (*See* potential
  well problems)
White ribbon effect, ICs, 429
Wire bonding, ICs, 476–78, **478**
Words or bytes in memory, 462
Work function of metals, 31,
  220–22
  junctions, 227–32

metal oxide transistor/
  MOSFET, 260, 262–64, **263**,
  272–73, **273**
Wurtzite lattice, 10

X
X-ray lithograph, 154

Y
Yield, 417

Z
Zener diodes, 194
Zener effect, 186–87, **187**, 486, 488

Zerbst plotting of capacitance,
  281, **282**
Zincblende lattice structures, 9, 11
Zinc (Zn)
  impurity energy levels,
    119–20, **119**
ZnS, 2, 112
  light-emitting diodes
    (LED), 392
  photoluminescence, 112, 114
ZnSe
  Gunn effect, 499
  light-emitting diodes
    (LED), 392
ZnTe
  light-emitting diodes
    (LED), 392
ZrO_2, 297

Capacitance: $C = \left| \dfrac{dQ}{dV} \right|$ (5–55)

Junction Depletion: $C_j = \epsilon A \left[\dfrac{q}{2\epsilon(V_0 - V)} \dfrac{N_d N_a}{N_d + N_a} \right]^{1/2} = \dfrac{\epsilon A}{W}$ (5–62)

Stored charge
exp. hole dist.: $Q_p = qA \displaystyle\int_0^\infty \delta p(x_n) dx_n = qA \Delta p_n \int_0^\infty e^{-x_n/L_p} dx_n = qAL_p \Delta p_n$ (5–39)

$I_p(x_n = 0) = \dfrac{Q_p}{\tau_p} = qA \dfrac{L_p}{\tau_p} \Delta p_n = qA \dfrac{D_p}{L_p} p_n (e^{qV/kT} - 1)$ (5–40)

$G_s = \dfrac{dI}{dV} = \dfrac{qAL_p p_n}{\tau_p} \dfrac{d}{dV}(e^{qV/kT}) = \dfrac{q}{kT} I$ (5–67c)

Long p^+-n: $i(t) = \dfrac{Q_p(t)}{\tau_p} + \dfrac{dQ_p(t)}{dt}$ (5–47)

MOS-n CHANNEL

Oxide: $C_i = \dfrac{\epsilon_i}{d}$ Depletion: $C_d = \dfrac{\epsilon_s}{W}$ MOS: $C = \dfrac{C_i C_d}{C_i + C_d}$ (6–36)

Threshold: $V_T = \underbrace{\Phi_{ms} - \dfrac{Q_i}{C_i}}_{\text{Flat band}} - \dfrac{Q_d}{C_i} + 2\phi_F$ (6–38)

Inversion: $\phi_s \text{ (inv.)} = 2\phi_F = 2\dfrac{kT}{q} \ln \dfrac{N_a}{n_i}$ (6–15) $W = \left[\dfrac{2\epsilon_s \phi_s}{qN_a} \right]^{1/2}$ (6–30)

$Q_d = -qN_a W_m = -2(\epsilon_s q N_a \phi_F)^{1/2}$ (6–32)   At V_{FB}: $C_{FB} = \dfrac{C_i C_{\text{debye}}}{C_i + C_{\text{debye}}}$

Debye screening length: $L_D = \sqrt{\dfrac{\epsilon_s kT}{q^2 p_0}}$ (6–25) $C_{\text{debye}} = \dfrac{\sqrt{2}\,\epsilon_s}{L_D}$ (6–40)

Substrate bias: $\Delta V_T \simeq \dfrac{\sqrt{2\epsilon_s q N_a}}{C_i}(-V_B)^{1/2}$ (n channel)   (6–63)